CW00730108

ADHESIVES AND SEALANTS

Basic Concepts and High Tech Bonding

Handbook of Adhesives and Sealants
Volume 1

Elsevier Internet Homepage - http://www.elsevier.com
Consult the Elsevier homepage for full catalogue information on all books, major reference works, journals, electronic products and services.

Elsevier Titles of Related Interest

Joining of Materials and Structures
Robert Messler
2004, 0-7506-7757-0

Fracture of Polymers, Composites and Adhesives II
J.G. Williams, A. Pavan, Bamber Blackman
2003, 0-08-044195-5

Adhesion Science and Engineering
Pocius
2002, 0-444-51140-7

Related Journals:

Elsevier publishes a wide-ranging portfolio of high quality research journals, encompassing the adhesives, composites, and polymer fields of materials science. A sample journal issue is available online by visiting the Elsevier web site (details at the top of this page). Leading titles include:

International Journal of Adhesion and Adhesives
Construction and Building Materials
Mechanics of Materials
Composites Science and Technology
Composites Part A: Applied Science and Manufacturing
Composites Part B: Engineering
Polymer
European Polymer Journal
Dental materials
NDT International
Reinforced Plastics

All journals are available online via ScienceDirect: www.sciencedirect.com

To contact the Publisher
Elsevier welcomes enquiries concerning publishing proposals: books, journal special issues, conference proceedings, etc. All formats and media can be considered. Should you have a publishing proposal you wish to discuss, please contact, without obligation, the publisher responsible for Elsevier's Material Science programme:

David Sleeman
Publishing Editor
Elsevier Ltd
The Boulevard, Langford Lane
Kidlington, Oxford
OX5 1GB, UK

Phone: +44 1865 843265
Fax: +44 1865 843987
E.mail: d.sleeman@elsevier.com

General enquiries, including placing orders, should be directed to Elsevier's Regional Sales Offices – please access the Elsevier homepage for full contact details (homepage details at the top of this page).

ADHESIVES AND SEALANTS

Basic Concepts and High Tech Bonding

Handbook of Adhesives and Sealants
Volume 1

SERIES EDITOR

PHILIPPE COGNARD
Versailles, France

2005

ELSEVIER

Amsterdam – Boston – Heidelberg – London – New York – Oxford
Paris – San Diego – San Francisco – Singapore – Sydney – Tokyo

ELSEVIER B.V.
Radarweg 29
P.O. Box 211, 1000 AE
Amsterdam, The Netherlands

ELSEVIER Inc.
525 B Street, Suite 1900
San Diego, CA 92101-4495
USA

ELSEVIER Ltd
The Boulevard, Langford Lane
Kidlington, Oxford OX5 1GB
UK

ELSEVIER Ltd
84 Theobalds Road
London WC1X 8RR
UK

First edition 2005

Library of Congress Cataloging in Publication Data
A catalog record is available from the Library of Congress.

British Library Cataloguing in Publication Data
A catalogue record is available from the British Library.

ISBN: 0-08-044554-3

Transferred to digital print 2007
Printed and bound by CPI Antony Rowe, Eastbourne

Contents

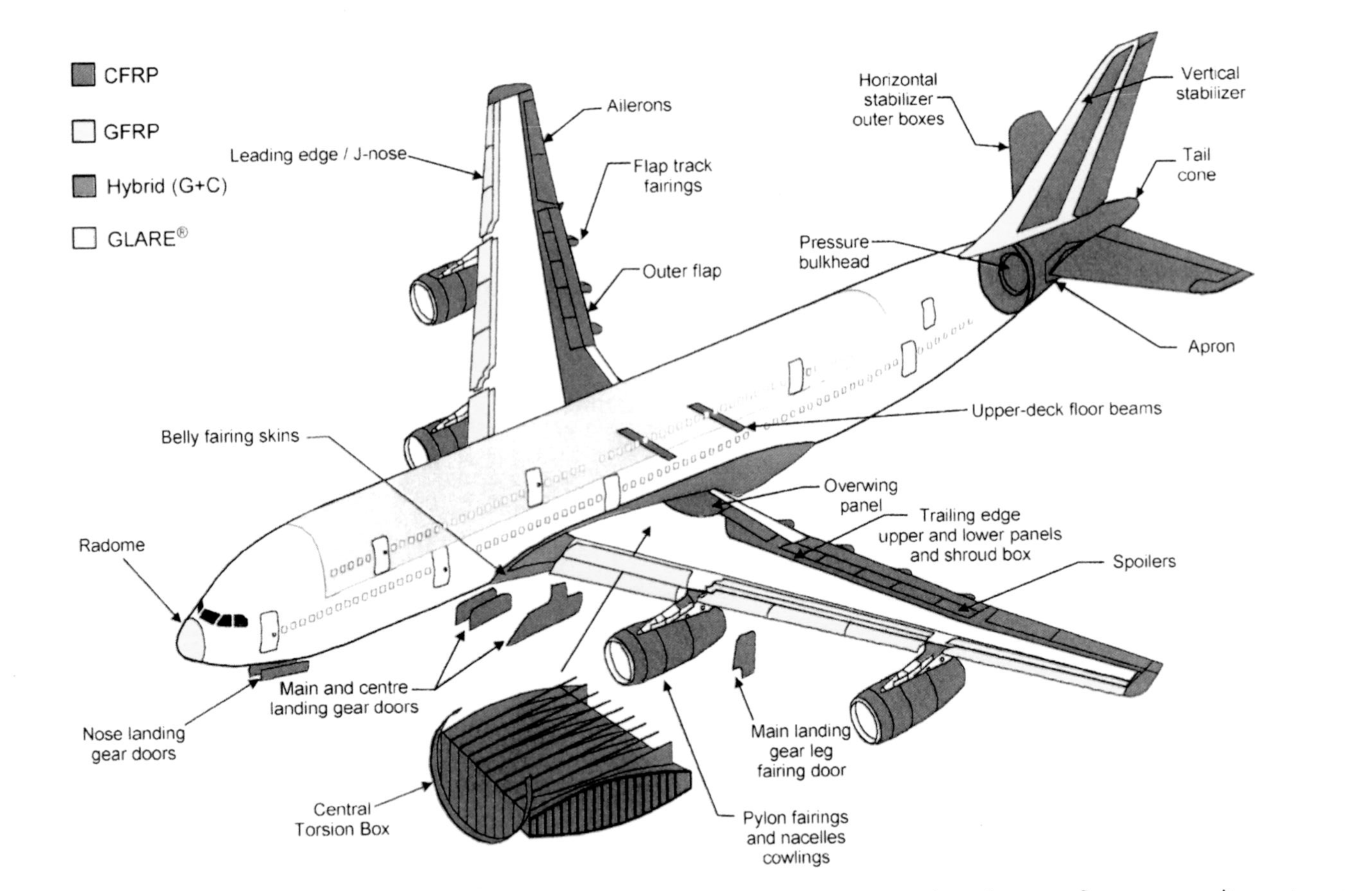

Plate 1: A380 composite materials application. Not shown: CFRP passenger floor panels and struts. *Some composite parts are bonded others are not*. Several metal parts are also bonded in modern civil and military aircrafts.

Plate 2: Electronic adhesives is now a huge market—400 million USD in 2004 worldwide.

(b)

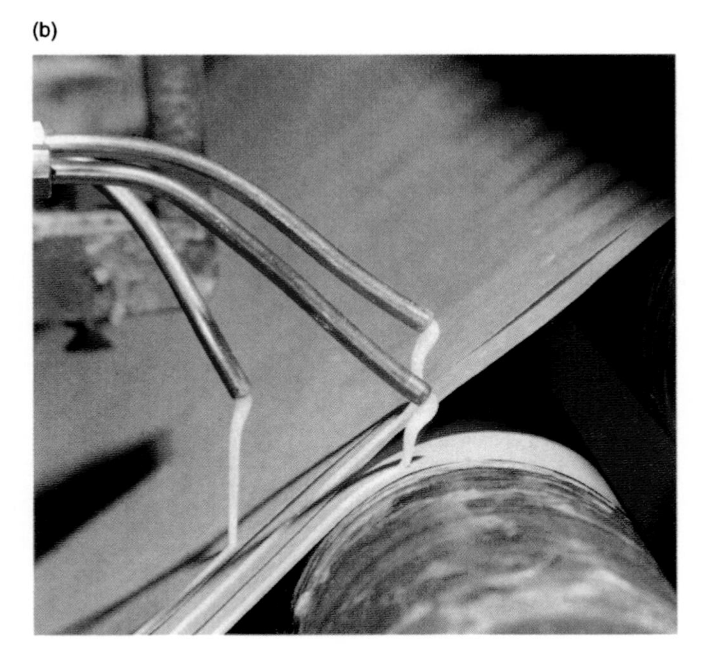

Plate 3: (b) A very simple way of applying an adhesive just by gentle pumping (manufacture of paper bags). Here the best would be to have a Newtonian adhesive.

Plate 4: Breaking down of honeycomb sandwich panel bonded with epoxy film. Honeycomb cells break down in different heights.

Plate 5: The various legal labels that are mandatory for risk warnings on chemical packagings/containers.

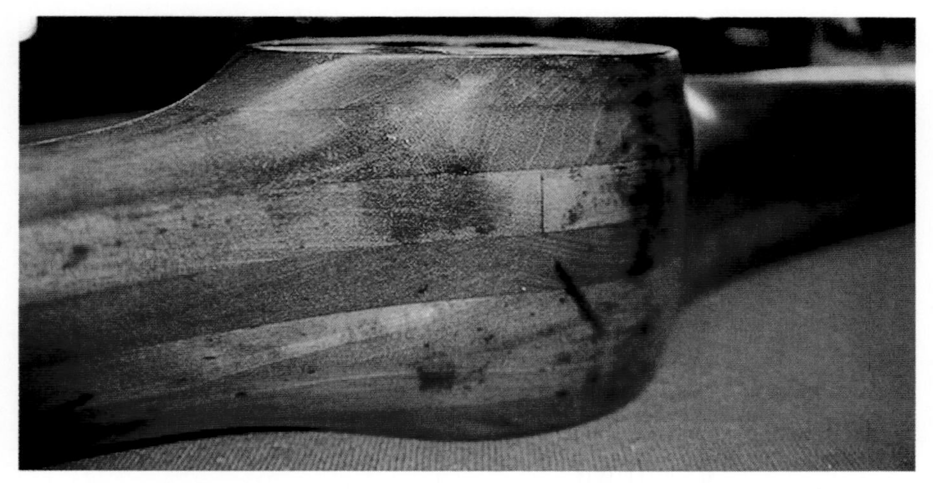

Plate 6: A 110-Lerhone bonded laminated propeller from a Nieuport scout of 1916–1920.

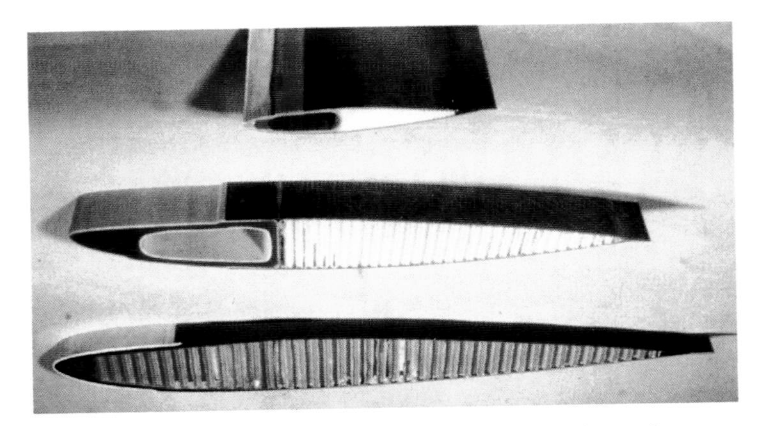

Plate 7: Bonded Westland helicopter rotor blade sections.

Plate 8: The original Redux film machine.

Plate 9: Typical press bonding shop.

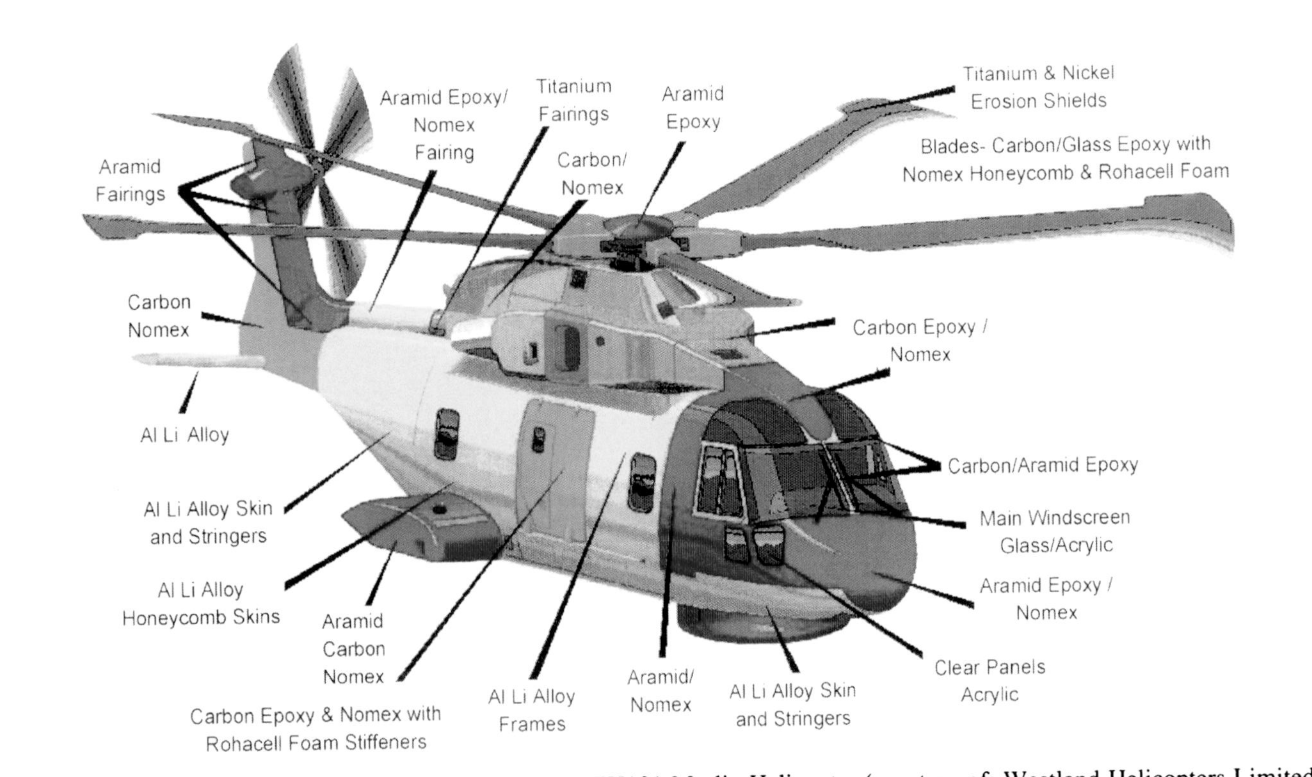

Plate 10: Schematic structure of Westland Helicopter's EH101 Merlin Helicopter (courtesy of: Westland Helicopters Limited).

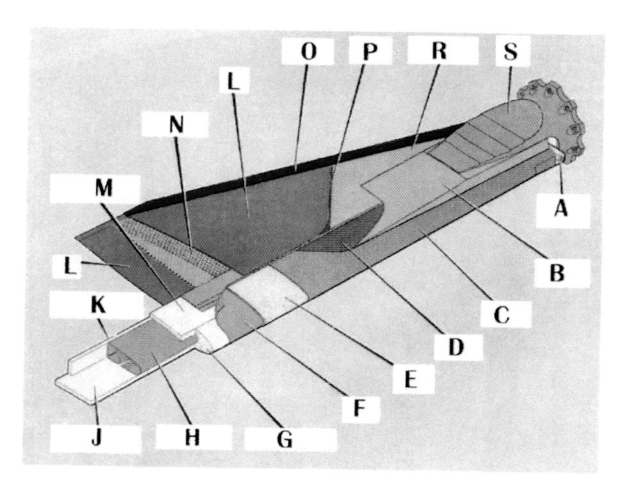

Plate 11: Schematic of the main rotor blade of Westlands' Sea King Helicopter (courtesy of: Westland Helicopters Limited), (A) BIM manifold, (B) Root doublers, (C) Erosion shield, (D) Packing piece, (E) Heater mat, (F) Outer wraps, (G) Uni-directional nose moulding, (H) Inner wraps, (J) Uni-directional sidewall slab, (K) Uni-directional backwall slab, (L) Skin, (M) Uni-directional sidewall slab, (N) Honeycomb, (O) Closing Channel, (P) Caulk, (R) Dummy skin, (S) Cuff.

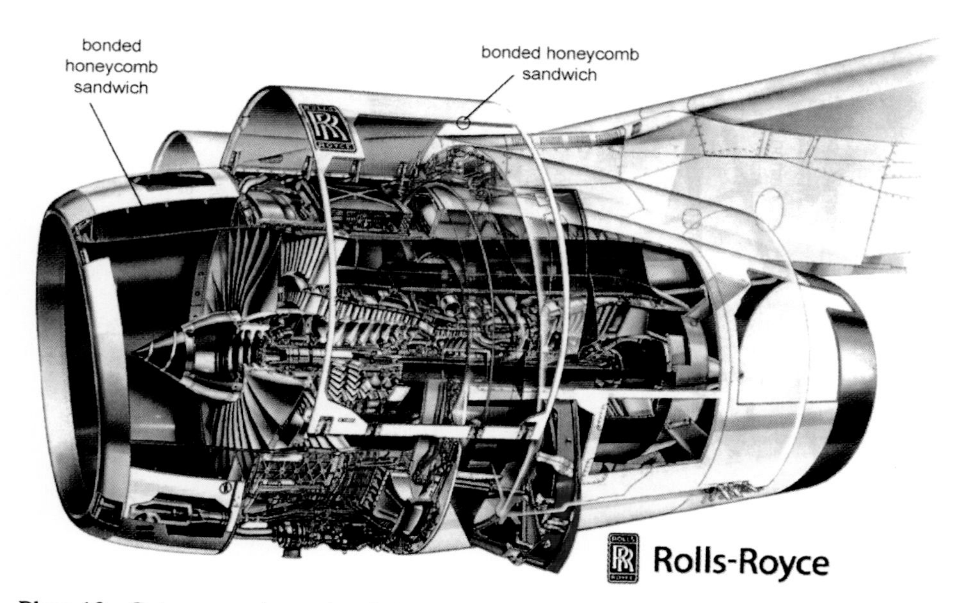

Plate 12: Cut-away schematic of the Trent 700 aero engine (courtesy of: Rolls Royce).

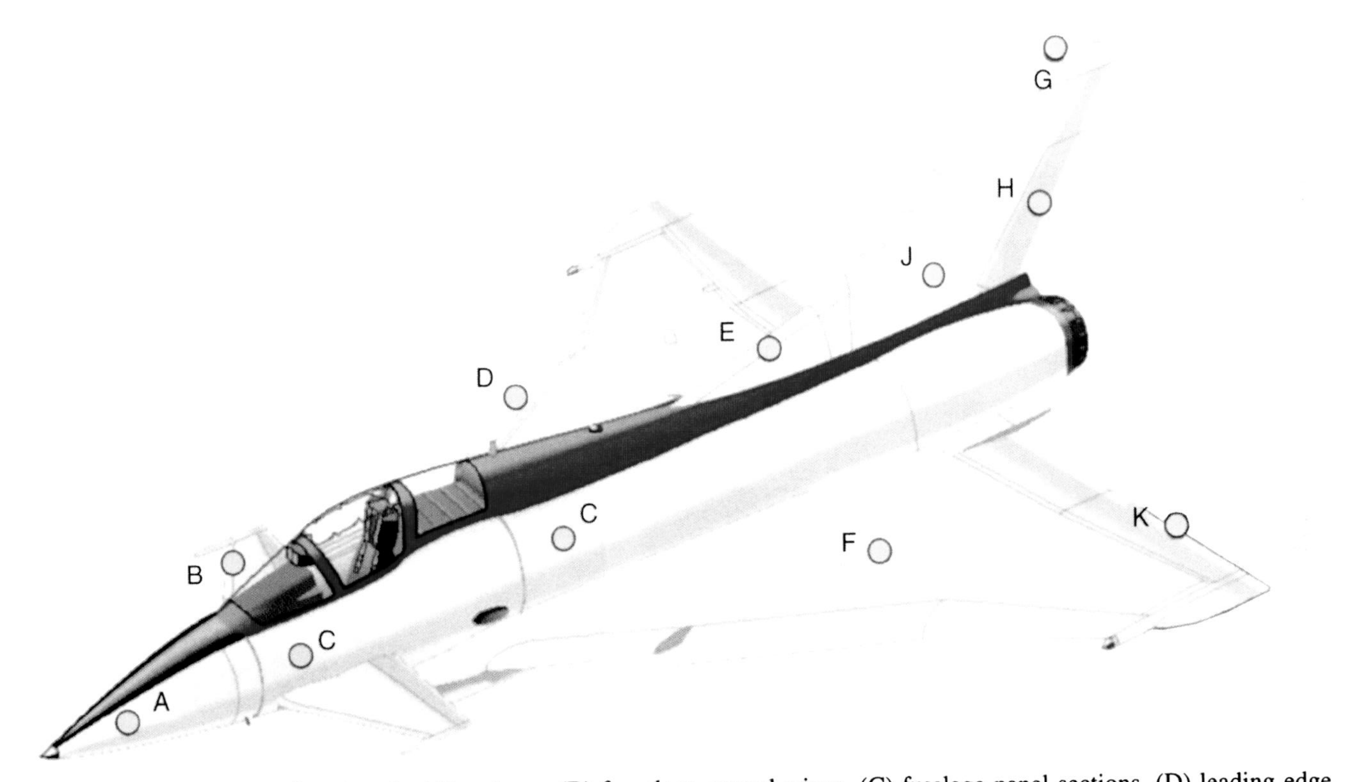

Plate 13: Generic military strike aircraft: (A) radome, (B) foreplane canard wings, (C) fuselage panel sections, (D) leading edge devices, (E) fin fairings, (F) wing skins and ribs, (G) fin tip, (H) rudder, (J) fin, (K) flying control surfaces.

Plate 14: Three generations of modern military aircraft. (a) Jaguar (Crown Copyright/MoD: Reproduced with the permission of the Controller of HMSO), (b) Tornado (Copyright: Roger Hadlow), (c) Typhoon (courtesy: Eurofighter at www.eurofighter.com).

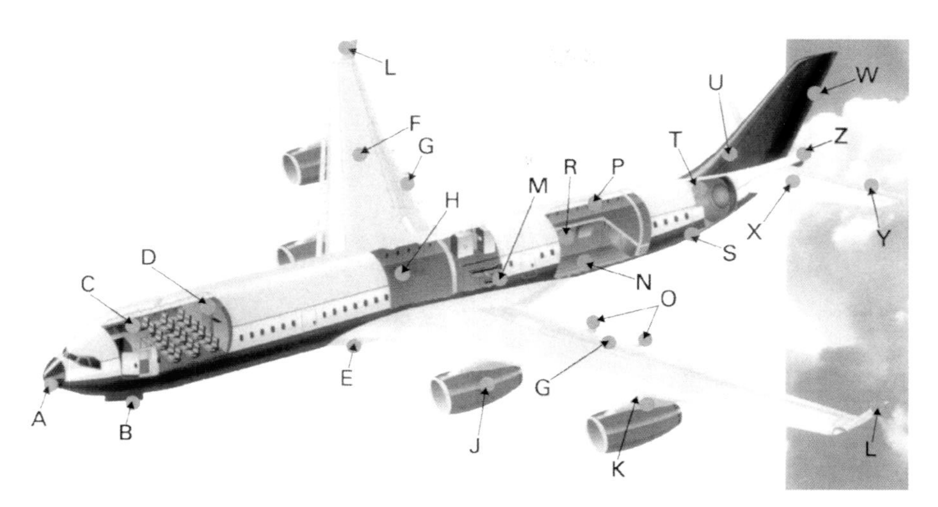

Plate 15: Generic commercial passenger aircraft: (A) radome, (B) landing gear doors and leg fairings, (C) galley, wardrobes, toilets, (D) partitions, (E) wing to body fairing. (F) wing assembly, (G) flying control surfaces: ailerons, spoilers, vanes, flaps and slats, (H) passenger flooring, (J) engine nacelles and thrust reversers, (K) pylon fairings, (L) winglets, (M) keel beam, (N) cargo flooring, (O) flaptrack fairings, (P) overhead stowage bins, (R) ceiling and sidewall panels, (S) airstairs, (T) pressure bulkhead, (U) vertical stabiliser, (W) rudder, (X) horizontal stabiliser, (Y) elevator, (Z) tail cone.

Plate 16: Lay-up of a typical Fibrelam panel.

(a)

(b)

Plate 17: MBDA missile systems: (a) Seawolf, (b) Rapier. (Copyright: MBDA Limited.)

Chapter 1

Introduction — How to Use This Handbook

Philippe Cognard, Series Editor

I strongly advise the reader to read this introduction where I explain how the Handbook should and can be used.

Many of the things that are used each day are bonded or assembled with adhesives, or sealed with sealants: your table and all the furniture, the foam cushioning of your seat, electronic parts in your computer, the double-insulated window, pressure-sensitive tapes and labels, your shoes, the food and other packaging, the carpet and floor coverings, the wallpaper, many parts in your car, many structures in airplanes, the PVC pipes in the kitchen, perhaps the tubes of your bicycle, the liner in some garments, shirt collars, the deflection yokes in your TV set, the metal base of electrical bulbs, children's toys, baby diapers and feminine napkins, and many other things — even this book! It would take too much time to list everything.

Assembly with adhesives is used by every industry and in many manufactured goods and products.

More recently, adhesives have entered other new fields such as dental care, textiles and garments, portable phones and even surgery!

Sealants are not as widely used, but today construction alone uses more than 1 million tons per year in Europe alone.

I believe that every engineer, designer, architect, chemist or technician in every industry will, at some point, need to use adhesives, sealants, or related products sometimes (for filling gaps, potting, repairs). Therefore, students in materials science, mechanical engineering, and technical universities, in general, should need to have a basic understanding of this field. This is the reason I have decided to compile this handbook.

1.1. Scope, Purpose and Contents of the *Handbook of Adhesives and Sealants*

Since the 1950s, many excellent books have been written on adhesives and sealants, often by American and European authors.

Handbook of Adhesives and Sealants
P. Cognard (Editor)

But these books usually focus either on the science and chemistry of adhesion and adhesives, on specific types of adhesives (such as pressure-sensitive adhesives, structural adhesives) or on industries that use adhesives and sealants, such as packaging, construction, metal bonding, etc.

I know of only two or three handbooks which are more or less comprehensive: one is well known by everybody in our industry for many years and was our "Bible" for all of us in the past — the Handbook of Adhesives. This was edited in 1962, 1980, and 1989 and written by the late Irving Skeist. It included some 46 chapters on the different chemical types of adhesives — a total of 800 pages.

Another is the Handbook of Adhesives and Sealants written by Paul Petrie and published in 2001 by McGraw-Hill, which also provides good general information.

These handbooks mostly covered chemistry of all types of adhesives and sealants, but did not concentrate on the industrial aspects and applications.

1.2. A Comprehensive Handbook

Our goal in publishing this handbook was very ambitious: I wanted to produce the most comprehensive handbook on adhesives and sealants ever published to address:

- *every scientific and technical issue* such as theories of adhesion, chemistry and physics of adhesives and sealants, technical characteristics, design and calculation of bonded or sealed parts, surface preparation before bonding or sealing, testing and standards (there are hundreds of standards altogether including those from the USA, Europe, Japan, and, of course, ISO international standards), methods of use, equipment for application, drying and curing, new curing techniques (there are now many, including the latest: UV curing, EB curing, microwaves, Joule effect) — in all, a total of 20 chapters.
- *every chemical type of adhesives and sealants*, thermoplastics and thermosetting, such as acrylics, different types of hot melts, engineering adhesives, epoxies, polyurethanes, heat-stable adhesives, formaldehyde adhesives (UF, PF, MF, and RF), water-, solvent-, rubber-based, vinyl and VAE, natural and renewable adhesives, as well as all the different types of sealants, and others — a total of more than 20 chapters. Each chapter needed to include: chemistry and physics of each type of adhesive or sealant, technical characteristics, various modes of curing, methods of use, standards, application techniques and equipment, the main uses in various industries, and lists of suppliers of adhesives and sealants and the raw materials.
- *every industry which uses (or could/should use) adhesives and sealants*. Almost all industries use or could use adhesives and sealants (and related products, for

surface preparation, potting, filling). With the help of the contributing authors, I will discuss the main users (in tons per year or turnover) from industries such as construction and decoration, packaging, woodworking and furniture, pressure-sensitive goods (tapes, labels, etc.), automotive, aerospace, electronics, transformation of metals, plastics and composites, footwear, disposable and sanitary items, abrasives, bonding of glass and ceramics, nonwovens, graphic arts, agglomerations of wood particles, fibers, chips and some recent developments which are still in their early stages such as dental care and repair, surgery and medical uses, textiles, etc. The survey of all utilizations of adhesives and sealants in all industries will account for a total of 40 chapters.

For each industry, we will explain — in detail — how, why, when, and where adhesives and sealants are, or may be, used for various assemblies: all the techniques, types of adhesives and sealants, materials, and applications that are available or possible in a given industry, together with case studies, examples, cost calculations, design tips, and suggestions for future applications.

Bonding and sealing are the two keywords to describe the scope of the handbook, but we will also study other jobs which are close to — and use the same techniques or products — such as filling, potting, jointing, caulking, assembling, formed in place gaskets, etc.

So in total I have identified some 80–90 chapters, a huge handbook of some 3,000 pages, which will be split into a series of 7 or 8 volumes — each one 350–500 pages — that will be published over a period of 4 years. This will make it more feasible and economical from the readers' standpoint (because they may buy only the volumes which deal with their business and the issues they have in mind) as every chapter will be published immediately after being written because the scientific and technical knowledge may become obsolete after 10 years or so.

Each chapter also provides lists of standards, testing methods, suppliers, and of course, an extensive bibliography for those who wish to learn more about specific, theoretical, scientific issues.

1.3. A Unique Feature of This Handbook Is the Fact that the Readers May Use it in Four Different Ways

General information, basic and scientific knowledge. Our readers may want to know some general rules of bonding such as design and calculation of structural adhesives, theory of adhesion, physics and chemistry of adhesives and sealants or they may need to select the equipment for applications. They may also wish to understand the physics and chemistry of the different types of adhesives and sealants.

All readers should read the several *general* chapters such as "Technical characteristics of adhesives and sealants", "Application equipment", "Physics and chemistry of adhesives and sealants", as these will provide the basic information required in order to understand other chapters.

by industries. Understandably, every reader will read the chapter devoted to their own industry, but they should also dig deeper in order to find similar techniques that could be transferred from another industry.

A good example is the possibility of the transfer of bonding techniques from aircraft manufacturing to other metal industries such as automotive, transportation, sandwich panels, and metalworking.

The aircraft industry was a pioneer in bonding, and started to bond structurally with wood in 1910–1920, aluminum during World War II, and composites in the 1980s.

Fig. 1 shows composite parts and many other parts that are bonded in the AIRBUS A380 — the largest aircraft in the world.

Thus, the high-performance adhesives which were developed during and after World War II, mainly epoxies, were then used later in automotive manufacturing. These were used for metal bonding in the 1950s in the assembly of steel stiffeners on hoods, at General Motors. Now composite parts are bonded in the car industry, with polyurethanes and structural acrylics, for instance (the first application was the bonding of exterior FRP panels on a steel skeleton in the Renault ESPACE, in 1983, with polyurethanes). More and more parts are bonded in a car and so each car needs several kilograms of adhesives at least and, of course, the same amount of sealants.

Another amazing adaptation is the use of cyanoacrylates, initially developed in the 1960s for very fast bonding of plastics, rubbers, and metals, but now employed in surgery, for fast suture of wounds. This began during the Vietnam War and is now used for domestic surgery!

Electronic parts are so tiny that only mini droplets of fast setting adhesives and sealants need to be used for their assembly and, although each piece of equipment uses only grams (or less) of adhesives, when multiplied by the huge number of units produced, it comprises a very big market in terms of Dollars or Euros.

And there are many other recent developments, such as do-it-yourself adhesives and sealants, laminating adhesives for flexible packaging, reinforcements in construction and civil engineering, many agglomerations (composite boards, etc.), dental repair, repairs on metal parts, etc.

Adhesives are now expected to enter other markets where they are not yet used: textiles and garments, medical products, replacing some welding operations, or railroad wagons and heavy industries.

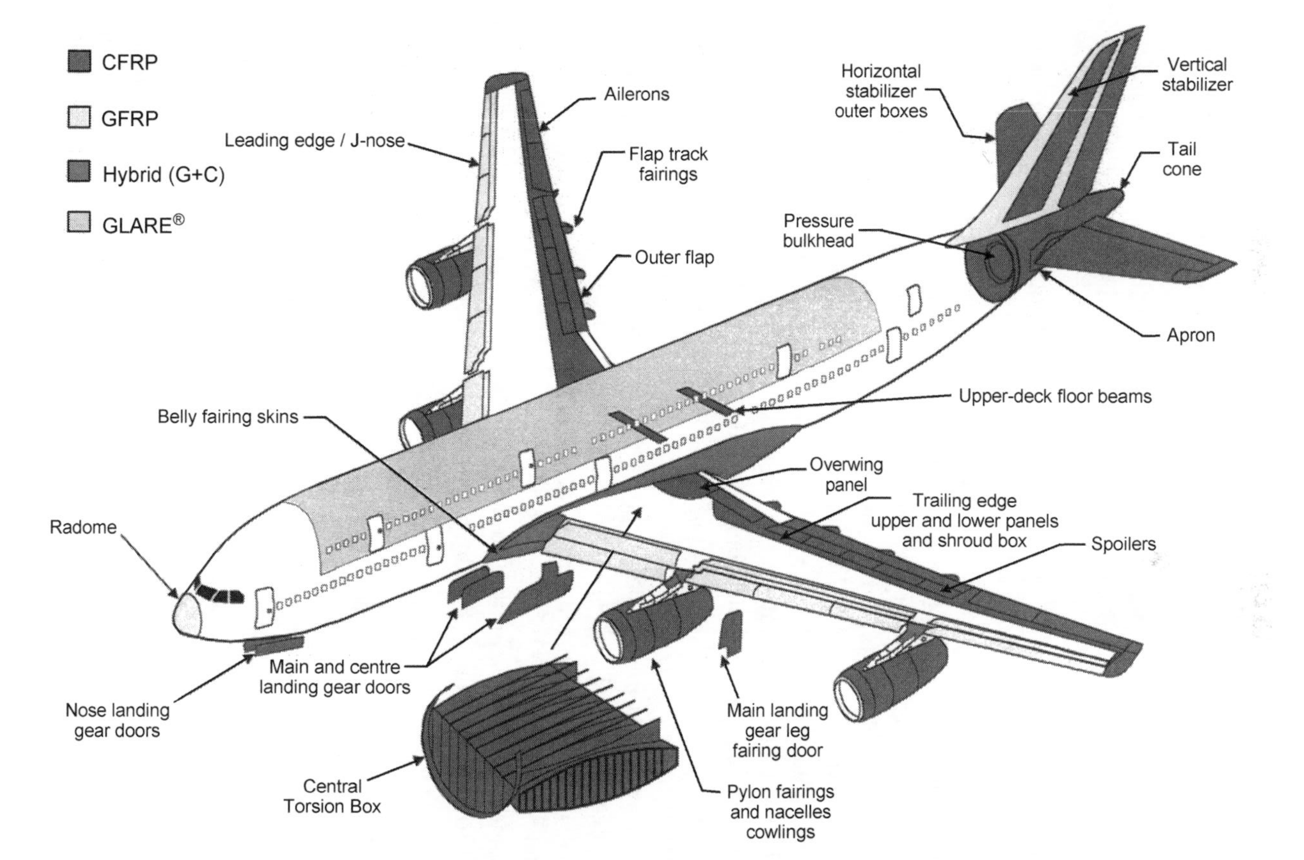

Figure 1: A380 composite materials application. Not shown: CFRP passenger floor panels and struts. *Some composite parts are bonded, others are not*. Several metal parts are also bonded in modern civil and military aircrafts. (Colour version of this figure appears on p. xi.)

I aim to give the reader many examples in differing industries, in order to suggest the smart *transfer of technologies*.

In order to complement their knowledge, the engineers who work with structural materials in automotive, aerospace, bonding of metals, plastics, and composites should also, of course, read the chapters: "Design and calculation of bonded parts", "Physics and chemistry of adhesion", "Surface preparation before bonding", "Metal bonding", "Bonding composites", and also "Epoxy adhesives", "Application equipment", etc.

by chemical types of adhesives and sealants. This is another way of addressing bonding techniques.

After reading some chapters on the various industries, the reader will need specific and detailed technical characteristics concerning the adhesives or sealants that they may consider for their own applications/end uses. For instance, those who wish to bond metal parts, those working in aircraft construction, automotive, and transportation, should read the chapters: "Epoxy adhesives", "Engineering adhesives", and maybe also "Heat stable adhesives".

Suppliers of adhesives, sealants and related products and product-oriented people, suppliers of raw materials, chemists, students, etc. should also read or scan through chapters in the sections dealing with chemical types.

by materials to be bonded. Adhesives are selected according to their required performances and also for the materials to be bonded. Metals, composites, plastics, and glass need structural adhesives, while paper, packaging, graphic arts, and pressure-sensitive goods use non-structural adhesives, based on quite different raw materials and techniques.

For instance, engineers in the automotive industries should also read the chapters: "Bonding metals", "Bonding composites", "Bonding rubbers", "Bonding plastics", etc.

The manufacturer of pressure-sensitive products should read the chapters: "Acrylic adhesives", "Styrenic hot melts", "Application equipment", "Testing and laboratory equipment", etc.

Thus, the readers should be able to navigate easily through the handbook and will be helped in several ways:

- in each chapter the authors and I have placed signs, footnotes, or detailed sentences which suggest which chapter to go to for more information on a given topic;
- a huge alphabetical index can be found in each volume as well as at the end of Volume 4 (midway) and Volume 8, with a total of some 2,500 significant keywords for all industries and all techniques or products (each industry has its own issues and its own keywords), and a detailed list of contents in each volume will help the reader find what they need. Each keyword will direct the reader to the many pages where this issue is addressed.

Therefore, the reader should be able to use the handbook in a variety of different ways.

Many chapters stand alone in their content, so that the reader can study them without necessarily having to refer to previous chapters.

1.4. A Comprehensive Book for Everybody

In this handbook, some chapters are scientific developments, which may require some technical or scientific university graduate background, but other chapters (generally those classified by industries) are both scientific and technological and can be read by anybody in a variety of job titles, businesses, or from different backgrounds. For example:

- industrial users such as engineers, technicians, scientists, designers in factories and laboratories, sales, purchasing, etc.
- construction, decoration and civil engineering companies, architects, contractors
- universities, professors, students, and also vocational high schools
- scientific and technical organizations, R&D organizations, trade organizations
- all material manufacturers and distributors, as their customers must assemble or seal their materials sooner or later, not only with mechanical fasteners but also increasingly with adhesives
- suppliers of all raw materials utilized for adhesives and sealants manufacturing, and production and testing equipment
- and, of course, all the adhesives, sealants, waterproofing equipment manufacturers and distributors.

I have designed the handbook so that it can be used for training students, engineers, or newcomers in any industry which use assembly, bonding and sealing, production engineering, etc.

1.5. All Authors Are High-Level Specialist Scientists, Engineers, or Chemists

The reader will get top quality, reliable information from people who have been working for many years in their given fields with good or high levels of responsibility.

We have gathered in excess of 65 authors, each being a well-known specialist in their given field of expertise.

1.6. List of Contents for Volumes 1–4

In order to allow the reader to plan acquisitions and reading, I have listed below the contents of the first volumes. A detailed list will, of course, be indicated on the Contents page at the beginning of each volume. I do, of course, reserve the right to add and alter contents of future volumes during the 4-year period in which this handbook is produced, but — while the reader is advised to check each volume's contents before purchase — the following should give a good idea of how the different books are to be produced.

Volume 1: Basic concept and high tech bonding:
Chapter 1: Introduction — how to use the handbook
Chapter 2: Technical characteristics of adhesives and sealants, by Philippe Cognard, Editor and Consultant
Chapter 3: Polyurethane adhesives and sealants, by scientists of Loughborough University, UK
Chapter 4: Surface preparation before structural bonding of metals and composites, by John Bishopp, Consultant in UK
Chapter 5: Aircraft and aerospace, pioneer for adhesive bonding, by John Bishopp
Chapter 6: Adhesives for electronics, by Guy Rabilloud, former General Manager Cemota

Fig. 2 shows one large application of adhesives in electronic goods that is now a very large market for high-performance adhesives and potting compounds. Fig. 3 shows an enlargement of the epoxy adhesive film on a bonded aluminum honeycomb.

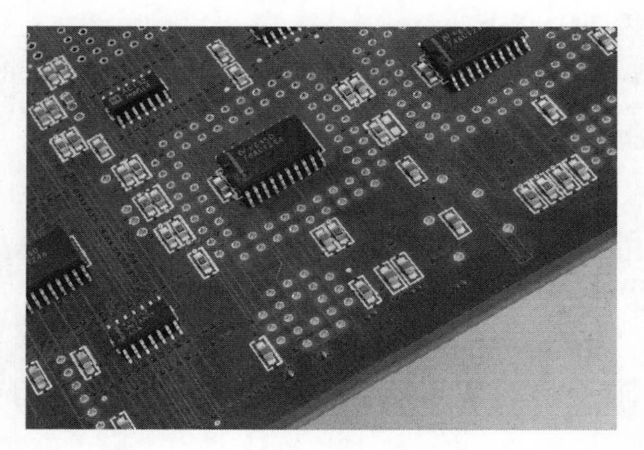

Figure 2: Electronic adhesives is now a huge market — USD400 million in 2004 worldwide. (Colour version of this figure appears on p. xii.)

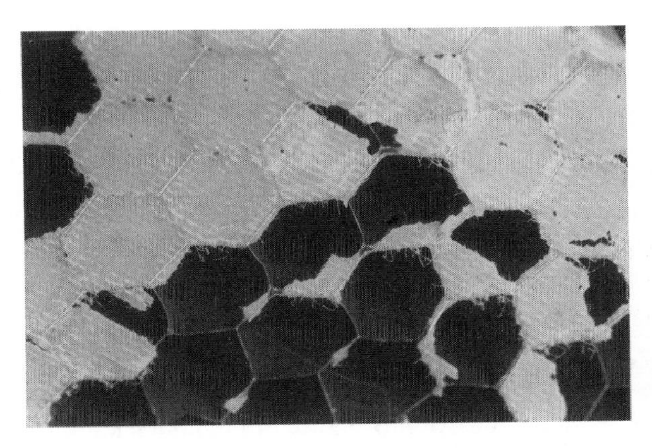

Figure 3: Close view of bonded aluminum honeycomb in a sandwich panel, showing (in white) the epoxy adhesive film after peeling of the aluminum sheet facing.

Several volumes will be published during 2005–2007 which contain chapters linked to the chapters in Volume 1, for instance: "Bonding metals", "Bonding composites", "Epoxy adhesives", "Bonding in automotive", "Structural adhesives", "UV curing", "Application equipment", and others.
Volume 2: General knowledge, application of adhesives, new curing techniques:
Chapter 1: Theory of adhesion, by John Comyn, Professor, De Montfort and experts from Loughborough University
Chapter 2: Application equipment for adhesives, by Philippe Cognard
Chapter 3: Design and calculation of bonded joints, by Richard Moulds, National Adhesives, UK
Chapter 4: UV curing, by Christian Decker, Mulhouse University, France
Chapter 5: Heat stable adhesives, by Guy Rabilloud
Chapter 6: Repairs of structural metal and composite parts, Dr Keith Armstrong, UK
Chapter 7: Flexible bonding, by Dr Burchardt, Sika Switzerland
Volume 3: Adhesives and sealants for construction, woodworking, etc.
Chapter 1: Furniture and woodworking adhesives, by Philippe Cognard
Chapter 2: Acrylic dispersion adhesives, by Dr Urban, Scientist, BASF, Germany
Chapter 3: Markets for construction and civil engineering adhesives and sealants, by Philippe Cognard
Chapter 4: Construction sealants, by Philippe Cognard
Chapter 5: Silicone sealants, by Andreas Wolf, Dow Corning Europe
Chapter 6: Thermosetting formaldehyde based adhesives, by Dr A. Pizzi
Volume 4: Adhesives for large volume applications and industries (Packaging, Graphic Arts, Hot melts, etc.)

Chapter 1: Markets for large volume applications, by Philippe Cognard
Chapter 2: Physics and chemistry, classification of adhesives and sealants, by Philippe Cognard
Chapter 3: Hot melt adhesives markets: graphic arts, woodworking, assembly by D. Grgetic, Nordson corp, USA
Chapter 4: EVA hot melt adhesives, by C. Laurichesse, Atofina
Chapter 5: Hot melt adhesives based on styrenic polymers, by Kraton Polymers
Chapter 6: Pressure-sensitive adhesives, by Luc Heymans
Chapter 7: Adhesives and sealants based on VAE emulsions, by Wacker company, Germany
Chapter 8: Adhesives for packaging and paper, bookbinding, by P. Cognard
Chapter 9: Polyurethane adhesives for laminating and packaging, by J. F. Lecam, Bostik-Findley

There will then be four more volumes that study many other types of adhesives and applications in all the other industries as well as the latest techniques available.

The reader will see that chapters have been written by many authors from several major industrial countries and companies, who will explain the various techniques and requirements of many different markets. It is also hoped that the handbook will be found useful throughout the world and especially in developing countries where adhesives, sealants and related techniques are increasingly being employed.

For instance, the chapters on footwear will be quite useful for the Chinese, Indian or Brazilian shoe manufacturers as today 60% of all shoes sold in the world are made in those countries. So, when compiling this handbook, consideration of their needs, the techniques and machines they use, the types of shoes they manufacture, etc. were taken into account.

The same theory applies to the woodworking industries in Far East Asia, Indonesia as well as the more developed European Nordic countries, and for the electronics industry in Japan and East Asia, etc.

1.7. Any Other Ideas or Suggestions?

As the handbook will be published as an 8-volume series, over a period of 4 years, I will have time to adjust its content and scope (as mentioned above). Therefore, if the reader feels there are other subjects that I — and the contributing authors — should study, please let me know and we will do our best to address these issues (for example, with regard to new and important applications, new chemical types, etc.).

My address is available from the Publisher, Elsevier, in Oxford.

I would also welcome more authors.

1.8. The Publisher — Elsevier

Elsevier is the leading publisher in the world for scientific and technical information. The group Reed Elsevier publishes thousands of scientific and technical journals, magazines and trade journals, in materials engineering, chemistry, physics, construction, medical sciences, etc. and also organises a number of very important trade shows such as BATIMAT, the biggest trade show for construction in the world, composite shows and as well as others.

1.9. Foreword to Chapter 5: "Aerospace, Pioneer in Structural Bonding"

I am sure this chapter will be of great importance for many of our readers.

Aircraft construction was the first "high tech" industry to use adhesives bonding as soon as the first airplanes left the ground, some 100 years ago, and now modern aircraft use large quantities of bonded metal and composite parts. If aircraft, which are subjected to high stresses and require 100% infallibility can be bonded, then everything else can also!

For readers who are involved in metal structural bonding, for automotives, metal working, mechanics, composites, etc. this chapter will provide a great deal of very valuable information about adhesives for metals and composites. It also shows how aircraft bonding developed step by step over the last 50 years from a historical perspective.

The reader should also refer to the chapter "Surface pretreatment for structural bonding" where the author, John Bishopp, provides detailed explanations on the various techniques. Good surface treatment before bonding is very important and makes the difference between a robust bond or a weak one.

John has worked for many years in *R&D* for CIBA GEIGY and HEXCEL, leaders in structural epoxy adhesives.

1.10. Foreword to Chapter 2: "Technical Characteristics of Adhesives"

This is a very important chapter because it provides all the definitions, all the technical characteristics of adhesives (QC control, use characteristics, physical, chemical and mechanical properties, durability, safety and cost) and testing methods, so that it is almost *mandatory* to read it before reading other chapters.

However, readers may save time by going directly to the definition and test method of each technical characteristic, when they need it.

The many illustrations show the testing methods and equipment (Fig. 4).

Hundreds of testing standards exist and users of adhesives should be made aware of a number of these.

The Series Editor, Philippe Cognard, has a comprehensive knowledge of adhesives gained after 35 years of experience in different leading roles within the industry.

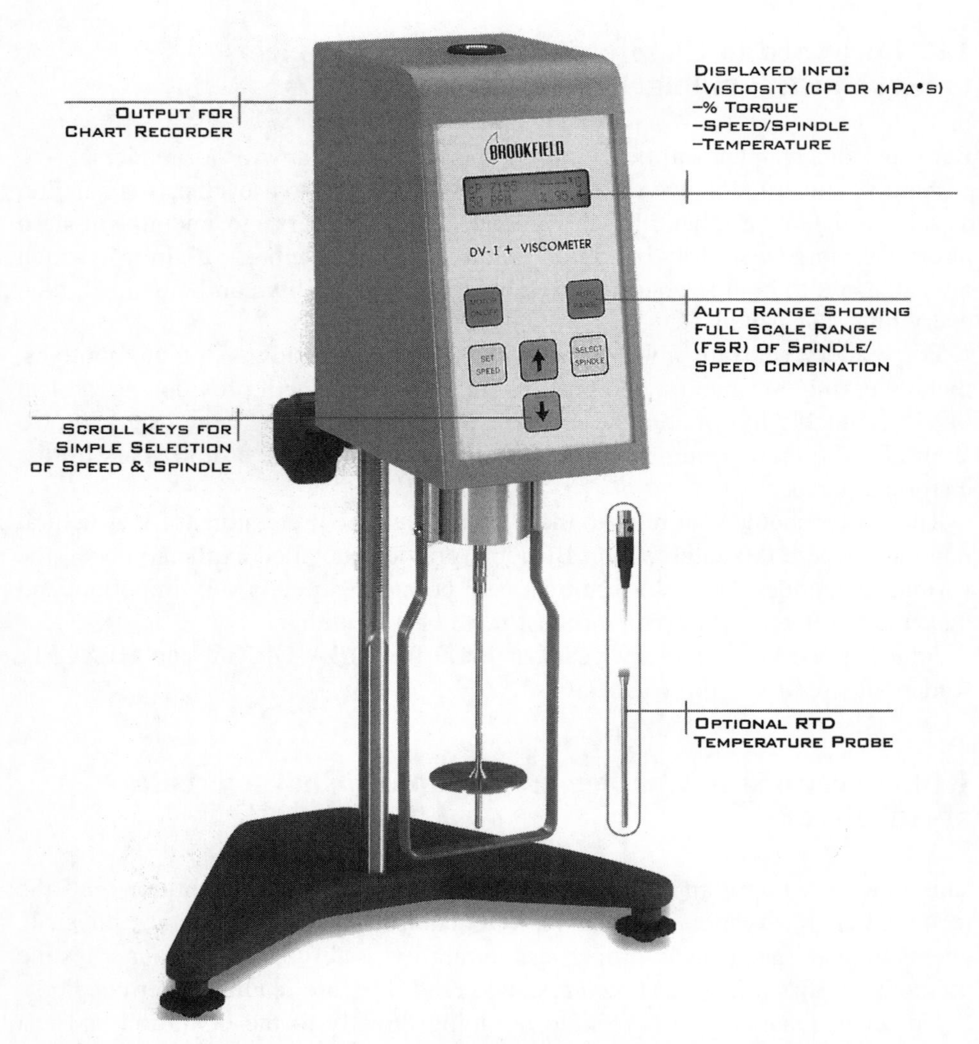

Figure 4: Testing equipment for adhesives (DV-1 + Viscometer of Brookfield).

1.11. Foreword to Chapter 3: "Polyurethane Adhesives and Sealants"

Polyurethane is the fastest growing chemical family in the adhesives and sealants industry, as polyurethanes enjoy a very versatile chemistry: One or two components, chemical curing, humidity curing or PUR reactive hotmelts, flexible, elastomeric or rigid adhesives, sealants, and potting compounds. All kinds of adhesives and sealants may be formulated with numerous different PU, polyols and isocyanate raw materials.

Polyurethane chemistry and adhesives and sealants is such a large and important topic that several chapters will be spent covering this subject: A general chapter in Volume 1 (written by a team of scientists from Loughborough University in the UK), PU sealants in Volume 3, PU flexible adhesives and sealants by Sika in Volume 2 and again in Volume 5 probably, and PU adhesives for laminating and packaging in Volume 4 by Bostik-Findley.

Readers (chemists, engineers, designers) from the various industries, and suppliers of adhesives, sealants, raw materials should read these chapters.

1.12. Foreword to Chapter 6: "Adhesives for Electronics"

These are very special adhesives that have specific properties (such as electrical properties).

This is a new, high tech market, where only grams of adhesive are used for each equipment. The number of pieces manufactured globally is so large (billions and more per year), and the adhesives of such high performance and high cost, that the worldwide market reached 800 million USD in 2003.

Thousands of companies/factories are now using electronic adhesives for computers, portable phones, Hi-fi systems, TV sets, domestic appliances, automobiles, aircraft, etc.

Guy Rabilloud, worked for years in this industry as General Manager of CEMOTA, a manufacturer of heat stable and electronic adhesives and also has written several excellent books and many patents on the subject.

You are now ready to learn more about adhesives and sealants, and I hope you will find many useful ideas in this Handbook.

Philippe Cognard, Editor
March 2005

In the following pages a list of suppliers corresponding to Volume 1 can be found.

1.13. Volume 1, Handbook of Adhesives and Sealants

1.13.1. List and Addresses of Suppliers

For Chapter 2, "Technical characteristics of adhesives and sealants", I have listed suppliers of testing laboratory equipment and sources of standards.

For the very large corporations, who have many companies and affiliates in several countries, I have provided only the corporate or main addresses. I cannot provide all of the address, phone, fax and internet details, and so the reader should seek this information from the local companies.

1.13.2. Laboratory and Testing Equipment

AMETEK/LLOYD Instruments	AMETEK TCI Division, 8600 Somerset Drive, Largo, FL 33773, USA. Tel.: 1 727 536 7831, fax: 1 727 539 6882, www.lloyd-instruments.co.uk (in America) and Lloyd Instruments Ltd, Forum House, 12 Barnes Wallis Rd, Segensworth East, Fareham, Hampshire PO15 5ST, UK. Tel.: 44 (0) 1489 486 339, fax: 44 (0) 1489 885 118 (in UK)
ASCOTT Analytical Equipment	Units 6 Gerard, Lichfield Road Industrial Estate, Tamworth, Staffordshire B79 7UW, UK. Tel.: 44 (0) 1827 318040, fax: 44 (0) 1827 318049, www.ascott-analytical.com
BROOKFIELD Engineering Laboratories	11 Commerce Blvd, Middleboro, MA 02346-1031, USA. Tel.: 1 508 946 6200, fax: 1 508 946 6262, www.brookfieldengineering.com
CAMBRIDGE Applied Systems	196 Boston Avenue, Medford, MA 02155, USA.

ChemInstruments Inc.	9349 Hamilton Dive, Mentor, OH 44060, USA. www.chemsultants.com
Fatigue Dynamics Inc.	969 Decker Road, Walley Lake, MI 48390-3217, USA. Tel.: 1 248 669 6100, fax: 1 248 624 3028
HAAKE GmbH	Dieselstrasse 4, D-76227 Karlsruhe, Germany.
INSTRON Corp	100 Royal St, Canton, MA 02021, USA.
KRUSS USA	9305 B Monroe Road, Charlotte, NC 28270, USA. www.krussusa.com
LABOMAT ESSOR	ZA Portes de Paris, 37 Bd A. France, 93200 Saint Denis, France. Tel.: 33 1 48096611, fax: 33 1 48099685, www.info@labomat.com
PERKIN ELMER Instruments	710 Bridgeport Av., Shelton, CT 06484, USA.
Q-PANEL	800 Canterbury Road, Cleveland, OH 44145, USA and Express Trading Estate, Farnworth, Bolton BL4 9TP, UK.
Testing Machines Inc.	Expressway Drive South, Islandia, NY 11749, USA and ADAMEL LHOMARGY, ZA de l'Habitat no 6, route d'Ozoir, 77680 Roissy en Brie, France. Tel.: 33 1 64409210, fax: 33 1 64409211, www.testingmachines.com
Tinius Olsen Testing Machines Co	Willow Grove, PA 19090, USA. www.TiniusOlsen.com
VOTSCH Industrietechnik GmbH	Beethovenstrasse 34, D-72336 Balingen-Frommern, Germany. http://www.v.it.com
ZWICK GmbH	August Nagel Str. 11, D-88079 Ulm, Germany. http://www.zwick.de

1.13.3. Polyurethanes, Raw Materials Suppliers, Adhesives and Sealants Suppliers

1.13.3.1. Raw Materials

ALBERDINGK BOLEY GmbH	Dusseldorferstrasse 53, 47829 Krefeld, Germany. Tel.: 90 21 51 528 0, fax: 90 21 51 57 36 43, alberdingk@alberdingk-boley.de
BASF AG	EDD/K-H201, D-67056 Ludwigshafen, Germany.
BAYER	Geshaftsfeld LS-M AM (Adhesives Materials) Gebaude F1/F46, D-41538 Dormagen, Germany.
DOW Chemical Company	PO Box 1206, Midland, MI 48642, USA. Tel.: 1 989 832 1560, fax: 1 989 832 1465, www.dow.com
DOW Europe SA	International Development Center, 13 rue de Veyrot, PO Box 3, 1217 Meyrin 2, Switzerland. Tel.: 41 22 719 4111, fax: 41 22 782 7666
HUNTSMAN Belgium	Everslaan 45, 3078 Everberg, Belgium. Tel.: 32 2 758 9211, fax: 32 2 759 5501, www.huntsman.com/pu/ac
HUNTSMAN International	2190 Executive Hills, Auburn Hills, MI 48326, USA. Tel.: 1 248 322 7300, fax: 1 248 3227303, www.huntsman.com/pu/ac
MERQUINSA	Gran Vial 17, 08160 Montmelo, Barcelona, Spain. Tel.: 34 93 572 1100, fax: 34 93 572 0934, info@merquinsa.com
ROHM and HAAS	100 Independence Mall West, Philadelphia, PA 19106, USA. Tel.: 1 215 592 3000, fax: 1 215 592 3021, poweratwork@rohmhaas.com

1.13.4. PU Adhesives and Sealants Manufacturers

BOSTIK-FINDLEY	Place de l'Iris, 92062 Paris la Defense, France. Tel.: 33 1 47969465, fax: 33 1 47969690, www.ato-findley.fr
ASHLAND Specialty Chemicals Co	PO Box 2219, Columbus, OH 43216, USA. Tel.: 1 614 790 3333, fax: 1 614 790 3206
DOW Automotive	1250 Harmon Road, Auburn Hills, MI 48326, USA.
HENKEL	Henkelstrasse 67, 40191 Dusseldorf, Germany. Tel.: 49 211 7 97 0, fax: 49 211 798 4008.
HENKEL-TEROSON GmbH	Henkel Teroson Str, 57, 69123 Heidelberg, Germany.
KOMMERLING Chemische Fabrik GmbH	Kweibruckerstrasse 200, D-66954, Pirmasens, Germany. Tel.: 49 6331 56 0, fax: 49 6331 56 22 26, www.koe-chemie.de
ROHM and HAAS	already cited above
SIKA Schweiz AG	Tuffenwies 16, Postfach, CH 8048 Zurich, Switzerland. Tel.: 41 1 436 4040, fax: 41 1 436 4564, www.sika.ch
UNIROYAL Adhesives and Sealants	2001 West Washington St, South Bend, IN 46628, USA.

Many other adhesives and sealants manufacturers also offer polyurethane adhesives and sealants, but I have limited this list to a few leaders in this field. A very large list of some 150 major adhesives and sealant manufacturers, located in all major countries, is provided elsewhere in this handbook.

1.13.5. Suppliers of Adhesives and Equipment for Aerospace

LOCTITE Aerospace, now called HENKEL Technologies	2850 Willow Pass Road, Bay Point, CA 94565, USA. Tel.: 1 925 458 8000, fax: 1 925 458 8030, www.loctite.com

HUNTSMAN Advanced Materials (formerly VANTICO)	Duxford, Cambridge CB2 4QA, UK. Tel.: 44 (0) 1223 493 000, fax: 44 (0) 1223 493 002, www.vantico.com/adhesives
HEXCEL USA	Dublin Blvd., Dublin, CA 94568-2832, USA. Tel.: 1 925 551 4900, fax: 1 925 828 9202
HEXCEL UK	Duxford, Cambridge CB2 4QD, UK. Tel.: 44 1223 833 141, fax: 44 1223 838808
STRUCTIL	18 rue Lavoisier, BP 10, 91710 Vert le Petit, France www.structil.com
3M USA	3M Center St, Saint Paul, MN 55144, USA. Tel.: 800-3M-HELPS, fax: 800 447 2053, www.3M.com
LONZA SpA	Via Enrico Fermi 51, I-24020 Scanzorosciate (Bergamo) Italy. Tel.: 39 035 652111, fax: 39 035 652799

1.13.6. Equipment (Autoclaves, etc.)

M. C. Gill Corp	4056 Easy St, El Monte, CA 91731, USA. Tel.: 1 626 443 4022, fax: 1 626 443 6094, www.mcgillcorp.com
Aeroform Ltd (Autoclaves)	Dawkins Road Industrial Estate, Poole, Dorset BH15 4JW, UK. Tel.: 44 (0) 1202 683 496, fax: 44 (0) 1202 675 957, www.aeroform.co.uk
Terruzzi Fercalx Spa (Autoclaves)	Viale Bianca Maria 31, I-20122 Milano, Italy. Tel.: 39 03 54879811, fax: 39 03 54879800, www.terruzzi.fercalx.com

1.13.7. Suppliers of Adhesives and Potting Compounds for Electronics

ABLESTIK	20021 Susana Road, Rancho Dominguez, CA 90221, USA. Tel.: 1 310 764 4600, fax: 1 310 764 2545
DELO Industrial Adhesives	Ohmstrasse 3, D-86899 Landsberg, Germany. Tel.: 49 8191 3204 2, fax: 49 8191 3204-144, www.DELO.de
ELECO Produits/EFD	125 Avenue Louis Roche, ZA des basses Noels, 92230 Gennevilliers, France. Tel.: 33 1 47924180, fax: 331 47922272, www.eleco-produits.fr
Emerson and Cuming	46 Manning Road, Billerica, MA 01821, USA.
Epoxy Technology Inc	14 Fortune Drive, Billerica, MA 01821, USA.
General Electric Silicones	Waterford, NY 12188, USA.
LOCTITE Corp	1001 Trout Brook Crossing, Rocky Hill, CT 06067, USA. Tel.: 800 562 0560, fax: 203 571 5465
3M	already cited above
National Starch and Chemicals Company	10 Finderne Av. Bridgewater, NJ 08807, USA.
HITACHI Chemical Company	PO Box 233, Mitsui Building, Shinjuku-ku, Tokyo 163, Japan.
TOSHIBA Chemical Corporation	Hankyu Express Building, 339, Shimbashi, Minato-ku, Tokyo 105, Japan.

1.13.8. Standardization Organizations

ISO, International Standardization Organization	1 rue de Varembe' Case Postale 56 CH-1211 Geneva 20, Switzerland. Tel.: 41 22 749 01 11, fax: 41 22 733 34 30

ASTM American Society for Testing Materials	1916 Race Street, Philadelphia, PA 19103, USA.
NASA (National Aeronautics and Space Administration)	USA
SAE (Society of Automotive Engineers also publishes standards adopted by aerospace industries)	400 Commonwealth Drive, Warrendale, PA 15096, USA.
TAPPI, Technical Association of the Pulp and Paper Industry	PO Box 105113 Atlanta, GA 30348-5113, USA. Tel: 770 446 1400, fax: 770 446 6947
ANSI, American National Standard Institute	1430 Broadway, New York, NY 10018, USA.
Federal Specifications and Standards	USA.
MIL	American military specifications, mostly used for aerospace industries, USA.
CEN, Comite Europeen de Normalisation	Brussels
BS, British Standards	London, UK.
DIN, Deutsche Industrien Normen	Berlin, Germany.
AFNOR, Association francaise de Normalisation	11 Av. Francis de Pressense, 93571 Saint Denis La Plaine cedex, France. Tel.: 33 1 41 62 80 00, fax: 33 1 49 17 90 00
JAS, Japan Standards	Tokyo, Japan.

I would like to remind the reader that at the end of each volume lists of suppliers (of raw materials, adhesives, sealants, laboratory and manufacturing equipment, application and curing equipment, information) are provided and each author has prepared both a bibliographic list and a list of relevant standards.

Chapter 2

Technical Characteristics and Testing Methods for Adhesives and Sealants

Philippe Cognard

Philippe Cognard is an *Ingénieur* of the Ecole Supérieure de Physique et Chimie de Paris–France's leading Physics and Chemistry college – and he received his Diploma in 1964.

He started his professional career in the USA the same year at the Pittsburgh Plate Glass Company moving to the Bloomfield Adhesives and Sealants Division in 1966 as a research and development chemist.

Over the years, he has held many top positions in leading adhesive and sealant companies around the world. These include:

- *Rousselot*, France's main developer and supplier of adhesives – later to become Elf Atochem – in various positions in the R&D, marketing and sales departments.
- *Weber et Broutin* – now *Saint Gobain* – dealing with building adhesives and mortars
- *Ato-Findley* – part of the *Elf Aquitaine Group* that became *Bostik-Findley* in the *Total Group* after the merger between *Elf* and *Total* – as Marketing Director for Adhesives.
- In 1996, he was appointed as a Director of the *Ato-Findley Adhesives company* in Guangzhou, China where he oversaw the launch, marketing and sales development in this country of *Ato-Findley* adhesives and sealants for application areas such as; construction, woodworking and furniture, packaging, laminating, footwear, and many others. This gave him a broad and thorough knowledge of the huge Chinese market, along with all R&D, manufacturing, marketing and sales activities employed by the company.

With more than 36 years of experience in the adhesives and sealants industry, he has also written and edited several related technical books and journal papers.

Keywords: Accelerated weathering tester; Acid value; Add-on (or consumption); Adhesion, adhesion to various substrates; Allergic effects; Application of adhesives; Ash content; ASTM; Bacteria resistance; Boeing wedge test; Brittleness; Brookfield viscometer; BS is British Standards; CEN (Comité

Handbook of Adhesives and Sealants
P. Cognard (Editor)

européen de Normalisation); Chemical resistance; Chromatography; Cleavage, cleavage resistance; Climbing drum peel test; CNAM; Cold resistance; Consumption; Contact with food; Corrosion resistance; Cost (of adhesive, sealants); Coverage; Creep; Cross-linking, density of; Curing; Degassing under vacuum; Density; Dermatitis; DIN is Deutsche Institut fur Normen; Drying, drying time; DSC (Differential calorimetry); Durability; Durometer; Dynamometer; EB curing; Elasticity modulus; Electrical characteristics; Environment; Epoxy equivalent; Expansion coefficient; Explosivity; Fatigue; Fatigue testing; Fire resistance; Flammability; Flash point; Flexibility; Food, contact with; Formaldehyde content; Freeze–thaw cycles; Gap filling; Gel time; Glass transition temperature; Handling and storage; Hardness; Heat resistance; High Frequency curing/drying; Honeycomb test; Hydroxyl value; Impact, impact resistance; Infrared analysis; ISO (International Standardization Organization); Isocyanate value; Lap shear test; Loop test tack measurement; MAK value; Mechanical properties; Mechanical resistance; MFFT is minimum film forming temperature; Mixing, mixing ratio; Newtonian; NF (French norms); Noxiousness; Occupational exposure; Oil resistance; Open time; Optical properties; OSHA; Peel, peel strength; Penetrometer; pH; Plasticizer migration, resistance; Poisson ratio; Pot life; Pressure; Price per kg, per liter; QC is quality control; Rheology; Rheometer; Ring and ball test, temperature; Rolling ball tack tester; RT is Room temperature, room temperature curing; Rupture, modes of; Safety data sheets (SDS); Safety regulations; Sagging, sag resistance; Setting time, setting speed; Shear rate; Solids content; Specific gravity; Stability of the adhesive (in tanks, etc.); Standards; Steps in bonding operation; Storage, storage conditions; Strain; Strain/stress curves; Stresses; Surface tension; Tack; Technical data sheets (TDS); Tensile shear strength; Tensile tester; T_g is Glass transition temperature; TGA (Thermal gravimetry analysis); Thermogravimetry; Thixotropy; Threshold limits; Toxicity, toxicological data; Transportation regulations; UV curing; UV spectrography; Vapor transmission; Viscometer; Viscosity, viscosity units; VOC (Volatile organic compounds); Waiting time; Waste disposal; Water resistance; Weathering, Weather-o-meter, QUV accelerated weathering tester; Wedge test; Wetting, wetting angle; Working temperature and humidity; Xenon test

2.1. General Comments, How Adhesives and Sealants Work

All A & S follow the same pattern for bonding operations:

- first they are delivered or prepared for bonding as a liquid or paste form, or as solid granulates, or sometimes as a film or powder,

- they are usually based on a polymer and resin blend, which must be spread evenly on the substrates to be assembled so that they must be a fluid during application and wet the substrates,
- then the parts are assembled together and the adhesive film must become solid either by drying or polymerization or curing through the use of a hardener, or by cooling for hot melts.

(We will study the theory of adhesion in Volume 2 and the various modes of setting and curing in another chapter of this Handbook.)

Consequently, 3 sets of technical characteristics may be measured on A & S, by the adhesive manufacturer or by the users:

- Properties of the liquid or paste before application, such as rheology, composition, solids, aspect, etc.
- Properties related to the method of use: how it is applied, how much should be used, how it will cure or dry or polymerize, etc.
- Properties of the cured adhesive after complete drying or curing, such as the mechanical, physical and chemical resistance of the bond or joint.

Of course, the technical characteristics depend on the types of adhesives – structural, nonstructural – the types of formulations – thermoplastic, drying type or hot melts, thermosetting, chemically curing and also the end uses: some end uses are very demanding such as high-performance structural adhesives for metal bonding, for aircrafts construction, etc., some are less demanding such as bonding paper or cardboard. But anyway, whatever the A & S, we will find out that they always require the user or designer to know some 20 technical characteristics at least for a proper utilization and for adapting it to its end use (refer to the list in Table 1).

We can identify here more than 70 different characteristics that may be useful to measure. This is an exhaustive list but, of course, for one single given application there might be only 15–25 characteristics to consider. However, the selection of an adhesive or sealant appears to be more complex than what many people believe. Also, it is important that the A and S manufacturers provide as much technical information as possible on their technical data sheets (TDS).

2.1.1. Reasons for Testing

There are several reasons for testing A & S:

1. to control the quality of an A & S and a bonding process, to check that the technical characteristics of the A and S meet the figures indicated in the TDS and also meet the customer's requirements,

2. to select an A & S adapted to: a given assembly, given materials to be bonded or sealed, given method of application and setting, and to all the requirements for bonded parts,
3. to confirm the effectiveness of a bonding process such as: surface preparation of substrates, application techniques, drying or curing conditions, durability and strength of the bond,
4. to investigate parameters that may bring differences in performances of the bond or may lead to failure.

Table 1: List of all the technical characteristics that may be measured on adhesives.

Before use:
Rheology and viscosity, consistency, thixotropy
pH
Solids content, ash content
Specific gravity
QC analysis: UV, IR spectra, chromatography, DSC, thermogravimetry
Acid value, epoxy equivalent, hydroxyl value, isocyanate value
Characteristics of use:
Viscosity and rheology
Wetting of the substrates (interfacial tension, critical surface energy)
Working temperature, minimum film forming temperature
Mixing ratio of two components A and S
Modes of setting, curing, polymerisation, drying
Pot life, gel time
Ring and ball temperature
Gap-filling properties, thickness of joint
Coverage or consumption
Stability of the adhesive in the applicator's tank
Sag resistance and flow
Waiting time before assembly
Tack
Open time
Pressure required
Possibility of radiation curing (UV, EB)
Setting time and setting speed
Storage conditions and storage life

(*Continued*)

Table 1: Continued

Mechanical properties of the cured adhesive:
- Adhesion to various substrates
- Tensile shear resistance
- Tensile resistance
- Peel strength
- Cleavage resistance, wedge test
- Curve strain/stress, elongation at break
- Impact resistance and resiliency
- Creep under constant load
- Hardness
- Flexibility

Physical characteristics of cured A and S
- Elasticity modulus and flexibility, Poisson ratio
- Heat resistance, T_g (glass transition temperature), ring and ball temperature
- Cold resistance, brittleness
- Expansion coefficient
- Electrical characteristics (resistivity/conductivity, dielectric constant, loss angle, ionic purity)
- Optical properties (refraction index, light transmission)
- Vapour transmission, waterproofing
- Degassing under vacuum

Chemical characteristics of cured A and S:
- Resistance to water and humidity, corrosion, salt spray
- Resistance to oils, grease and plasticizers
- Resistance to chemicals (acids, alkalis, solvents, etc.)
- Density of cross-linking

Durability in various life conditions, fatigue resistance

Safety characteristics:
- Flammability, flash point
- Toxicity, noxiousness
- VOC, environment position, occupational exposure, threshold limits
- Allergic effects, etc.
- Formaldehyde content
- Contact with food

Cost:
- Price per kg, litre, gallon, sq m or sq ft
- Specific gravity, coverage or consumption

2.1.2. Test Methods, Standards

Each characteristic may be measured by one or several methods of testing.

For adhesives we have numbered some 300 standard test methods in the US alone plus 200 in Europe, which apply to some 50 different technical characteristics.

The leading industrial countries have their own test methods, standards and specifications:

- In USA, there are ASTM standards (Committees C 24 and D 14), SAE, US Federal specifications, MIL specifications (for military applications such as aerospace), Society of Automotive Engineers (SAE), several professional organizations have also enacted standards, for instance, Technical Association for Pulp and Paper Industries (TAPPI), Pressure Sensitive Tape Council (PSTC) and others.
- In Europe, we have the German DIN standards and norms, the French AFNOR, the British BS, and a few other (Spanish, Swedish, etc.), but a huge effort has been made since 1985 to systematize all these test methods into European norms enacted by the Comité européen de Normalisation (CEN), in Brussels and Paris.
- There is also the international standard organization (ISO) in Geneva, which is trying to systematize on a worldwide level and has already issued a large number of international standards test methods.
- Some professional organizations have also developed their own specific standards such as Association des fabricants europeens de rubans adhésifs (AFERA) for pressure-sensitive tapes, the European association for pre-adhesed labeling industry (FINAT), TAPPI, etc.
- Japan and China have also their own standards for A & S.
- Let us finish this long list by indicating that some very important industry leaders have also established specific test methods and their own specifications, for instance, BOEING and AIRBUS in aircraft construction or all the major automotive manufacturers.

2.1.3. Sealants Standards

Many technical characteristics or test methods are similar for adhesives and sealants, for instance, viscosity, specific gravity, shear strength. But the test methods usually differ because sealants have different properties and different functions from adhesives: sealants must usually fill a large gap (several mm), they must remain flexible and much softer in order to accept movements, while adhesives must provide a stronger and harder bond.

In the present chapter, we will study all the technical characteristics and test methods of adhesives, but only 20% of these characteristics and test methods are suitable for sealants. Therefore, there will be another chapter devoted to the characteristics and test methods of sealants; also in the various chapters which will study sealants there will be sections related to their testing. We suggest our readers to refer to these chapters in Volumes 3, 5 and 6, such as "Silicone sealants", "Construction sealants", "Industrial sealants", etc.

2.1.4. General Methods of Testing

There are several general standards such as terminology, lists of technical characteristics of adhesives, methods of preparing test specimen, sampling methods, etc. These standards exist in CEN, ASTM and ISO (refer to the annex "List of standards" at the end of Volume 5), for instance, ASTM D 907 Terminology of Adhesives.

2.2. Technical Characteristics of A & S before and during Use

2.2.1. Quality Control (QC) Tests

When the supplier manufactures the A & S, it mixes or polymerizes a base polymer with other raw materials: solvents, water, resins, rubbers, tackifier, plasticizers, fillers, hardeners, catalysts, additives, etc. In order to ensure that this mixing has occurred properly in the right proportions, and that the finished A & S meets its specifications (as indicated on its TDS), the manufacturer usually measures several significant characteristics.

2.2.1.1. Rheology and Viscosity

Rheology is the study of the change in form and the flow of matter, embracing elasticity, viscosity and plasticity. Viscosity is the internal friction of a fluid, caused by molecular attraction, which makes it resistant to flow. This friction becomes apparent when a layer of fluid is made to move in relation to another layer. The greater the friction, the greater the amount of force required to cause this movement, which is called shear. Shearing occurs whenever the fluid is moved or distributed, as in pouring, spreading, mixing, spraying, etc. Highly viscous fluids require more force to move than fluid ones.

In Fig. 1, parallel planes of fluid of equal area A are separated by a distance dx and are moving in the same direction at different speeds V_1 and V_2. The force required to maintain the difference in speed is proportional to the difference in speed through the liquid or "velocity gradient"

$$F/A = \eta\, dv/dx$$

where η is a constant for a given material, which is called its viscosity.

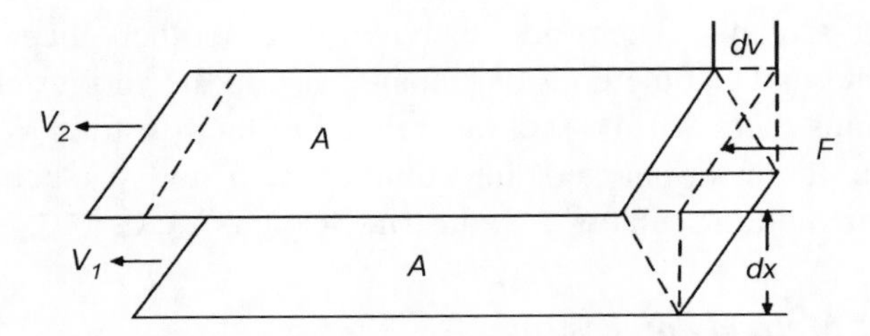

Figure 1: Relative movements of the layers of a fluid.

The velocity gradient dv/dx is a measure of the change in speed at which the intermediate layers move with respect to each other. It describes the shearing effect in the fluid and it is called "shear rate". It will be written as S hereunder.

The term F/A indicates the force per unit area required to produce the shearing action. It is referred to as "shear stress" and will be symbolized by "F1". Its unit of measurement is in dynes per square centimeter (dynes/cm^2).

So now the equation becomes:

$$\text{Viscosity} = \eta = F1/S = \frac{\text{shear stress}}{\text{shear rate}}$$

Viscosity is usually expressed in pascal seconds (Pa sec) or millipascal seconds: a material requiring a shear stress of 1 dyne/cm^2 to produce a shear rate of one reciprocal second has a viscosity of 100 Pa sec.

2.2.1.1.1. Different behaviors of fluids

Fluids may have different rheological behaviors:

Newtonian fluids

Newtonian fluids are represented in Fig. 2A and B: viscosity is constant when the shear rate varies. Thus at a given temperature, the viscosity will remain constant regardless on which viscometer model, spindle or speed is used to measure it.

But most fluids are non-Newtonian.

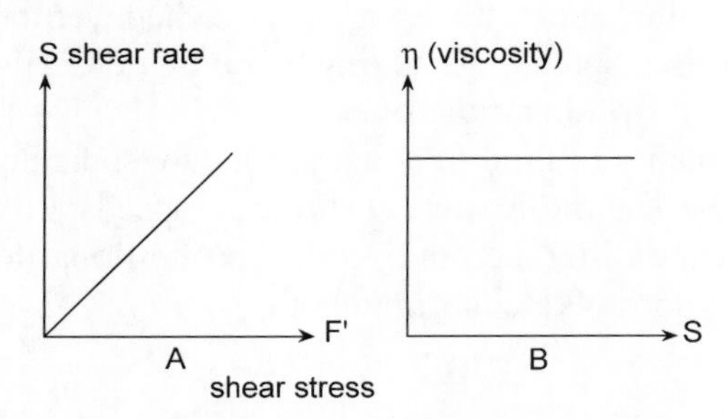

Figure 2: Newtonian fluids.

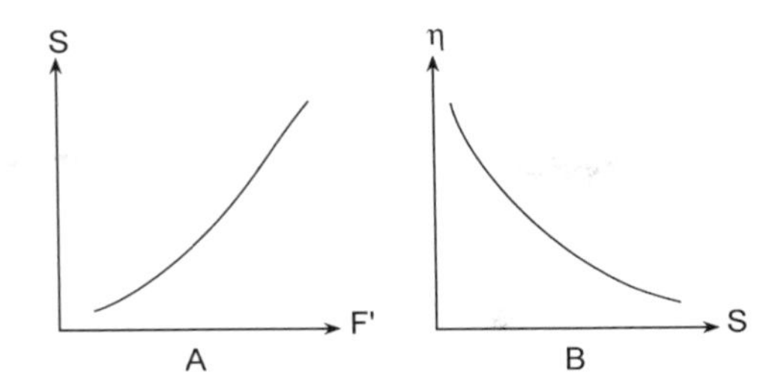

Figure 3: Pseudoplastic fluids.

Non-Newtonian fluids
Here the relationship $F1/S$ is not a constant — when the shear rate varies, the shear stress does not vary in the same proportion or even necessarily in the same direction. The viscosity of such fluids will therefore change as the shear rate varies. Thus, the experimental parameters of the viscometer model, spindle and speed all have an effect on the measured viscosity. This measured viscosity is called the apparent viscosity and is accurate only when explicit experimental parameters are adhered to and indicated in the recorded measure.

There are several types of non-Newtonian flow behavior, the most common types being the following:

Pseudoplastic Refer to Fig. 3A and B. The most common pseudoplastic types are some water-based paints, adhesives, emulsions and dispersions. When you mix the material, it becomes more fluid in a way similar to yoghurt.

Dilatant Refer to Fig. 4A and B. It is rarer than pseudoplasticity, but it may be observed in fluids containing high levels of deflocculated solids, such as clay slurries and sand/water mixtures: we all have observed on the beach that if you press slowly your foot into the sand/water mix it will penetrate, but if you hit it strongly it will resist penetration. This may also be the case of sand/cement mortar.

Plastic Refer to Fig. 5A and B. This type of fluid will behave as a solid under static conditions; a certain amount of force must be applied to the fluid before

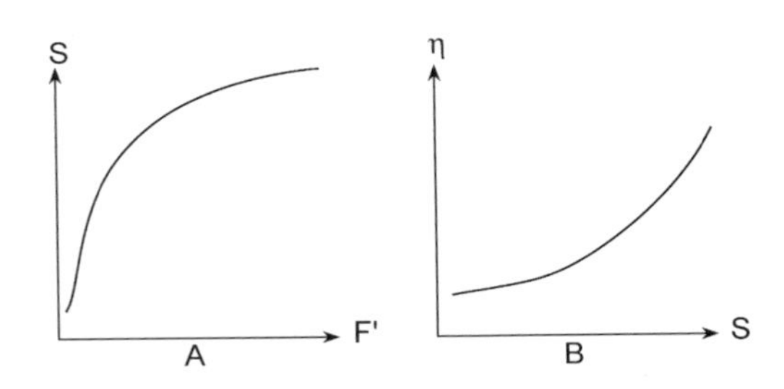

Figure 4: Dilatant fluids.

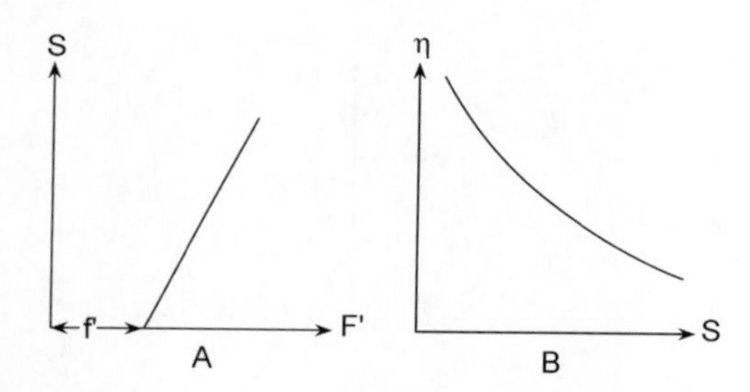

Figure 5: Plastic rheology: f = yield value.

any flow is started, this force f is called yield value. Tomato ketchup is an example of this type: it will not pour from the bottle until the bottle is strongly shaken. Once the yield value is exceeded and the flow starts, plastic fluids may display Newtonian, pseudoplastic or dilatant behavior.

Thixotropy and rheopexy Now let us consider what happens when time is considered. Some fluids will display a change in viscosity with time under conditions of constant shear rate. Thixotropic fluids show a decrease in viscosity with time when subjected to constant shearing, as indicated in Fig. 6. Rheopexy is just the opposite (Fig. 6). Rheopectic fluids are very rare and rheopexy would be quite detrimental for A and S! Thixotropy is frequently observed in materials such as greases, building adhesives, paints and it is done purposely by the formulator so that the adhesive will not sag when applied gently on a vertical surface but by brushing or rolling it will become more fluid and spread easily. Both thixotropy and rheopexy may occur in combination with any of the previously discussed flow behavior, or only at certain shear rates. The time element is variable: under constant shear, some fluids will reach their final viscosity in a few seconds, while others may take up to several days.

Hysteresis cycles: When subjected to varying rates of shear, a thixotropic fluid will react as illustrated in Fig. 7: the up and down curves do not coincide because

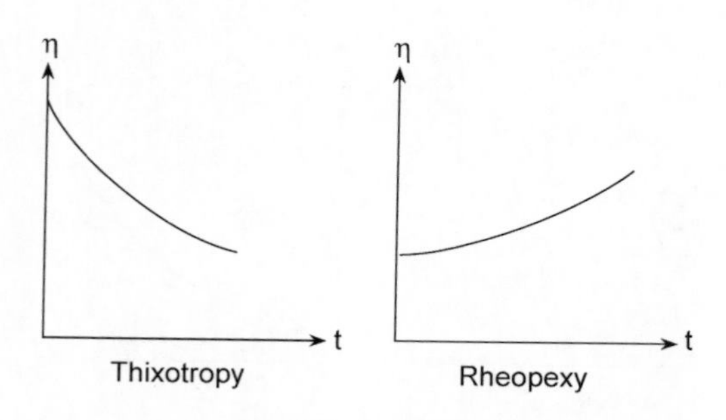

Figure 6: Thixotropy and rheopexy.

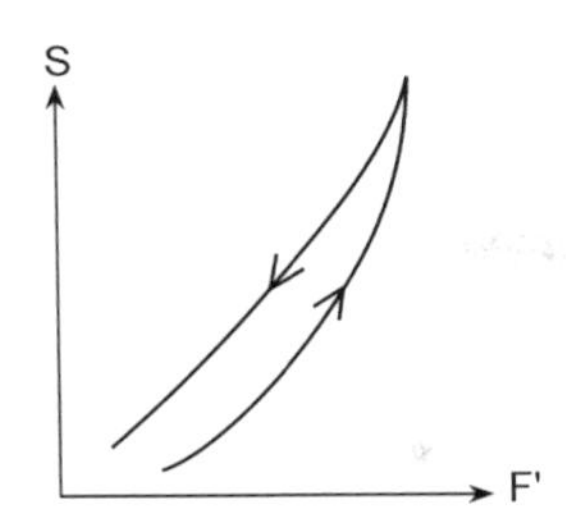

Figure 7: Hysteresis cycle of a thixotropic adhesive.

the fluid viscosity decreased with increasing time of shearing. Such an effect may or may not be reversible: some thixotropic fluids, if allowed to stand undisturbed for a while will regain their initial viscosity (for example, yogurt and thixotropic emulsion adhesives), while others never will.

The rheological behavior of a fluid can of course have a marked effect on viscosity measurement.

Laminar and turbulent flow: The concept of viscosity implies a laminar flow in which the movement of one layer of fluid past another does not cause a transfer of matter from one to the other, and viscosity is just the friction between these layers. Laminar flow exists until a certain speed beyond which a transfer of mass occurs: this is called turbulence. Molecules pass from one layer to the next one and dissipate energy in the process. Thus a larger energy input is necessary to maintain this turbulent flow.

Turbulent flow is used for the mixing operation during the manufacture of the A and S, in order to mix the different raw materials of the formulation. We will study this later in the chapter "Manufacturing techniques and equipment for adhesives and sealants".

A simple way for measuring viscosities of fluid adhesives is by using a special cup such as a Ford cup or ISO/DIN/NF/BS cup. The time necessary to empty the cup through its bottom hole is recorded (in seconds). Tables provide correspondence between cup measurements and viscosity in mPa sec.

2.2.1.1.2. Factors that affect the rheological properties

Temperature
Viscosity decreases when temperature rises. This is quite noticeable for the hot melts adhesives, as shown in Fig. 8.

Shear rate
Because the majority of fluids are non-Newtonian it is important to know the effect of shear rate. For instance, it would be impossible to pump a dilatant fluid through

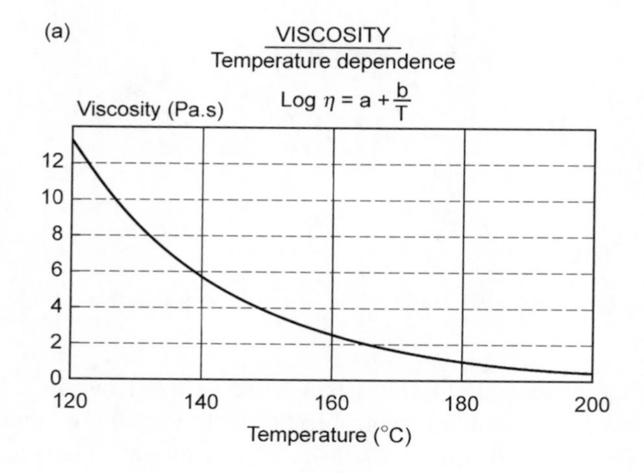

(b)

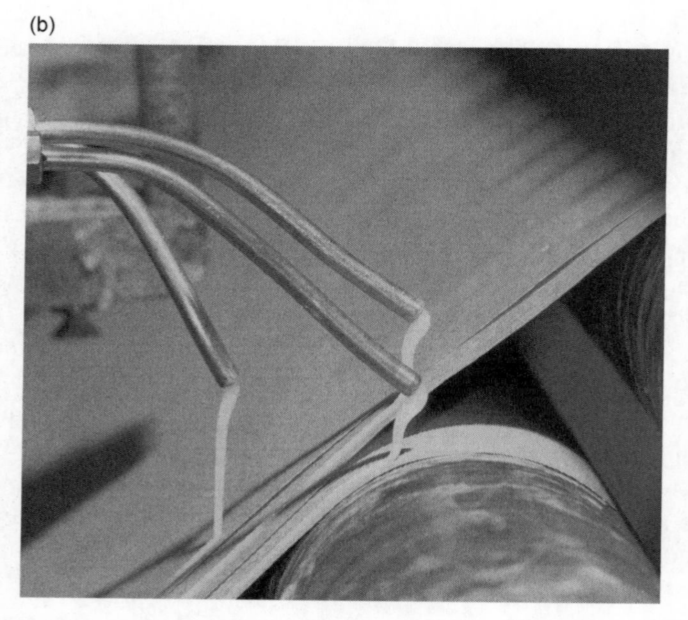

Figure 8: (a) Viscosity/temperature dependence. (b) A very simple way of applying an adhesive just by gentle pumping (manufacture of paper bags). Here the best would be to have a Newtonian adhesive. (Colour version of this figure appears on p. xii.)

a high-rate system where it may become almost solid inside the pump! When a material is to be subjected to a variety of shear rates during use, it is very important to know its viscosity at the various shear rates. It is always useful to measure viscosities at different shear rates in order to detect rheological behavior that may have an effect on processing or use.

Table 2 provides examples of shear rates found in A and S application with various equipments and manufacturing. (Source: Brookfield, USA).

Measuring conditions
The condition of a material during measurement of its viscosity can have a large effect on the result.

- the viscosity measurement techniques should be adhered to. Variables such as viscometer model, spindle/speed combination, sample size and temperature all affect the viscosity of the material.
- The sample material may be sensitive to the ambient atmosphere (oxidation, drying…).
- The sample must be homogeneous.

Time
As we have discussed, time elapsed under conditions of shear affects thixotropic materials. Many materials will undergo changes in viscosity during the chemical reaction, and also during storage of reactive A and S.

Viscometers: The main supplier is BROOKFIELD and we show several types of viscometers in Figs. 9 and 10. Other suppliers are HAAKE Germany.

Rheometers: Allow to measure the yield value. For instance, Brookfield SST 2000 may be used to measure viscoelastic behavior of pseudoplastic and thixotropic products.

Methods of measurement for viscosity: There are several standards:

- EN 12092 (2000) is a European standard
- ASTM D 2556: test method for apparent viscosity of adhesives having shear rate dependent flow
- ASTM D 1084: test method for viscosity of adhesives, ISO…

2.2.1.1.3. Resistance to penetration, consistency

For very thick products such as sealants, bituminous products, caulks, mortars, instead of viscosity it is more meaningful to measure the penetration of a needle into the product under a given weight and during a certain period of time, according to ASTM D 1321, D 1916, D 4950, DIN 51804, etc. (penetrometer).

2.2.1.2. pH

pH is just an indication; it can only be measured on water dispersions or solutions; it depends on the type of raw material; for instance, acrylic emulsions have alkaline pH while vinyl emulsions have acidic pH. It is used only as a quality control tool: the general idea when you manufacture an adhesive which is

Table 2: Examples of shear rates found in A and S application and manufacture.

Situation	**Sedimentation of powders in a liquid**	**Leveling**	**Extrusion**	**Dip coating**	**Mixing, stirring**	**Pipe flow**	**Spraying brushing**	**Milling fillers in a fluid**	**High speed coating**
Range of shear rate	10^{-6}–10^{-4}	10^{-2}–10^{-1}	10^{0}–10^{2}	10^{1}–10^{2}	10^{1}–10^{3}	10^{0}–10^{3}	10^{3}–10^{4}	10^{3}–10^{5}	10^{5}–10^{6}
Application	Paints, mortars	Paints	Sealants, heavy pastes	Fluid adhesives	Manufacturing	Pumping	Adhesives	Manufacturing	Roll application

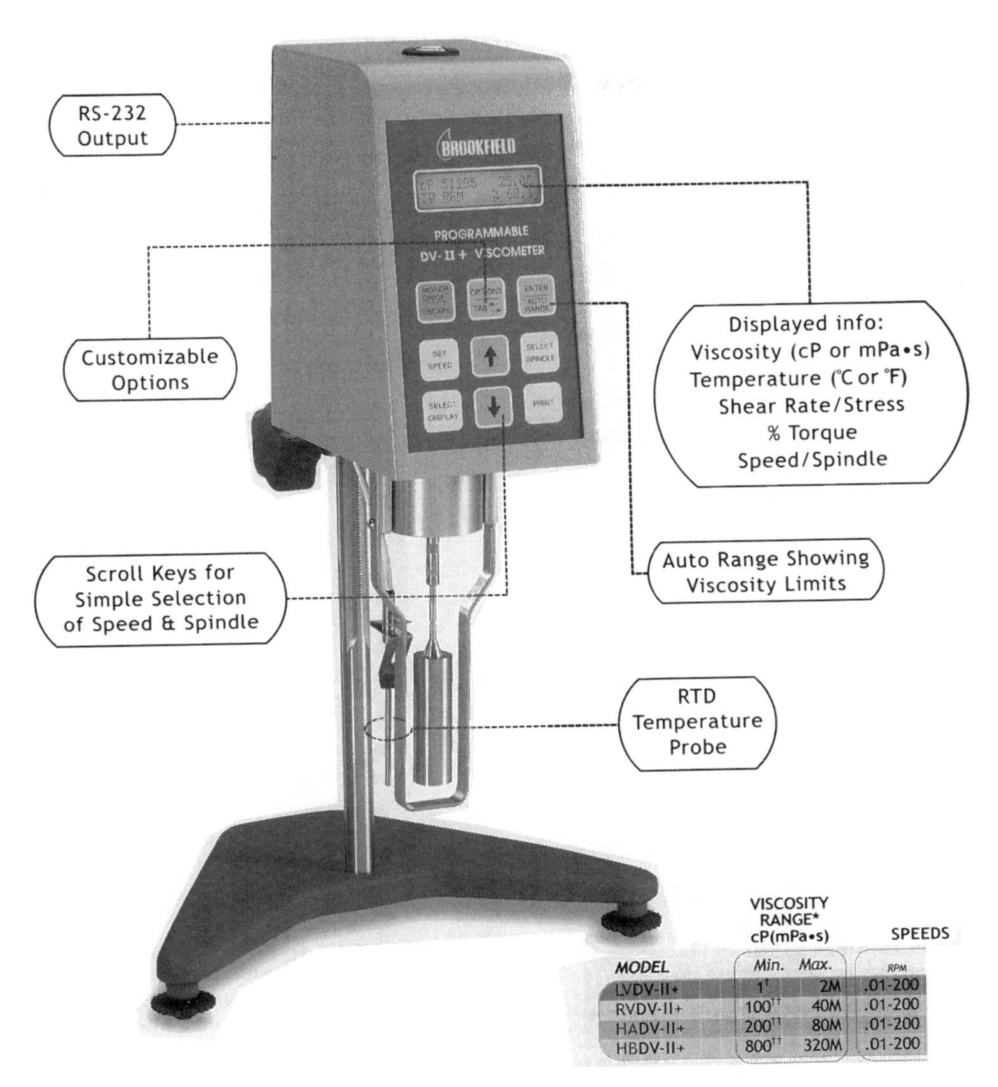

MODEL	VISCOSITY RANGE* cP(mPa•s) Min.	Max.	SPEEDS RPM
LVDV-II+	1†	2M	.01-200
RVDV-II+	100††	40M	.01-200
HADV-II+	200††	80M	.01-200
HBDV-II+	800††	320M	.01-200

Figure 9: Brookfield Viscometer DV-II +. (By courtesy of Brookfield, USA.)

a mixture of several raw materials is to control a set of 3 or 4 technical characteristics, for instance, viscosity, pH, specific gravity and another one appropriately selected; if all these characteristics are acceptable and if the worker has weighed correctly every raw material, there is a very high probability that the formulation has been followed completely.

pH may be measured according to European standard EN 1245 (1998), with a pH meter.

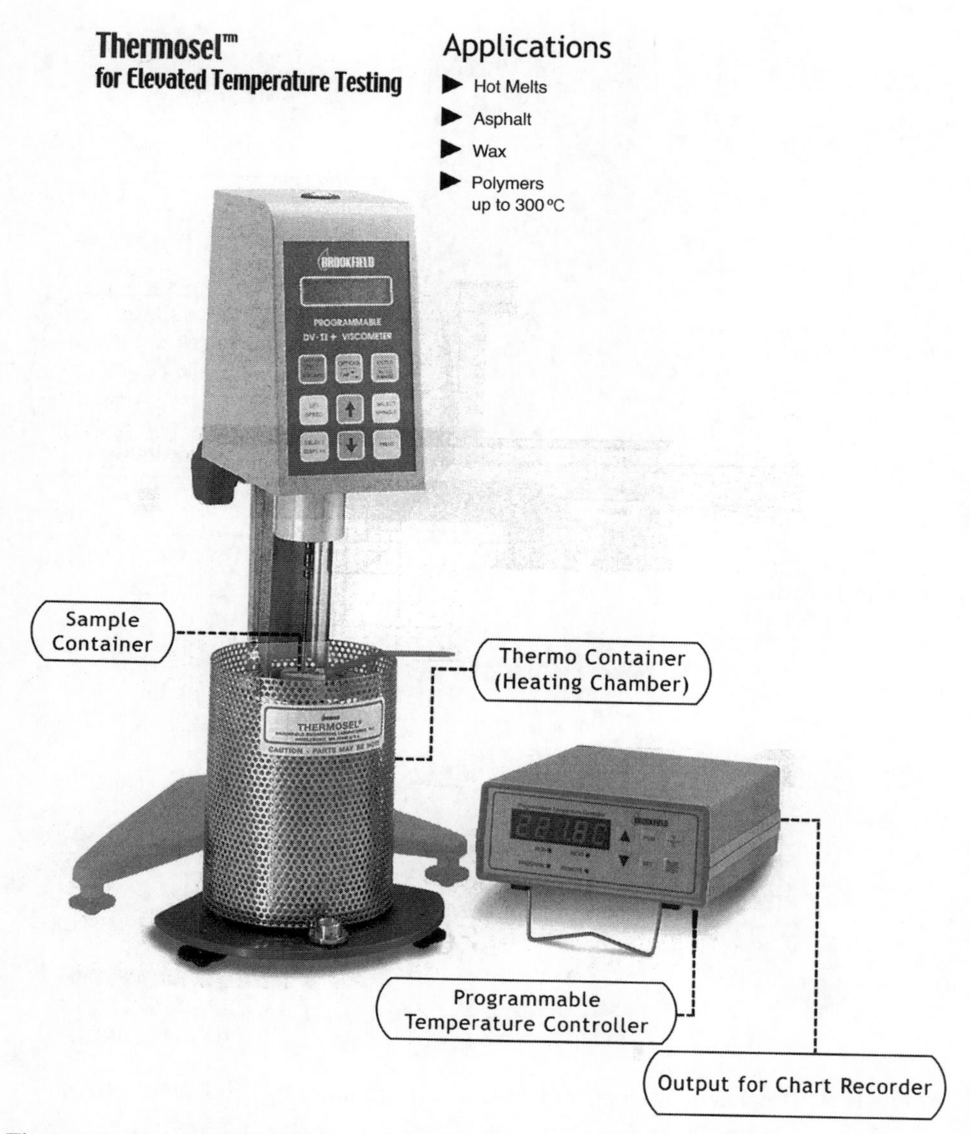

Figure 10: Brookfield Thermosel Viscometer for elevated temperature measurement of viscosity. (By courtesy of Brookfield, USA.)

2.2.1.3. Solids Content

Many adhesives and some sealants are dispersions or solutions of polymer, resin and other raw materials in a carrier liquid that may be water or solvents.

The dry solids measure the ratio of the active part of the adhesive to the total weight:

$$\text{DS\%} = \frac{\text{weight base polymer} + \text{weight of resin} + \text{weight of additives, fillers, plasticizers}}{\text{total weight of formulation}}$$

or: $100 - \text{DS} = \%$ of liquid carrier. The liquid carrier will evaporate during use, and therefore it has no economical value for the user.

Examples of dry solids:

- solvent-based adhesives (rubber based, etc.): 17–50%
- vinyl white glues: 45–70%
- UF, PF, RF (formaldehyde glues): 60–70%
- Solvent-based sealants (butyl, etc.): 80–85%
- Hot melts: 100%
- Epoxy adhesives: 100%
- Silicone sealants: 95–100%

Dry solids are measured for quality control. DS are also useful to determine how much water or solvent should be evaporated during drying and setting of the adhesive or sealant. In some cases, low solids content is a way for some manufacturers to reduce the cost of a water-based or solvent-based adhesive. If solids are too low, there might be not enough active adhesive in the glue line, which may become starved with voids or absence of adhesive between the two parts.

For some sealants, such as butyl or acrylic based, the solids content ranges from 75 to 85%, so that there is a shrinkage during drying, which may cause stresses and eventually problems in the joint. The solids content should be as high as possible. High quality and high elasticity sealants such as silicones, PUR, have always 100% DS.

Measurement of dry solids: The nonvolatile solids content is measured by weighing 2 g of A/S in a special aluminum cup to an accuracy of 0.01 g, then heating and drying the A/S until a constant weight is obtained (for instance, 30 min at 100–150°C according to adhesive type and evaporation rate of the liquid carrier). The solids content is the ratio of the sample weight after and before drying.

Many test methods are available for the determination of solids content:

- ASTM D 1489: nonvolatile content of aqueous adhesives,
- ASTM D 1582: nonvolatile content of phenol, resorcinol and melamine adhesives,
- CEN 827 (1995): determination of dry solids of adhesives,
- ISO,
- EN 542 (1995): determination of dry solids.

2.2.1.4. Specific Gravity or Density

Specific gravity of raw materials varies:

- from 0.8 to 1 for the solvents,
- from 1.1 to 1.5 for resins and polymers,
- from 1.2 to 2.0 for the mineral fillers,
- from 1.0 to 1.3 for rubbers,

so that specific gravity of A and S may vary from 0.8 to 1.5. Specific gravity may be measured according to EN 542 for paste adhesives and EN 543 for powders and granulate adhesives.

Measurement of specific gravity is useful for QC and also to evaluate the A/S consumption:

The weight W of A/S needed to fill a joint is: $W = V \times d$, where V is the volume of the void between the 2 parts and d the density of the A/S. It is customary in most countries to indicate consumption or add-on in g/m^2, but in fact in would be better to measure it in volume, and this is what is done in USA where the coverage is measured often in sq ft/gallon (area that may be covered with one US gallon of adhesive), but also in lbs per sq ft.

If an adhesive contains a large amount of fillers (such as calcium carbonate, calcium sulfate, sand that are dense) its specific gravity will be high, probably higher than 1.2 or 1.3, and consequently the consumption will be higher. Manufacturers' TDS should always indicate the density of A and S.

2.2.1.5. Ash Content, Mineral Content

This is a way to measure the content of mineral fillers and other raw material that do not burn. Ash content is measured by weighing accurately a few grams of adhesive in a refractory crucible and then burning it completely at high temperature: 500–900°C according to the needs. At these temperatures all the resins, polymers and organic compounds are burnt, their carbon atoms are also burnt and become CO_2 gas which evaporates. Only the mineral fillers will be left after complete combustion for 1 h. Calcium carbonate will also decompose at 900°C: $CO_3Ca = CO_2 + CaO$, and only CaO will be left so that combustion at 900°C is useful to measure the calcium carbonate content by difference between residue at 500°C and residue at 900°C. Mineral fillers being much cheaper than the other components of the A and S, a high ratio of mineral fillers may be a way to lessen the cost of the adhesive, and such a high ratio may lower the performances of the A and S. However, fillers may also increase cohesion, gap filling power, and also allow one to control the viscosity because they absorb a part of the polymer at the surface of their particles thereby thickening the A or S.

EN 1246 (1998) and ASTM D 5040 standards are used to measure the ash content.

2.2.1.6. Aspect, Color, Delivery Form

A and S may have different forms when delivered:

- liquids or pastes of different colors: white glues (vinyls, acrylics, VAE), beige or yellow glues (polychloroprene, UF), brown resorcinol–formaldehyde glue,
- solid granulates or pellets or rods for hot melts, white or beige colors,
- powders (cement mortars, hot melt powders),
- films with or without a carrier (thickness from 0.1 to 0.5 mm), such as some epoxies or heat sealable films,
- pressure-sensitive films and tapes,
- sealants tapes or other preformed shapes.

The color is given by the darkest resins contained in the adhesive (and, very rarely, by a specific pigment or carbon black contained in some rubber-based formulations). Color will almost always darken when the A & S is exposed to high heat and oxidation, and this is a way to evaluate the heat resistance of hot melts. It will yellow after exposure to sunlight, UV and oxidation. Color may be measured with the Gardner scale.

2.2.1.7. Spectrographic IR, UV, Chromatographic Analysis

These analyses are of course the best QC tests, because they provide a "finger print" of the formulation: a reference spectrum printed with the right formulation is used and all the spectra of all batches should match this reference perfectly. Each raw material will give its own absorption peaks, and the formulator should get all the spectra of the various raw materials in order to be able to find it in a formulation. Refer, for instance, to Fig. 11, which provides the spectrum of an epoxy resin, showing all the specific peaks of absorption.

Big repertories of spectra of many chemicals and raw materials are published by suppliers of equipment (Perkin Elmer, Cambridge Instruments, etc.). The experienced chemist can read the spectra and tell the types of polymers or rubbers or resins contained in the formulation and the reactive sites. Thus, this is also a way to analyze an unknown or competitive product, although one needs first to separate the various raw materials by classic laboratory techniques.

2.2.1.8. Differential Calorimetry (DSC)

This analysis technique measures the difference of energy absorbed by a product and by a blank reference when subjected to a controlled increase in temperature: during this increase in temperature, the product sample undergoes some physical

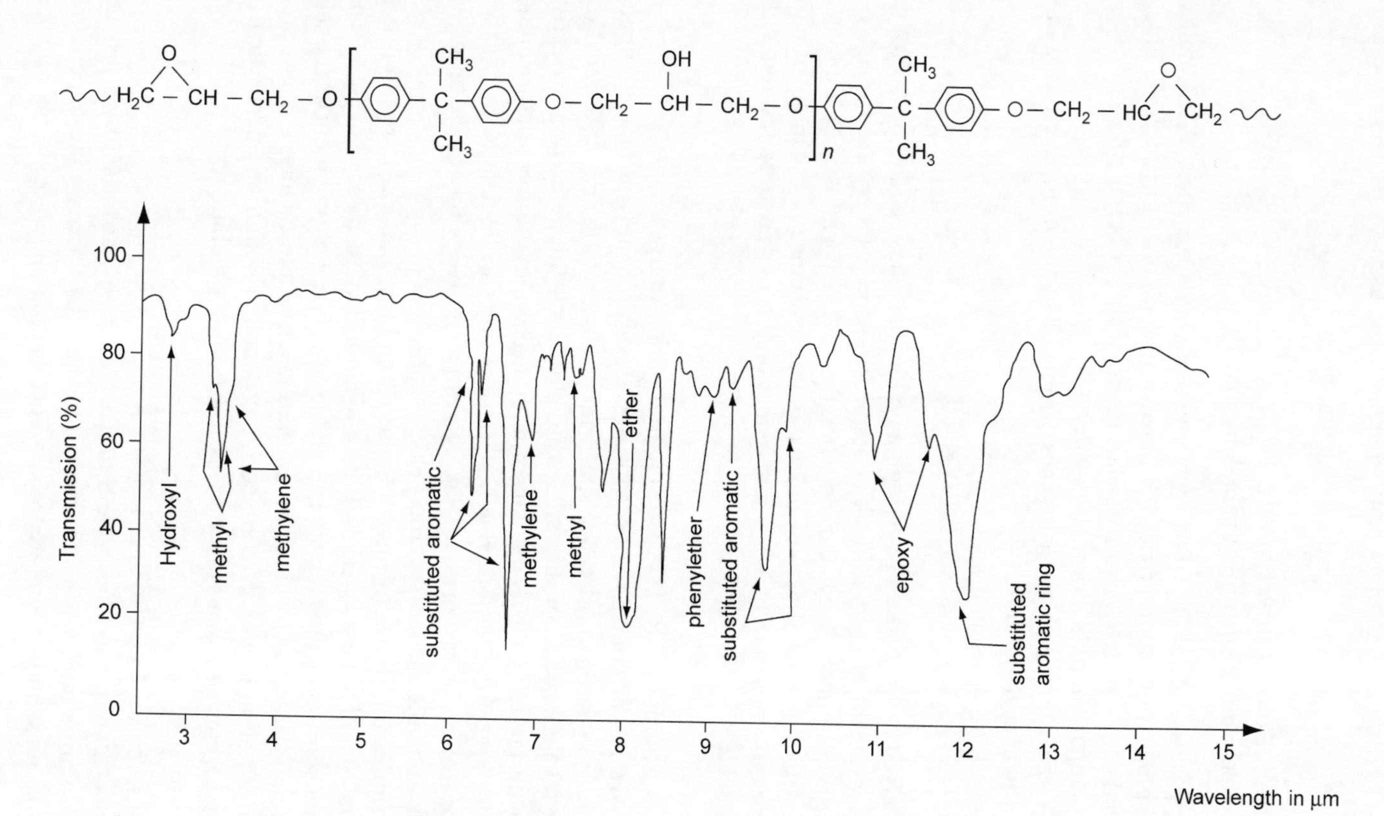

Figure 11: Reading the Infrared spectrum of an epoxy resin, the various chemical groups are detected by the absorption peaks.

transformation, chemical reaction that may absorb or generate energy. For instance, polymerization or oxidation decomposition give exotherms while melting, evaporation, etc., are endothermic.

DSC may be used:

- on nonpolymerized adhesive for QC control: for instance, Fig. 12 shows the glass transition temperature T_g and an exothermic peak of polymerization at 147°C,
- on polymerized adhesive in order to measure the percentage of polymerization, the T_g of the cured adhesive. When the cure reaction has been completed there is no more variation of the enthalpy.

2.2.1.9. Thermal Gravimetric Analysis (TGA)

In this technique, the weight of a sample is measured during a controlled increase in temperature.

For more information about these analyses, please refer to the chapter "Testing methods for adhesives and sealants" later in this Handbook (refer to the list of contents).

2.2.1.10. Acid or Hydroxyl Value, Epoxy Equivalent Isocyanate Content

According to the chemical type of A and S, the supplier will measure these values in order to determine the reactivity and the mixing ratio between part A or resin and part B or hardener.

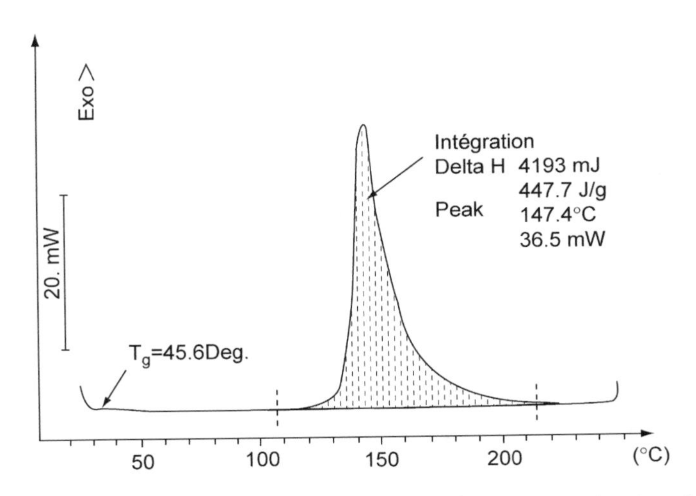

Figure 12: DSC analysis on an epoxy adhesive before polymerization (AV119 from ciba geigy/Vantico) the exothermic peak of polymerization at 147°C allows to compute enthalpy of the reaction and the range of curing temperature.

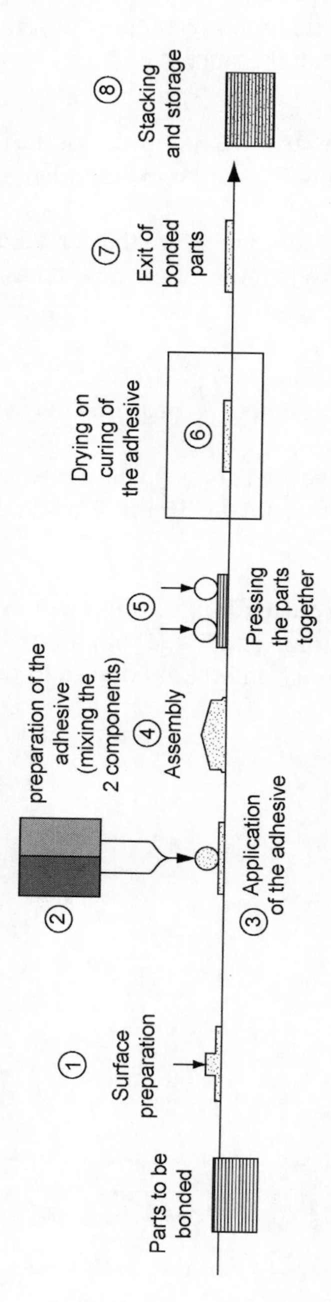

Figure 13: Successive steps of a bonding process.

Several standard methods exist:

- to measure hydroxyl value: EN 1240 (1998),
- for isocyanate content: EN 1242 (1998),
- for epoxy equivalent: refer to the chapter "Epoxy adhesives",
- for acid value: ASTM D 1994, determination of acid number of hot melt adhesives.

2.2.2. Properties Related to the Method of Use

In this section, we will study only the bonding properties of adhesives. For sealants, the methods of use are different because they are not supposed to give a strong assembly but instead to fill a gap and seal against humidity, water, etc., so that their technical characteristics are different. We will study their specific properties in the various chapters devoted to sealants in this Handbook, in Volumes 3, 5 and 6 (refer to list of contents).

Every bonding process involves 8 or 9 different steps (Fig. 13):

1. surface preparation: cleaning of substrates, surface preparation by mechanical, physical or chemical techniques,
2. preparation of the adhesive: mixing the 2 components, diluting, melting the hot melts,
3. application with various systems: hand tools, rollers, spray gun, extrusion, etc.
4. pre-drying if necessary (with water- or solvent-based adhesives),
5. assembly,
6. pressing together the parts with various equipments: hot or cold plates presses, rollers, clamps, etc.
7. drying, curing or polymerization of the adhesive in hot air oven, hot presses, autoclaves or with UV or HF curing,
8. exit of bonded parts, stacking, storage before shipment.

To each step are associated some technical characteristics of the adhesives, and adhesives must be adapted to the bonding process and bonding equipment as we will see hereunder.

Table 3 lists the bonding process steps, the corresponding requirements for the adhesives and the relevant characteristics. We will now study each of these "characteristics of use".

2.2.2.1. Viscosity and Rheology

Viscosity and rheology must be adapted to the application system: some need very fluid adhesives, for instance, spray guns need viscosities from 100 to 1000 mPa sec, needle applicators for electronics need also very fluid adhesives in order to flow through the fine diameter of the needle.

Table 3: Bonding steps, requirements and corresponding characteristics of adhesives.

Bonding steps	Requirements	Characteristics of adhesives
Surface preparation	Should be compatible with substrates and with the adhesive	– Adhesion to the substrates – Selection of adequate primer
Preparation of the adhesive	Mix properly the two components Respect mixing ratio	– Mixing ratio – Pot life of the mix
Application of the adhesive	Application must be performed: – at the speed of the equipment; – as an even coat, with the required coverage	– Viscosity – Wetting of materials – Coverage, consumption – Stability of the adhesive in the tanks – Melting temperature of the hot melts
Transfer of parts from application to assembly	After application, adhesive must remain sticky until assembly	– Waiting time – Open time
Assembly	Some adhesives can maintain parts together; others cannot and require a mechanical fixture (clamps, press, frames, etc.)	– Tack – Wetting of second substrate
Pressing and hardening (time, pressure, temperature)	Adhesives must dry or cure during this time	– Mode of setting – Speed of setting, drying, curing – Time, pressure, temperature for curing
Exit of bonded parts	The bond should not fall apart; adhesive must have already some strength	– Tack and hot cohesion – Speed of setting
Finishing, storage	Assemblies should resist finishing operations	– Complete setting time – Mechanical resistance of bond after setting
Transportation	May create mechanical or thermal stresses	– Resistance of adhesive to environment, handling, heat, moisture

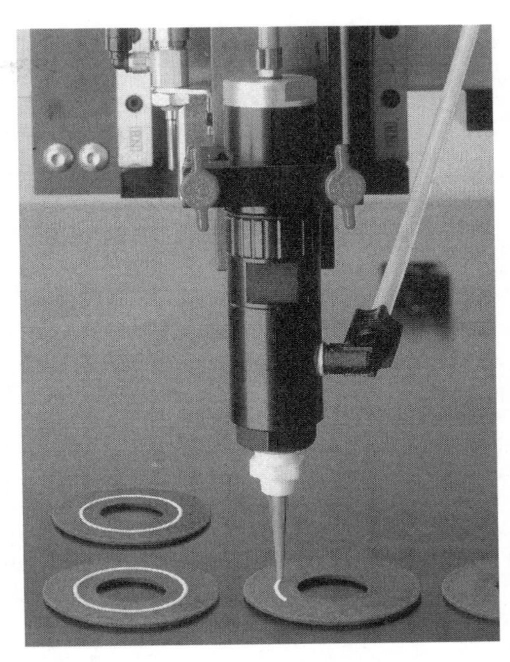

Figure 14: A piston valve EFD applies here automatically a bead of white glue of medium viscosity — beads of glue do not sag. (Source EFD USA/Dosage 2000 France).

Other applicators require higher viscosities and nonsagging A and S, for instance:

- extrusion guns used to apply epoxy or PU mastics for automotive bodies in white bonding need a high viscosity paste – from 50,000 to 500,000 mPa sec – so that the adhesive will fill the gap between parts up to 1 or 2 mm thickness without sagging,
- Fig. 14 shows a piston valve application of a medium viscosity glue (20,000 mPa sec) at high speed,
- construction and industrial sealants have also high viscosities, or consistency,
- mortars used in construction for thick joints – up to 10 or 15 mm thick – are also high viscosity products, from 100,000 to 500,000 mPa sec, or in better words high consistency, in order to resist sagging and flowing out of the joint.

Thixotropy may also be useful, for instance, in order to apply an adhesive onto a wall with a hand roller, without dripping. Refer also to Section 2.2.2.11: Sag resistance.

2.2.2.1.1. Viscosity versus temperature, hot melts

For all A and S, viscosity decreases when temperature rises, but this factor is more important for hot melts as indicated in Figs. 8 and 15. At 20°C hot melt

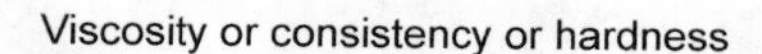

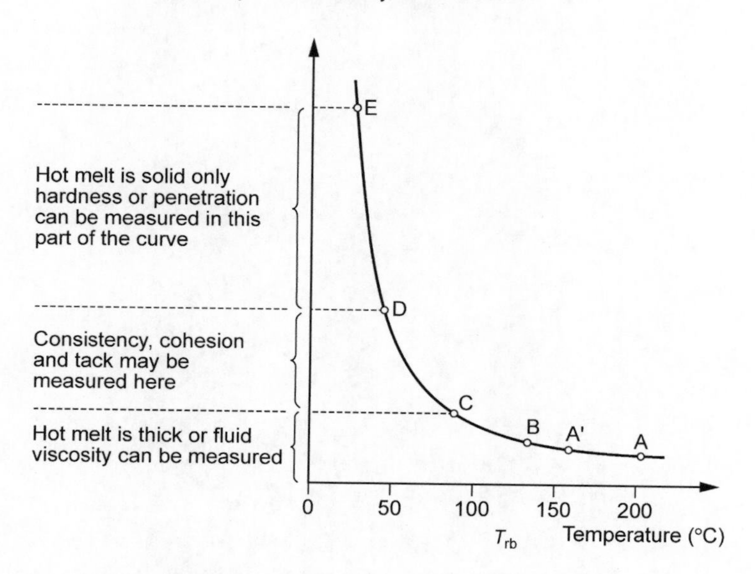

Zone AB The hot melt adhesive is liquid, its viscosity increases when temperature lowers - it can be applied only between A and A'.

Zone BC The adhesive "congelates" at high speed. T_{rb}(ring and ball temperature) gives an indication of "congelation" (or solidification) point.

Zone CD Development of cohesion and tack. The hot melt sets and starts to hold the parts together.

Zone DE Development of hardness and cohesion. The hot melt has reached now its full mechanical performances.

Figure 15: Hot melts: viscosity, consistency and hardness versus temperature.

adhesives are solid. Then they will gradually soften when temperature increases to 50–70°C, and they become "liquid" or fluid when temperature reaches 100–150°C according to types and formulations. Hot melts being high polymers do not have a sharp melting point as some other chemicals: instead they go through a progressive softening until they become more or less liquid, as shown in the figure. In order to get an indication of the change of state from solid to a paste, formulators use a test called "ring and ball" (Fig. 16). The equipment is gradually heated, for instance, 2°C per minute, until the steel ball falls through the ring immersed into the hot melt. The recorded temperature is called ring and ball temperature.

On the curve of Fig. 15, the recommended application temperature is located between A and A', which corresponds for many hot melts to temperatures ranging

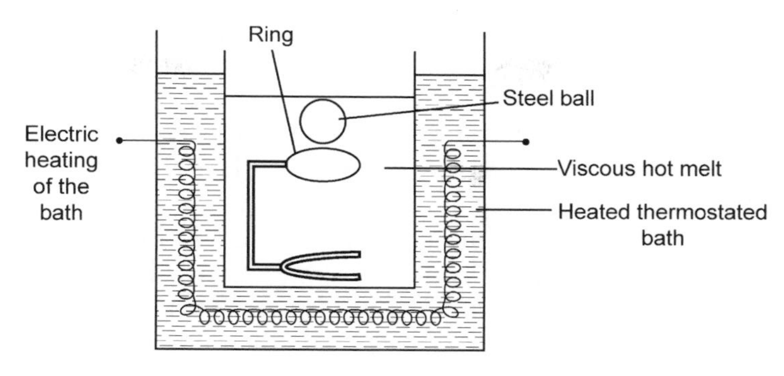

Figure 16: Ring and ball test apparatus for hot melt adhesives.

from 140 to 200°C. At these temperatures the hot melts are fluid (from 1000 to 5000 mPa sec) and can be applied on substrates by several means: hot melt guns, hot rolls, etc., and it will spread on/wet the substrates.

After application, the HM adhesive will follow the curve in the reverse way: from 200 to 100°C it will thicken quickly, become "solid" again below the ring and ball temperature and it will eventually become strong enough to hold together the parts when its temperature falls below 60 or 50°C.

2.2.2.2. Wetting of Substrates

Proper wetting of substrates is mandatory for a good adhesion (refer to the chapter "Theory of adhesion" in Volume 2).

Wetting of the substrates depends on several factors:

1. The angle of wetting depends on the type of adhesive (or sealant) and the nature of the substrate. Wetting follows the Young equation (refer to Fig. 17):

$$\gamma_{LV} \cos \theta = \gamma_{SV} - \gamma_{SL}$$

where γ_{SV} is the interfacial tension of the solid material in equilibrium with the fluid vapor, γ_{LV} is the surface tension of the fluid material in equilibrium with its vapor, γ_{SL} is the interfacial tension between the solid and liquid materials.

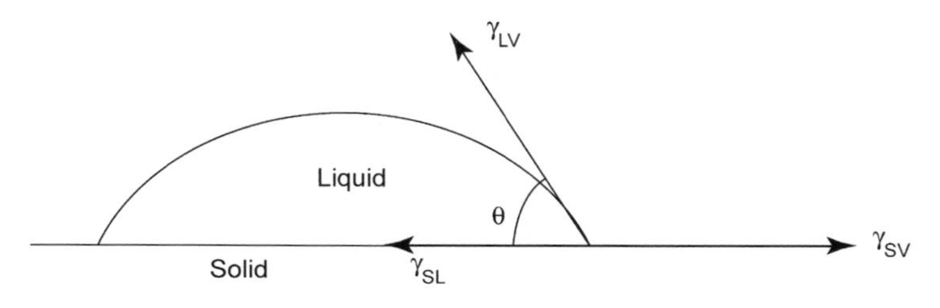

Figure 17: Wetting of a solid by a liquid.

Complete wetting occurs when $\theta = 0$ or $\cos\theta = 1$, i.e. when

$$\gamma_{SV} \geq \gamma_{SL} + \gamma_{LV}.$$

Therefore, wetting is good when the substrate surface tension γ_{SV} is high and its critical surface energy is high, and when the surface tension of the wetting liquid is low. Thus low-energy polymers and adhesives will wet easily high-energy substrates like metals, but most adhesives do not wet low-energy surfaces like polyethylene ($\gamma_c = 31\ \mathrm{mJ/m^2}$) or Teflon ($\gamma_c = 18\ \mathrm{mJ/m^2}$).

European standard EN 828 (1998) indicates how to measure wetting properties by measuring the contact angle and the critical surface tension of a solid surface.

2. Wetting depends also on the viscosity. Viscosity should be low enough so that the adhesive will spread on the surface easily. We can understand that a good wetting may be difficult to obtain with a sealant because sealants usually have high viscosity. In order to improve their wetting properties, sealant formulations should include a wetting agent. However, silicone sealants have very low surface tension of 21–22 mN/m and can wet many different substrates.

3. The pressure applied on the adhesive will force it to spread and penetrate into porous, fibrous materials and into the roughness of the surfaces. This is why parts to be bonded should always be pressed together with as much pressure as possible.

Fig. 18 shows the difference between a good wetting and a bad one. Probably 20% of bond failures result from poor wetting.

2.2.2.3. Working Temperature and Humidity

Ambient temperature and humidity have an impact on drying and setting speed. Most A and S are designed for utilization around 20°C and relative humidity (RH)

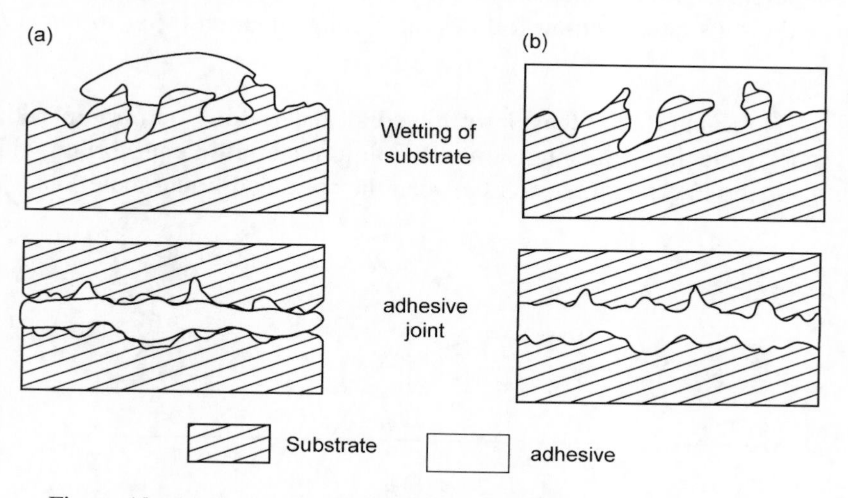

Figure 18: (a) Bad wetting. (b) Excellent wetting of all the surfaces.

between 40 and 80%, in most countries. If temperature is much higher, the A/S may start to dry on its surface (also called skinning), thus reducing the open time (see later, Section 2.2.2.14) and the wetting of the other substrate. This is especially true for solvent-based A/S.

Dew point: If ambient humidity is high there will be a problem with solvent-based adhesives formulated with fast drying solvents (such as acetone, hexane, ethyl acetate, etc.): the solvent will evaporate quickly, this will cool down the surface of the adhesive and the ambient humidity will condense on this surface, as a dew. It will be visible as a blush and a matt surface. Mini droplets of water condensed on the surface will prevent, for instance, contact polychloroprene adhesives to transfer to the other coated surface and the bond will be very poor. Thus solvent blend should be adapted to the weather conditions in the country where the adhesives are used.

2.2.2.3.1. Minimum film forming temperature

Low ambient temperature is quite detrimental to water-based emulsion adhesives. When the temperature is too low, the dispersed polymer particles will not coalesce or fuse together, and it will prevent the formation of a strong continuous film of adhesive. Instead, a powdery, cracked film will occur if the temperature is lower than a required minimum. There is a minimum film-forming temperature (MFFT) below which there will be no film formation and consequently no bonding. This temperature may be measured with several test methods: ASTM D 2354 and ISO 2115, by using specific equipments such as the Sheen Instrument minimum film temperature bar (Fig. 19): a microprocessor controlled stainless steel plate is cooled at one end and heated at the other. The sample to be tested is laid down at 75 μm film thickness and 25 mm width. After 45–90 min, a clearly defined coalescence zone will be visually obvious, and the temperature at this point will be recorded as MFFT.

For instance, emulsion vinyl glues display MFFT from +2 to +10°C. MFFT can be adjusted by proper formulation: for instance, plasticizers and solvents may help to lower the MFFT a certain extent (refer to the chapters related to Water-based emulsion adhesives in the subsequent volumes).

Freezing during application or during setting: If freezing conditions occur at these times, the ice crystals will totally prevent bonding. The A/S should be scraped and the bonding job done again after the temperature rises to a minimum of 10°C to be safe.

For all these reasons, bonding and sealing should be done always at temperatures ranging from +10 to +35°C. In very hot or very humid countries special formulations and extra care are mandatory.

Sealants reacting with ambient humidity: Some sealants and adhesives (one component PU, MS polymers) cure by a chemical reaction with ambient humidity.

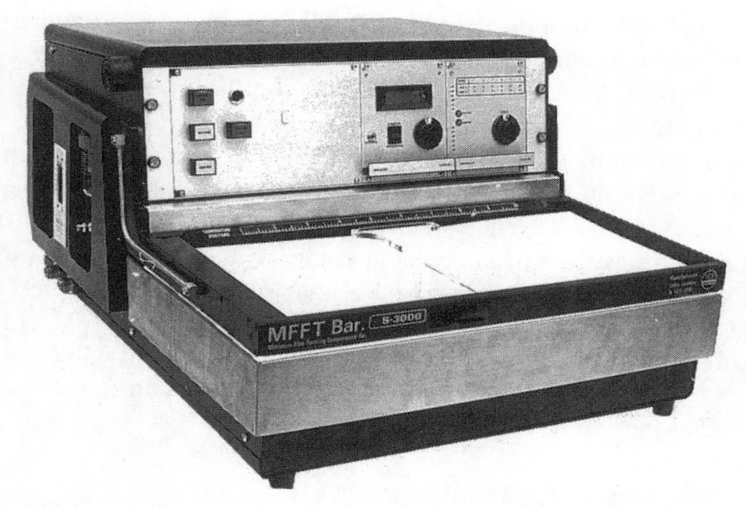

Figure 19: Minimum film forming temperature bar from Sheen Instruments Ltd England; on the left of the plate, cooled zone; on the right, heated zone.

This is the only case where humidity is an advantage. They may also react with the humidity contained in the substrates.

2.2.2.4. Mode and Conditions of Setting

After application of the liquid or paste A/S, it must harden; this is called setting or curing. Table 4 shows all the different modes of setting, by drying, curing, chemical reactions, etc. Modern physics and chemistry offer many different modes of setting, including the latest ones discovered such as UV curing, HF (high frequency) or MW (micro waves), electron beam (EB) polymerization. In the chapter "Physics and chemistry of A/S" we will study in detail all the available modes of setting and curing. Many characteristics of A/S depend on the mode of setting. It is thus very important for the users to understand how setting/drying or curing will occur during the bonding process, and the TDS should always indicate it precisely, but unfortunately this is almost never done!

We will now study each of these parameters and related technical characteristics.

2.2.2.5. Mixing and Proportioning 2 Components Adhesives

The 2 components should be preferably mixed immediately prior to application so that the chemical reaction is not yet started and it leaves time to apply the A/S.

Proportions of the mix are quite variable according to types of A and S:

- some formulations require only the addition of a minor amount of hardener or catalyst (1–10%), for instance, UF and MF resins, use of isocyanate hardeners with polychloroprene or PU adhesives, Thiokol/polysulphide sealants, etc.

Table 4: Different modes of setting for A and S and setting parameters.

Types of A/S	Modes of setting	Setting parameters	Examples
Water based A/S	– Water evaporation before assembly – Absorption of water by substrate(s)	– Film thickness – Absorption of substrates – Temperature, humidity – Link between water and polymer	– Vinyl, acrylic emulsions – Other water based (PU emulsions)
Solvent-based adhesives	Evaporation and absorption of solvents	– Evaporation rate of solvents – Absorption of substrates – Temperature	– Solvent-based PU, PVC adhesives
Contact adhesives applied on both substrates	Complete drying followed by immediate bond by contact under pressure	– Same as solvent based + – Minimum pressure 5 bar	– Polychloroprene adhesives
2 components thermosetting A and S	Chemical reaction between 2 components at room or elevated temperature or with a catalyst	Mixing ratio – Reactivity between components – Time and temperature	– Epoxies, PU, struct. acrylics – UF, MF, PF, RF
1 component thermosetting adhesives	Chemical reaction triggered by heat	Chemical reactivity – Temperature, pressure, time	– Epoxies, PU, UF, MF, PF
Hot melt adhesives	Cooling of adhesive after application	– Cooling speed – Viscosity = F (temperature)	– EVA, PA, PU hot melts
Reactive hot melts one component	Combine both: cooling + curing	– Cooling speed – Chemical reactivity	– PU reactive hot melt
Adhesives which react with humidity in the ambient air	Chemical reaction between water and polymer	– % humidity in the air and in the substrates – Temperature	– PU A/S, MS polymers – PU reactive hot melts
UV and EB curing	Chemical reaction triggered by radiations	– Wavelength, power	– UV and EB curing PU, acrylics

- other adhesives and sealants are formulated in order to have a ratio, which is easier to measure and control. For instance, some epoxies, acrylics or PU adhesives may be mixed at a 1:1 ratio or at least the quantities of part A and part B should be of the same order, for instance, 60:40 or 70:30.

According to the type of chemical reaction between the 2 components, A and S may be more or less tolerant to a variation of the ratio; in most cases the mixing ratio should be respected within a 5% tolerance.

Fig. 20 shows an example of a mixing and metering machine for 2 components A and S.

2.2.2.6. Pot Life, Gel Time

These characteristics are related only to 2 components A and S. Pot life is the time during which the adhesive may be used after the 2 components have been mixed. After mixing, the chemical reaction between the 2 components starts and the mix start to thicken so that after some time the adhesive becomes too thick to be spread and to wet and stick to the substrates.

Pot life varies with:

- chemical reactivity of the 2 components.
- temperature: when temperature increases pot life is reduced. A general and approximate rule is that pot life will be halved when temperature increases by 10°C.
- quantity mixed: when the quantity of adhesive mixed increases, pot life will be reduced because most chemical reactions are exothermic and the heat generated, larger with a larger quantity, will in turn speed up the reaction.

Example: For a given epoxy adhesive let us suppose that the pot life is 1 h at 20°C for a 1 kg mix. At 35°C it will be only 25 min and if the mixed quantity is 5 kg it may be only 10–15 min. For many chemical reactions, the speed of the reaction doubles when temperature increases by 10°C.

The pot life of 2 components adhesives may be measured according to ISO 10364 (1993) or ASTM.

Gel time: This is a way to measure the pot life and the chemical reactivity. There are different methods of testing. One method is by measuring the viscosity from time to time after mixing. When viscosity becomes too high the mix is no longer usable.

Automatic gel timers record the increase of viscosity versus time, according to BS 2782, for instance.

Several standards exist to measure pot life and gel time, for instance, ISO 10364 or ASTM.

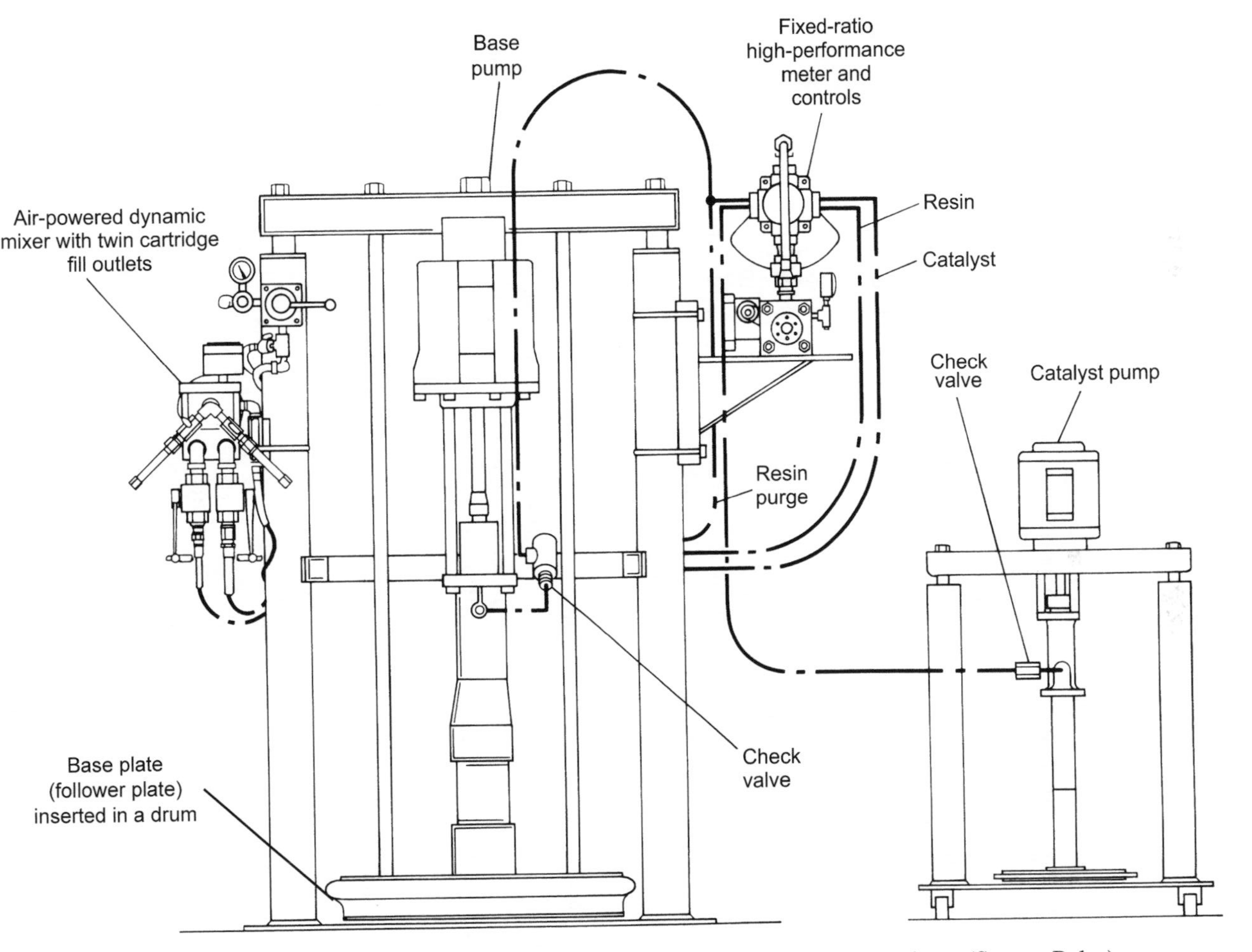

Figure 20: Fixed-ratio meter/mix machine for 2 components adhesives or sealants (Source Pyles).

2.2.2.7. Application of the A and S, Tools and Equipment

This will be explained in detail in the chapter "Application equipment for adhesives" in Volume 2. In this section, we will only explain the characteristics of A and S that are related to the application equipment.

The selection of application equipment depends on several characteristics:

- viscosity: some applicators need very fluid adhesives such as spray guns in order to break the stream of adhesive into fine particles; others need higher viscosities, for instance, guns for sealants and caulks. Each type of applicator requires a given range of viscosity,
- thickness of the joint, quantity of A and S needed: when the gap between parts to be assembled is large (1 mm or more), for instance, for sealants, the application equipment must deliver a large quantity per minute, without sagging,
- one or two components, pot life: 2 components A and S should be mixed and used much before the end of pot life. It is even better to mix the 2 components immediately after they leave the applicator; some equipment allow it by external mixing, or by separate applications of resin and hardener (resin being applied on one substrate and hardener on the other one as is done for resorcinol–formaldehyde adhesives in wood industries).

2.2.2.8. Gap-filling Properties

Usually adhesives should be used in low thickness, from 0.05 to 0.25 mm in order to lower stresses (refer to the chapter "Theory of adhesion" in Volume 2). But some adhesives allow one to fill larger gaps from 0.25 to 2 or 3 mm.

In order to fill large gaps the adhesive should meet the following requirements:

- solids content should be 100% or very close to that level so that when the adhesive will cure it will not release volatile compounds and it will not shrink or develop internal stresses. Thus, gap filling adhesives are usually cured by a chemical reaction (1 component thermosetting or 2 components),
- in most cases, a flexible adhesive will be better than a rigid one, unless a very rigid assembly is necessary, for instance, for concrete to concrete bonding,
- curing should not give evaporation of volatile compounds (resulting from chemical reaction) because this would cause pressure in the joint, stresses and shrinking.

Gap-filling properties are measured as the largest gap that may be filled by the adhesive without reducing mechanical characteristics and without cracking. ASTM D 3931 is used to determine the strength of gap-filling adhesive bonds in shear by compression loading.

A and S used for gap filling:

- epoxies may fill gaps up to 2 or 3 mm, for instance, to bond concrete to concrete in civil engineering (segmented bridges),
- PU, polyesters and structural acrylics are also used for thick joints up to several mm thickness, for instance, for boat building,
- formulations similar to sealants formulation, based on PU, thiokols or MS polymers, may also be used as flexible adhesives, and they may fill large gaps, up to 5 mm,
- mortars may fill gaps up to 2 cm, for instance, in construction (Fig. 21).

Formation of a meniscus: When bonding honeycomb to a flat surface, the contact area between the honeycomb cells and the flat surface is small. Thus, it is necessary to increase it as much as possible in order to increase the mechanical resistance. The way to do it is to use an adhesive that will be very fluid during application and/or curing, with an adequate surface tension so that it will climb on the walls of the cells as indicated in Fig. 22. The pictures of Fig. 23 show what happens when the bond is broken: in some places the wall cells break 1 or 2 mm above the adhesive film, which is good, in other places the walls of the cells break immediately near the adhesive surface.

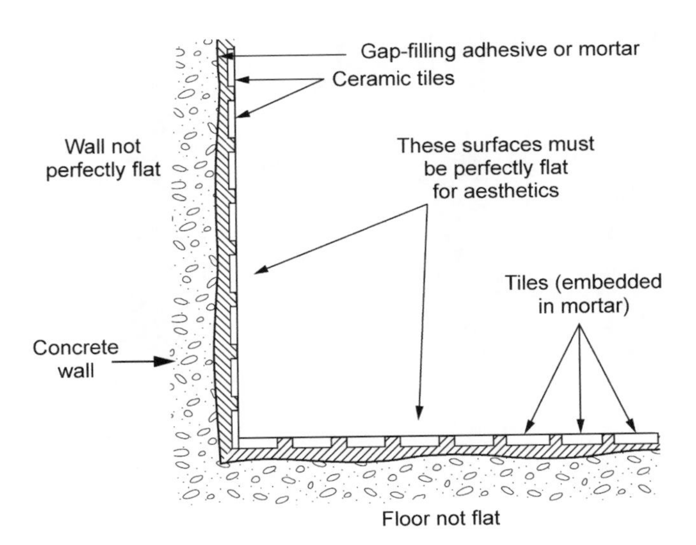

Figure 21: Gap-filling effect of a ceramic tile adhesive or mortar: although wall and floors are not perfectly flat the surface of tiles must be perfectly flat. The adhesive on mortar must fill the gaps between wall/floor and the back of tiles.

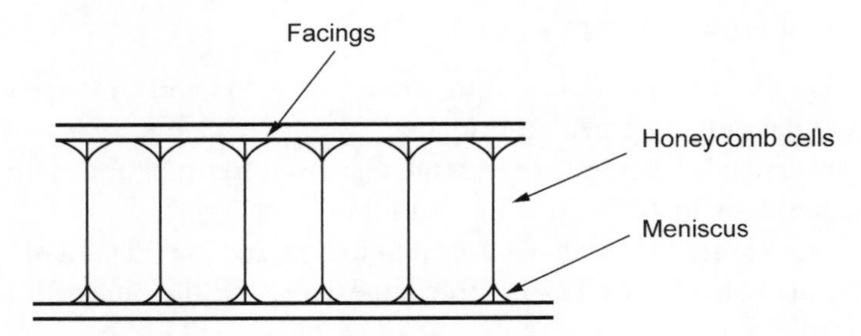

Figure 22: Formation of the meniscus on honeycomb.

Figure 23: Breaking down of honeycomb sandwich panel bonded with epoxy film. Honeycomb cells break down in different heights. (Colour version of this figure appears on p. xiii.)

2.2.2.9. Coverage or Consumption or "Add-on"

This is the optimum amount of adhesive that should be applied on substrates in order to get good bond and strength. European companies measure it in g/m^2, American in sq ft per gallon or pounds per sq ft. ASTM D 898 and 899 are used to measure applied weight per unit of dried or liquid adhesives. Coverage depends on the volume that should be filled by the adhesive. If there is not enough adhesive to fill the gap, the bond strength will be too low. If there is too much, the drying may be too slow, there will be a waste of adhesive, and moreover the bond strength may be lower because stresses may develop in the glue line under tension or shear.

Measurement of coverage: Take a paper or plastic film of a given area, weigh it accurately, apply the adhesive and weigh again immediately (or after complete drying if the goal is to measure dry consumption); record the weigh difference for a 1 m^2 area. Simple notched gauges (Fig. 24) with notches ranging from 10 to 1000 μm depth are quite convenient to measure the thickness of the wet film of adhesive.

2.2.2.10. Stability of the Adhesive in the Applicator's Tank

- Hot melt adhesives must stay in the melting tank for a whole working day at the application temperature. At these elevated temperatures, hot melts may start to degrade by oxidation or burning (charring). They will become darker, burned particles may clog the guns and properties will be damaged.

  The heat stability is usually measured by following the darkening of color after increasing periods of time (from 1 h to several days) at different

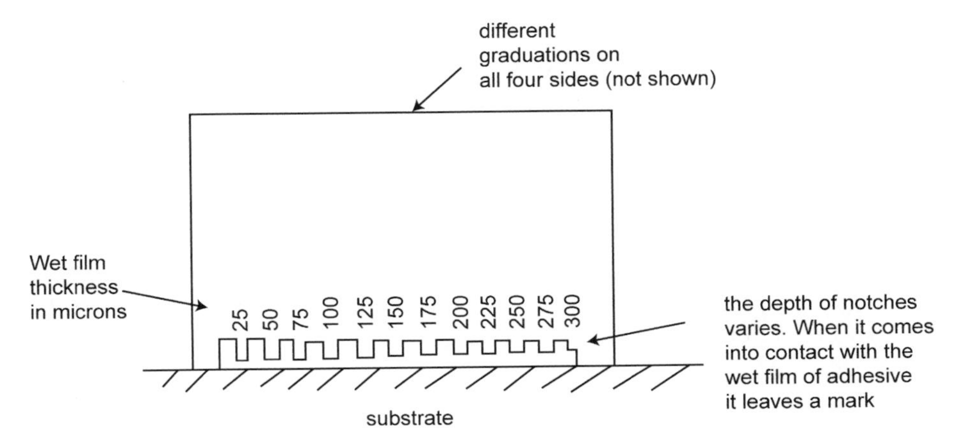

Figure 24: Metal notched gauges to measure the thickness of a wet film of adhesive (or paint).

temperatures. The color chart may be supplied by the adhesive manufacturer and the color evaluated with a Gardner scale of colors. EN ISO 10363 (1995) is used to measure the heat stability of hot melt adhesives.

- Drying types of adhesives (water based and solvent based) may start to dry by water or solvent evaporation in the tanks or on the rolls by the effect of agitation, and some latex-based adhesives may become unstable and coagulate after a long agitation. Mechanical stability refers to the ability of an emulsion to withstand agitation, shearing during pumping, working on rollers or other systems involving high shear. This is important on high speed roller machines in packaging, laminating, graphic arts, paper bonding.

A practical test is the best way to assess this mechanical stability:

Five hundred grams of the adhesive are poured in the tank of a laboratory roll applicator; the roll distance is adjusted to the expected gap, for instance, 0.1 mm and the rolls are started at high speed: 100 rpm for the pick up roll and 400 rpm for the applicator roll. After 2 h, note the viscosity change, possible coagulation, foaming, etc. A similar test may be run with a laboratory mixer and a 500 g beaker, by running the lab mixer at 2000 rpm for 2 h. (After these tests, one may also test the adhesive by running an actual bond test and checking for any loss of performance.)

2.2.2.11. Sag Resistance

Sag is related to the rheology of the A and S. EN ISO 14678 is used to measure the sag resistance of adhesives. ASTM D 6005 measures slump resistance of carpet adhesives, for instance, (there are also other standards to measure sag of sealants: refer to the chapters related to sealants).

2.2.2.12. Waiting Time before Assembly, Pre-drying

This applies only to water- or solvent-based adhesives that set by drying. After application on the substrates, water or solvents start to evaporate at the ambient air, and also to be absorbed by substrates if these are porous. The adhesive thickens, becomes tacky or sticky and eventually develops enough cohesion so that it becomes possible to assemble the parts without risk of parts moving or sliding. The waiting time is the time that the operator must wait before assembly in order to let some water or solvent evaporate and to develop enough tack (refer to definition in Section 2.2.2.13).

Waiting time depends:

- on the type of adhesive: solvent-based adhesives containing fast evaporating solvents (acetone, MEK, alcohol, ethyl acetate, etc.) need only seconds if the adhesive is applied on one side only or a few minutes if adhesive is applied on both parts to be bonded. Other adhesives containing slow evaporating solvents or water may need from 3 to 20 min. Contact polychloroprene adhesives which are always applied on both parts (double application or 2 ways technique) and are used frequently on impervious materials need 10 min waiting time; after this time they develop a very strong "grab" or tack by simple contact between the 2 films under 5 bars pressure.

- on the stiffness of the materials to be bonded: for instance, to bond a stiff film or plate a high tack is required so that this material will not move or separate from the other part, if the parts are not pressed or clamped together during drying of the adhesive.

2.2.2.13. Tack

This is the ability of the adhesive coat to "grab" the other part and prevent it from moving immediately after assembly. For pressure-sensitive adhesives, it is the ability to stick immediately after application of the tape or label. Tack develops during waiting time and also it may be provided by a thick and sticky adhesive so that there will be a suction effect, for instance, in mastics. Tack may be measured with many different test methods according to the industries. We will just mention a few of these methods.

Tack of pressure-sensitive adhesives: We will study pressure-sensitive adhesives in a special chapter later in Volume 4.

Pressure sensitive adhesives (PSAs) are thermoplastic adhesives, which after application and drying or setting, remain sticky for very long periods of time (several years). They are used to manufacture PS tapes and labels.

Tack of PSA is measured by several tests:

- Rolling ball tack tester: refer to Fig. 25 and to ASTM standard D 3121.
- Loop test: refer to Fig. 26; a loop of PSA tape with PS adhesive outside is placed to contact a stainless steel clean plate, so that the contact area is of

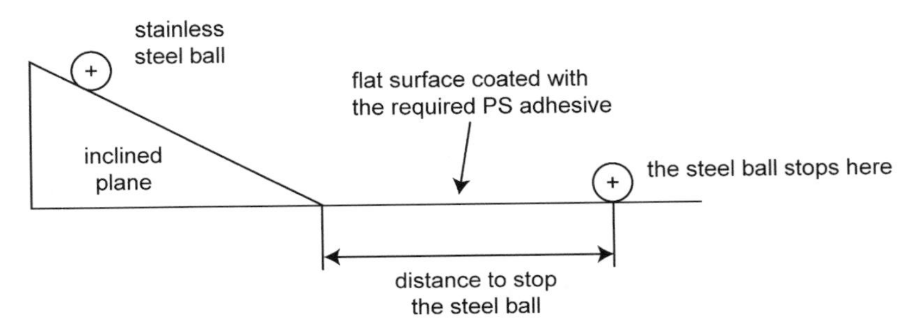

Figure 25: Measurement of PSA tack by rolling ball tack tester.

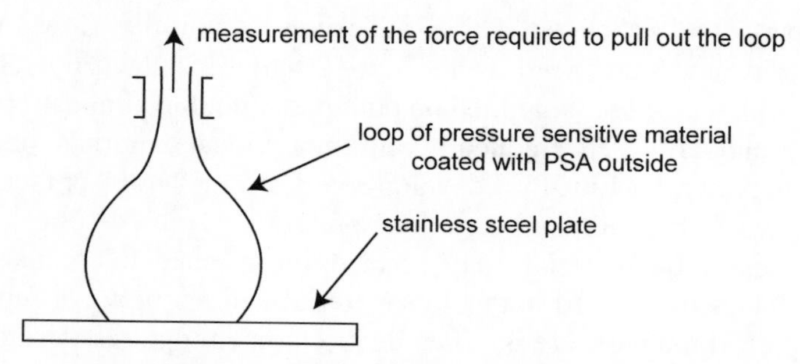

Figure 26: Measurement of tack by the loop test.

a few cm^2; after a few seconds the loop is pulled upward to separate it from the steel plate and the force required is recorded, in grams per cm width of the tape.
- D 2979: test method for pressure-sensitive tack of adhesive using an inverted probe machine.

2.2.2.14. Open Time

With almost all adhesives, after the adhesive has been applied on one substrate, it starts to thicken, either by starting the chemical reaction for reactive adhesives, or by evaporation for water- and solvent-based adhesives, or by cooling down for hot melts. After some time, the adhesive becomes too viscous or too dry and it is no longer able to transfer to/and wet the other substrate. This maximum waiting time is called open time.

Examples:

- For two components adhesives, the open time is related to the gel time; to be safe it is advisable to assemble the parts within an open time equal to half the gel time.
- Hot melts cool down very quickly after application and for the EVA or polyamide hot melts the open time is only a few seconds.
- Vinyl emulsions for wood and paper bonding have open times from 30 sec to several minutes.

Fig. 27 shows the two time limits, waiting time and open time; the working time WT is equal to OT − WT. It should be adapted to the speed of the bonding operations and the production line.

Measurement: After coating the adhesive on one substrate, pieces of the second substrate are laid on the surface of the adhesive from time to time, at 1 sec, or 1 or 5 min intervals. Later after some drying or curing, these pieces are pulled away and the open time is the maximum time one can wait and still get a good transfer of the adhesive and a good bond. (To be safe acceptable transfer should be at least 90%.)

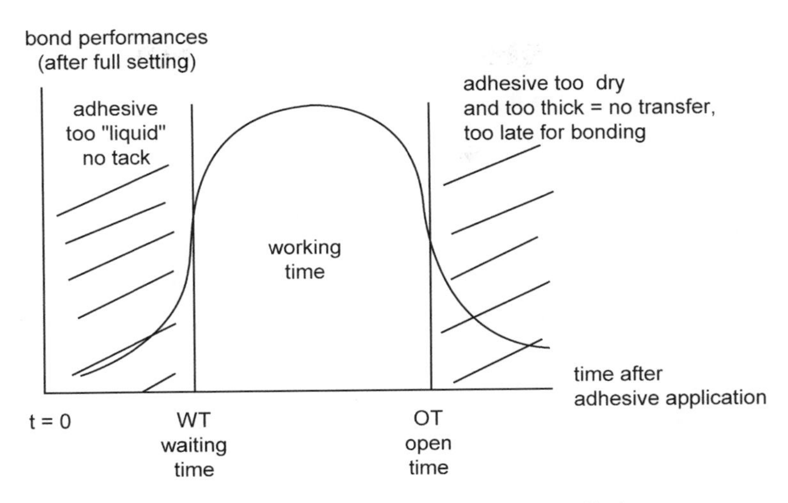

Figure 27: Open time and waiting time limits.

2.2.2.15. Pressure Required

During drying or curing, the parts must be pressed together firmly for several reasons:

- the adhesive film and the parts should not move during drying or curing,
- in case some volatile compounds should evaporate during drying or curing (gases from condensation, water, solvents), pressure will prevent parts to be separated and counter effect the gas pressure,
- pressure will force the adhesive to penetrate into the roughness of the substrates.

According to the chemical types of adhesives and their mode of curing, the required pressure may vary from mere contact under low pressure to 1 bar up to 12 bars. For drying or curing conditions, temperature, time and pressure are indicated by the adhesive supplier.

2.2.2.16. Setting Time and Setting Speed

Adhesives acquire their final mechanical and physical properties only after they harden and become cohesive, according to the different modes of setting indicated in Table 2.

- a first level of setting or curing will provide some strength, enough for parts to be handled and further machined (by drilling, cutting, painting, etc.),
- later a complete 100% setting/curing will develop.

The curves in Fig. 28 show some different types of setting and/or curing. Setting speeds are very different according to the types of adhesives, mode of setting, temperature, chemical reactivity.

Examples:

- hot melts set in seconds after assembly, through fast cooling,
- cyanoacrylates set in 30 sec,
- UF, MF, RF, vinyl wood glues, one component heat curing epoxies, all provide 50–80% of final mechanical strength immediately after hot pressing and cooling,
- two components RT curing epoxies or PU take several hours for partial set,
- a special case is provided by the reactive PU hot melts that work in 2 steps: applied at 120°C, they set in seconds after cooling, this gives some strength already, but later the chemical reaction of the PU provides within hours a much higher mechanical strength.

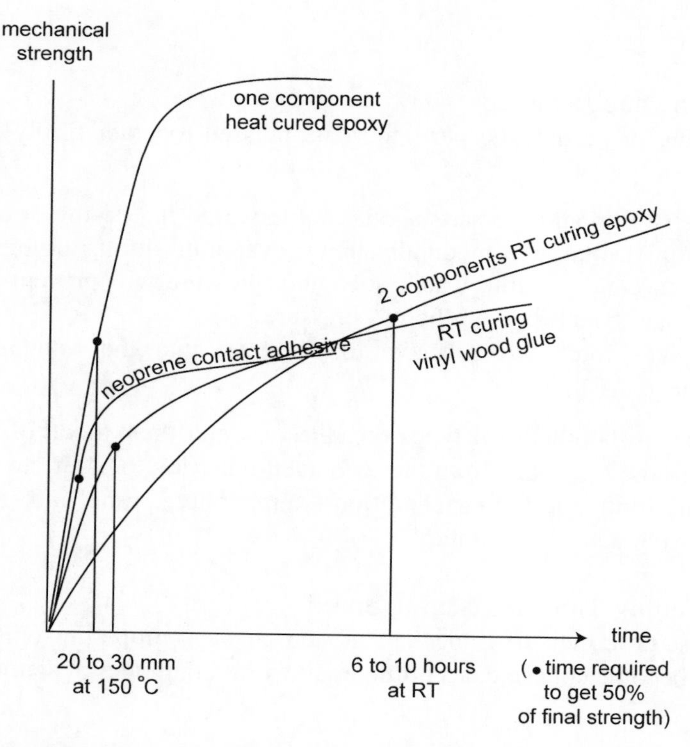

Figure 28: Time for drying/setting/curing for various types of adhesives (for other specific curves please refer to the chapters dealing with other chemical types of adhesives). • = 50% of final strength.

2.2.2.17. Storage and Storage Life

According to their mode of setting and curing, adhesives have different storage lives:

- water- and solvent-based adhesives that set by drying may be stored in closed, original containers for a fairly long time, 1 year at least,
- hot melts may be stored several years because nothing will happen in the granulates at room temperature,
- two components adhesives may usually be stored for 1 year or more in separate and closed packaging at room temperature, but some hardeners or catalysts such as isocyanates have a shorter storage life (6–9 months), and should be stored in very tight containers because they react with ambient humidity,
- one component thermosetting adhesives such as epoxies, PU, etc., may have a storage life limited to 6–9 months at room temperature: they react quickly at elevated temperature but they also react slowly at room temperature (most chemical reactions have a speed that doubles when temperature increases by 10°C so that 9 months at 20°C is equivalent to 1 month only at 50–60°C),
- when they contain fillers, liquid adhesives may settle during storage, and should be remixed thoroughly just before use.

Temperature during storage: Some very reactive adhesives such as epoxy one component film adhesives or some cyanoacrylates should be stored in refrigerators at − 10 or − 20°C before utilization in order to increase storage life. At these temperatures they may be stored several months, while at 20°C they could only be stored a few weeks. Suppliers always indicate the shelf life of their A and S.

2.2.2.18. Stability to Freeze–Thaw Cycles

It is important to know what happens if the A/S has been subjected to conditions near freezing. EN 1239 is used to measure stability to freeze–thaw cycles.

2.3. Properties of Cured/Dry Adhesives

2.3.1. Mechanical Characteristics

2.3.1.1. Adhesion to Various Substrates

Adhesion of an adhesive A to a material M depends on the couple A–M, for instance:

- epoxy and PU adhesives will bond/adhere to many materials: all metals, many types of plastics, wood, glass, composites, ceramics, foams (and also to

cardboard, paper, textiles, etc., but these latter materials are not strong and do not need such high performance, structural and high cost adhesives, and thus the epoxies are used only for structural bonding). The good adhesion of epoxies and PU to many materials is due to the structure of their molecules, which contain polar groups (refer to the chapters: "Theory of adhesion", "Epoxy adhesives" and "Polyurethane adhesives").
- on the other hand, vinyl emulsion adhesives bond/adhere to fibrous materials such as wood, cardboard, paper, textiles, but they do not bond to metals, plastics or glass and other impervious materials.

We will not explain here why a given adhesive should stick to a given material, because this is explained in the chapter "Theory of adhesion" in Volume 2 in the light of wetting, surface energy, adsorption, work of adhesion, electrostatic, diffusion, covalent bonds and van der Waals forces, and it is also discussed in all the chapters dealing with the various chemical families of adhesives.

We will only indicate how to measure adhesion through mechanical testing of the bonds.

We will not describe the laboratory equipment and the detailed test methods that may be used for all the tests hereunder, because this will be done later in the chapter titled "Methods of testing adhesives, sealants and their joints" in Volume 7.

In the chapter " Selection of an adhesive", we provide charts indicating which types of adhesives should be used to bond a material M1 to a material M2. The only way to measure adhesion to a given substrate is to prepare samples of actual bonded parts and test it by various mechanical breakdown tests. Basically, bonded joints may be stressed and eventually broken in 4 different modes: tensile shear, tensile force, peel and cleavage. These modes are shown in Fig. 29.

These test methods have 3 purposes:

- When the bond breaks, if the failure is adhesive to one of the substrates, the breaking load provides an evaluation of the adhesion to this substrate.
- If the failure is cohesive, the breaking load gives the internal strength of the cured adhesive.
- In both cases, the breaking load provides an evaluation of the mechanical resistance of the bond, and this value may be later utilized for the design and calculation of the joint, as we will see in the chapter dealing with these issues (in Volume 2).

There are many standards/test methods that are used for these mechanical tests, because the size and shapes of bonded samples, the forces involved, the speed of pulling and loading, the stiffness of the materials, the specific conditions of testing vary with the materials to be bonded, the industries and applications, etc.

We will describe here the main testing methods. Other specific methods will be described in the chapters dealing with the various industries, for instance, in "Pressure sensitive adhesives", "Electronic adhesives" because these industries have their own tests methods. Suppliers of testing equipment are listed in the appendix.

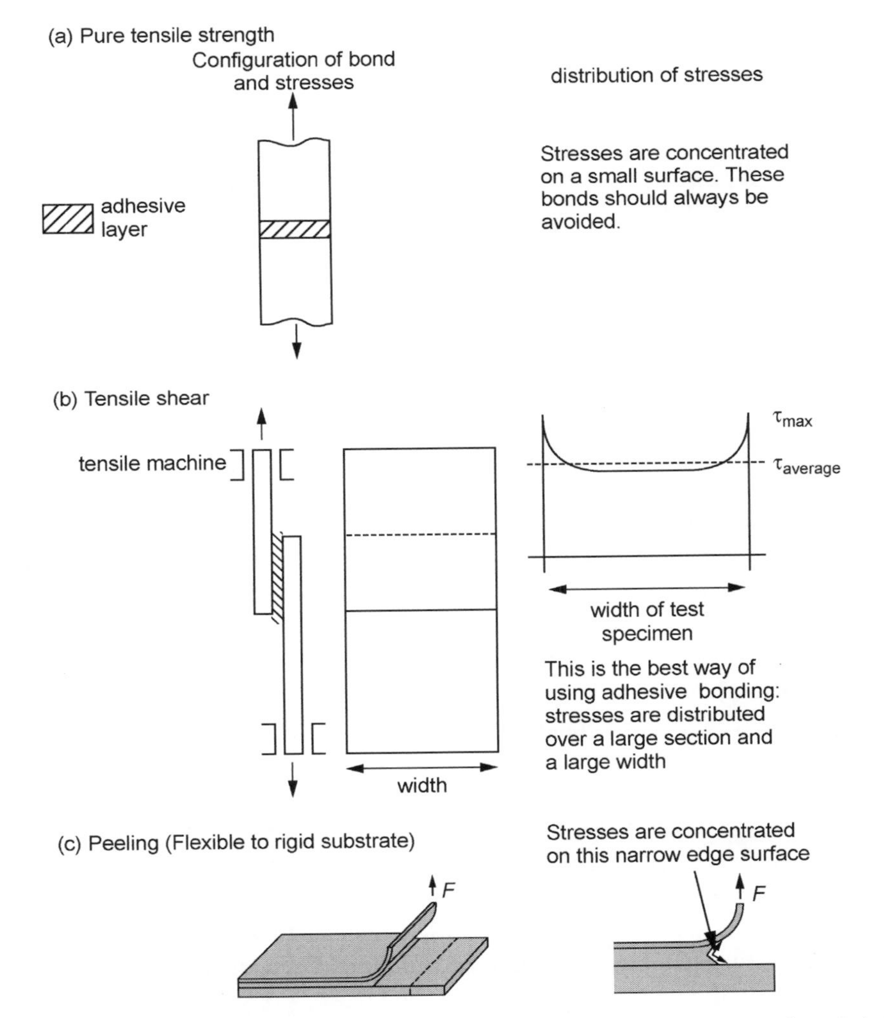

Figure 29: Different modes of mechanical stresses on adhesives (or sealants) joints. Modes of rupture of adhesives and sealants bonds.

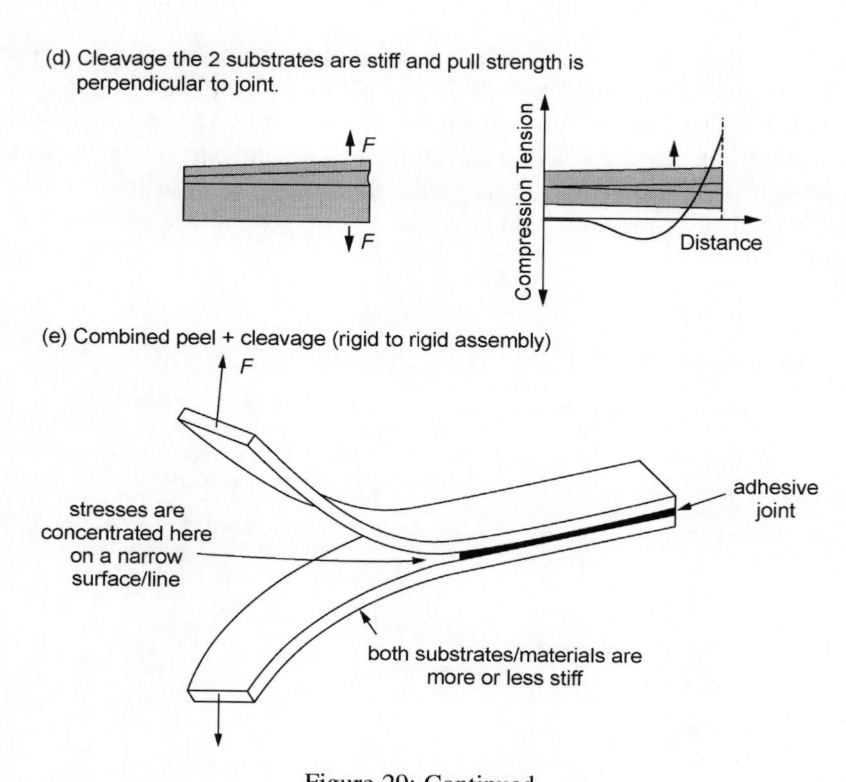
(d) Cleavage the 2 substrates are stiff and pull strength is perpendicular to joint.
F
F
Compression Tension
Distance
(e) Combined peel + cleavage (rigid to rigid assembly)
F
adhesive joint
stresses are concentrated here on a narrow surface/line
both substrates/materials are more or less stiff

Figure 29: Continued.

Tensile shear resistance

This is one of the most important tests because adhesive bonds show a good resistance to tensile shear stresses and it is always recommended to load the bonded parts in tensile shear mode in order to get the highest resistance (refer to the chapters: "Design and calculation of bonded joints" in Volume 2 and "Bonding metals" in Volume 5).

The thickness and the mechanical resistance of the test specimen should be close to that of the actual parts to be bonded, so that it will not interfere with the tests results and it will duplicate the real parts. Test specimens are prepared according to the standard test method (refer, for instance, to Fig. 31) including the proper surface preparation, then the samples are bonded and cured or dried according to the adhesive or customer specifications.

A certain period at 20 or 23°C must elapse before testing, in order to standardize the test specimen. Then the test specimen is clamped in a tensile machine, such as the one shown in the colour section, and pulled with a constant rate of increase of the pull force, until it breaks, and the forces are recorded. The mode of rupture, adhesive or cohesive or mixed, is also carefully noted (Fig. 29b).

Modes of ruptures	Désignations
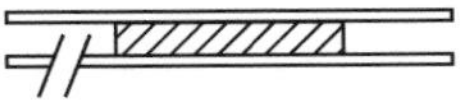 Rupture of one substrate	RS
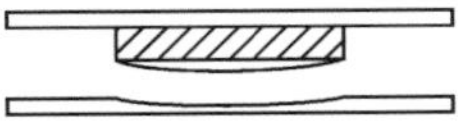 Rupture in one of substrates by delamination	RSD
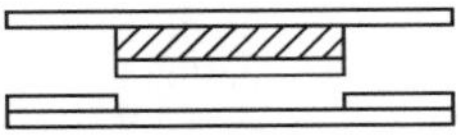 Rupture in the coating of one substrate (I)	RCS
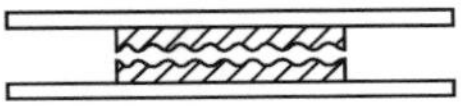 Cohesive rupture	RC
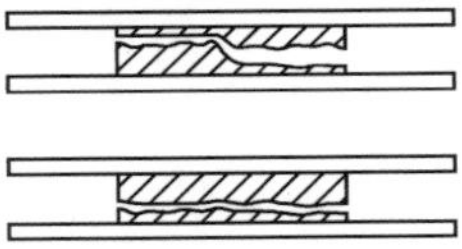 Cohesive rupture near the surface of the substrate	RC NS
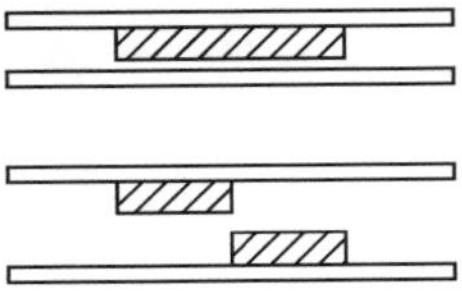 Adhesive rupture (into the adhesive film)	RA
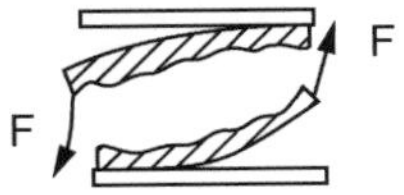 F F Cohesive rupture by peeling	RCP

Figure 29: Continued. The coating may be a paint or varnish or a primer or a surface preparation.

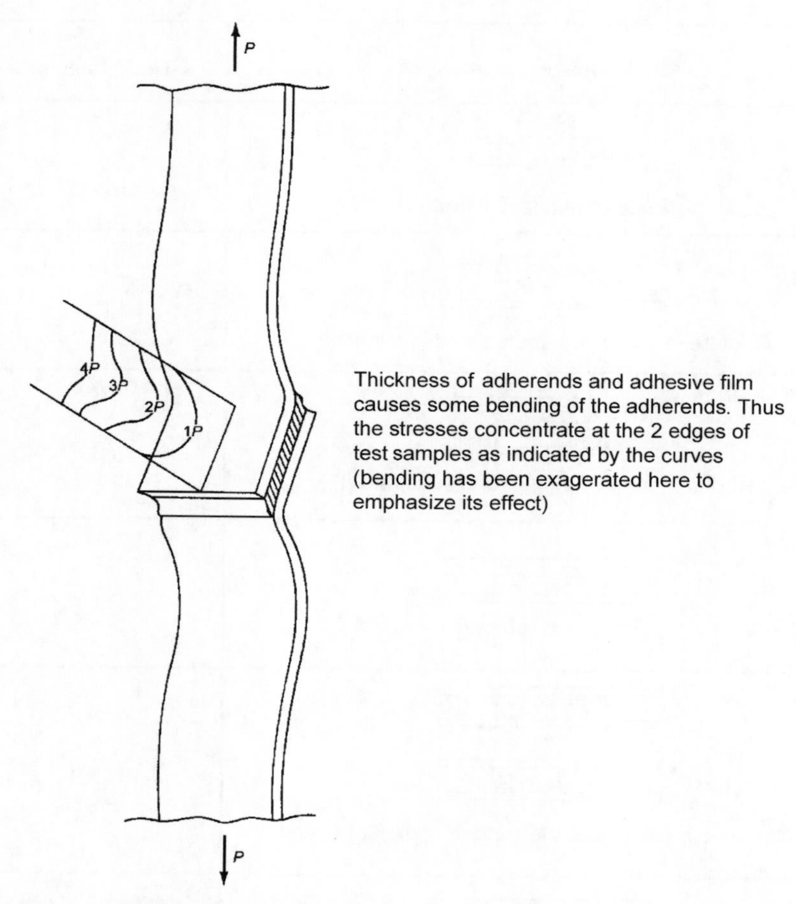

P
4P
3P
2P
1P
Thickness of adherends and adhesive film causes some bending of the adherends. Thus the stresses concentrate at the 2 edges of test samples as indicated by the curves (bending has been exagerated here to emphasize its effect)
P

Figure 30: Deformation of a lap shear specimen showing stresses distribution when the load varies.

There are several test methods available:

- ASTM: D 1002 for metals, D 3163 and 3164 for plastics and D 5868 for composites,
- D 3163: test method for determining the strength of adhesively bonded rigid plastic lap shear joints in shear by tension loading,
- DIN 53283, 55284,
- EN 1465 (1995) and ISO 4587 (1995): determination of tensile shear resistance of rigid–rigid assemblies, single lap samples,
- ISO 6237 for wood to wood adhesives.

Test specimens should be sufficiently thick and rigid, or else they would bend as indicated in Fig. 30, some peeling effect would occur and the pull values would be lower or even much lower if the adhesive was very rigid.

For this reason there are also test methods for double lap shear such as ASTM D 3528 (for generic substrates), and for off set notched lap shear such as ASTM D 5656 (for metals) and D 3163, D 3164 and D 3165 (for generic substrates). Refer to Fig. 32.

When adhesives are heat cured, if bonded materials are different and have different coefficients of expansion this difference will cause during cooling some stresses and the pull values will be lower, mostly if the adhesive is very rigid.

Examples:

- Structural adhesives such as epoxies may be tested on metal or composites or plastic specimens. The usual size of specimen is indicated in Fig. 31. On metal substrates, epoxies give values between 15 and 40 MPa according to the metal, the surface preparation and the specific epoxy adhesive. Because the metal resistance is higher than these values, the metal specimens will not break (unless they were too thin and then the test should be performed again with thicker samples).
- Wood to wood bonds made with vinyl white glue are also tested with double lap shear test specimen (Fig. 32), in order to avoid cleavage effects at the ends of the joint. In this case the tensile shear resistance will be from 4 to 10 MPa and some wood fiber tear will occur.

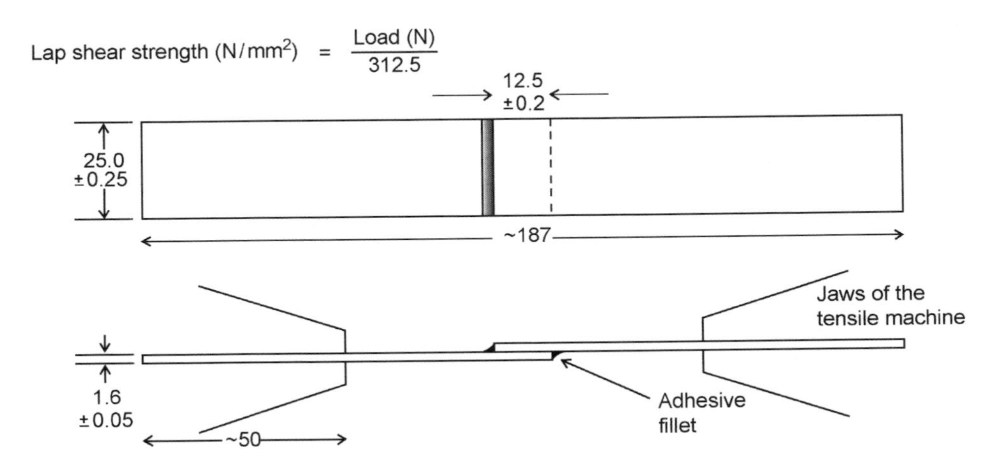

Figure 31: Tensile shear test of tensile shear specimen (DIN 55285).

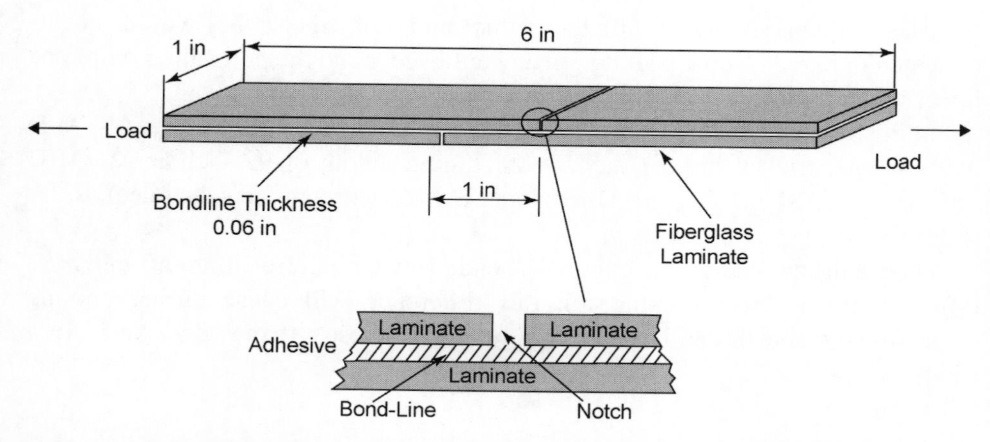

Figure 32: Double Lap Shear (reduce peel stress): ASTM D3528 (generic substrate), Offset or Notched Lap Shear (reduce peel stress), ASTM D5656 (metals), ASTM D3165 (generic substrate).

- Cardboard, textiles, thin plastic films are not tested in this way because they would break before the adhesive and the test would not be significant. These fragile and flexible materials and bonds are usually tested by a peel test or a creep test, as we will see hereunder.

Tensile test (tension perpendicular to joint)
Pure tensile stresses are not good for adhesives because the bonded surface is too small to get good values. However, in some very special cases the user may need to know this value, for instance, for the bond between a stiffener and a plate. The ISO standard 6933, ISO 6922 (1987) or CEN 26922 (1993) may be used to measure the tensile strength of bonded assemblies perpendicularly to the bond line, for metal to metal assemblies.

The test will be performed as for the tensile shear resistance but with the required configuration of the joints. The bonded parts should be aligned very carefully as is explained in the above standards, because if they are not there will be some cleavage effect and as the adhesives do not resist to cleavage the pull value will be low, and this is the reason why the bonded joints should never be loaded in pure tensile mode. For the same bonded area, the tensile value will be lower than the tensile shear value.

This tensile test may be used for measuring the modulus and the Poisson coefficient (read below paragraph Section 2.3.1.2 for the definitions).

Also metal to honeycomb bonds need to be tested in tensile load. The aircraft industry uses EN 2243-4 or the American MIL A 25463 and MIL 401B for this test (Fig. 33).

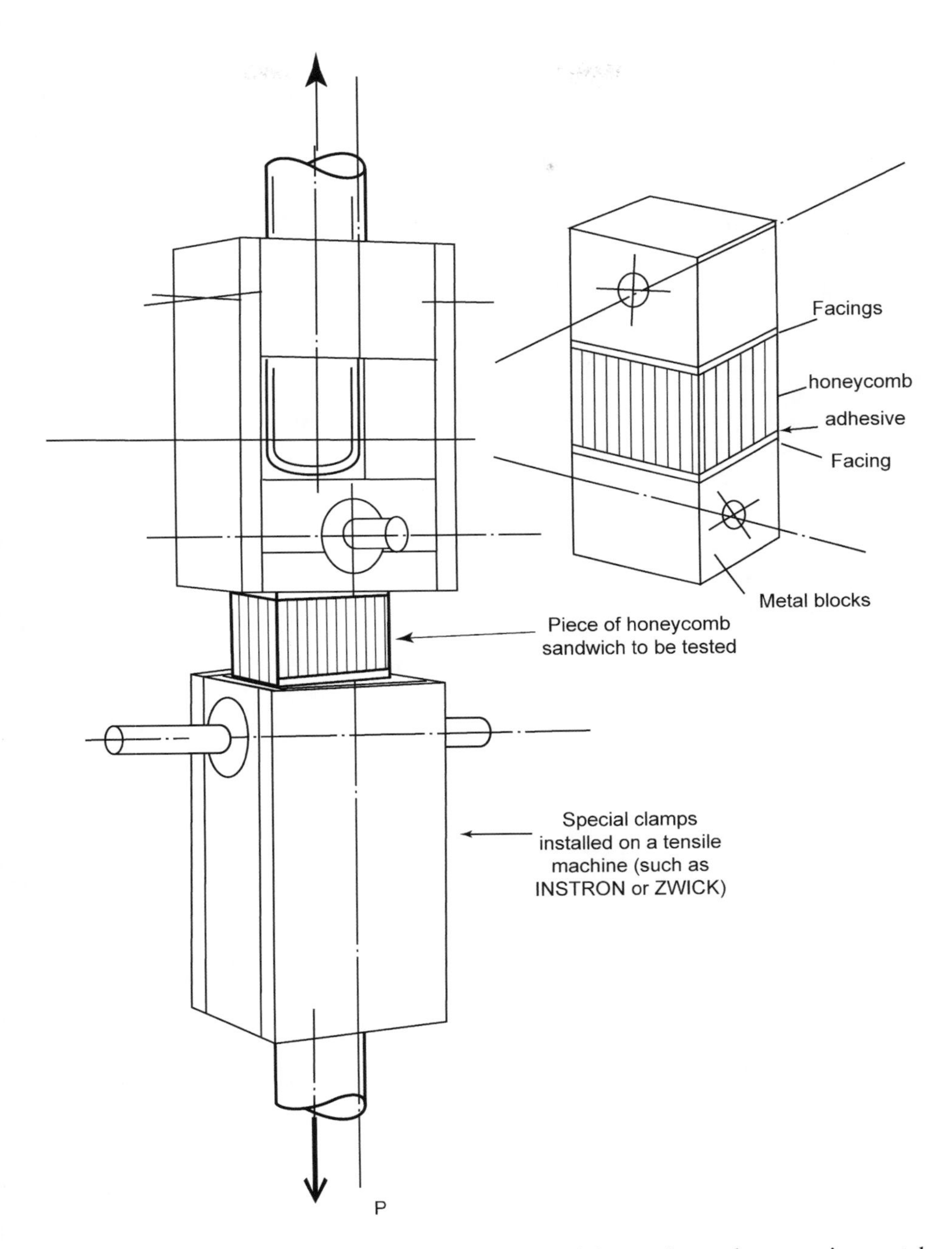

Figure 33: Pure tensile test on honeycomb sandwich panel — clamps axis must be perfectly aligned in order to avoid any cleavage effect.

Peel strength

This test may be performed only when at least one of the materials is pliable/flexible (such as paper, plastic films, textile, thin metals, etc.) and when both materials can withstand the pull force without tearing or breaking.

One of or both material is (are) peeled off in a tensile machine, under a constant speed of separation, which is usually much higher (300–500 mm/min) than for the tensile shear test.

Here again there are many different test methods according to the materials to be tested, the requirements of the industries where bonding is used, the types of

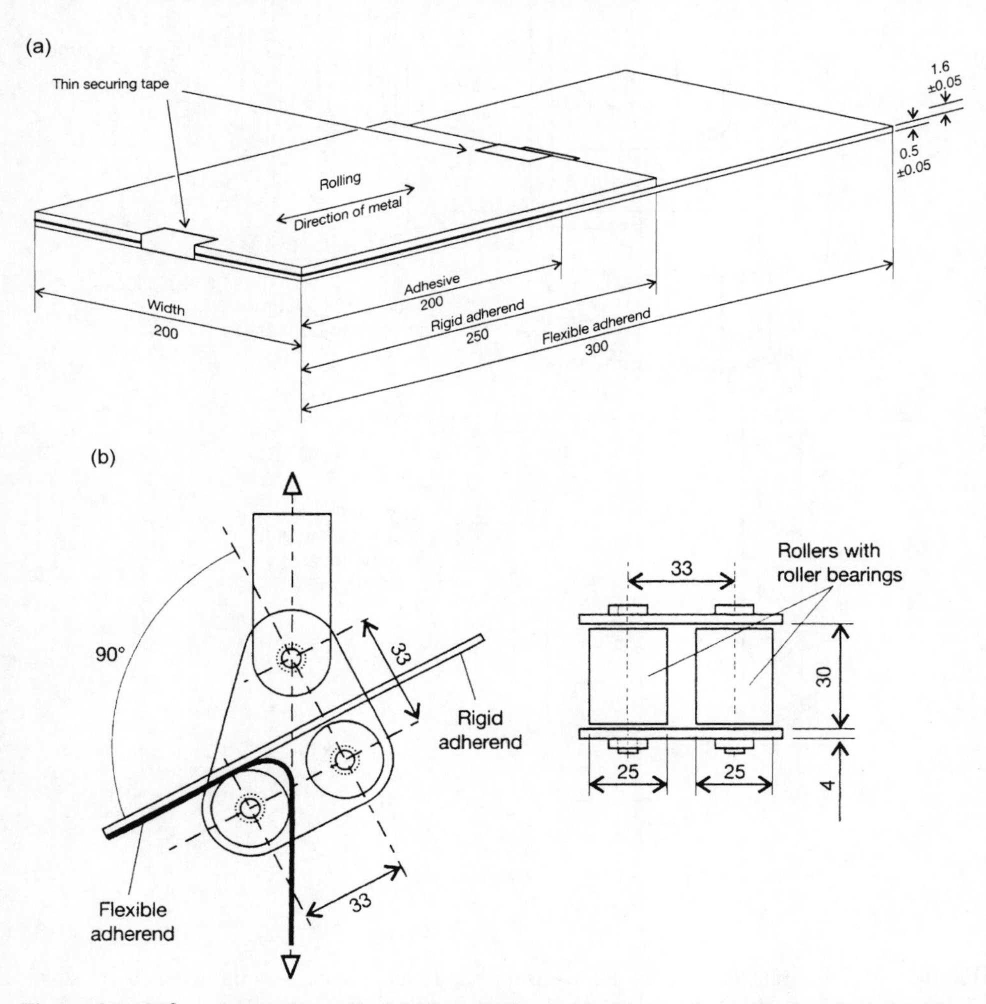

Figure 34: 90° peel test also called floating roller peel: (a) panel prior to bonding (cut into 25 mm strips after bonding); (b) peel test apparatus (all dimensions in mm).

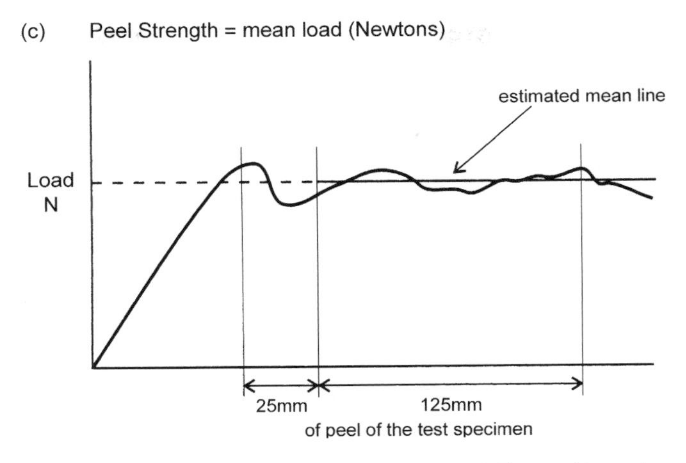

Figure 34: Continued. (c) Typical trace of a 90° metal/metal peel.

bonded parts to be tested, etc. We will review the most frequently used test methods hereunder.

- 90° peel: ISO 8510-1 December 1990 or EN 28510-1 are used to measure the peel strength of a flexible material bonded to a rigid one (Fig. 34),
- 180° peel: ISO 8510-2 or EN 28510-2, for flexible to rigid assemblies (Fig. 35a),
- another 180° peel is called T peel for flexible to flexible assemblies, it is performed according to ISO 11339 (1993),
- 180° peel may also be measured according to ASTM D 903 (refer to Fig. 35b),
- EN 1464: Determination of peel resistance for structural adhesives with floating rollers equipment.

In aircraft construction, sandwich panels made of metal facings bonded onto honeycomb are usually peeled with the climbing drum peel test, according to EN 2243-3 (Fig. 36) or ASTM D 1781 or MIL A 83376.

For all these tests, the peel strength is measured in N (Newton)/cm width.

The peel strength varies according to the types of adhesives. They are very low (200–800 g/cm) for PSAs, up to 100 N/cm for PU adhesives used for shoe soles attachment (peel strength of shoe soles attachment may be measured according to ASTM D 2558).

Peel strength measurement is always performed on pressure sensitive adhesives (PSAs) where it is the most important test.

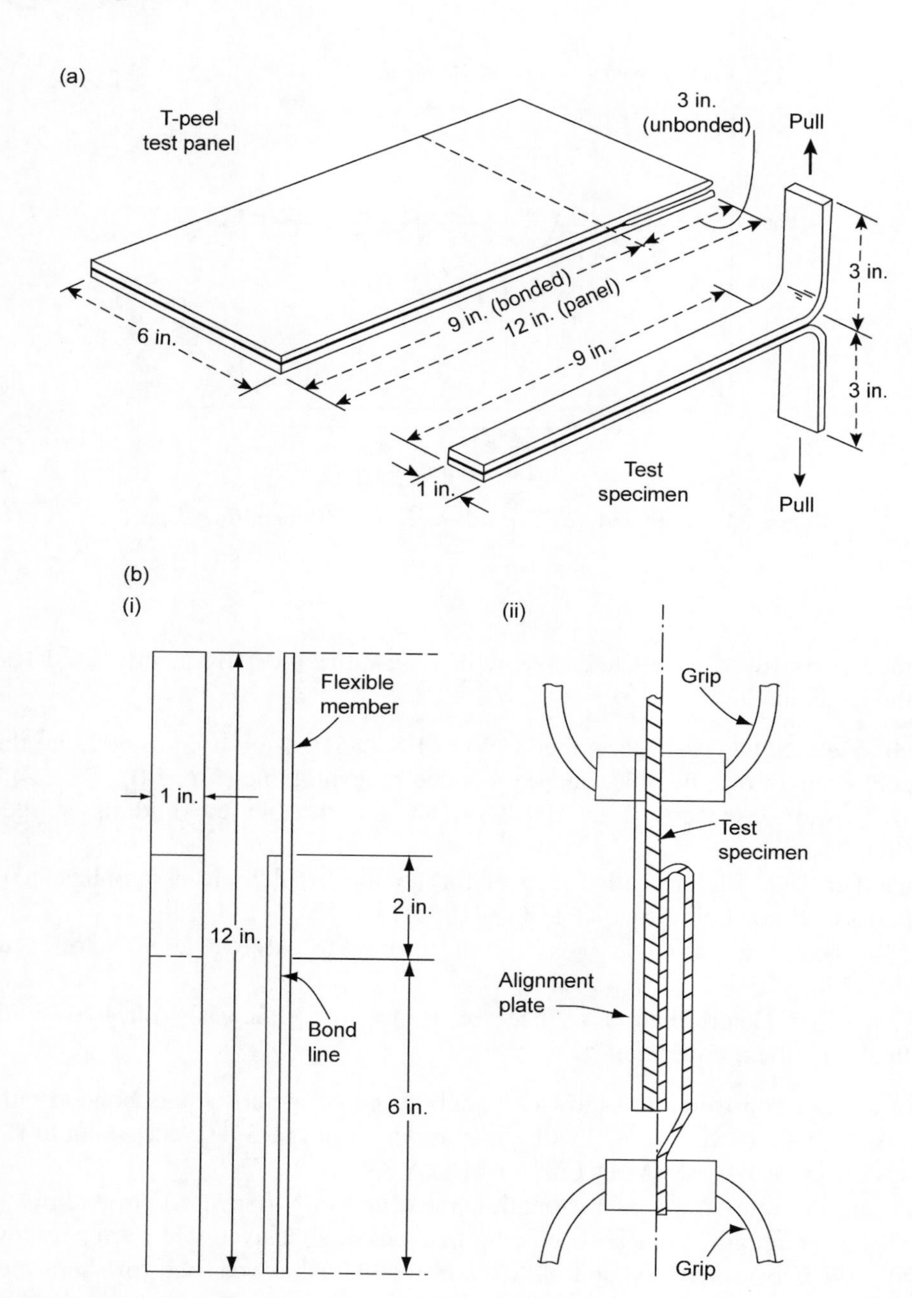

Figure 35: (a) Test panel and specimen for T-peel (from ASTM D 1876) (both substrates are flexible); (b) 180° peel test specimens: (i) specimen design, (ii) testing equipment for flexible to rigid substrate bond. (From ASTM D 903.)

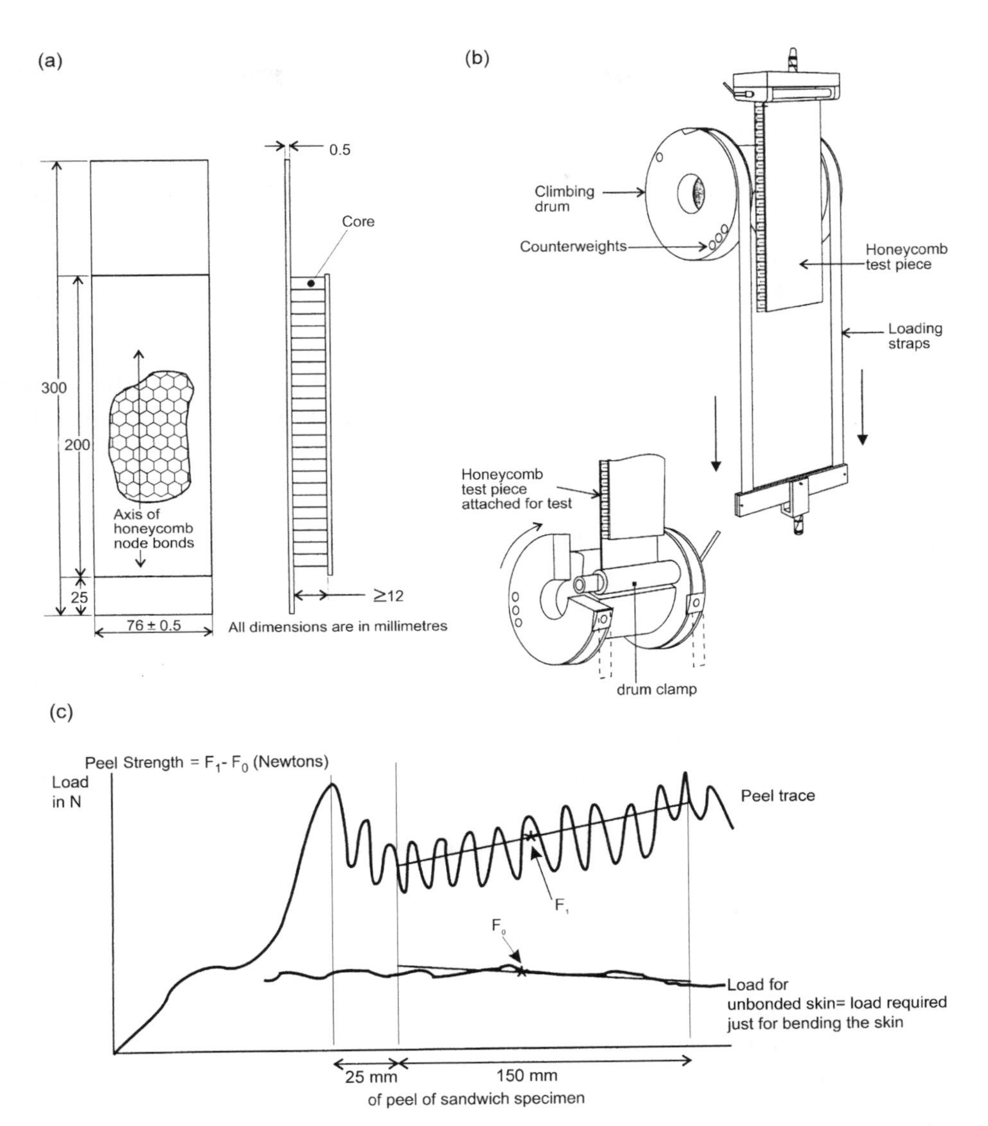

Figure 36: Climbing drum peel test for honeycomb sandwich panels: (a) specimen configuration; (b) climbing drum apparatus; (c) typical trace of climbing drum peel test.

Examples:

- PSAs for tapes and labels display low peel strengths from 200–800 g/cm width, because they need to be peeled off easily from substrates, without leaving traces. Thus, their peel strength must be lower than the paper or plastic film on which they are coated,

- epoxy adhesives, even though they are structural adhesives, have fair peel strength from 20 to 60 N/cm width, because they are quite rigid and therefore do not have a very high resistance to peeling.
- some flexible and tough adhesives (for instance, some PU adhesives) may have high peel strength combined with fairly good tensile shear resistance, which may be a very good combination for some applications. This will be studied in the chapter: "Flexible structural bonding" in Volume 2,
- PU or polychloroprene adhesives used for bonding shoe soles may reach 100 N/cm in peel.

These values and the phenomenon of peeling of adhesives and sealants are explained in the chapter "Theory of adhesion" and also in other chapters of the Handbook.

Cleavage, wedge test, Boeing test

This kind of stress is bad for adhesives, generally speaking, because adhesive bonds do not have good resistance to cleavage. Cleavage can only occur with rigid substrates.

The resistance to cleavage is measured by a wedge test, as shown in Fig. 37 according to ISO 10354 standard: durability of parts bonded with structural adhesives, wedge test, may be used, similar to the Boeing wedge test. In this test, a wedge is pushed inside the joint and the crack propagation is measured.

This test is very important in metal bonding and aerospace industries, where it is used not only to measure the adhesion and mechanical resistance of the joints and adhesives, but also to test the efficiency of the preparation of surfaces before bonding.

In the BOEING wedge test, the test specimens are immersed in water at 50°C for progressive periods of time, and the length of the crack is measured after each period with a microscope. If the surface preparation was not good enough, then the crack will progress quickly and far, and the aircraft constructors have their own specifications for this test that they perform with various surface preparations such as PAA (phosphoric acid anodization), CAA (chromic acid anodization), etc.

2.3.1.2. Curve Stress/Strain, Shear and Tensile Modulus

On a sample of cured adhesive, it is very useful to plot the curve stress/strain which usually looks like the curve in Fig. 38. The tensile stress $S = F/A$ in which F is the force applied by the pulling machine and A is the cross-sectional area of the sample. If L_o is the original length of the sample and L the length after a certain amount of tension is applied, the engineering tensile strain or elongation is:

$$e = \frac{L - L_o}{L_o}$$

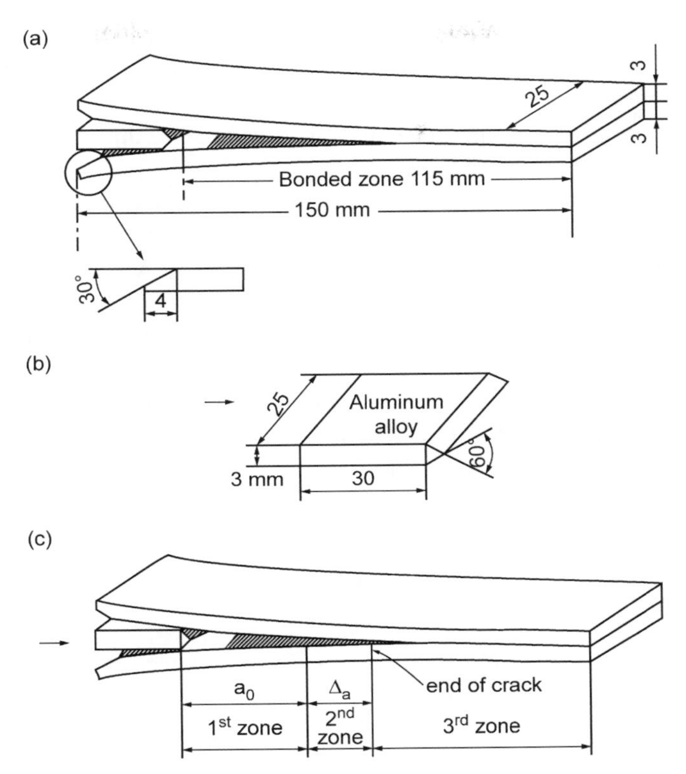

Figure 37: Cleavage test also called crack propagation (ISO 10354) or BOEING wedge test: (a) test specimen (all dimensions in mm); (b) wedge; (c) Test specimen after the test: wedge has been pushed inside the test specimen (at a speed of 1 mm/min, for instance); original opening was a_o, which increased by Δa so that crack propagation was Δa (Δa is measured very accurately with a microscope).

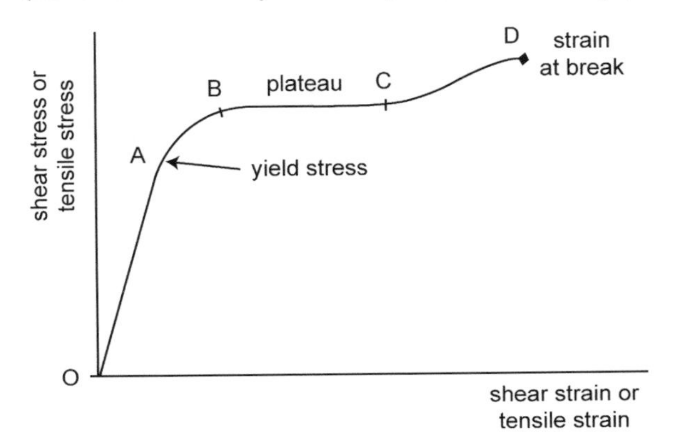

Figure 38: Stress–strain curve.

The slope of the initial part of this curve from O to A is very important to consider in order to classify various materials, and we define the tensile modulus or Young modulus as being $E = S/e$ if this part of the curve is linear. It is expressed in MPa. Materials that have a high Young modulus are stiff or rigid and materials with low Young modulus are flexible and "elastic".

In the zone from O to A, the curve is linear and the material is elastic; A marks the yield point. After this point, the material becomes nonelastic and it is plastically deformed. Plastic deformation means that now the material absorbs energy during further deformation. The plateau from B to C is a deformation without further increase of stresses: such a plateau exists for some materials, which display neck in or draw down deformation or plastic flow. After point C, the material cannot withstand stresses and eventually will break in D. The stress at breaking point is also called ultimate tensile strength. The tensile modulus gives an evaluation of the elasticity of the adhesive.

Poisson ratio: When materials are subjected to a tensile strength, they stretch and become thinner in cross section. The Poisson ratio is defined as:

$$\nu = \frac{\Delta r}{r_o} / \frac{\Delta l}{l_o}$$

where r is the radius of a cylindrically shaped tensile specimen at a given stress, r_o is the original radius and $\Delta l/l_o$ is the tensile strain (Fig. 39). It measures the ratio of lateral strain to the tensile strain, in the elastic region. For isotropic materials (those which have the same properties in any direction) Poisson has predicted that this ratio should be close to 0.25 as it is indicated for some materials in Table 5.

All these values and characteristics may be measured on halter specimens with a tensile machine that records both the stresses and the deformations. They are very important to know when we want to calculate joints as we will see in chapter "Design and calculation of bonded joints" in Volume 2.

2.3.1.3. Impact Resistance and Resiliency

When bonding structural and rigid materials such as metals and composites, it is important in some industries and applications (automotives, ships at sea which

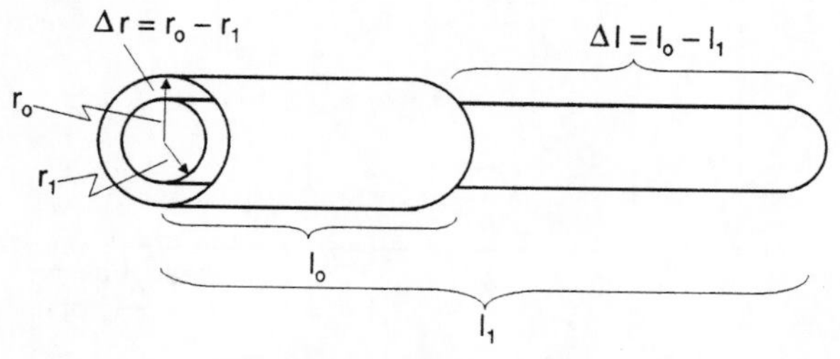

Figure 39: Poisson ratio.

Table 5: Tensile modulus and Poisson ratio for some materials and adhesives.

Material	**Young's modulus (Pa)**	**Poisson ratio**
Aluminum	7×10^{10}	0.33
Mild steel	2.2×10^{11}	0.28
Silicon	6.9×10^{10}	
Glass	6×10^{10}	0.23
Polymethyl methacrylate	2.4×10^{9}	0.33
Polycarbonate	1.4×10^{9}	
Low-density polyethylene	2.4×10^{8}	0.38
Natural rubber	2×10^{6}	0.49

must withstand the shocks of the waves, machine tools) to know how the bonds will resist various impacts.

Fig. 40 shows the impact test also called IZOD impact test according to GM 9571P standard or similar ASTM D 950 standard test method. There is also a Charpy impact test. Impact resistance will be measured in Joules, a measure of the energy required to break the bond. Impact resistance may be measured also according to EN/ISO 9653.

Rigid adhesives tend to have a low resistance to impact and much work has been done in the last few years to promote some flexible structural adhesives such as

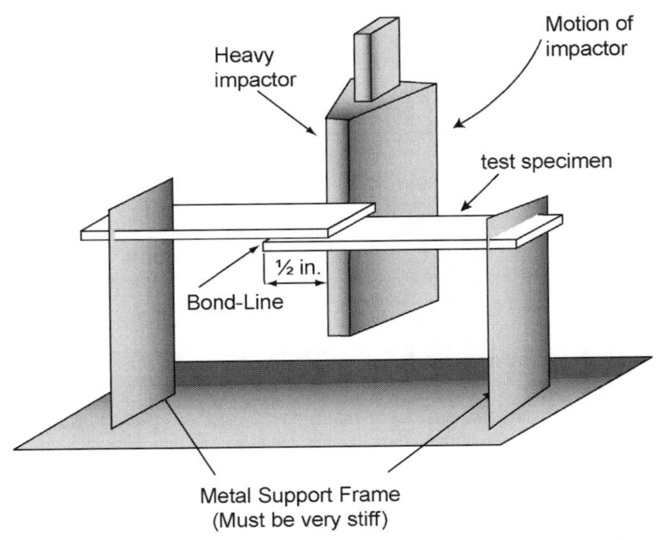

Figure 40: Schematic of Izod shear impact sample in frame showing direction of impact load (ASTM D950).

acrylics and polyurethanes in order to improve impact resistance of structural adhesives as we will see in other chapters such as "Flexible structural bonding", "Automotive adhesives" and "Transportation adhesives" in subsequent volumes.

2.3.1.4. Creep under Constant Load

In many applications and industries, it is important to know the behavior of bonds under a constant stress, inferior to the breaking load, but measured sometimes at elevated temperature.

In many cases, the adhesive bond must withstand continuous loads, and all adhesives tend to creep more or less under continuous loads.

For instance, cardboard boxes bonded with hot melt adhesives must withstand the weight of the packaged goods and also elevated temperatures during transportation. It could be from 50 to 70°C: at these temperatures the hot melt starts to soften, very slightly, but this is enough to observe a slow creep of one part towards the other, until the bond may eventually fail. The same may happen on a segmented bridge made of concrete hollow segments bonded with epoxies: the epoxy adhesive should have a total creep resistance, at the service temperature, and we know that some epoxies, although they give high shear resistance will start to creep at temperatures as low as 40–50°C, under the heavy weight of the concrete blocks.

The creep test is performed as follows (Fig. 41): bonded parts are submitted to a dead load *W*, at the required temperature (in a chamber at this temperature) and the creep (or deformation of the joint) is measured, with a microscope, every hour or day or month, until failure when the bottom part falls (a system records that moment, for instance, by breaking or opening an electric circuit, starting an alarm). An operator will record the dead load used, the creep in mm per day, and the time to failure, and also the temperature of testing.

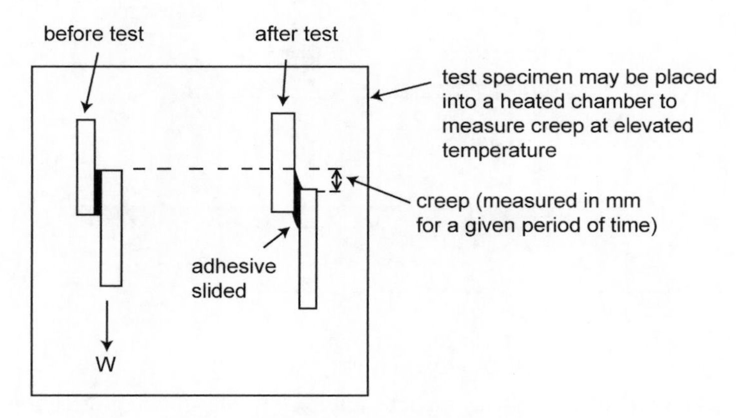

Figure 41: Measurement of creep.

Depending on the adhesive, the loads and testing conditions, the time required for a measurable deformation might be very long, as is the case for the concrete segmental bridge.

Instead of dead loads, one can use calibrated springs. Special fixtures are used for that.

There are several test methods to measure creep:

- ASTM D 2294 utilizes a spring loaded apparatus to maintain a constant stress. Once loaded the elongation of the lap shear specimen is measured by observing at regular periods of time the separation of fine marks with a microscope.
- ISO 15109 (1998) Determination of time to rupture of bonded joints subjected to a static load.

Fig. 42 shows a typical curve of creep versus time for an elastic adhesive or for a viscoelastic sealant. When the load is removed there will be a relaxation of stress and a gradual recovery of strain with time, and also some irreversible strain.

2.3.1.5. Hardness

Hardness of an adhesive is not important in itself, but for the sealant it may be important because it may be damaged if it is too soft, for instance, when the sealant joint is on the floor and should withstand traffic. Hardness is measured in a classic way as rubber hardness with a Shore durometer. There are of course relationships between hardness, flexibility, stiffness, etc. Hardness may also be used sometimes as an indication of the degree of cure.

2.3.1.6. Flexibility

In the past, for structural and semi-structural bonding, adhesives technologists favored the strong adhesives which had very high tensile shear resistance, for

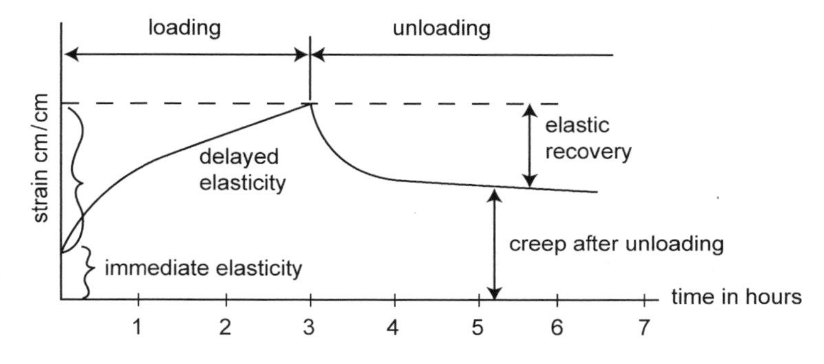

Figure 42: Permanent strain after creep test of a viscoelastic sealant.

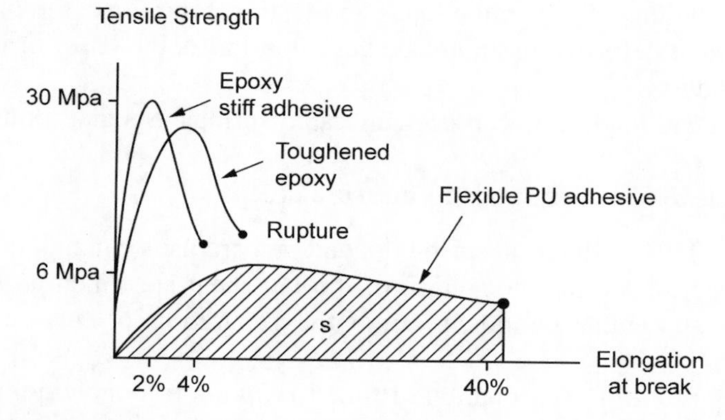

Figure 43: Comparison between rigid and flexible adhesives: the work required to break the bond is proportional to the surfaces under the curves (s).

instance 30 MPa or more such as epoxies. But recently, some suppliers remarked that a very stiff, hard adhesive, having a 30 MPa tensile shear but which would accept only a few percent elongation at break, could break quickly when submitted to movements, while an elastic adhesive allowing 30% or more elongation at break and a few MPa in shear could need more work to break the bond, as indicated in Fig. 43.

This is due to the fact that the work of adhesion (refer to chapter: "Theory of adhesion") is proportioned to the surface below the stress–strain curves.

This explains that adhesives suppliers first developed in the seventies the so-called toughened epoxies, flexibilized by addition of a small amount of cross-linkable rubber, and then recently the company SIKA pioneered the so-called "flexible structural bonding" in which they use PU adhesives derived from their experience in PU sealants, to get semi-structural bonds, for instance, for transportation and automotive or naval uses.

Fig. 43 shows that a flexible, semi-structural PU adhesive giving only 6 MPa tensile strength may require greater work to break the bond than a high-tensile strength structural epoxy. The reader should refer to the chapters: "Flexible structural bonding" by SIKA, and "Epoxy adhesives" by 3M, in following volumes of this Handbook.

2.3.2. Physical Characteristics

2.3.2.1. Elasticity Modulus and Flexibility

We already defined it. Here we will consider it in relationship with the expansion coefficient of the materials to be bonded. If we have 2 materials with different coefficients of expansion, the adhesive film or sealant joint will be stressed when the temperature will change, and this is especially important during the heat curing and cooling of the adhesive: some permanent stresses may result from heating, hardening and then cooling of the adhesive. Therefore, it may be desirable to use a slightly flexible adhesive, which will be able to follow these movements without developing too much internal stresses. Some products need a very flexible adhesive, for instance, shoe soles that should withstand some 100,000 flexions during their whole life without bond failure (and in this case, it is aggravated by the fact that the shoe should also withstand walking into water).

2.3.2.2. Heat Resistance

Adhesives are based on various polymers which have very different behaviors when submitted to heat. All adhesives will soften when the temperature reaches their softening point, but some adhesives will start to soften at 60°C such as the thermoplastic adhesives (vinyls, EVA hot melts), others at 70–80°C (neoprene and rubber based); the thermosetting adhesives will not soften until 90–150°C.

Adhesives, based on polymers, do not melt sharply at a given temperature like some chemicals, but instead they soften gradually when the temperature rises, as is shown in Fig. 15.[1] Before the temperature reaches the softening zone, the adhesive becomes softer and may creep under moderate loads.

The heat resistance of the various adhesives will be indicated in the chapters dealing with these adhesives in this Handbook. Heat resistance may be evaluated by measuring mechanical properties at various temperatures. For example, the standard ASTM D 2295 allows one to measure the strength properties of adhesives in shear by tension loading at elevated temperatures. Heat resistance may be also measured by DSC analysis as indicated in Fig. 44

Heat exposure of bonded parts varies greatly according to the applications and industries, for example:

- outside exposure in Construction will subject adhesives and sealants to temperature cycles from −20 in winter up to 70–85°C according to climates (or even more than 80°C on black absorbing surfaces in very hot countries).
- aircraft adhesives are classified in several grades: those which should withstand up to 80°C, those that withstand temperatures from 80 to 150°C for most uses, and some high temperature adhesives used for supersonic military aircrafts and near the engines which must resist to 260°C. Epoxy adhesive can withstand temperature from 80 to 150°C, epoxy-novolac may go as high as 200°C and some very special adhesives such as BMA (bismaleimide) and cyanate esters may resist up to 260°C service temperature.

[1] Editor's note: In Volume 2 of this Handbook, our readers will find a very important and comprehensive chapter titled 'Heat Stable Adhesives' by Guy Rabilloud, which provides much new information of such adhesives that may resist up to 300 or 400°C, for instance for aerospace, mechanics, machines or appliances.

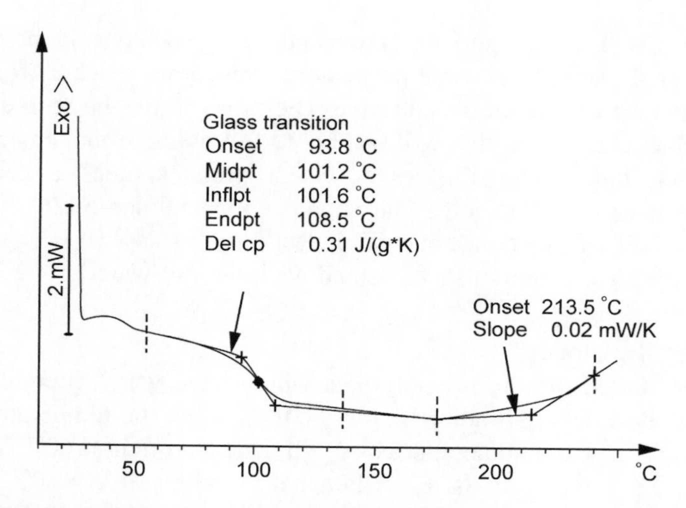

Figure 44: DSC Analysis on polymerized epoxy adhesive. It shows: $T_g = 101°C$; beginning of thermal degradation at 213°C.

- packaging adhesives should withstand the temperature during transportation which could reach 60°C in normal climates inside the containers, or even more in hot countries.
- in the automotive industry, there are also several classes of adhesives regarding their heat resistance. Under the hood applications require service resistances up to 150°C or even more which requires heat stable adhesives such as the very best modified epoxies. Under the windshield and back window, the temperature may reach 80°C or more in hot countries.

2.3.2.3. Cold Resistance

Cold temperatures will harden the A and S, so that they will become more brittle, less flexible, and thus more sensitive to peel and cleavage effects. Therefore, the adequate test will be to submit the A and S joints to the expected low temperatures, and measure peel strength, shear strength resistance to cleavage and impact resistance at ambient temperature and at low temperatures, and decide what percentage loss would be acceptable.

Some adhesives that are already hard and brittle at room temperature (for instance, some hard and densely cross-linked epoxies, UF, cyanoacrylates, etc.) will not tolerate low temperatures below 0°C or much below.

2.3.2.4. Thermal Expansion Coefficient

It is preferable that the coefficient of expansion of the adhesive be of the same order as those of the substrates, so that stresses will not develop if the service temperature ranges for over a large span. But this can be compensated for by using a more flexible adhesive that will tolerate some minor differences of movements resulting from variations of temperatures.

2.3.2.5. Other Specific Properties

There are also other specific characteristics which are required only for special applications such as:

- electrical characteristics (dielectric constant, resistivity, loss angle, etc.), for instance, ASTM D 2739 is used to measure the volume resistivity of conductive adhesives,
- thermal characteristics (T_g or glass transition temperature, expansion coefficient, thermoelasticity),
- optical properties (refraction index, transmission of light, etc.), water transmission and waterproofing,
- ionic purity for electronic applications.

These very specific properties will be defined and studied in the relevant 40 chapters of this Handbook, which are devoted to the specific industries. For instance, in this Volume 1, chapter 6 "Adhesives for electronics" indicates all the specific characteristics of these adhesives (electrical properties, etc.).

2.3.3. Chemical Characteristics

Here we will study the effects and resistance of adhesives and sealants to various chemicals, which may modify or degrade the products.

2.3.3.1. Resistance to Water and Humidity

Water will often degrade the adhesives in different ways:

- It may penetrate into the adhesive by a slow absorption following the Fick law, and then soften it slowly, this happens with many adhesives,
- it may also penetrate between the adhesive and the substrates and destroy the adhesion, this is also frequent,
- water and humidity may also penetrate into porous substrates, swell it and cause movements that are detrimental.

Therefore, water resistance tests must always be performed if the bonds may be subjected to water or humidity during the product life. There are many different tests for water resistance, for instance:

- bonds between porous materials (wood, paper) may be immersed and soaked with water, either at ambient temperature or at elevated temperature, for a given period of time, and their mechanical resistance may be measured before and after this immersion to see its effect.

For instance, the water resistance of adhesives used for the assembly of wood windows is tested according to the EN standard EN 204 with the following test and cycles.

Beechwood test samples are bonded and fully dried. Adhesives are then classified in 4 classes as follows:

- E1 adhesives require only a dry resistance of 10 N/mm^2.
- E2 should withstand a 3 h immersion in water at 20°C followed by drying at room temperature and the test requires a minimum initial resistance of 10 N/mm^2 and after drying 8 N/mm^2. These requirements correspond to occasional exposure in kitchens and bathrooms.
- E3 adhesives must withstand the following cycles:
  - 7 days at RT,
  - followed by 7 days at RT + 4 days immersion in water at 20°C,
  - followed by 7 days at RT + 4 days immersion in water at 20°C,
  - followed by 7 days drying at RT

and the resistance requirements are: initial resistance 10 N/mm^2, wet resistance 2 N/mm^2 and resistance after drying 8 N/mm^2. These requirements correspond to outside windows and doors, kitchen and bathroom equipments.

- E4 adhesives should withstand the following cycles:
  - 7 days at RT,
  - 7 days at RT + 4 days immersion in water at 20°C,
  - 7 days at RT + 6 h in boiling water and 2 h in water at 20°C,
  - 7 days at RT + 6 h in boiling water and 2 h in water at 20°C,
  - 7 days drying at RT.

And the resistance requirements are: initial 10 N/mm^2, wet 4 N/mm^2 and after drying 8 N/mm^2. This corresponds to outside exposure to rain, humidity, etc.

- Metal to metal bonds may be also subjected to immersion in hot water at 50°C, for example in order to speed up the phenomena, and the effect of water may be measured with the wedge test (such as the Boeing wedge test that we described above): the propagation of the initial crack with time of immersion will show the degradation effect of water, and also at the same time it will show evidence of the importance of a good preparation of the metal surface before bonding; if the surface preparation was not good, then the water will affect the bond faster and the crack will grow faster and further.
- In order to know whether humidity has penetrated into the adhesive film, a DSC test may be performed.

- It is also possible to test the resistance to humidity by submitting bonded samples to an accelerated weathering tester such as the QUV spray tester, which combines water spray and condensation plus UV light to simulate sun rays. We will study this in the "Weathering" section hereunder.
- In the automotive industry, some car manufacturers are using the "wet patch" or wet cataplasma: they place samples of bonded parts in contact with cotton impregnated of water, put it into a plastic impervious bag and then store it at a given temperature for a given period of time. They measure mechanical resistance (such as tensile shear strength) before and after this test, and they have their own rules of experience in order to decide the percentage loss of properties, which is acceptable for a given application.
- Other industries have their own specific tests: for instance, textile and garments industries have their own tests to simulate washing resistance of bonded parts in washing machines, other industries may have their specific tests for "tropicality" (of packaging for instance) in very humid and hot climates.
- Fig. 45 shows a very special test for the evaluation of cathodic delamination in seawater, which combines the effects of load and environment.

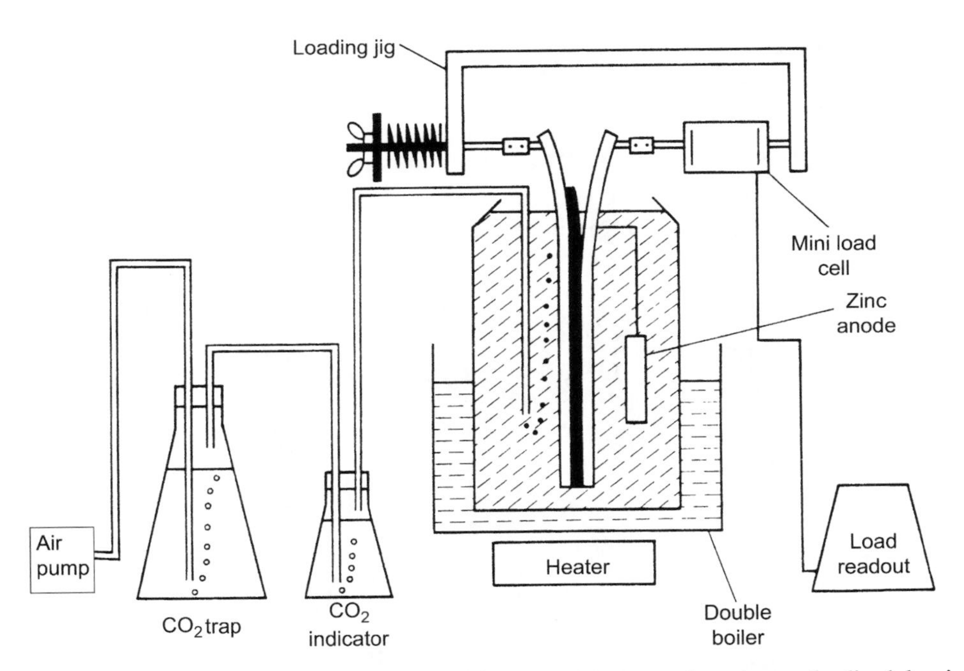

Figure 45: Spring-loaded double cantilever beam specimen undergoing cathodic delamination in seawater. The spring lowers the rate of decay of G with debond distance.

- Construction sealants have also their own water resistance tests and durability tests. Refer to the four chapters dealing with sealants.
- Double insulated windows need a specific test to measure waterproofing properties of A/S.

2.3.3.2. Resistance to Oils, Greases and Plasticizers

All these products may penetrate into the adhesive film and soften it by penetrating between the molecules of adhesive. Fig. 46 shows what happens when a plasticizer or oil migrates into a film of thermoplastic adhesive: the film swells, it is softened, this degrades the adhesion and eventually the bond will break.

The best test is to simulate what happens in the real product life by immersing bonded parts into these products at a slightly elevated temperature, for instance 60°C, in order to speed up the degradation. It is always assumed that an increase in temperature by 10°C will double the speed of many chemical and physical reactions, and therefore it is quite common to perform accelerated tests at elevated temperatures. But of course the laboratory chemist should also use their experience in order to correlate an accelerated test and a real long-term exposure.

Resistance to oils and greases is important not only for applications in mechanics, but also for automotives, aircraft (resistance to some fluids), packaging of fat materials, bonding of fat leather in footwear, etc.

Resistance to plasticizers is a problem when bonding plasticized PVC or rubber. The plasticizer may migrate slowly:

- to the interface and create a soft layer which will after some time separate the adhesive from the substrate (Fig. 46),
- and/or into the adhesive film and soften it.

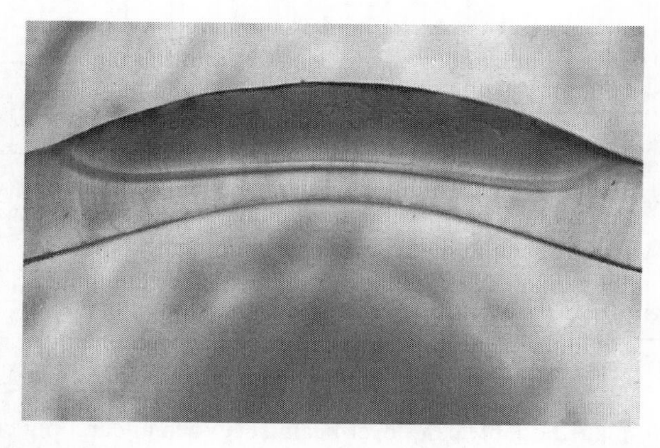

Figure 46: Migration of oil or plasticizer into new adhesive film. (Source BASF.)

In industries that are bonding plasticized materials, there are some specific tests to measure for instance peel strength before and after accelerated aging test. The A and S suppliers, who know the raw materials included in their formulations, can tell quickly whether a given adhesive should be sensitive or not to oils, greases and plasticizers. Some adhesives are sensitive to oils (vinyl emulsion glues), some are not (epoxies, anaerobics, etc.).

2.3.3.3. Resistance to Chemicals (Acids, Alkalis, Solvents)

Many applications require some resistance to chemicals: for instance bonding of chemical equipment, automotive (resistance to some fluids, bonding batteries), packaging (resistance to some aggressive liquid foods such as tomato juice, oils and fats, etc.), or even furniture (resistance to the solvents of varnishes), clothes (during dry cleaning), electronics (resistance to etchants, etc.), or decoration (resistance to cleaning agents). Here again the best test is to immerse bonded parts into the aggressive chemical and measure the bond strength before and after this test.

2.3.3.4. Density of Cross-linking

For the chemist this may give good indications regarding resistance to water, oils, plasticizers: the more the polymer is cross-linked, the more the chances are that it will be resistant to penetration of other fluids and consequently to softening. Also a highly cross-linked polymer will probably give a higher resistance to heat. However, some purely thermoplastic polymers have a very crystalline structure with strong links between the molecules chains and these polymers have a high resistance to heat or to water or other chemical agents.

For all these chemical properties, the A and S supplier will consider the chemical type of the base polymer: there are general rules in chemistry that tell what are the expected performances of a given polymer.

2.3.3.5. Degree of Cure

DSC analysis allows one to measure the degree of cure of thermosetting and 2 components adhesives. Fig. 47 shows that at 150°C for 30 min this epoxy adhesive was totally polymerized. At the other temperatures residual enthalpy proves that polymerization was not finished.

2.3.4. Durability

2.3.4.1. Durability Tests, Accelerated Aging

Users want to know or forecast how long will a bond last under a given set of service life conditions. The best tests should combine all the risks of degradation

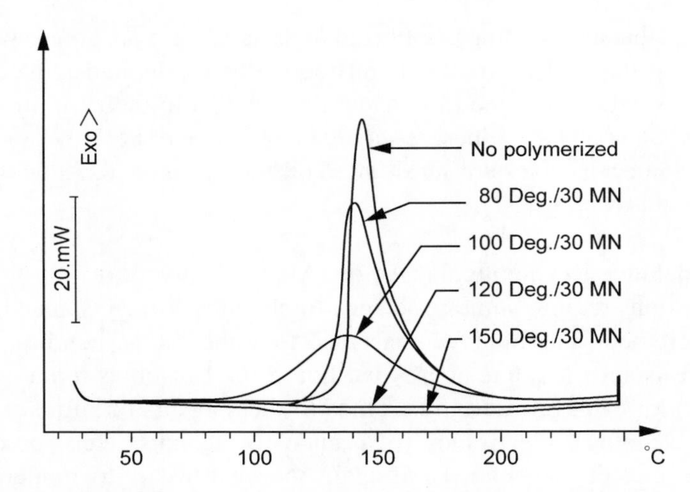

Figure 47: DSC Analysis on epoxy adhesives polymerized at various temperatures during 30 min. It shows that at 150°C/30 min the polymerization was completed. At the other temperatures, residual enthalpy proves that polymerization was not finished.

during actual use. This is why lab equipment manufacturers have designed some specific equipment to measure durability under various conditions.

An important issue is: How long will the bonded product last outdoors when subjected to temperature cycles (heat and cold), humidity cycles, sunlight, etc. Let us mention here the QUV accelerated weathering tester from Q panel lab products, USA. It reproduces the damages caused by sunlight, rain and dew, by exposing samples (of bonded parts or pieces of A and S alone) to alternating cycles of light and moisture at elevated temperatures. In a few days or weeks, the QUV reproduces the damages that occurs over months or years outdoors.

Fig. 48 shows this equipment which uses:

- UV lamps with various wavelengths spectra and energy to simulate sunlight,
- dew condensation or spray to simulate rain and variation of humidity. Hot vapor maintains the chamber at 100% RH and at elevated temperature.

We may also mention the Weather-o-Meter which is similar. Other equipment can simulate corrosion with a salt spray, or simulate the aggressions from other agents such as chemicals.

Several standards exist for measuring durability, for instance D 3762: test method for adhesive bonded surface durability on aluminum (wedge test).

2.3.4.2. Fatigue Resistance

Fatigue resistance is also another big issue. Several equipments are used to subject bonded parts to alternate fatigue cycles such as the one in Fig. 49. Fatigue

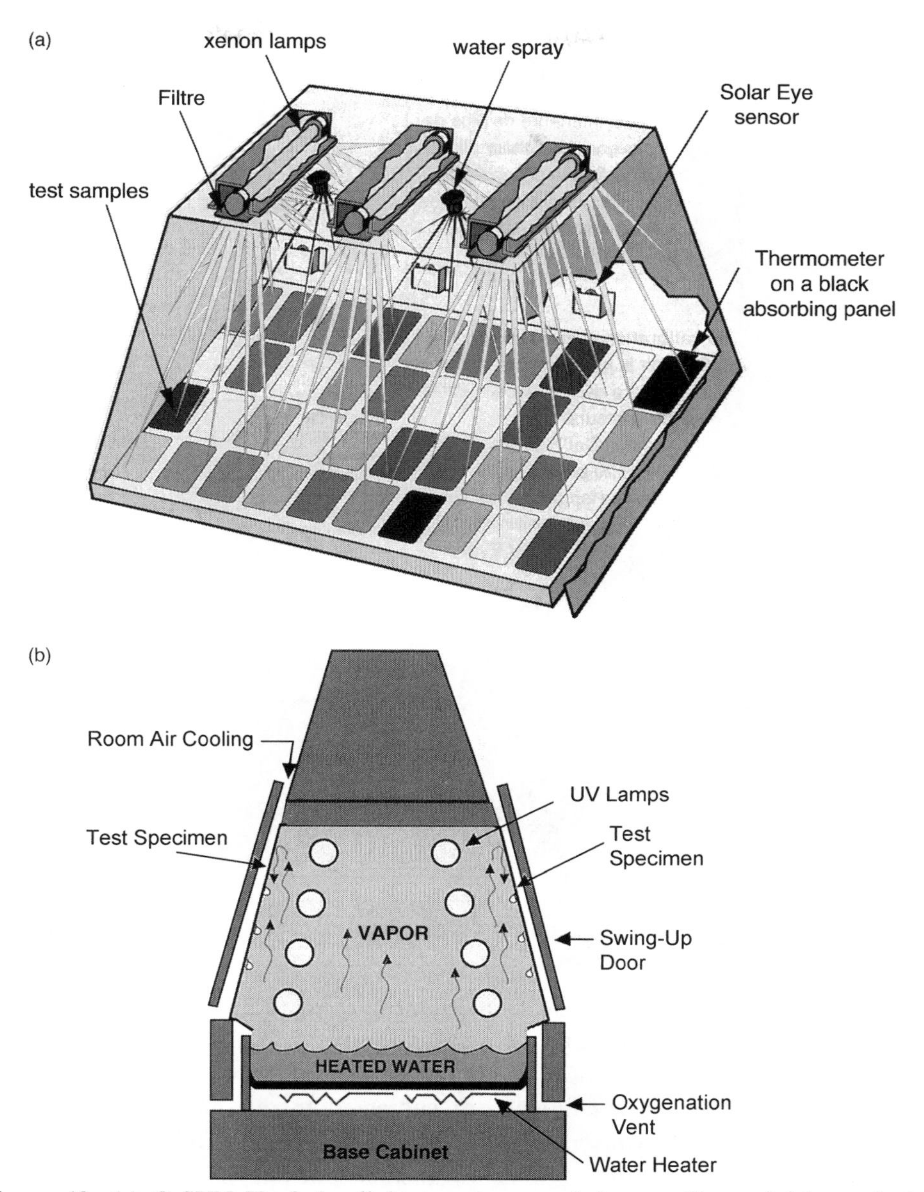

Figure 48: (a) Q-SUN Xe 3 (sunlight + water spray) (source Q panel lab products); (b) QUV cross-section during condensation period (humidity + UV light).

Figure 49: One kind of fatigue resistance test on a bonded specimen. The machine flexes the test specimen to a certain extent for a very large number of alternate flexions.

resistance may be measured for structural adhesives according to standard ASTM D 3166: test method for fatigue properties of adhesives in shear by tension loading (for metal to metal bonds) and other tests, such as ISO 9664.

In these tests, bonded specimens are subjected to alternate stresses of for instance 20% of the maximum tensile shear resistance, and this cycle may be repeated 100 000 or one million times, for instance, or until the bond breaks, and here again mechanical resistance is measured before and after the test.

2.3.4.3. Attack by Bacteria, Roaches

The adhesive may be attacked while it is stored in its packaging: this may be measured by ASTM standard D4783: tests methods for resistance preparation to attack by bacteria, yeast and fungi. The dry film of adhesive in the joint may be also degraded and there are also standards for that such as D 1382: susceptibility of dry adhesive film to attack by roaches.

2.3.5. Safety Characteristics

Here we must consider many factors:

- the various risks and dangers: flammability, volatile organic compounds (VOCs), noxiousness, toxicity, allergic dermatitis, explosion,
- some of these risks and dangers occur during manufacturing of the A and S, transportation and storage, others occur during the utilization of the A and S,
- bad smell,
- food and drugs compatibility,
- protection of the environment.

Another difficulty is the fact that the laws and regulations differ between Europe and USA, and also differ with some developing countries like China or India or Eastern countries where there is a lack of strong laws and regulations.

2.3.5.1. Flammability

In the past, many adhesives and some sealants used to contain flammable solvents which could be flammable or even explosive for given ratio of solvent to air, and/or toxic or noxious by breathing. Since the eighties, solvent use has been reduced and solvent-based adhesives have been replaced by hot melts, water-based adhesives or curing adhesives, all of which do not contain solvents. But there is still a large use of solvents in some countries and some industries, for instance, in shoe manufacturing in the main producing countries such as China, India and Brazil, and solvents are still used also in Western countries in footwear, rubber bonding, for PVC pipes bonding, contact polychloroprene adhesives, etc.

Therefore, we provide hereunder some basic information and we suggest to readers to contact the safety organizations in their own countries (because regulations differ from one country to another) when they need more information. Almost all solvents are highly flammable except the chlorinated solvents such as trichorethylene, trichlorethane, etc., but these have another drawback: they are toxic. Flammability is measured by the flash point which is the minimum temperature at which the vapors of an adhesive will start to burn, giving a short "flash" of fire. There are several test methods for measuring the flash point: closed cup, open cup, Abel cup, Pensky Mertens; readers should refer to detailed regulations and standards for more information.

Flammability is indicated by the red label (example in Fig. 50) and there are several classes: very highly flammable, highly flammable, etc., according to the

Figure 50: Label for flammable products (transportation regulations).

flash points. For instance, ethyl acetate, with a flash point of −4°C in closed cup and −7°C in open cup is labeled easily flammable. Its limits of explosion are from 2% by volume, mixed with air, up to 11.5%. Boiling point is 77°C at atmospheric pressure so we can see that the flash point has nothing to do with the boiling temperature.

2.3.5.2. Explosivity

If the ratio of solvent to ambient air is within a given range the mix is explosive in the presence of fire or electric sparks. In areas where these solvents are used all the equipment and electric equipment should be spark proof, and the solvent vapors should be eliminated by exhaust fans (but now in some countries it is unlawful to release large quantities of noxious solvents in the ambient air so that very large users may need some system for recycling of solvents, for instance those who coat very large surfaces of pressure-sensitive tapes and labels or laminate plastic films for flexible packaging).

In order to warn of these dangers, flammable products must be labeled as indicated in Figs. 50 and 51, for transportation and utilization. The various labels indicate the class of danger.

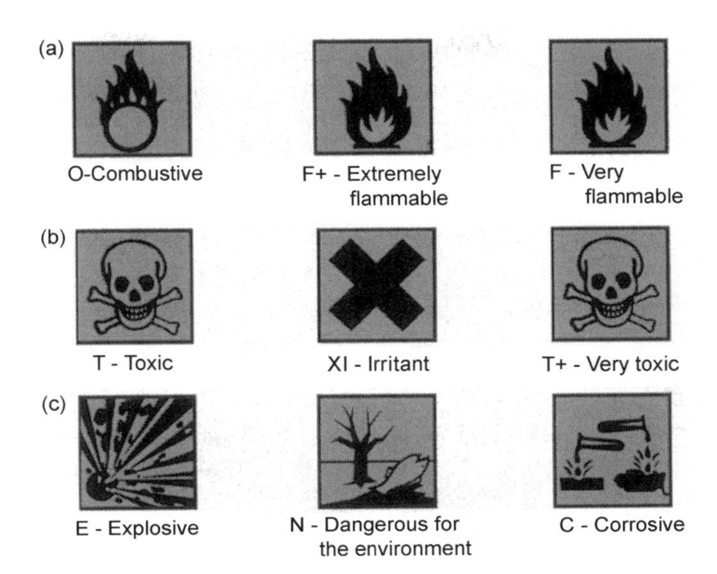

Figure 51: The various legal labels that are mandatory for risk warnings on chemical packagings/containers. (Colour version of this figure appears on p. xiv.)

2.3.5.3. Noxiousness and Toxicity, Threshold Limit Values

Solvents such as toluene, benzene, methanol, chlorinated solvents, etc. and some chemicals (such as isocyanates, formaldehyde, etc.) are harmful to health, either by breathing or by ingestion. Here again there are several regulations, related to the use of a number of solvents and chemicals, and all the dangerous products must be labeled according to European Community laws or US laws. Fig. 51 shows an example of such danger labels.

In Europe, dangerous chemicals are listed in the "European Directive 67/548/CEE, for classification, packaging, labeling of dangerous chemical substances", which was started in 1967 and modified 28 times since, so that today we have the 29th amendment dated April 2004. This very long list provides for all chemicals their CAS number and EINECS number, the class of danger, the labels and danger warning sentences which must be affixed on each container/package, and some more information.

In the USA and outside the EC there are other regulations, but the various developed countries have the same understanding of the potential dangers of some chemicals.

In workshops, there is a maximum amount of solvent vapors which is acceptable, and these values, called "threshold limit values for occupational exposure to chemicals" are given by tables which may differ from one country to another. However, toxicologists in all developed countries all agree to list more or less the same chemicals in their lists, but the threshold limit values may differ.

For instance in Europe, the Directives 91/322/CEE of May 29, 1991, and 96/94/CEE of December 18, 1996 indicate the following threshold values:

- methanol: 200 parts per million (ppm) or 260 mg/m^3, for an 8 h exposure, and the French CNAM indicates:
- ethyl acetate: 400 ppm,
- ethyl acrylate: 5 ppm,
- methanol: 200 ppm,
- toluylene diisocyanate: 0.01 ppm only,
- *n*-hexane: 50 ppm,
- methyl methacrylate: 100 ppm is 20 times more than ethyl acrylate (beware of general assumptions, all acrylates are not the same),
- methyl ethyl ketone: 200 ppm,
- di butyl phthalate: only 5 mg/m^3,
- toluene: 100 ppm,
- trichlorethylene: 75 ppm.

In USA, the American Conference of Governmental Industrial Hygienists (ACGIH, 1330 Kemper Meadows Drive, Cincinnati, OH 45240-1634, USA) publishes each year the "Treshold limit values for chemical substances and physical agents and biological exposure index", which are frequently used in other countries that do not have their own regulations.

Let us mention also the Occupational Safety and Health Administration of USA lists (OSHA), which publishes legal rules such as OSHA final rule, air contaminants — permissible exposure limits, title 29, Code of federal regulations, part 1919–1000. In Germany, the Deutsche Forschungsgemeinschaft DFG publishes each year the "MAK and BAT values, maximum concentrations at the workplace and biological tolerances values for working materials".

Sometimes the threshold values differ between countries. For instance, for the methyl-2-cyanoacrylate the American ACGIH gives an average threshold value of 0.2 ppm, while the French CNAM and German MAK values are of 2 ppm. All depends on how the various countries perceive the danger of chemicals and it depends also on the legal and Health administrations.

2.3.5.4. Volatile Organic Compounds (VOCs) Regulation

VOCs have been defined as: "all organic compounds having a vapor pressure higher than 0.01 kPa at 20°C". Emissions of VOCs may have effects on health (of users, manufacturers) and the environment. More than 25% of all VOCs released in the atmosphere are solvents coming from paints, varnishes and other products such as adhesives, cleaning agents, etc.

The European Community has enacted the Directive of March 11,1999 which limits the use of solvents in order to protect the environment. Conformity of existing plants and installations must be obtained before October 30, 2005. Generally speaking, when the exhaust exceeds 2 kg/h the maximum VOC emission is 110 mg/m^3 expressed as carbon content. This regulation concerns various industries: application of PSAs on tapes and labels, manufacture of paints, varnishes, adhesives and inks, various laminating operations, footwear manufacturing, painting of cars and other vehicles, surfaces cleaning and other large users of adhesives.

For all types of risks and dangers, our readers should consult the safety regulations and Safety and Health agencies in their countries, for detailed information.

All A and S manufacturers and suppliers must warn all their customers about all these potential risks and dangers, by several means:

- labels and warning sentences on all packaging, according to the various rules and regulations,
- by providing materials safety data sheets (MSDS). These are now mandatory in most countries: they list all dangers and risks of a given product, which usually needs several pages in order to refer to all regulations. Refer to Section 2.3.5.6 and Table 6.

2.3.5.5. Allergy and Dermatitis

Some products may cause allergies or dermatitis, such as cement, epoxy resins and polyamide hardeners, etc. The safety data sheets also warn about these risks.

Since 1985, adhesives manufacturers in all developed countries strived to reduce all these risks by several ways:

- replacing solvent-based adhesives by water-based and hot melt adhesives,
- eliminating the dangerous solvents and chemicals, with new formulations,
- and at least warning carefully about all the remaining risks by detailed labeling and safety data sheets.

2.3.5.6. Materials safety data sheets

In order to safeguard all users of chemicals, most developed countries, governments have required the manufacturers to provide to all users MSDS incorporating all the relevant legal information and warnings.

In Europe, the Directive EEC 91/155/EEC was issued in 1991: the SDS must contain 16 sections, and Table 6 lists these 16 potential hazards. SDS must be issued by manufacturers even when they feel that the product is harmless because there is always at least one risk, it is the ecological one, because for instance adhesives cannot be thrown into rivers! In practice, for many A and S, provided they are handled with proper care, the potential risks

Table 6: Contents of European standard material safety data sheet, as defined by EC Directive 91/155/EEC.

	Subject	Contents
1	Labelling of substance/ preparation and manufacturer's designation	Proprietary name, intended use, manufacturer's name and address
2	Composition/information about constituents	Chemical description, hazardous, constituents (incl. Chemical Abstracts number, concentration hazard symbols, risk and safety phrases)
3	Potential hazards to human health and the environment	Designation of hazards, special hazard warnings
4	First-aid measures	General advice, action to be taken following inhalation, skin contact, eye contact, ingestion
5	Action in case of fire	Suitable extinguishing agents, special hazards posed by the product itself or by gases and vapours released during combustion, special protective equipment needed for fire-fighting
6	Action in the event of accidental release	Precautionary measures designed to protect exposed persons, environmental safeguards, clean-up procedures
7	Handling and storage	Safety measures designed to prevent fire and explosion, storage specifications for bulk/mixed storage, additional storage recommendations
8	Limiting exposure and personal protective equipment	Constituents that need to be monitored in terms of their maximum workplace concentrations, personal safety equipment (respirators, gloves, goggles, protective clothing)
9	Physical and chemical properties	Appearance, safety-related data
10	Stability and reactivity	Conditions to be avoided, dangerous reactions, thermal degradation and hazardous decomposition products
11	Toxicology data	Sensitization, known effects of human exposure (in cases of skin or eye contact, inhalation or ingestion)
12	Ecological data	Information on possible environmental hazards (contamination of water, rivers soil and air)

(*Continued*)

Table 6: Continued.

	Subject	Contents
13	Disposal of waste	Disposal of product and soiled packaging
14	Carriage and movement of goods	Classification for transport by road, rail, air and sea
15	Legal requirements	Labelling in accordance with national and international regulations (e.g. EC-Directive 88/379/EEC on hazardous substances, toxicity classification, water pollution classification, etc.)
16	Miscellaneous points	

are less than may appear to some people, but a small number of A and S do represent a real risk.

Conclusion

I am sure that after reading this chapter, with its long list of some 75 different technical characteristics and properties, our readers are now able to read and understand any paper, book or the subsequent chapters written on adhesives and sealants. I regret that these things may look complicated, but remember that for any given bonding problem or any choice of adhesives, you will probably need to know only 15 to 25 of these characteristics!!

Chapter 3

Chemistry of Polyurethane Adhesives and Sealants

D.M. Segura, A.D. Nurse, A. McCourt, R. Phelps and A. Segura*

PUR-Technology Centre of Excellence, Wolfson School of Mechanical and Manufacturing Engineering, Loughborough University, Loughborough, Leicestershire, LE11 3TU, UK

Diana Segura received her BSc (2002) in chemical engineering from Los Andes University, Bogota, Colombia, and she is a PhD student at the PUR Technology Centre of Excellence at Loughborough University, United Kingdom. Her areas of interest are in polyurethane chemistry, and artificial intelligence for product formulation. Her current research work is in the area of case-based reasoning for formulating polyurethane materials using fuzzy logic, genetic algorithms, and neural networks.

Andrew D. Nurse completed his PhD in 1990 at the University of Sheffield. His postdoctoral research was performed at University of Sheffield until 1992. He has been a senior lecturer in stress analysis in the Department of Mechanical Engineering, Loughborough University, since 1993. His research interests include new manufacturing processes for plastics, knowledge-based engineering, and inverse analysis. He is director of a Centre of Excellence in polyurethanes.

Keywords: Additives; Aliphatic; Aliphatic isocyanates; Amine-isocyanate reaction; Aromatic isocyanates; Blocked isocyanates; Chain extenders; Construction (application in); Glass transition temperature; Hydroxyl-isocyanate reactions; Hot melt reactive PU; Isocyanates; MDI; Moisture curing PU; One-component PU; Packaging (application in); Plastic and composites (application in); Polyester; Polyether polyols; Polyisocyanates; Sealants; Silane PU hybrids; TDI; Testing of PU adhesives; testing of PU sealants; Testing standards; TPU thermoplastic PU; Transportation (application in); Two-component PU; Water-borne PU

3.1. Introduction

Polyurethane adhesives and sealants are part of the polyurethane family legacy that emerged in 1937 when Bayer et al. pioneered the polyaddition polymerisation

*Corresponding author. Tel.: +44(1509)-227554; fax: +44(1509)-227648.
E-mail address: A.D.Nurse@lboro.ac.uk (A.D. Nurse).

Handbook of Adhesives and Sealants
P. Cognard (Editor)

reaction of polyisocyanates with di- or polyfunctional hydroxyl or amine compounds and with that the practical uses of isocyanate compounds. Since then it has been discovered that by varying the nature of the polyols and the isocyanate components, a wide range of properties can be obtained [1].

This chapter introduces readers to the fundamentals of polyurethane chemistry, the basics of polyurethane adhesives, their applications and test standards. The chapter will also highlight relevant developments that have led formulators to create and tailor-make many polyurethane (PUR)[1] adhesives and sealant products.

3.2. Polyurethane Chemistry

In 1849, nearly 90 years before Bayer et al.'s developments, the actions of isocyanates were discovered by Charles-Adolphe Wurtz who was one of the first chemists to react aliphatic isocyanates with hydroxyl compounds to obtain urethane compounds. Since then, it has been discovered that urethane chemistry comprises the reactions that isocyanates undergo with any active hydrogen compound. Isocyanates will react with any compound containing hydrogen atoms attached to a nitrogen atom [2].

Polyurethanes include polymers containing a significant number of urethane groups, regardless of the composition of the rest of the molecule [3]. For instance, a typical polyurethane may contain, in addition to urethane groups, aliphatic and aromatic hydrocarbon, ester, ether, amide, and urea groups.

Polyurethanes are the most versatile of all polymers. Their applications include diverse types of foams, (soft and rigid), coatings, adhesives, sealants, and elastomers. Although the number of chemicals is small, the molecular weight of the reactants and the method of polymer formation can be varied widely to meet the desired properties of the final product. There have been many books published on the subject of polyurethane chemistry [3–12].

This section of the chapter will present the basic reactions found in the chemistry of polyurethane compounds, such as the reaction of isocyanates with polyols, water, and amines. The reactions of isocyanates with urethanes, ureas, and amides are also of significant importance in polyurethane chemistry as they will lead to an increase in materials choice.

For those readers who need to review the fundamental aspects of organic chemistry, there is a quick refresher available [13]. As for reviewing the reactions and preparation of isocyanates more deeply, exhaustive studies on the chemistry of organic isocyanates have also been carried out [2,3,14].

[1] According to international standards (ISO, DIN, etc.), the only correct abbreviation used for the term polyurethane is "PUR". However, the form "PU" commonly used throughout the world, although not officially standardised, will also be used in this volume.

Table 1: Reactivity degree of active hydrogen compounds.

Hydrogen compound	Formula	Reactivity degree
Primary aliphatic amine	$R{-}NH_2$	Highly reactive
Secondary aliphatic amine	$R_2{-}NH$	
Primary aromatic amine	$Ar{-}NH_2$	
Primary hydroxyl	$R{-}CH_2{-}OH$	
Secondary hydroxyl	$R_2{-}CH{-}OH$	
Tertiary hydroxyl	$R_3{-}C{-}OH$	
Phenol	$Ar{-}OH$	
Water	$H{-}O{-}H$	Less reactive

3.2.1. Reactions of Isocyanates

Isocyanates will react with all compounds containing hydrogen atoms attached to a nitrogen atom [2]. There are four basic reactions chemists employ to make polyurethanes. The reaction of isocyanates with hydroxyl groups to produce urethane is the primary reaction. The reaction of isocyanates with amines yields urea; and the reactions of isocyanates with urea and urethane produce biurets and allophanates, respectively.

3.2.2. Reaction of Isocyanates with Alcohols

Isocyanate groups react with polyfunctional active hydrogen compounds to give high molecular weight polyurethane products. One of the most important reactions of isocyanate compounds is with di- or polyfunctional hydroxyl compounds, e.g. hydroxyl terminated polyesters or polyethers. The functionality of the hydroxyl-containing compound as well as of the isocyanates can be increased to three or more to form branched or cross-linked polymers. Different degrees of reactivity are expected from different compounds as shown in Table 1. This is also affected by the steric hindrance of either the isocyanate or the active hydrogen compounds.

The reaction proceeds at ambient temperatures without the use of catalysts. Reactivity is higher for primary alcohols, decreasing for secondary, tertiary, and aromatic alcohols (Fig. 1).

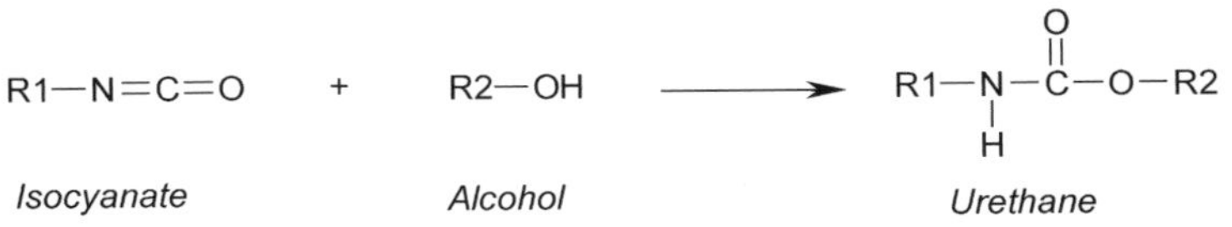

$$\underset{\textit{Isocyanate}}{R1{-}N{=}C{=}O} \quad + \quad \underset{\textit{Alcohol}}{R2{-}OH} \longrightarrow \underset{\textit{Urethane}}{R1{-}\underset{H}{\underset{|}{N}}{-}\overset{O}{\overset{||}{C}}{-}O{-}R2}$$

Figure 1: Reaction of isocyanates with alcohols; R1 and R2 stand for an aromatic or aryl group.

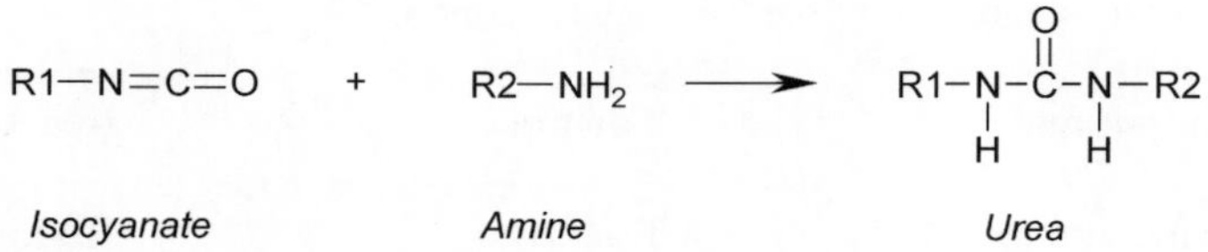

$$R1{-}N{=}C{=}O \quad + \quad R2{-}NH_2 \longrightarrow R1{-}NH{-}C(=O){-}NH{-}R2$$

Isocyanate *Amine* *Urea*

Figure 2: Reaction of isocyanates with primary amines.

3.2.3. Reaction of Isocyanates with Amine

In addition, isocyanates react with amines usually at 0–25°C yielding urea. Primary aliphatic amines react most quickly followed by secondary aliphatic amines and aromatic amines (Figs. 2 and 3).

3.2.4. Reaction of Isocyanates with Water

In general, isocyanates have a strong affinity to water, which makes them difficult to store. This reaction is not desirable in applications such as structural adhesives and sealants [15]. Isocyanate reacts with water giving carbamic acid, an unstable compound which spontaneously decomposes into a primary amine and carbon dioxide, together with a subsequent urea formation (Fig. 4).

3.2.5. Allophanate and Biuret Formation

Isocyanates can react, in excess conditions, with urea and urethane to produce allophanate and biuret compounds. The first reaction is quicker as seen in Table 2 with a kinetic constant (K) 30 times higher for the reaction of phenyl isocyanate with urea than with urethane at the same temperature [14, p. 339]. In polymers containing both urea and urethane groups in roughly the same concentrations, branching is introduced principally by biuret formation. These reactions will occur more readily at higher temperatures and it can be seen that the products, allophanates and biurets, are in equilibrium with the starting materials, isocyanates and active hydrogen compounds (Figs. 5 and 6).

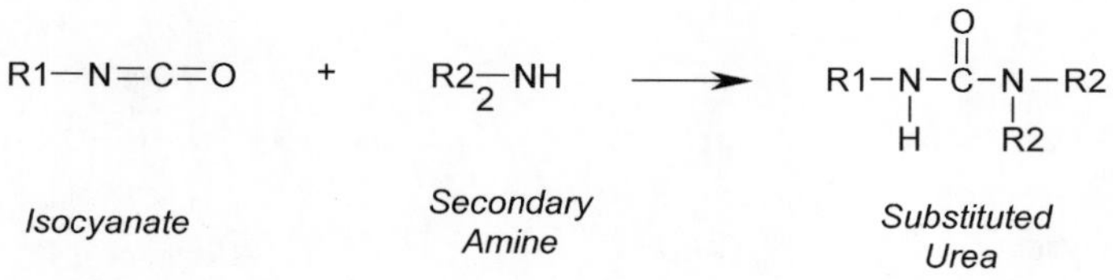

$$R1{-}N{=}C{=}O \quad + \quad R2_2{-}NH \longrightarrow R1{-}NH{-}C(=O){-}N(R2){-}R2$$

Isocyanate *Secondary Amine* *Substituted Urea*

Figure 3: Reaction of isocyanates with secondary amines.

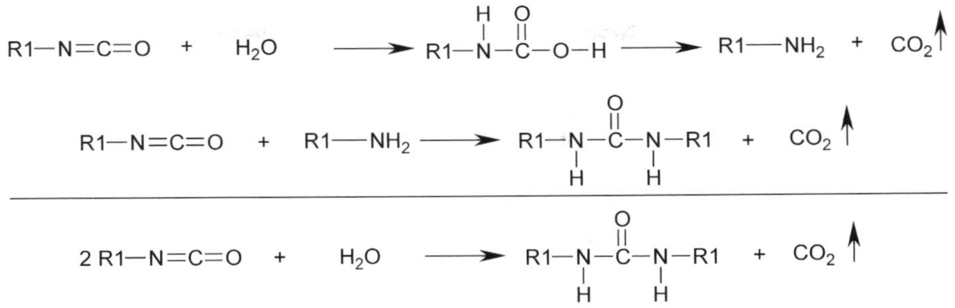

$$R1{-}N{=}C{=}O + H_2O \longrightarrow R1{-}NH{-}C(=O){-}O{-}H \longrightarrow R1{-}NH_2 + CO_2\uparrow$$

$$R1{-}N{=}C{=}O + R1{-}NH_2 \longrightarrow R1{-}NH{-}C(=O){-}NH{-}R1 + CO_2\uparrow$$

$$2\,R1{-}N{=}C{=}O + H_2O \longrightarrow R1{-}NH{-}C(=O){-}NH{-}R1 + CO_2\uparrow$$

Figure 4: Reaction of isocyanates with water and polyurea formation.

3.3. Common Polyurethane Raw Materials

Common isocyanates used as building blocks for polyurethane adhesives include aromatic and aliphatic isocyanates. Commercial aromatic isocyanates include principally toluene diisocyanate (TDI), 4,4′-diphenylmethane diisocyanate (MDI), and polymeric MDI.

3.3.1. Aromatic Isocyanates

In the majority of reactions, mainly with active hydrogen compounds, aromatic diisocyanates are much more reactive than their aliphatic equivalents.

3.3.1.1. TDI

One of the most important monomers used in the polyurethane industry is TDI. TDI is usually supplied as a mixture of the 2,4- and 2,6-isomers. The products available vary from >98% 2,4-TDI to a 65/35 and 80/20 2,4/2,6-isomer mixture (Fig. 7).

Table 2: Relative rates of reactions of aromatic isocyanates and substituted ureas and urethanes.

Isocyanate	**Urea**	**Urethane**	**Temperature (°C)**	**$K{\cdot}10^4$, l/mole sec**
Phenyl isocyanate	$C_6H_5NHCONH_2$	–	60	3.7
Phenyl isocyanate	$C_6H_5NHCONH_2$	–	100	32
Phenyl isocyanate	$C_6H_5NHCONH_2$	–	140	48
Phenyl isocyanate	–	$C_6H_5NHCOOC_2H_5$	140	18
o-Tolyl	–	$C_6H_5NHCOOC_2H_5$	140	0.1

Source: Ref. [14].

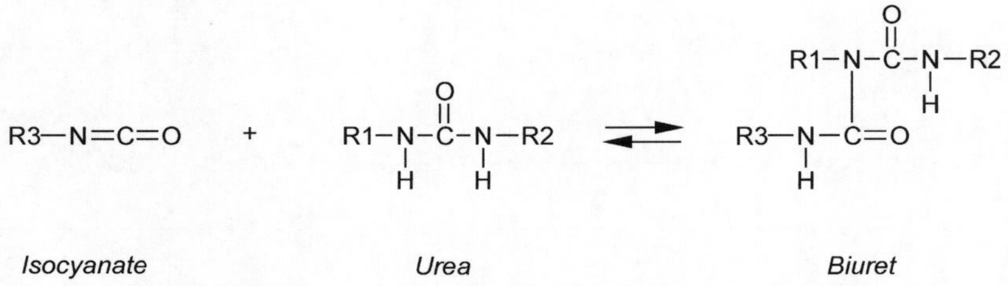

Figure 5: Biuret formation.

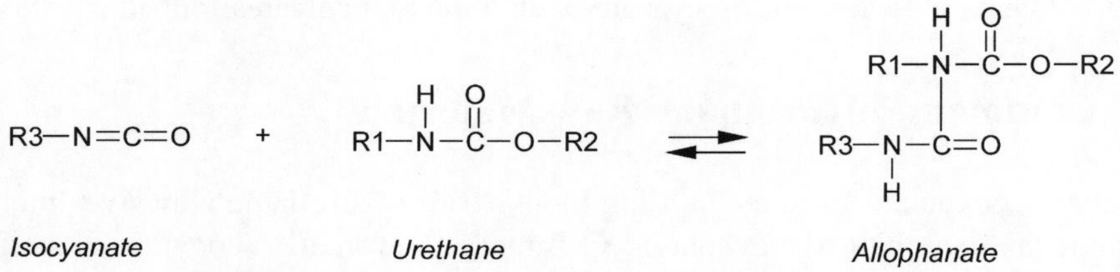

Figure 6: Allophanate formation.

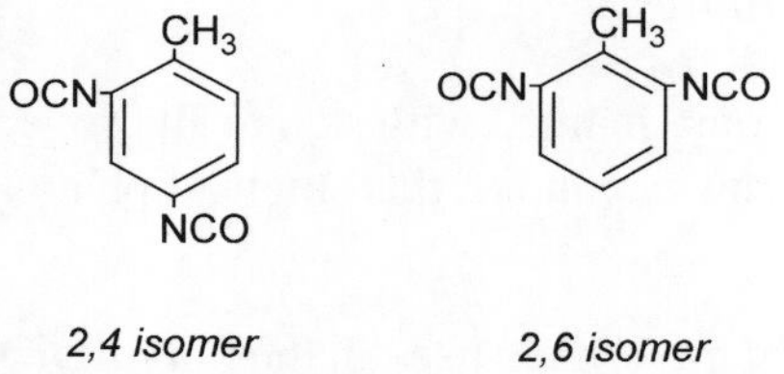

Figure 7: Toluene diisocyanate isomers.

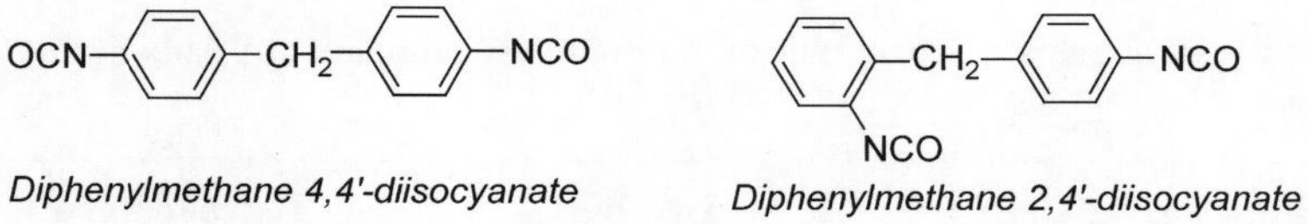

Figure 8: MDI monomers.

3.3.1.2. MDI

MDI is one of the monomers more widely used in the polyurethane industry. MDI is preferred over TDI because of its significantly low vapour pressure and the usually higher performance polymers that can be produced. MDI is produced as a variety of isomers and oligomers by phosgenation of the condensation product of formaldehyde and aniline (Fig. 8).

The symmetrical 4,4-isomer is isolated from this mixture by a distillation process resulting in solid monomer at room temperature. Pure MDI has a functionality of 2, i.e. there are two reactive groups per molecule, whilst polymeric and liquid (modified) MDI with average functionalities from 2.1 up to about 3 is available as a mixture of the 2,4- and 4,4-isomers.

3.3.2. Polyisocyanates

Polyisocyanates are high molecular weight resins such as prepolymers, adducts, and isocyanurate trimers or blends of them. They contain very low levels of monomeric diisocyanate which reduce the concerns associated with industrial hygiene and handling of polyurethane adhesives and sealants. MDI-based polyisocyanates are commonly referred to as "polymeric MDI". They are oligomer mixtures of 2-ring MDI (2,4- and 4,4-isomers) of the type shown in Table 3. These materials are supplied as low viscosity liquids containing no solvent, and are used to prepare high solids (high content of solids) or solvent free sealants, coatings and caulks [16]. For most sealants, either TDI or MDI polyisocyanate types are the materials of choice.

3.3.2.1. Blocked Isocyanates

Another group of polyisocyanates commonly used are blocked isocyanates. A blocked isocyanate is an isocyanate which has been reacted with a material, e.g. monofunctional alcohols or amines, to prevent its reaction at room temperature, with compounds that conventionally react with isocyanates, but will permit that reaction to occur at higher temperatures [17]. Less commonly used terms to describe them are "capped", "heat latent", "thermally liable", "masked", and "splitters". Their chemistry has been extensively reviewed by Wicks et al. [17–20].

Blocked isocyanates are used extensively in many CASE[2] applications mostly for technical and economic reasons when the presence of free isocyanate must be eliminated, due to the potential toxic hazards associated with their use, and heat curing is possible. Blocked polyisocyanates are also used to crosslink both solvent-borne and water-borne resins offering a wide spectrum of possible formulations (refer to Sections 3.5.2.2 and 3.5.2.3).

They are blended with polyols, pigments, additives and fillers to improve flow. These mixtures are stable at room temperature whilst at higher temperatures (from 120 to 250°C), the blocking group volatilise either regenerating totally polyisocyanate functionality or remaining at least partially with the finished

[2] CASE = coatings, adhesives, and sealants.

urethane. Common blocking agents are methyl ethyl ketoxime (MEKO) or 2-butenone oxime,[3] phenol,[4] and ε-caprolactam.[5]

There are several variables which affect the rate and extent of the unblocking reaction such as:

- the structure of the blocking agent and the polyisocyanate used,
- the structure of R,
- presence of catalysts,
- effects of solvents, and
- temperature.

For instance, blocked polyisocyanates based on aromatic polyisocyanates dissociate at lower temperatures than those based on aliphatic ones. Generally, the dissociation temperatures of blocked polyisocyanates based on commercially utilised blocking agents decrease in the order: alcohols > ε-caprolactam > phenols > methyl ethyl ketoxime > active methylene compounds.

3.3.2.2. Liquid MDI

In the presence of phosphorus-containing catalysts, MDI monomer can react with itself to produce carbodiimide, which in turn reacts with MDI in excess to form MDI uretoneimine. This compound reduces the melt point, so that a liquid material results at temperatures above 10–15°C. These products have an average functionality of over 2 (Fig. 9).

Another liquid form of MDI, crude MDI or polymethylene phenylene isocyanate (often described by its name PAPI® polymeric MDI products) [21] has a lower cost than MDI. However, disadvantages lie in its dark colour and low functionality.

3.3.3. Aliphatic Isocyanates

Principally, the major aliphatic polyisocyanates comprise hexamethylene diisocyanate (HDI), isophorone diisocyanate (IPDI), and dicyclohexylmethane 4,4′-diisocyanate (H_{12}-MDI). Aliphatic isocyanates are used if a non-yellowing sealant is required. Table 3 contains a list of the most common isocyanates used in adhesives and sealants.

Some companies and their isocyanate series trademarks are displayed in Tables 4 and 5. More detailed and up-to-date information is readily available by contacting the original vendor publications. Additional information can be found in the Specialchem4adhesives raw material database [23].

---

[3] CAS No. 96-29-7.
[4] CAS No. 108-95-2.
[5] CAS No. 105-60-2.

Table 3: Common aliphatic isocyanates.

Isocyanate	**Formula**	**Features**	**CAS[22][a]**
HDI 1,6-Hexamethylene diisocyanate	$OCN-(CH_2)_6-NCO$	$C_8H_{12}N_2O_2$, is also known as HMI, 1,6-hexamethylene diisocyanate and 1,6-diisocyanatohexane. It is pale yellow liquid made by the phosgenation of hexamethylenediamine	822-06-0
IPDI 3-Isocyanatomethyl - 3,5,5-trimethylcyclohexyl isocyanate	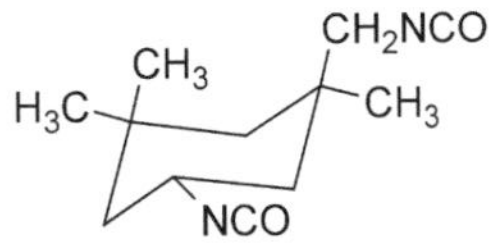 H_3C, CH_3, CH_2NCO, CH_3, NCO	$C_{12}H_{18}N_2O_2$ is a colourless to yellow liquid	4098-71-9
H_{12}-MDI bis (4-isocyanatocyclohexyl) methane	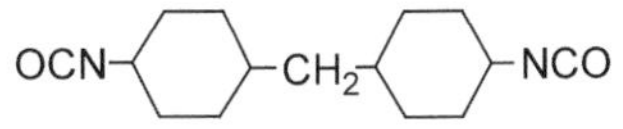 OCN, CH_2, NCO	$C_{15}H_{22}N_2O_2$, is also known also as hydrogenated MDI (HMDI or H_{12}MDI), and bis (4-isocyanatocyclo-hexyl) methane-H_{12} MDI, among others. It is a useful material for the production of hydrolytically stable polyurethanes [16]	5124-30-1

[a] The CAS register number is a unique identifier for chemical substances that has no chemical significance but provides an unambiguous way to identify a molecular structure.

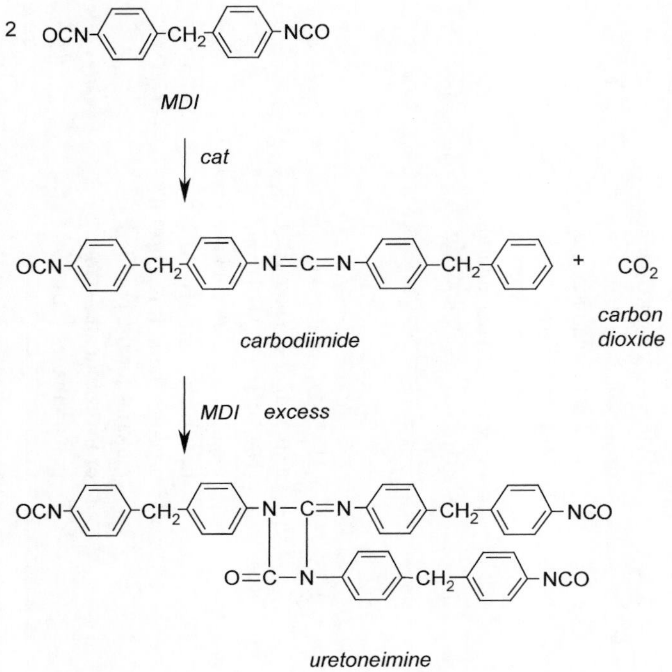

Fig. 9: Catalyst preparation of liquid MDI. (Source: [11].)

3.3.4. Polyols

A variety of different polyols are used in the production of different specifications of polyurethane adhesives and sealants. Polyols are either polyesters or polyethers. In most prepolymers the polyols will consist mostly of both diols and triols.

Table 4: Isocyanates series trademarks for polyurethanes adhesives and sealants.

Company	Polyisocyanates series trademarks
BASF/Elastogran	LUPRANATE
Bayer AG	DESMODUR, MONDUR
Cytec	TMI
DOW Chemical	ISONATE, PAPI, VORANATE
Huntsman	SUPRASEC
Lyondell	SCURANATE, TDI-80
Merquinsa	DISPERDUR
Rhodia	TOLONATE AT

Table 5: Principal polyisocyanate products used for polyurethane adhesives and sealants.

Product	**Type**	**Supplier**
Lupranat® ME	Isocyanate–MDI	BASF/Elastogran
Lupranat® MI	Isocyanate–MDI	BASF/Elastogran
Lupranat® MIP	Isocyanate–MDI	BASF/Elastogran
Lupranat® MM 103	Isocyanate–MDI–modified MDI	BASF/Elastogran
Lupranat® MP 105	Isocyanate–MDI–modified MDI	BASF/Elastogran
Desmodur® 2460 M	Isocyanate–MDI–monomeric MDI	Bayer
Desmodur® 2460 M	Isocyanate–MDI–monomeric MDI	Bayer
Desmodur® DA-L	Isocyanate–HDI, hexamethylene diisocyanate	Bayer
Desmodur® E 14	Prepolymer–Isocyanate–TDI based	Bayer
Desmodur® E 15	Prepolymer–Isocyanate–TDI based	Bayer
Desmodur® E 21	Prepolymer–Isocyanate–MDI based	Bayer
Desmodur® E 22	Prepolymer–Isocyanate–MDI based	Bayer
Desmodur® E 23	Prepolymer–Isocyanate–MDI based	Bayer
Desmodur® I	Isocyanate–IPDI, isophorone diisocyanate (cycloaliphatic)	Bayer
Desmodur® IL EA	Prepolymer–Isocyanate	Bayer
Desmodur® L 75	Prepolymer–Isocyanate–TDI based	Bayer
Desmodur® T 100	Prepolymer–Isocyanate–TDI based	Bayer
Desmodur® T 80 P	Prepolymer–Isocyanate–TDI based	Bayer
Desmodur® VH 20	Isocyanate–MDI	Bayer
Desmodur® VL	Isocyanate–MDI	Bayer
Desmodur® VL 50	Isocyanate–MDI	Bayer
Desmodur® VL 51	Isocyanate–MDI	Bayer
Desmodur® VL R 10	Isocyanate–MDI	Bayer
Desmodur® VL R 20	Isocyanate–MDI	Bayer
Desmodur® W	Isocyanate–MDI	Bayer
Desmodur® XP 7144	Prepolymer–Isocyanate–MDI based	Bayer
Mondur® 1441	Isocyanate–MDI	Bayer
Mondur® MA 2902	Isocyanate–MDI–modified MDI	Bayer
Mondur® MA 2903	Prepolymer–Isocyanate–MDI based	Bayer
Mondur® XP 7143	Isocyanate–MDI, diphenylmethane diisocyanate	Bayer
	Isocyanate–TDI, toluene diisocyanate	
TMI®	Isocyanate–unsaturated aliphatic isocyanate	Cytec
Isonate™ 125MCJ	Isocyanate–MDI–pure MDI	Dow Chemical
Isonate™ 143L	Isocyanate–MDI–modified MDI	Dow Chemical
Isonate™ 143LJ	Isocyanate–MDI–modified MDI	Dow Chemical

(*Continued*)

Table 5: Continued.

Product	Type	Supplier
Isonate™ M 124	Isocyanate–MDI–monomeric MDI	Dow Chemical
Isonate™ M 125	Isocyanate–MDI–monomeric MDI	Dow Chemical
Isonate™ M 125P	Isocyanate–MDI–monomeric MDI	Dow Chemical
Isonate™ M 143	Isocyanate–MDI–modified MDI	Dow Chemical
Isonate™ 181	Prepolymer–Isocyanate–MDI based	Dow Chemical
Isonate™ M 304	Isocyanate–MDI–modified MDI	Dow Chemical
Isonate™ M 309	Isocyanate–MDI–modified MDI	Dow Chemical
Isonate™ M 340	Prepolymer–Isocyanate–MDI based	Dow Chemical
Isonate™ M 342	Prepolymer–Isocyanate–MDI based	Dow Chemical
Papi™ 135	Isocyanate–MDI–polymeric MDI	Dow Chemical
Papi™ 27	Isocyanate–MDI–polymeric MDI	Dow Chemical
Papi™ 2940	Isocyanate–MDI–polymeric MDI	Dow Chemical
Papi™ 580N	Isocyanate–MDI–polymeric MDI	Dow Chemical
Papi™ 901	Isocyanate–MDI–polymeric MDI	Dow Chemical
Papi™ 94	Isocyanate–MDI–polymeric MDI	Dow Chemical
Voranate™ M 220	Isocyanate–MDI–polymeric MDI	Dow Chemical
Voranate™ M 229	Isocyanate–MDI–polymeric MDI	Dow Chemical
Voranate™ M 2940	Isocyanate–MDI–polymeric MDI	Dow Chemical
Voranate™ M 580	Isocyanate–MDI–polymeric MDI	Dow Chemical
Voranate™ M 590	Isocyanate–MDI–polymeric MDI	Dow Chemical
Voranate™ M 595	Isocyanate–MDI–polymeric MDI	Dow Chemical
Voranate™ T-80	Isocyanate–TDI, toluene diisocyanate	Dow Chemical
Suprasec® 1000	Isocyanate–MDI–pure MDI	Huntsman
Suprasec® 1004	Isocyanate–MDI–modified MDI	Huntsman
Suprasec® 1007	Prepolymer–isocyanate–MDI based	Huntsman
Suprasec® 1100	Isocyanate–MDI–pure MDI	Huntsman
Suprasec® 1306	Isocyanate–MDI–pure MDI	Huntsman
Suprasec® 1400	Isocyanate–MDI–pure MDI	Huntsman
Suprasec® 1412	Prepolymer–Isocyanate–MDI based	Huntsman
Suprasec® 2004	Isocyanate–MDI–modified MDI	Huntsman
Suprasec® 2008	Prepolymer–isocyanate–MDI based	Huntsman
Suprasec® 2010	Prepolymer–isocyanate–MDI based	Huntsman
Suprasec® 2018	Prepolymer–isocyanate–MDI based	Huntsman
Suprasec® 2020	Isocyanate–MDI–modified MDI	Huntsman
Suprasec® 2021	Prepolymer–isocyanate–MDI based	Huntsman
Suprasec® 2023	Prepolymer–isocyanate–MDI based	Huntsman
Suprasec® 2029	Isocyanate–MDI–modified MDI	Huntsman
Suprasec® 2030	Prepolymer–isocyanate–MDI based	Huntsman
Suprasec® 2034	Prepolymer–isocyanate–MDI based	Huntsman
Suprasec® 2049	Prepolymer–isocyanate–MDI based	Huntsman

(*Continued*)

Table 5: Continued.

Product	Type	Supplier
Suprasec® 2050	Prepolymer–isocyanate–MDI based	Huntsman
Suprasec® 2054	Prepolymer–isocyanate–MDI based	Huntsman
Suprasec® 2058	Prepolymer–isocyanate–MDI based	Huntsman
Suprasec® 2059	Prepolymer–isocyanate–MDI based	Huntsman
Suprasec® 2060	Prepolymer–isocyanate–MDI based	Huntsman
Suprasec® 2061	Prepolymer–isocyanate–MDI based	Huntsman
Suprasec® 2069	Prepolymer–isocyanate–MDI based	Huntsman
Suprasec® 2085	Isocyanate–MDI–polymeric MDI	Huntsman
Suprasec® 2090	Isocyanate–MDI–Modified MDI	Huntsman
Suprasec® 2211	Isocyanate–MDI–Modified MDI	Huntsman
Suprasec® 2214	Isocyanate–MDI	Huntsman
Suprasec® 2234	Prepolymer–isocyanate–MDI based	Huntsman
Suprasec® 2332	Prepolymer–isocyanate–MDI based	Huntsman
Suprasec® 2385	Isocyanate–MDI–Modified MDI	Huntsman
Suprasec® 2386	Isocyanate–MDI–Modified MDI	Huntsman
Suprasec® 2388	Isocyanate–MDI–Modified MDI	Huntsman
Suprasec® 2408	Prepolymer–isocyanate–MDI based	Huntsman
Suprasec® 2419	Prepolymer–isocyanate–MDI based	Huntsman
Suprasec® 2495	Isocyanate–MDI	Huntsman
Suprasec® 2496	Isocyanate–MDI	Huntsman
Suprasec® 2497	Isocyanate–MDI	Huntsman
Suprasec® 2642	Isocyanate–MDI	Huntsman
Suprasec® 2644	Prepolymer–isocyanate–MDI based	Huntsman
Suprasec® 2645	Isocyanate–MDI–Modified MDI	Huntsman
Suprasec® 2647	Isocyanate–MDI	Huntsman
Suprasec® 3051	Isocyanate–MDI	Huntsman
Suprasec® 5005	Isocyanate–MDI–polymeric MDI	Huntsman
Suprasec® 5025	Isocyanate–MDI–polymeric MDI	Huntsman
Suprasec® 5030	Isocyanate–MDI–polymeric MDI	Huntsman
Suprasec® MPR	Isocyanate–MDI–Pure MDI	Huntsman
Scuranate® T80	Isocyanate–2,4/ 2,6 TDI mixture	Lyondell
TDI 80-Type 1	Isocyanate–2,4/2,6 TDI mixture	Lyondell
Tolonate® AT	Isocyanate–TDI, Tris (6-isocyanatohexyl) isocyanurate	Rhodia
Disperdur® 444-20	Isocyanate–aliphatic polyisocyanate	Merquinsa

The polyesters as well as polyethers are high molecular weight materials prepared from monomers. As with all polymeric products, polyols have an average molecular weight; this varies from 200 to 10,000 depending on the application.

3.3.4.1. Polyethers

Polyether polyols based upon polyoxypropylene polyols are often the polyols of choice for polyurethane sealants. They are manufactured by the base-catalysed addition of propylene oxide to propylene glycol or di-propylene glycol as initiator. Other initiators are those displayed in Table 6. This reaction yields predominately secondary hydroxyl groups. Higher incidence of primary hydroxyl groups can be attained by reacting the homopolymer with ethylene oxide (EO) to form the block copolymer.

3.3.4.2. Polyesters

Polyesters include various classes of high molecular weight substances obtained generally by polycondensation of multifunctional carboxylic acids and hydroxyl compounds. Polyesters contain the ester group –O–CO as the repeating unit in the chain. Typically, polyester polyols offer abrasion resistance and adhesion promotion while polyether polyols provide low-temperature flexibility and low viscocity.

For adhesive applications, polyesters produced from polyalkylene phthalate or adipates are preferred because they produce high strength and modulus. These properties are not required for sealants. Speciality polyesters such as caprolactone polyols are used to enhance performances in a wide range of applications.

3.3.4.3. Other Polyols

Other polyols used in polyurethane CASE applications improve mechanical properties and prepare systems for specific applications; some of them are explained as follows.

3.3.4.3.1. Hydrophobic polyols

This type of polyol is used to extend polyurethanes with percentages of mineral oil and to improve resistance to hydrolysis, acids, and bases. Olefinic polyols such as hydroxy-terminated polybutadiene (Poly BD) are commonly used. An example is shown in Fig. 10, where the value of n is 57–65 and the functionality is 2.4 and equivalent weight is about 1260 [24, p. 27]. Another hydrophobic polyol is castor oil. It is a triglyceride which contains a non-saturated acid as the acid component, ricinoleic acid, which has a free hydroxyl group used as the alcohol component for polyurethanes.

3.3.4.3.2. Silanol modified polyols

Hydroxyl-containing silicon material with the hydroxyl groups attached directly to a silicon atom are combined together to react to form silanol-functionalised urethane prepolymers that can cure at ambient temperatures in the presence of moisture to form polysiloxane polyurethanes. For instance, physical blends of a polypropyleneoxide polyol intermediates and an organic-silicone block copolymer-based polyol intermediates have been prepared to improve elongation and reduced

Table 6: Common initiators for polyether polyols.

Initiators	**Structure**[a]	**Formula**	**Functionality**	**CAS**
Water	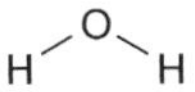	H_2O	2	7732-18-5
Ethylene glycol	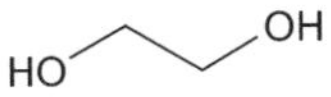	$C_2H_6O_2$	2	107-21-1
1,2 propylene glycol	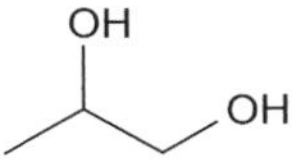	$C_3H_8O_2$	2	57-55-6
Glycerine	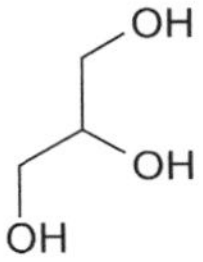	$C_3H_8O_3$	3	56-81-5
Trimethylol propane	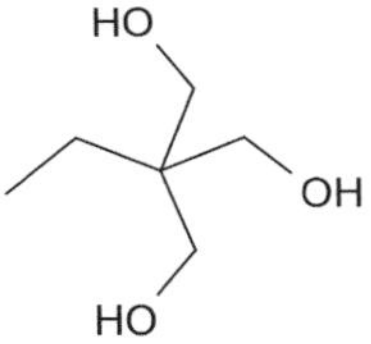	$C_6H_{14}O_3$	3	77-99-6
Triethanol amine	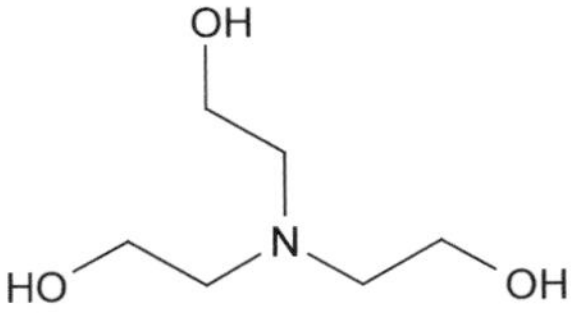	$C_6H_{15}NO_3$	3	102-71-6
Pentaerythritol	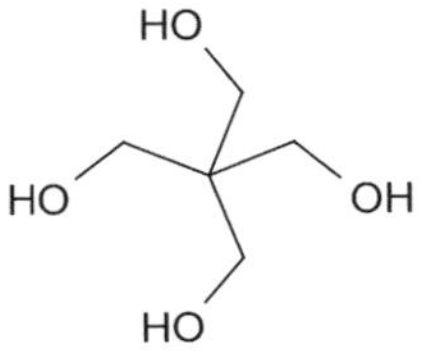	$C_5H_{12}O_4$	4	115-77-5

[a] To convert a line structure into a structural formula, place a C atom at the end of each line and at each line intersection, then add enough –H connections to give each C atom 4 connections.

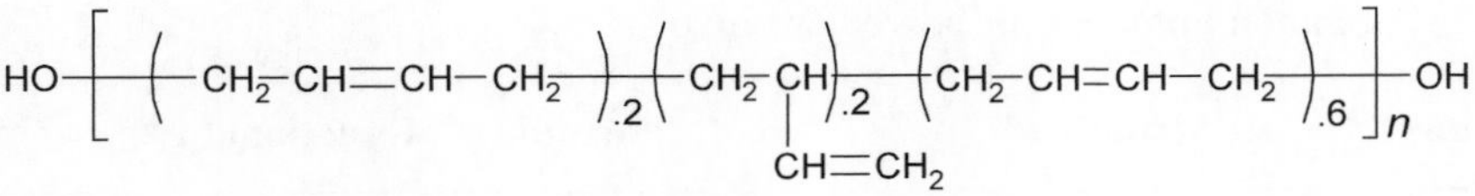

$$HO-\left[\left(CH_2-CH{=}CH-CH_2\right)_{.2}\left(CH_2-\underset{\displaystyle CH{=}CH_2}{CH}\right)_{.2}\left(CH_2-CH{=}CH-CH_2\right)_{.6}\right]_n-OH$$

Fig. 10: Hydroxy-terminated polybutadiene.

modulus of polyurethane sealants when compared to conventional urethane compositions [25]. Several patents show the actual state of the art [25,26].

3.3.4.4. Chain Extenders

Chain extenders are usually difunctional compounds, such as glycols, diamines or hydroxylamines used in polyureas and polyurethane/ureas. Table 7 shows common hydroxylated compounds used as chain extenders or crosslinkers.

Some companies and their polyol series trademarks are displayed in Tables 8 and 9. More detailed and up-to-date information is readily available by contacting the original vendor publications.

Table 7: Hydroxylated chain extenders and crosslinkers.

Compound	**Structure**	**Formula**	**Molecular weight**	**CAS**
Ethylene glycol	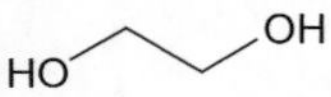	$C_2H_6O_2$	62.0682	107-21-1
Diethylene glycol	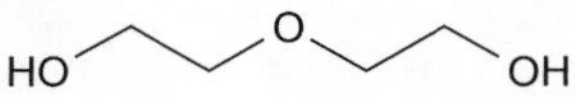	$C_4H_{10}O_3$	106.1212	
Propylene glycol	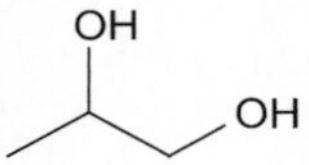	$C_3H_8O_2$	76.095	57-55-6
Dipropylene glycol	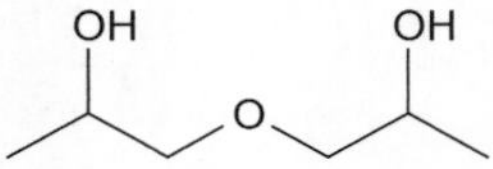	$C_6H_{14}O_3$	134.1748	25265-71-8
1,4-Butanediol	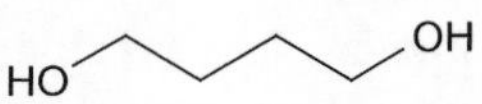	$C_4H_{10}O_2$	90.12181	110-63-4
2-Methyl-1,3 propanediol	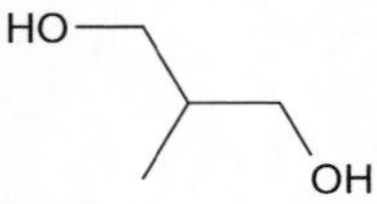	$C_4H_{10}O_2$	90.12181	2163-42-0

(*Continued*)

Table 7: Continued.

Compound	Structure	Formula	Molecular weight	CAS
Water	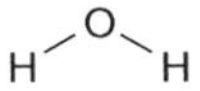 H–O–H	H_2O	18.0152	7732-18-5
Hydrazine	$H_2N—NH_2$	H_4N_2	32.045	302-01-2
Ethylenediamine	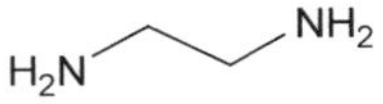 H_2N–CH₂CH₂–NH_2	$C_2H_8N_2$	60.0986	107-15-3
1,4-Diamino-cyclohexane	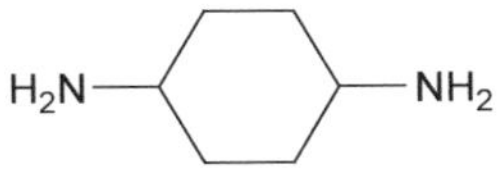 H_2N–C₆H₁₀–NH_2	$C_6H_{14}N_2$	114.19	2615-25-0
Isophorone diamine	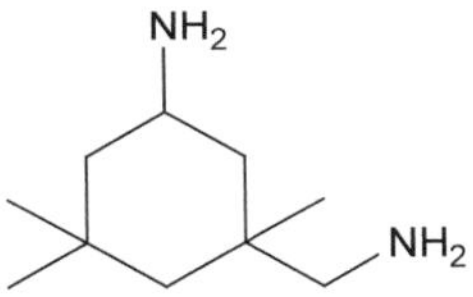 NH_2, NH_2	$C_{10}H_{22}N_2$	170.2972	2855-13-2

Table 8: Polyols series trademarks for polyurethane adhesives and sealants.

Company	Polyol series trademarks
BASF	LUPRANOL, LUPRAPHEN
Bayer AG	ARCOL POLYOL, DESMOPHEN
C. P. Hall	URETHHALL
Crompton-Uniroyal Chemical	FOMREZ
DOWChemical	TONE, VORANOL
DuPont	TERATHANE
Huntsman	DALTOREZ
P.A.T. Products	DIEXTERG214
Reichhold	POLYLITE
Repsol	ALCUPOL
Rokra-Kraemer	ROKRAPOL
Scandiflex	CORDAFLEX
Shell Chemicals	CARADOL
Solvay	CAPA
Stepan	STEPANPOL
Uniqema	PRIPLAST

Table 9: Principal polyols used for polyurethane adhesives and sealants.

Product	Polyol Type	Supplier
Lupranol® 1301	Polyether	BASF-Elastogran
Lupranol® 2001	Polyether	BASF-Elastogran
Lupranol® 2031	Polyether	BASF-Elastogran
Lupranol® 2032	Polyether	BASF-Elastogran
Lupranol® 2043	Polyether	BASF-Elastogran
Lupranol® 2090	Polyether	BASF-Elastogran
Lupranol® 2095	Polyether	BASF-Elastogran
Lupranol® 2100	Polyether	BASF-Elastogran
Lupranol® VP9272	Polyether	BASF-Elastogran
Lupranol® VP9289	Polyether	BASF-Elastogran
Lupraphen® 8002	Polyols	BASF-Elastogran
Lupraphen® 8004	Polyester	BASF-Elastogran
Lupraphen® 8100	Polyester	BASF-Elastogran
Lupraphen® 8101	Polyester	BASF-Elastogran
Lupraphen® 8103	Polyester	BASF-Elastogran
Lupraphen® 8104	Polyester	BASF-Elastogran
Lupraphen® 8105	Polyester	BASF-Elastogran
Lupraphen® 8106	Polyester	BASF-Elastogran
Lupraphen® 8108	Polyester	BASF-Elastogran
Lupraphen® 8109	Polyester	BASF-Elastogran
Arcol Polyol 11-34	Polyether	Bayer
Arcol Polyol E-351	Polyether	Bayer
Arcol Polyol E-648	Polyether	Bayer
Arcol Polyol LG-56	Polyether	Bayer
Arcol Polyol LHT-42	Polyether	Bayer
Arcol Polyol PPG-1025	Polyether	Bayer
Arcol Polyol PPG-2000	Polyether	Bayer
Arcol Polyol PPG-2025	Polyether	Bayer
Arcol Polyol PPG-3025	Polyether	Bayer
Arcol Polyol PPG-4000	Polyether	Bayer
Desmophen F-2035	Polyester	Bayer
Desmophen S-1011-35	Polyester	Bayer
Desmophen S-1072-30	Polyester	Bayer
Desmophen S-1074-30	Polyester	Bayer
Desmophen® 5034 BT	Polyether	Bayer
Desmophen® 550 U	Polyether	Bayer
Desmophen® C 200	Polyester	Bayer
Multranol® 9181	Polyether	Bayer
Tone™ 0201 Polyol	Caprolactone	Dow Chemical

(*Continued*)

Table 9: Continued.

Product	Polyol Type	Supplier
Tone™ 0210 Polyol	Caprolactone	Dow Chemical
Tone™ P737 Polymer	Polyester	Dow Chemical
Voranol™ 1010 L	Caprolactone	Dow Chemical
Voranol™ 2000 L	Polyether	Dow Chemical
Voranol™ 2070	Polyether	Dow Chemical
Voranol™ 2100	Polyether	Dow Chemical
Voranol™ 2110	Polyether	Dow Chemical
Voranol™ 2110-TB	Polyether	Dow Chemical
Voranol™ 2120	Polyether	Dow Chemical
Voranol™ 220-028	Polyether	Dow Chemical
Voranol™ 220-056	Polyether	Dow Chemical
Voranol™ 220-094	Polyether	Dow Chemical
Voranol™ 220-110	Polyether	Dow Chemical
Voranol™ 220-260	Polyether	Dow Chemical
Voranol™ 222-029	Polyether	Dow Chemical
Voranol™ 222-056	Polyether	Dow Chemical
Voranol™ 230-056	Polyether	Dow Chemical
Voranol™ 230-112	Polyether	Dow Chemical
Voranol™ 230-238	Polyether	Dow Chemical
Voranol™ 230-66	Polyether	Dow Chemical
Voranol™ 232-027	Polyether	Dow Chemical
Voranol™ 232-028	Polyether	Dow Chemical
Voranol™ 232-034	Polyether	Dow Chemical
Voranol™ 232-035	Polyether	Dow Chemical
Voranol™ 235-056	Polyether	Dow Chemical
Voranol™ 2471	Polyether	Dow Chemical
Voranol™ 800	Polyether	Dow Chemical
Voranol™ CP 1055	Polyether	Dow Chemical
Voranol™ CP 260	Polyether	Dow Chemical
Voranol™ CP 300	Polyether	Dow Chemical
Voranol™ CP 3055	Polyether	Dow Chemical
Voranol™ CP 3355	Polyether	Dow Chemical
Voranol™ CP 4055	Polyether	Dow Chemical
Voranol™ CP 450	Polyether	Dow Chemical
Voranol™ CP 4655	Polyether	Dow Chemical
Voranol™ CP 4755	Polyether	Dow Chemical
Voranol™ CP 6055	Polyether	Dow Chemical
Voranol™ CP 755	Polyether	Dow Chemical
Voranol™ EP 1900	Polyether	Dow Chemical
Voranol™ EP 2001	Polyether	Dow Chemical

(Continued)

Table 9: Continued.

Product	Polyol Type	Supplier
Voranol™ P 1010	Polyether	Dow Chemical
Voranol™ P 2000	Polyether	Dow Chemical
Voranol™ P 400	Polyether	Dow Chemical
Voranol™ RA 640	Polyether	Dow Chemical
Voranol™ RA 800	Polyether	Dow Chemical
Voranol™ RH 360	Polyether	Dow Chemical
Voranol™ RN 490	Polyether	Dow Chemical
Daltorez™ P133	Polyester	Huntsman
Daltorez™ P312	Polyester	Huntsman
Daltorez™ P315	Polyester	Huntsman
Daltorez™ P321	Polyester	Huntsman
Daltorez™ P345	Polyester	Huntsman
Daltorez™ P355	Polyester	Huntsman
Capa2043	Caprolactone	SolvayCaprolactones
Capa2054	Caprolactone	SolvayCaprolactones
Capa2085	Caprolactone	SolvayCaprolactones
Capa2100	Caprolactone	SolvayCaprolactones
Capa2101A	Caprolactone	SolvayCaprolactones
Capa2200	Caprolactone	SolvayCaprolactones
Capa2200A	Caprolactone	SolvayCaprolactones
Capa2205	Caprolactone	SolvayCaprolactones
Capa2302	Caprolactone	SolvayCaprolactones
Capa2302A	Caprolactone	SolvayCaprolactones
Capa2304	Caprolactone	SolvayCaprolactones
Capa2402	Caprolactone	SolvayCaprolactones
Capa3050	Caprolactone	SolvayCaprolactones
CapaHC1060	Caprolactone	SolvayCaprolactones
CapaHC1100	Caprolactone	SolvayCaprolactones
CapaHC1200	Caprolactone	SolvayCaprolactones

3.3.5. Catalysts

The most commonly used catalysts in polyurethanes are tertiary amines. They promote isocyanate reactions which will occur at moderate temperatures, i.e. reaction with alcohols, water, and carboxylic acids. However, the tertiary amines are not strong catalysts for the reactions of isocyanates and isocyanate derivatives at elevated temperatures [14]. Strong catalysts for these reactions are the strong bases, e.g. NaOH, NaOR, and R_4NOH.

With aromatic isocyanate resins, the formation of the urethane linkage can be promoted by a number of metals in the form of organometallics and/or salts of organic acids. Tin compounds such as dibutyl tin dilaurate and tin (II) octoate are particularly effective, having superseded the more toxic lead equivalent components.

Synergistic effects of tin and amine catalysts are of technical importance and are widely studied principally because of the differences in reactivity. Metal catalysts are usually employed in systems based on the slower reacting aliphatic isocyanate adducts [27].

Standard moisture cure catalysts used principally in adhesives are:

- *Tin catalysts*. Dibutyltin dilaurate (DBTDL), dibutyltin diacetate (DBTDA);
- *Amines*. Morpholine derivatives, tertiary amines;
- Bismuth catalysts, which are increasingly replacing mercuric catalysts.

Both blocked aromatic and aliphatic polymers use latent curatives such as ketimines and oxazolidines; however, they do have some inherent disadvantages. The use of ketimines and aldemines often produces products that have a tendency to yellow upon exposure to sunlight, and that take a longer time to achieve complete cure and retain slow evaporating ketones or aldehydes. Oxazolidine modification can also result in some yellowing and reduced chemical resistance to some acids [26].

Some companies and their catalysts series trademarks are displayed in Table 10. More detailed and up-to-date information is readily available by contacting the original vendor publications.

Table 10: Catalysts series trademarks for polyurethanes.

Company	Catalysts series trademarks
Air Products and Chemicals Inc	DABCO, POLYCAT
Cosan Chemical Co	COCURE, COSCAT, COTIN
Enterprise Chemical Corp	QUINCAT
Kao Corporation	KAO LIZER
Merk & Co	METASOL
Rohm and Hass Co	DMP
Th Goldschmidt Co	KOSMOS, TEGO
Texaco Chemical Co	TEXACAT
Toyo Soda Co	TOYOCAT
Union Carbide Corp.[a]	NIAX
Witco Chemical Corp	FOMREZ, FOMREZ UL

[a] Union Carbide Corp. is a wholly owned subsidiary of The Dow Chemical Company since 2001.

3.3.6. Solvents

Solvents are used principally to reduce the viscosity and improve surface wetting of the substrate. Although the trend is to reduce or eliminate solvents for most polyurethane sealants, the solvent content must be low (0 to $<10\%$) in order to maintain properties after cure otherwise the polymer will develop stresses at the interface, leading to debonding. Typical solvents are:

- mineral spirits;
- odourless mineral spirits;
- xylene;
- solvent naphtha;
- glycol ether acetates.

Water absorbency and active hydrogen group containing solvents must be avoided with two packed isocyanate containing adhesives.

3.3.7. Plasticisers

Plasticisers act under the same phenomenon as solvents, increasing the free volume of the polymer but without producing complete dissolution. Both are governed by the same laws of solubility. However, each plays a different role. While solvents serve mostly as viscosity modifiers, plasticisers modify the curing properties of the sealant, softening and lowering the glass transition temperature.

Common plasticisers in polyurethane sealants are:

- butyl benzyl phthalate;
- diundecyl phthalate (DIDP);
- dioctyl phthalate (DOP);
- dipropylene glycol dibenzoate;
- hydrocarbon extenders are also used as internal plasticisers;
- free polyol in low isocyanate index PUs.

3.3.8. Additives

Depending on the application, several additives are added to the co-reactant or curative component having active hydrogen containing groups. The main types are:

- UV stabilisers;
- UV absorbers;
- adhesion promoters;
- pigments and colourants;
- mildewcides and fungicides;

- pigments and extenders;
- levelling agents;
- thickening agents;
- air release agents;
- antioxidants;
- bituminous extenders;
- suspending agents;
- antiskinning agents;
- surfactants;
- rheological modifiers.

3.4. Structure Property Relationships

From Section 3.3, it was shown there are a large number of monomers and oligomers available for polyurethanes. It is often said that if cost was not of concern, then urethane-based polymers could be tailored to replace most polymers for applications that did not demand too high a service temperature [28]. Polyurethanes, besides adhesives and sealants applications, can be found as foams (rigid, flexible, micro-cellular), elastomers, and encapsulants which differ slightly in raw materials. However, processing parameters and additives make feasible a diverse synthesis of this material.

As was discussed in Section 3.2, chemistry variables are introduced in the polyurethane structure mainly by chain extenders and cross-linker agents, and polyols and to a less extent by isocyanates. Variables introduced by polyols and chain extenders are:

- type of monomer precursor;
- molar mass variation, i.e. 300–10,000;
- functionality;
- use of polyol blends of different molar mass;
- pendant groups or bulky substituents in the main chain.

Polyurethane's technology aims to study the interaction between those variables to create different morphologies which interact differently, changing by various extent the mechanical properties and performance for all types of PU materials.

3.4.1. Hard–Soft Segment Theory

According to the theory [29], soft segments are derived from the polymeric polyol and hard segments from the diisocyanate chain extended with low molecular

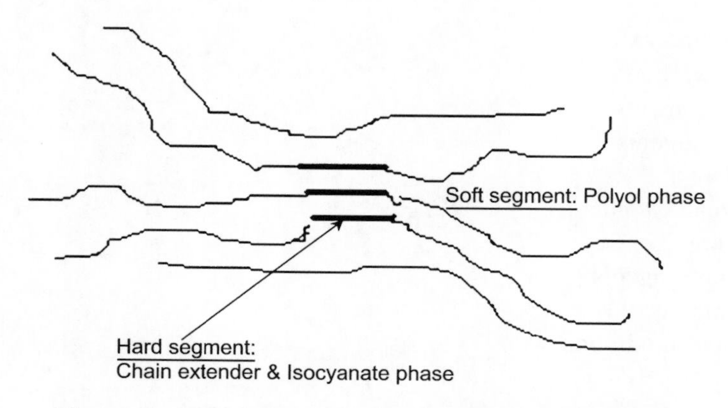

Figure 11: Morphology of polyurethanes. (Source: [30].)

weight diol or diamine. Refer to Fig. 11. Hard segments are plastic-like domains that form in a continuous rubbery phase (polyol) and may be partially crystalline. They are formed in two ways:

1. Reaction of isocyanate-terminated prepolymer with water produces urea — plastic phase.
2. Reaction of polyisocyanate with chain extenders
   - diols — ethylene glycol, butanediol, etc.
   - diamines — diethyl toluenediamine, ethylene diamine, etc.

Different physical forces are present between segments. Strong secondary bond interactions such as hydrogen bonding between the polar groups of hard segments are present acting like reinforcing "filler particles". Less extensive interaction forces occur between the non-polar groups of soft segments. Hydrogen bonding, the bond in which a hydrogen atom is associated with two other atoms, is particularly important in polyurethanes. This type of bond occurs between two functional groups in the same or different molecules.

In addition to hydrogen bonding, secondary-bond forces lead to the aggregation of separate particles into solid and liquid phases; they are not of great importance for stable chemical compounds. However, many physical properties such as surface tension and frictional properties, miscibility and solubility are determined to a large extent by intermolecular forces. Three types of forces acting between molecules are recognized, dipole, induction, and dispersion forces. Occasionally, the term *van der Waals forces* is applied to the dispersion forces alone.

The usual thermodynamic incompatibility of these two types of segments leads to a phase separation, leading to a structure consisting of hard segments and soft segment domains. This segregation gives rise to a micro-nonuniform

Table 11: Advantages and disadvantages of hard segments.

Advantages	Disadvantages
Tensile and overlap shear strength increases Raises the hardness and modulus (stiffness) Peel strength may increase (to a point) Can have good modulus at moderate temperatures and excellent low temperature impact properties	Overlap shear strength is not as high as the reverse morphology (rubber-modified plastic) Thermal stability – not as good (function of thermal stability of the urethane, not the morphology)

structure in which hard segment domains are dispersed in a soft segment matrix. The extent of phase separation is dependent on the level of association between hard segments, and also on the degree of compatibility between hard segments and soft segments.

Some advantages and disadvantages of hard segments [30] are given in Table 11.

Another important factor is that of crystallinity. Some factors that encourage and discourage crystal growth are given in Table 12.

The same factors that encourage and discourage crystal growth in hard segments apply for soft segments (polyol). For polyols, enhancement of crystallinity plays an important role for both urethane adhesives and sealants according to the hard–soft segment theory, some of them are given in Table 13 [30].

Table 12: Factors that encourage and discourage crystal growth in both hard and soft segments.

Encourage	Discourage
Symmetrical structure	Non-symmetrical structure
Aromatic rings	Hetero-atoms (O, S, etc.)
Even number of carbon atoms	Odd number of carbon atoms
No side chain substituents (encourage chain packing)	Contains side chain substituents (discourage chain packing)

Table 13: Importance of soft segment crystallinity on both urethane adhesives and sealants.

For urethane adhesives	For urethane sealants
Fast green strength – crystallisation provides the fast strength after applying the adhesive, e.g. water-borne urethane adhesives, solvent-borne urethane adhesives, curing hot melt adhesives May improve tensile properties	Crystallisation of soft segment usually not desirable, but may provide additional resistance to permanent set, e.g. tensile set, compression set (below crystallisation temperature) Amorphous polyols can produce lower viscosity adhesives and sealants, very important for high solids adhesives and sealants

3.5. Classification of Polyurethane Adhesives and Sealants

A recent survey carried out in 2000 showed that the total polyurethane consumption reached nearly 8.5 million tonnes [4] and for formulated polyurethane CASE products was approximately 3.1 million tonnes, with total isocyanate demand considered as 900,000 tonnes and total polyol demand accounting for around 1.5 million tonnes [31].

According to Tom Mach Industry Analyst Frost & Sullivan [32], polyurethane sealants are enjoying strong levels of growth. In North America, it is expected that the highest CAGRs increase (Compound Annual Growth Rate) in this sector will come from polyurethane sealants, at 4.1 percent. This growth principally stems from:

- Continued growth of polyurethane foam sealants in the DIY and construction sectors;
- Increasing importance of other chemistries in the polyurethane family, such as polyureas; and
- Growing acceptance of polyurethane sealants as a replacement of older technologies, such as polysulfides for marine applications.

In addition, Tom Mach points out that there is a fragmentation in the sealants market due to the fact that there are a large number of types of chemical sealants which lack of consistency between suppliers as to what is or is not a sealant [32]. For instance, some products called sealants are in fact adhesives and vice versa.

A sealant is defined as a liquid, paste or foam material, that, when applied to a joint or orifice, forms a tight seal against liquids or gases and can withstand the movements of the joint. Moreover, caulks are generally used where elastomeric properties are needed, while sealants are materials that are used where elastomeric as well as structural strength properties may be needed. Nevertheless, those terms are used interchangeably to denote compounds that are used to fill and seal a joint, such as to fill in cracks, crevices or gaps in structural units and to fill spaces between building units. Their primary function is to protect the sealed surfaces against weather conditions, i.e. temperature, moisture, and sunlight exposure conditions, in addition to absorbing shear, compression, and extension stresses.

Additionally, the classification of a product that has both adhesion and sealing capabilities is obscure. For this reason, the classification does not have a well-defined demarcation line. In some respects, the physical properties of urethane adhesives differ from those of urethane sealants. In line with that, this section presents the main types of polyurethane sealants and adhesives, their characteristics, advantages and disadvantages according to the form in which they are found, i.e. 100% solids, solventborne, waterborne, and the cure characteristics whether one or two-component material.

3.5.1. Nomenclature

From a chemical point of view, "one-component" (1K) and "two-component" (2K) terms are often misleading, in some cases erroneous. On the one hand, one-package reactive systems in fact use at least two constituents which will form a polymeric system on the substrate [33]. Under this heading, for instance isocyanate-terminated polyurethanes with a relatively high molecular weight (prepolymer) and rather low remaining isocyanate content that cure with atmospheric pressure and resins containing blocked isocyanates groups in combination with reactants are included.

On the other hand, components in two-package systems are typically much more reactive at ambient conditions and are mixed just prior to application. For instance, highly cross-linked systems until relative high solid contents that cure at low temperature can only be achieved by the two-package application mode [33]. A review of processing requirements and physical properties of one and two-component urethane adhesives is presented elsewhere [34].

A comparison of one and two-component adhesives is listed in Table 14 [34].

Polyurethane adhesives are preferred in a vast range of applications for a number of reasons [35]:

- They can interact with most substrates through polar interactions (e.g. hydrogen bonding), mainly due to polar sites on the urethane and urea groups.
- They can form covalent bonds with substrates that have active hydrogen atoms (for reactive adhesives).

Table 14: Comparison of one- and two-component urethane adhesives.

One component	Two component
Chemistry is limited to room temperature stable packages	Unlimited chemistry
Very long open times	Limited open time, variable from second to hours
No mixing	Must be meter mixed
Simple dispensing equipment	Complicated equipment, sometimes very sophisticated
No flushing required	Flushing and cleaning needed after some predetermined time
Minimum surface preparation	Best results obtained with primer to clean and prepare the surface
Complete bond line heating and fixturing for the cure cycle	Spot heating for curing to fixture the part

- They effectively wet the surface of a number of substrates (plastics, wood, metal, glass...).
- Their relatively low molecular weight/small molecular size allows them to permeate porous substrates (for reactive adhesives).
- Through molecular composition the adhesive stiffness, elasticity, and crosslinking can be tailored to suit specific needs.

Some requirements which are common to both polyurethane adhesives and sealants are [24, p. 23]:

- low viscosity;
- low volatile organic compounds (VOC);
- low cost;
- with TDI, low free monomer;
- compatibility with a broad variety of substrate materials.

Commonly, adhesives require high tensile strength while sealants cure at low modulus and high elongation capacity.

3.5.2. Classification by General Type

As previously stated, classification of polyurethanes does not have a well-defined demarcation line. However, in order to include the principal urethane adhesive types in this section, they will be explained as depicted in Table 15. Following

Table 15: Classification of systems.

Class	Chemistry (mechanism)	Types	Applications
Non-reactive	TPU & PUDs[a] (carrier evaporation)	Solvent-borne, water-borne, hot-melt	Contact adhesives
Reactive: one component	Isocyanates (moisture cure)	Solvent-borne, solvent-free	Crosslinker, wood binding
	Isocyanates prepolymers (moisture cure)	Solvent-borne, solvent-free	General purpose adhesives, sport floors sealants
Reactive: two component	NCO + Amine (polyurea)	Solvent-free	Roof coatings
	NCO + Polyol (polyurethane)	Solvent-borne, water-borne, solvent-free	Auto refinish and OEM coatings, sealants, encapsulants, flexible packaging adhesives, synthetic mortar
	Blocked NCO + Polyol (polyurethane)	Powder stoving	Can and coil coatings, maintenance coatings, adhesives for metals
Reactive: other	Urethane acrylates (radiation cure)	Solvent-free	Wood coatings, adhesives for GRP (glass reinforced plastics)

Source: Ref. [36].
[a] TPU, thermoplastics polyurethane; PUD, polyurethane dispersion.

this, they are grouped according to the *form* by which they are found, as 100% solids, dissolved in organic solvents or dispersed in water.

3.5.2.1. 100% Solids Polyurethane Adhesives

Polyurethane reactive hot melt adhesives (PURhma) consist of an isocyanate-terminated polyurethane or high molecular weight hydroxyl-terminated polyurethane which is solid at ambient temperature, but which melts at low temperature. They bond through the physical process of rapid solidification from the melt as well as through chemical reaction with ambient moisture. As a result of this curing reaction, PURhma develop temperature and chemical resistance exhibiting a degree of stiffness as well as toughness.

These materials have the advantage of being relatively easy to apply. However, melt adhesives are typically not reactive in nature and, therefore, do not develop sufficient strength and sufficient heat and chemical resistance for certain applications [37]. Advantages of hot melt curable adhesives over traditional liquid curing adhesives are:

- their ability to provide "green strength" upon cooling prior to cure and
- they provide adhesives of very low cross-linking density and thus high levels of flexibility and toughness.

The so-called "modifiers" have been used are in an attempt to improve the balance of processing, thermal and mechanical attributes of PURhma. Conventional modifiers include:

- rosin glycerol ester;
- polycaprolactone diol; and
- terpene phenolic resins.

Most of these modifiers have not provided a viable solution to the formulators' needs that can provide a useful balance of processing, thermal and mechanical attributes to PURhma. Recent attempts have been made to confer such desirable compositions [37,38].

3.5.2.2. Solvent-Borne Polyurethane Adhesives

Solvent polyurethane adhesives have been of great economical and technical importance because of their easy processing and high performance. They have been used traditionally for the bonding of rubber, leather, textiles, metal, paper, wood, and plastics including highly plasticised polyvinyl chloride.

These adhesives consist of high molecular weight (approximately 100,000) dissolved in a solvent (refer to Section 3.3.6). They are formed by the reaction of high molecular weight polyester diols with a diisocyanate.

3.5.2.3. Water-Borne Polyurethane Adhesives

Aqueous polyurethane dispersions are not new; they have been available since the early 1970s. However, environmental concerns and impending legislation are in favour of non-solvent containing materials and reduction of VOCs making these systems commercially available nowadays. Performance of new aqueous dispersions has reached the point that they exceed the performance of some solvent-based systems [16]. These urethane dispersions can also often be processed with the same or similar equipment as solvent systems. They may be used as the sole resin or blended with other resin systems, such as acrylics.

These adhesives are high molecular weight polyurethanes dispersed in water (PU dispersions, or PUDs). They bond through cure by chemical reaction as well as through the physical process of drying evaporation [6]. This means that the water carrier is eliminated during use, leaving the precipitated and coalesced polymer to form the adhesive bond [35].

One-component water-borne polyurethane systems can be derived from polyurethane dispersions or blocked polyisocyanates (refer to Section 3.2.1). Blocked isocyanates are added to the co-reactant resins providing one-component systems with excellent shelf life. This type of adhesives is principally used for non-porous materials and the bonding of unlike metals such as aluminium to steel, and stainless steel to mild steel. They are also useful in bonding some of the high pressure laminates such as those based on phenolics and melamine [39]. Systems based on water-borne blocked polyisocyanate crosslinkers and suitable water-borne polymers approach the performance levels previously obtained only by solvent-borne systems [16].

Hydrophilic or salt forming groups are incorporated into the polyurethane backbone to make the polymer self-dispersible. These non-reactive polyurethane dispersions are in aqueous form. Anionic forms based on pendant acid functional groups are the most common.

Other pendant groups frequently used in PUDs may include:

- dimethylol propionic acid (DMPA);
- sulphonic acid pendant groups;
- cationic forms based on tertiary amines in the backbone or pendant;
- non-ionic surfactants built into the polymer.

Preparation of water-borne polyurethane dispersions is done by emulsification of hydrophobic polyurethanes in water with the aid of protective colloids or suitable external emulsifiers. An example of the formation of a polyurethane dispersion is shown below [40]. Excellent reviews on water-borne polyurethanes can be found elsewhere [41,42] (Fig. 12).

3.5.3. Other Curing Characteristics

Cure characteristics of one-component systems categorise adhesives as moisture curable, UV curable, powder blocked isocyanates, etc., and for two-components systems as either polyol-isocyanate reactive curable or polyol-amine reactive curable.

3.5.3.1. Moisture Curable Polyurethane Adhesives

The curing principle for moisture-curing CASE polyurethane is described by the reaction of isocyanate groups with water. Polyisocyanate/polyol combinations

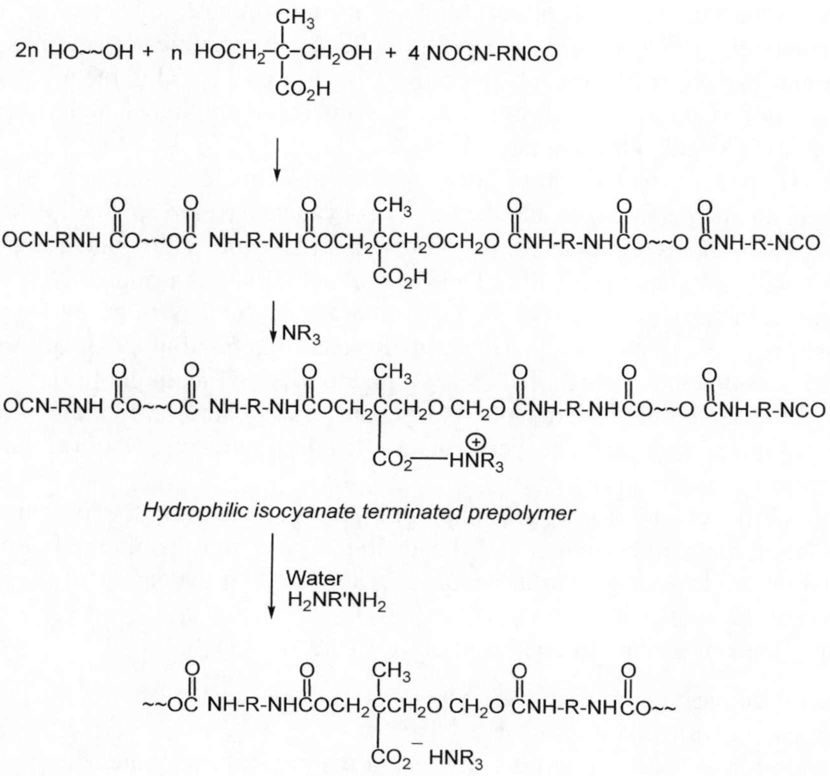

$$2n\ HO{\sim\sim}OH + n\ HOCH_2\text{-}C(CH_3)(CO_2H)\text{-}CH_2OH + 4\ NOCN\text{-}RNCO$$

$$\downarrow$$

$$OCN\text{-}RNH\overset{O}{\overset{\|}{C}}O{\sim\sim}O\overset{O}{\overset{\|}{C}}\ NH\text{-}R\text{-}NH\overset{O}{\overset{\|}{C}}OCH_2C(CH_3)(CO_2H)CH_2OCH_2O\ \overset{O}{\overset{\|}{C}}NH\text{-}R\text{-}NH\overset{O}{\overset{\|}{C}}O{\sim\sim}O\ \overset{O}{\overset{\|}{C}}NH\text{-}R\text{-}NCO$$

$$\downarrow NR_3$$

$$OCN\text{-}RNH\overset{O}{\overset{\|}{C}}O{\sim\sim}O\overset{O}{\overset{\|}{C}}\ NH\text{-}R\text{-}NH\overset{O}{\overset{\|}{C}}OCH_2C(CH_3)(CO_2^{-}\text{—}\overset{\oplus}{H}NR_3)CH_2O\ CH_2O\ \overset{O}{\overset{\|}{C}}NH\text{-}R\text{-}NH\overset{O}{\overset{\|}{C}}O{\sim\sim}O\ \overset{O}{\overset{\|}{C}}NH\text{-}R\text{-}NCO$$

Hydrophilic isocyanate terminated prepolymer

$$\downarrow \text{Water}, H_2NR'NH_2$$

$${\sim\sim}O\overset{O}{\overset{\|}{C}}\ NH\text{-}R\text{-}NH\overset{O}{\overset{\|}{C}}OCH_2C(CH_3)(CO_2^{-}\ HNR_3)CH_2O\ CH_2O\ \overset{O}{\overset{\|}{C}}NH\text{-}R\text{-}NH\overset{O}{\overset{\|}{C}}O{\sim\sim}$$

Figure 12: Aqueous dispersion of polyurethane-urea.

with an excess of isocyanate groups (prepolymers) crosslink with atmospheric moisture to give insoluble higher molecular weight polyurethanes/polyureas (recall the reaction of isocyanate with water, Fig. 4).

The properties of moisture-curing one-component systems are principally determined by the nature of the particular base isocyanates used. For example, one-component coatings based on aliphatic polyisocyanates (HDI, IPDI, DESMODUR W diisocyanate) generally need longer drying times than those based on aromatic isocyanates (TDI, MDI). In addition, drying times depend on both the temperature and the amount of atmospheric moisture present [16]. Humidity affects curing, formation of bubbling, loss of properties, potential shelf life, storage issues, and manufacture.

Some of the advantages of moisture curable CASE polyurethanes are that there are no solvent emissions, and an improvement in properties due to the formation of urea groups.

Single component moisture curable sealants are widely used. They provide liquid and gaseous barriers in diverse applications including bonding of dissimilar materials, sealing of expansion joints, weatherproofing, and perimeter sealing (sealing around doors, windows and other building components), amongst others.

3.5.3.2. Silane Polyurethane Hybrids

Silane polyurethane hybrids are urethane-based polymers which have been end-capped with reactive silane groups. Urethane-based and silicone-based sealants are two major, single component sealant technologies useful in many applications. For instance, they are used for sealing and bonding cement-containing compounds, metals, plastics, and glass.

Urethane-based sealants improve rheological and mechanical properties, and adhere well to a variety of substrates as do silicone-based sealants. However, urethane-based sealants tend not to accumulate dirt and dust and are easier to compound than silicone-based sealants. For this reason, hybrid sealants based on moisture-curable hydrolysable urethane prepolymers have been proposed [43]. Silane polyurethane hybrid sealants are systems that maximise the beneficial features of each of the urethane-based and silicone-based technologies, whilst minimising the undesirable characteristics.

The moisture-curable prepolymers are made by compounding moisture-curable alkoxylane functional polyether urethane prepolymers with rheological modifiers, adhesion promoters, oxidative stabilisers, plasticisers, and cure catalysts. Stuart [43] described the preparation and processing of sealants from silylated polyether urethane prepolymers (refer also to Section 3.3.4.3) prepared from endcap precursors containing dialkyl maleates.[6]

Some of their advantages are:

- completely non-gassing;
- no unreacted monomeric isocyanate;
- quick curing.

However, they involve a high cost and prepolymer synthesis can be required.

3.5.3.3. Amine Cured Polyurethanes

Primary and secondary amines react quickly with isocyanates yielding urea (Section 3.2.3). The inherent high reactivity of isocyanate, particularly aromatic isocyanate, with amine containing materials is a disadvantage of using an amine-cured binder. Accordingly, systems formed using amine-cured binders often react

[6] The term "silylated" or "endcapped" is applied to the resulting polymer which comes from the termination of pendant isocyanate groups with the amino group of the amino alkylalkoxysilane, and the molecule used to terminate the isocyanate groups is referred as to "endcap".

so quickly that application can be accomplished only with specialised plural component spray equipment which represents a high cost increase as compared to other systems. In addition, amine cured systems have a tendency to yellow upon exposure to sunlight, which prevents their use in applications where colour stability is required.

3.6. Applications of Polyurethane Adhesives and Sealants

3.6.1. Adhesives

In today's competitive markets it is often necessary to use complex structures made of different layered materials in order to achieve the desired combination of properties for particular applications.

It has been found that one effective way to make strong and durable laminates is through the use of adhesives. Bonding of such composite materials has several advantages over other joining methods. For example, when welding and soldering, the surfaces are permanently changed by thermal stresses whilst by using adhesives the strains are dissipated over the whole surface without creating local stresses. Also, by bonding surfaces the pieces are not weakened as happens with riveting, nailing, sewing or screwing methods. The little weight that an adhesive adds to a composite structure is also a great advantage [6].

Polyurethane adhesives are produced in many grades such as one-component, two-component, dispersion and solvent based, and hot melted for use in different application areas. PUR adhesives have a good adhesion to wood, metal and plastics and find, therefore, end use in many industrial applications. Some applications outlined by sector are presented in this section.

3.6.1.1. Construction

Some reactive polyurethane adhesives such as one- and two-component ones are particularly suitable for wood bonding. They are currently being used for the bonding of parquets and other wood flooring, mainly because of their strength and resistance. They are also being used for bonding of heavy duty rubber floor in commercial areas with heavy foot traffic [24].

Other common applications for the building and construction industry include foam PUR adhesives for sub-flooring, drywall, plywood, foamboard, fibreboard, wallboard, and brick. The specific features that make PUR adhesives suitable for those applications are their adherence to wet or frozen timber, their fast initial grab, high shear strength, their superior bonding and their anti-flammable properties amongst others. There are also some premium grades suitable for mirrors, ceramic tiles, ABS plastics, bath surrounds, concrete, and masonry.

PUR adhesives can be produced in many grades and can even be customised for special applications.

Table 16 presents several commercial construction polyurethane adhesives from Maris Polymers® outlining their type, properties, and common applications [44].

3.6.1.2. Packaging,[7] Woodworking

Most food packaging is now being made of two or more different films (refer to the chapter "PUR Adhesives for Flexible Packaging" in Volume 4). For products like pastries or liquid foodstuffs, efforts are being made to develop transparent packaging materials and some film composites based on PE, cellophane, PP, and PET have proven themselves to be appropriate. However, an efficient transparent adhesive that meets various government health regulations for being in contact with food was necessary. Suitable formulated polyurethane adhesives were found to be appropriate because in addition to good strength and no-discolouration, they have sufficient elasticity and good performance at temperatures of food freezing [6].

Depending on their use and characteristics, solvent or solventless polyurethane based adhesives are available and can also be customised for special packaging requirements. Some typical uses for one-component PUR adhesives are for adhesion of shopping bags, printed films, and adhesives for snack food bags. These are especially suitable for these applications mainly because they are solvent free, moisture curable, and easy to apply.

Other custom-modified two-component PUR adhesives are used for the adhesion of PVC to Al sheets for medicine cases [45].

Profile wrapping is another major application for hot melt PUR adhesives (HMPUR). Some wrapping applications include the bonding of printed papers and elasticised vinyls onto MDF profiles (refer to the chapter "Adhesives for Wood and Furniture" in Volume 3). These applications require some of the characteristics of the PUR adhesives such as low application temperatures and low viscosity [46].

3.6.1.3. Transportation[8]

Many of the new materials for the transportation sector – carbon fibre, GRP, plastics, light alloys, metals, as well as glass – cannot be joined or may not operate to their full potential with rigid fixing technology. The advantages of PUR

---

[7] A detailed study of PU adhesives for flexible packaging is provided in Volume 4 by the chapter "PU adhesives for laminating" by J-F LECAM.

[8] For a detailed study of PU adhesives for transportation, automotive, etc., the reader may refer to the chapters "Elastic Bonding and Sealing" by SIKA in Volume 2 and "Flexible Structural Bonding" also by SIKA in the next volumes of this Handbook.

Table 16: Commercial construction polyurethane adhesives.

Product	Adhesive type	Applications	Properties
MARISTICK ® 1000 (Maris Polymers ®)	Semi-rigid, one–component, without solvents	Adhesion of wood and ceramics	Provides strong mechanical and chemical resistance
MARISTICK ® 1070	Semi-rigid, one-component, with solvents	Adhesion of wood and ceramics	Provides strong mechanical and chemical resistance
MARISTICK ® 1700	Semi-rigid, flame retardant, two component, without solvents	Production of metal-rockwool sandwich panels	Strong mechanical and chemical resistance, flame retardancy
MARISTICK ® 1702	Semi-rigid, two component, without solvents	Production of sandwich panels made of ABS or PVC. Polystyrene foam blocks	Provides elasticity of hard rubber, excellent adhesion without shrinkage, cures by crosslinking even at very low temperatures
MARISTICK ® 1705	Semi-rigid, two–component, without solvents	Used for the production of sandwich panels made of metal or wood with rockwool. Polystyrene foam	Provides elasticity of hard rubber, excellent adhesion without shrinkage, cures by crosslinking even at very low temperatures
MARISTICK ® 1750	Semi-rigid, two -component, without solvents	Adhesion of metal, wood and ceramic tiles on horizontal and vertical surfaces	Provides elasticity of hard rubber, excellent adhesion without shrinkage, cures by crosslinking even at very low temperatures
Subfloor polyurethane adhesive from M-D Co	Premium grade quick curing PUR foam adhesive	Subfloor, plywood, gypsum, foamboards, block, fibreboard, drywall, wallboard, brick, hardboard	

Source: Ref. [44].

adhesive bonding include movement capability, noise dampening, vibration reduction, and improved durability as well as the ability to cope with large variations in bondline thickness.

One example of high performance polyurethane adhesives is the bonding in windscreens in cars and commercial vehicles. The use of this kind of adhesives provides a fast, reliable and watertight installation as well as adding structural support [47].

PUR adhesives are also used for the bonding of interior materials of automobiles such as door trims, door centre panels, headlining and dash boards that are composed of at least two or more materials. Other polyurethane adhesives such as INSTA-GRIP™ [48] are used to attach fibreglass liners to trailers. The required properties are their initial adhesive strength, and heat resistance as well as their easy and quick application.

Bonding of plastic retainers to wood-fibre panels, fixing air vents beneath instrument panels or reinforcing door panels with side-impact-protection parts also requires the fast-setting and durable properties with high resistance to mechanical stress and heat of PUR adhesives.

The bonding of side walls of vehicles is another important application for PUR adhesives (hot melts). This kind of application can be very demanding and only PUR adhesives offer the resistance to temperatures from -40 to 160° F to which these vehicles are exposed. The typical construction of the side walls include a multi-layer lamination with PUR adhesives to bond the fibre glass-reinforced plastic, two layers of lauan plywood, a layer of EPS foam and interior finish panel with decorative paper. Reichhold's Ever-Lock® 2U255 adhesive and Ever-Lock® 2U246 provide a long open time (LOT) and yet a high level of creep resistance at relatively high and low plant temperatures [46].

Sika Corporation also offers a great variety of adhesives and sealants for the transportation sector and it has developed products used by manufacturers of trailers, trucks, buses, cars, agricultural equipment, van conversions, and specialty vehicles. The main application areas in these sectors are direct glazing, roof bonding, body panel bonding, floor bonding, and flooring. Figure 13 shows chequer plate flooring using both Joint Sealant Sikaflex®221 and Adhesive Sikaflex® 252 or 254.

3.6.1.4. Marine

One- and two-component semi-rigid PUR adhesives are widely used in marine installations. They are suitable for strong wood to wood and wood to metal adhesions, particularly, for the production of sandwich panels made of extruded polystyrene with metal sheets or rockwool/metal. Their high strength, elasticity performance, high resistance to water, seawater, and sewage water make them the perfect bonding material for the marine sector [44] for several

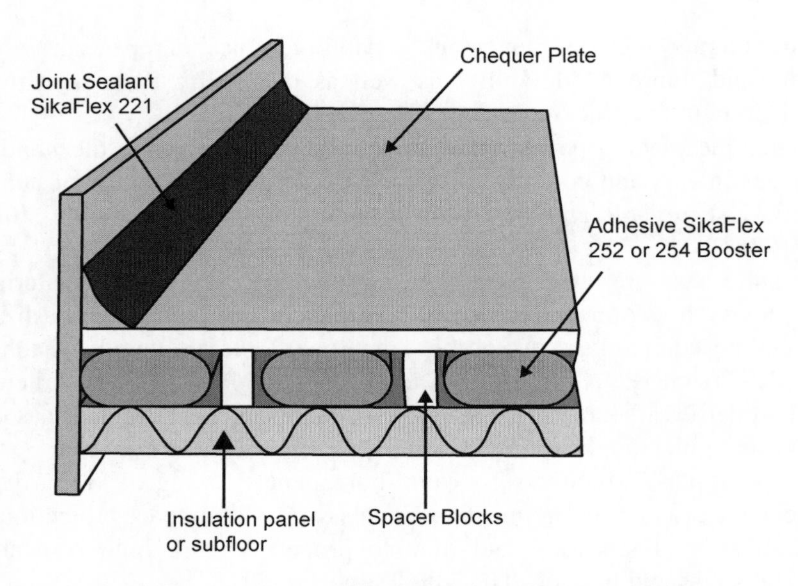

Figure 13: Chequer Plate Flooring. (Source: [49].)

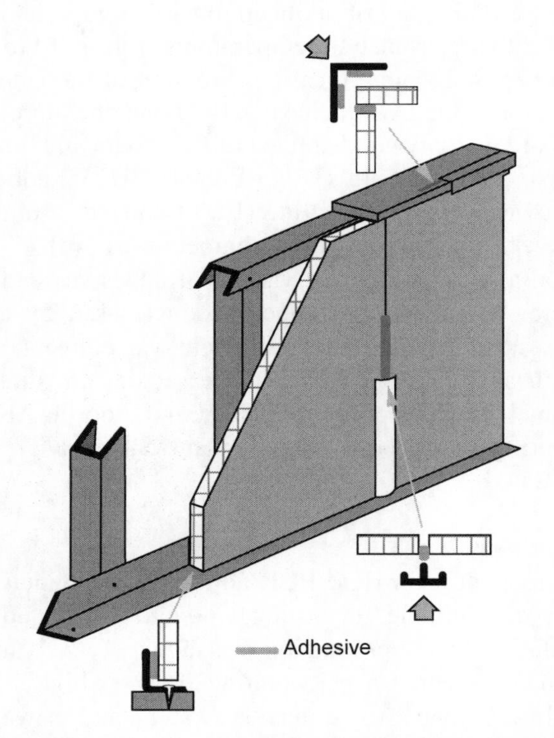

Figure 14: Sikaflex® 291 and 292 to assemble panels and profile trims. (Source: [49].)

applications. Some other two-component adhesives with flame retardant properties are widely used at the manufacture of division panels for ships and governmental offices.

Figure 14 shows the assembling and bonding of panels and trims for the interior decoration of large cruise ships with Sikaflex®291 and 292.

Another kind of commercial semi-rigid, thixotropic, two-component PUR adhesive with no solvents provides strong mechanical and chemical resistance and it is commonly used for the adhesion of ceramic tiles on metal, wood, and concrete on vertical and horizontal surfaces at the marine sector (Table 17).

There are several PUR-based adhesives created for specific functions within the marine sector such as the boat maintenance adhesives from Sika (Table 18) and Huntsman (Table 20). Amongst its products there is a one-component UV grade PUR adhesive that has been specially formulated for a marine environment where resistance to direct sunlight is required. It is commonly used in building areas such as hatches, portholes or other transparent parts. Another one-component moisture cured with high thixotropy and high strength adhesive is particularly suited for bonding of floor or replacing fasteners in the manufacture of boats. Also from Sika [49] there is an LOT one-component marine grade adhesive for use above and below the waterline. Its moisture cured and non-sag properties make it

Table 17: Marine applications.

Marine applications
Levelling, bonding and caulking teak decks
Bonding timber components
Bonding anti-slip deck coverings
Bedding and sealing of fittings and hardware
Flybridge bonding
Bonding of rub rails and fenders
Deck-to-hull bonding
Keel-to-hull joints
Bonding and sealing organic “glass” windows
Bonding and sealing mineral glass windows
Sealing of sacrificial anodes
Bonding of decorative panels and work surfaces
Bonding of deck panels and feature decks
Sealing of high-UV-risk areas
Bonding of lightweight internal partitions
Bonding of anti-slip plates for engine rooms

Table 18: Sika adhesives for various applications.

Product	Adhesive type	Applications	Properties
Sikaflex®-291	One-component, polyurethane-based sealing/adhesive compound	All-purpose grade is used for general marine adhesive sealing applications	Low viscous, medium modulus,
Sikaflex®-292	One-component, PUR-based system	Used for a wide range of bonding applications, deck-to-hull bedding of chainplates, through-hull fittings and toe rails	Thixotropic, high modulus, structural
Sikaflex®-295 UV	Fast-curing, one-component polyurethane-based adhesive	Bonding and sealing of windows and portholes. Suitable for all types of organic (PU, PMMA) window panels	UV resistant, fast curing, flexible, high performance
Sikaflex®-296	One-component PUR-based adhesive	Bonding windows and portholes. Suitable for all types of mineral glass. Its high degree of UV resistance also allows the use of the system as a weatherproof sealant	Fast-curing, one component, flexible, high performance
Sikaflex®-852 FR	One component PUR-based sealing and bonding adhesive	Marine flame-retardant sealing and bonding	Flame-retardant properties

suitable for light bonding of items identified for future removal such as bedding of deck hardware, hatches, port lights, etc.

3.6.1.5. Electronics

PUR adhesives are also widely used in the assembly of electronic devices for bonding of various materials. They are suitable for this application because of their room temperature curing, their durability, and resistance to thermal shock loads as well as vibration. The main requirement for adhesives to be used in this industry is their compatibility with the electronic components.

Using non-compatible adhesives to bond electronic components will cause:

- a corrosive effect on electronic components;
- the absorption of adhesive by the active layers of electronic components;
- a negative effect on electrical isolation;
- disruption of components.

In the electronics market, PUR adhesives compete with other joining methods such as mechanical fasteners, brazing, welding, soldering, and thermocompression bonding. The major uses of these adhesives are the bonding of embossed or die-stamped printed circuit boards, bonding of copper foil to dielectric material, fabrication of electronic coils and surface mount conductive adhesion. PUR adhesives are commonly employed in these applications because their flexibility and toughness up to 125°C and their good bond strength to a variety of surfaces.

A common type of adhesive for electronics is the hot melt PUR adhesives (HMPURs) which have the necessary high temperature resistance, the fast production speeds and the viscoelastic properties of an elastomer and are found in audio speakers, electronic enclosures, and the installation of sound dampener systems [50].

3.6.1.6. Footwear[9]

One very important property of PUR is its resistance to water. It will not break down in wet conditions even after much flexing. This feature, therefore, makes PUR adhesives ideal materials for footwear that needs to be water resistant. It has been demonstrated that specially developed PUR adhesives can be used to create a waterproof bond between waterproof lining and waterproof insole materials. PUR adhesive has been used successfully to seal the stitched seam on strobel lasted shoes.

PUR adhesives can be used to bond a wide variety of materials in the shoe industry. Featuring the one-way approach, PUR adhesives have been used to cost

[9] For a detailed study of footwear PU adhesives our readers should refer to the chapter "Footwear Adhesives and Bonding Techniques" by CTC Centre Technique du Cuir in this Handbook.

effectively improve productivity in such diverse applications as laminated trims to PVC Wellington boots, to inserting socks into PU sandals. Both applications require permanent bonding in extreme service conditions [51].

The sole attaching application also needs the PUR adhesive due to its high initial strength, its high heat resistance and cold flexibility, excellent water resistance and fast chemical crosslinking as well as fast setting.

3.6.1.7. Textiles and Fibres*

There has been a great interest in this area of application for PUR adhesives. They are currently been used for textile coating and some products have been developed for bonding very soft low-surface-energy materials such as Teflon, and PUR films for active-wear applications. The main characteristics that make PUR suitable for textile coating are its low viscosity Newtonian behaviour that allows rapid machining at low temperature, great hydrolysis resistance for fabrics to be washed several times and contain reactive flame retardants.

Some HMPUR adhesives such as the Jowat AG's Jowatherm-Reaktant® have been developed to provide a permanent bond for lingerie with excellent optical and tactile properties. Jowathern-Reaktant® is also widely used to laminate beach wear textiles due to their UV resistance, pleasant textile feel. It is also used for the manufacture of breathable sport and leisure clothing as well as protective garments (refer to the chapter "Adhesives Bonding in Textiles and Garments").

PUR hot melt adhesives are highly suitable for these applications and for the multilayer compound which consists of inner and outer textile layer with a central membrane open to diffusion. Some of Jowatherm-Reaktant hot melt adhesive types are used for several purposes (Table 19) [52].

3.6.1.8. Tanks and Pipes

Rigid polyurethane foams form strong adhesive bonds, which allows them to bond effectively with a wide range of building facings. The adhesion is so strong that the bond strength is usually higher than the tensile or shear strength of the foam. The low thermal conductivity of the foam allows it to be used as thermal insulators for pipes and its chemical resistance allows it to be used in chemical tanks. Some formulated systems from Huntsman are being developed to meet the specifications of the pipe industry, including oil and gas, district heating and cooling, chemical plants and manufacturing plants. Two of such systems are the Rubinate® and Rubitherm®.

PUR rigid foams are excellent insulate adhesives and sealants because of their efficient retention of heat, high level of compression and shear strength, processability, excellent adhesion, resistance to extreme temperatures, low water permeability, lightness and chemical resistance.

---

*Editor's note: Readers may refer to the chapter "Bonding Textiles and Garments" in a subsequent volume of this Handbook.

Table 19: Jowat's hot melt adhesive products range.

Product	Application	Properties
Jowatherm-Reaktant® 603.08	Manufacture of textile compounds in the automotive sector, the textile processing industry and for bonding special textiles	For general use, high initial strength
Jowatherm-Reaktant® 603.78	Manufacture of textile compounds in the automotive sector, the textile processing industry and especially in applications for the medical area	Low viscosity, resistance to constant washing, steam sterilisation resistance (good hydrolysis resistance)
Jowatherm-Reaktant® 603.88	Manufacture of textile compounds in the automotive sector, the textile processing industry. Specially for water repellent textiles	Low viscosity, suitable for spray application

3.6.1.9. Plastics and Composites*

Adhesives are increasingly being used to replace traditional fastening methods such as welding, bolts and rivets, in industrial applications. For structural and engineering applications polyurethane adhesives are widely used. These adhesives are used to bond many different material types including ceramics, metals, glass, plastics, and composites; some of them discussed previously. General plastic and composite bonding principles are found elsewhere [53].

Polyurethane adhesives currently used to bond most thermosets, including glass fibre reinforced plastics (GRP), sheet moulding compound (SMC), glass fibre reinforced epoxy (GRE), and thermoplastic composites including ABS, PVC, acrylics, polycarbonates, etc., are pre sented in Table 20.

Applications of Araldite®2000 include bonding of different substrates such as bonding of rubber/rubber and rubber/composite (Fig. 15), glass/aluminium interfaces (Fig. 16), bonding of navigation equipment screens into housings (acrylic to ABS) (Fig. 17) and bonding whip handles to horse whips (rubber onto fabric and carbon fibre) (Fig. 18), among others.

---

*Editors's note: For a detailed study of adhesives and sealants for plastics and composites, refer to the two chapters dealing with these materials in the next volumes of this Handbook.

Table 20: Huntsman advanced materials Araldite ®2000 PU adhesives.

Araldite®	Bonding	Properties
2018	Bonds well to most thermo plastics, including GRP/SMC/GRE/thermoset composites/ABS/PVC/ acrylics/polycarbonates, etc.	Flexible polyurethane, pale opaque, light handling strength 4 h, excellent impact resistance, room temperature curing, very flexible, lightly thixotropic liquid PU, ideal for large areas of bonding and/or a long working life
2026	Bond of transparent substrates, polycarbonate, PMMA and glass	Transparent, light handling 1 h, excellent chemical, water, and impact resistance, ideal for use where transparent or invisible joints are required, UV resistant, room temperature curing, flexible
2027	Ideal for SMC and GRP	Beige, light handling 1.5 h, good water and impact resistance, flexible, excellent gap filling capability, for primer less SMC bonding

Figure 15: Bonding rubber/rubber and rubber/composite. Bonding of the rubber layer (15 cm long) to the upper surface of the surf board tail which modifies the tail kick according to the amount of stress. (Source: [54].)

Figure 16: Assembly of inserts into neck of glass bottles. Glass/aluminium. Good adhesion to glass and aluminium together with water-like clarity. (Source: [54].)

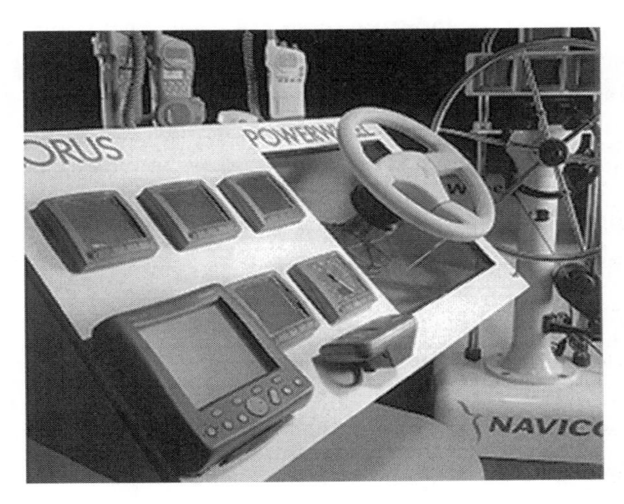

Figure 17: Bonding of navigation equipment screens into housings. Acrylic to ABS. Must resist vibration and be resistant to water. (Source: [54].)

Figure 18: Bonding whip handles to horse whips. Bond rubber onto fabric and carbon fibre. Client required product to be flexible, and weather and impact resistant. (Source: [54].)

3.6.2. Polyurethane Sealants Applications[10]

3.6.2.1. Construction

Polyurethane sealants are excellent for sealing dynamic moving joints such as pre-cast concrete, tilt-up panels, window and door perimeters, expansion, control, curtainwall joints and general construction applications. As an example, the Hardcast® PT™-304 Polyurethane sealant [55] is a one-part, ready to use, permanently flexible, fast moisture cure, non-sag, paintable, multi-purpose construction sealant and adhesive. PT-304 has high bonding and strength capabilities, with outstanding elasticity and exhibits high recovery making it excellent for sealing joints subjected to dynamic and thermal movement. Other commonly used PUR sealants are the Sonneborn® Sonolastic products listed in Table 21 [56].

The OSI® PL® [57] sealants are also widely used for general construction applications presented in Table 22.

3.6.2.2. Transportation

One supplier of raw materials to the adhesive and sealant industry for the automotives sector is Bayer. The polyurethane sealants created from these raw materials are used to prevent leaks, reduce wind and noise, and protect against corrosion amongst other applications.

[10] PU and other types of sealants will be studied in much details in the chapters "Construction Sealants" by Philippe Cognard in Volume 6 of this Handbook, and in the chapter "Industrial Sealants", later.

Table 21: Sonneborn® Sonolastic's polyurethane sealants for construction applications.

Product	Sealant type	Applications	Properties
Sonolastic 150	One-part polyurethane based	For use on substrates such as aluminium, concrete, masonry, wood, and stone. It can be used in applications such as wet glazing, curtain wall construction, expansion wall joints, EIFS, panel walls, precast units, aluminium and wood window frames, fascia, and parapets	Premium-grade, high performance, silyl-terminated, polyurethane-based, non-sag elastomeric sealant. It is a low-modulus formula, possessing extreme joint movement capability. + 100/− 50%
Sonolastic NP1 polyurethane sealant	One-part, moisture cure gun grade polyurethane sealant	For expansion wall and EIFS joints, curtain wall construction, panel walls, precast units, and perimeter window sealing	USDA approved, joint movement capability ± 25%, weather resistant, wide temperature application range

(Continued)

Table 21: Continued.

Product	Sealant type	Applications	Properties
Sonolastic NP2 polyurethane sealant	Two-part, chemical cure gun grade polyurethane sealant	For expansion wall and EIFS joints, curtain wall construction, panel walls, precast units, and perimeter window sealing	Elastomeric, movement capability of $\pm 50\%$, resistant to weather, air-borne pollutants, and chemicals, UL listed
Sonolastic SL1 self-levelling polyurethane sealant	One-part, self-levelling polyurethane sealant	For pavements, and decks	Abrasion and puncture resistant and elongation and recovery properties
Sonolastic SL2 self-levelling or slope-grade polyurethane sealant	Two-part, self-levelling or slope-grade polyurethane sealant	For expansion joints in pavements, decks, or other concrete applications	Abrasion and puncture resistance as well as elongation and recovery properties
Sonomeric 1 bitumen-modified polyurethane sealant	One-part, low modulus, self-levelling bitumen-modified polyurethane sealant	For horizontal joints in metal and masonry that are subject to sunlight, and jet fuel, e.g. airport runways, highways, bridges, and industrial concrete floors	Weather resistant, jet fuel resistant, exceptional elongation, chemical resistant

Table 22: OSI® PL® polyurethane sealants for building applications.

Product	Description
The PL® polyurethane door, window and siding sealant	Ideal for installing or repairing door and window frames, bonding with wood, vinyl, metal, aluminium, stucco and other materials, sealing applications on all types of siding
PL® polyurethane roof and flashing sealant	Seal cracks between wood and metal or flashing around chimney bases, wells, etc. Excellent for shingle tabbing and sealing cracks in gutters, skylights waterproof/ weatherproof
PL® polyurethane concrete and masonry sealant	Perfect for sealing cracks or gaps in concrete or masonry structures, holds its grip in all types of weather. Textured surface looks like cured concrete seals out radon migration
PL® polyurethane self-levelling concrete crack sealant	Produces a flexible long-lasting seal in expansion joints in concrete floors and decks. Use it on sidewalks, pavements, decks, and other concrete structures where abrasion and puncture resistance are required. When cured, it resists deterioration caused by weather, stress, movement, traffic, and water

Direct glazing is a common application in the automotive industry since the polyurethane sealant can perform the important functions of adhesion of glass directly into the metal frame, cope with the vibration and movement and still remain weather tight. The Sikaflex® -255FC is a one-component compound from Sika which, because of permanent elasticity and excellent adhesion to metal, glass, and many other materials, is ideal for the automotive industry.

Seam sealing is another application which is familiar to all bodyshops, but the attention that has been paid to this apparently minor detail has made a significant difference to vehicle life. We all know the speed with which water and road salts can creep into seams and cause corrosion. Effective polyurethane sealing will prevent this and is essential for warranty coverage and resale value [49].

3.6.2.3. Tanks and Pipes

A polyurethane sealant must be resistant to bacteriological attack, to abrasion and be weather tight in order to seal tanks and pipes. Emer-Seal 200 from Parchem is a polyurethane sealant for water retaining structures. It is used for sealing movement and static joints especially in applications likely to be subjected to biological degradation such as sludge digestion tanks, sewerage and water treatment plants, filtration and aeration tanks, water reservoirs, marine installations, and also most building applications such as tanks and pipes.

When compared to adhesives, sealants are generally low modulus/high movement capability materials. Sealants generally do not have to be cohesively strong; however, they should adhere strongly to the jointing surfaces and be capable of allowing the designed movement of the joint and maintaining a water and weather-tight seal.

According to Sika the most versatile and widely used sealants are those derived from the polyurethane family. A vast range of properties may be gained from these, thus offering the option of products tailor made for the different application requirements.

Today, suppliers have developed one- and two-component polyurethane systems, although the two-component systems are not widely used. One-component systems offer faster curing times which make them more suitable to meet most customers' needs.

Once cured, polyurethanes offer tough, elastic sealants with a range of strengths and movement capability and higher strength bonding adhesives for direct glazing and other more structural applications. The following section gives an overview of such applications.

3.6.2.4. Marine

3M manufactures a wide array of maintenance products specifically designed to perform in the harsh marine environment. Some of its high-performance polyurethane adhesive/sealants become tack free in 48 h, and completely cure in 5–7 days with no shrinkage. The polyurethane seal is extremely strong, retains its strength above or below the water line, and stays flexible allowing for structural movement. Stress caused by shock, vibration, or swelling is effectively absorbed. As an example, the 3M® 5200 Fast Cure sealant [58] has excellent resistance to weathering and salt water and works to seal fibreglass hulls to deck joints, wood to fibreglass, portholes and deck fittings, etc.

The PL® Polyurethane Marine Sealant is another marine-grade sealant from OSI® for joints and boat hardware, skylights, and hatches. It can be used on wood, fibreglass, and metal to provide a watertight, weather-resistant, long-lasting seal.

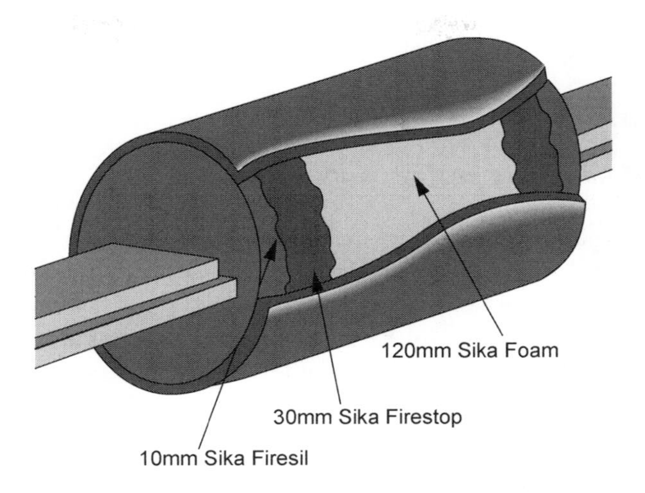

Figure 19: Sika® Cable Duct System. (Source: [49]).

3.6.2.5. Electronics and Electrical Equipment
Sealants in the electronics industry are primarily used to bond, seal, and insulate, and dissipate component damaging heat. Fig. 19 shows cable duct systems with Sika sealants.

3.7. Testing and Standards of Polyurethane Adhesives and Sealants[11]

Polymeric adhesives and sealants are finding increased use in aerospace and automotive applications. This is due to the benefits they offer enabling the manufacture of structures with high strength yet low weight. Readers are referred to the chapter "Sealants for Aerospace" in this Handbook.

There are many test methods in use to indicate the "strength" of bonded and sealed joints. Results from these tests are used by manufacturers to optimise formulation and processes, and by end users in the selection of an adhesive or sealant system.

There are a number of challenges associated with the testing of adhesive and sealant systems. The performance of these systems is never solely due to material properties. Often, the results of a test are a function of the test rather than the system being tested. This can be due to a number of variables including the

[11] Refer also to the Chapter "Technical Characteristics of Adhesives and Sealants", by Philippe Cognard, Chapter 2 of this Volume 1.

viscoelasticity of the system which is very dependent on strain rate and other test parameters.

3.7.1. Tack of Adhesives and Sealants

One property of interest in many applications is tack. Tack can be described as a measure of the ability of a system to instantaneously form a bond to a substrate using light contact pressure.

Tack is dependent not only on the material properties but also on the bonded surface properties. Other influences include contact pressure, the duration of contact, the rate of separation of the surface, the mechanical stiffness of the system and the test environment [59].

3.7.2. Green Strength Adhesives

As previously stated, pressure sensitive adhesives develop stronger bond properties over time. The minimum strength for a material to be strong enough to adhere the part, but not so strong that it cannot be disassembled and refitted is referred to as "green strength". In general, the requirement is that the bond should be strong enough to adhere the part, but not so strong that it cannot be disassembled and refitted.

3.7.3. Peel Strength Adhesives and Sealants

The peel strength of a system is defined as the force required to progressively separate the two parts of a bonded system over the adherend surfaces. Peel tests are quite widely used to evaluate adhesive performance of adhesive bonded joints, particularly with thin, flexible adherends. Various configurations, including climbing drum and floating roller, have been developed to provide particular test method characteristics. The T-peel test is a simpler arrangement which can be used on symmetrical joints [60]. Peel tests apply to both adhesives and sealant systems; however, there are differences in the test regimes (Fig. 20).

Specimens are usually peeled at an angle of 90 or 180°. The values resulting from any of the test methods mentioned can differ considerably; it is, therefore, important to specify the test method employed. The climbing drum and floating roller tests produce more consistent results due to the fact that the angle of peel is constant. Other factors that may influence the peel strength of adhesives are the viscoelastic properties and thickness of the adhesive. Elastomeric adhesives, resulting in thicker bondlines generally result in higher peel strengths, whilst with

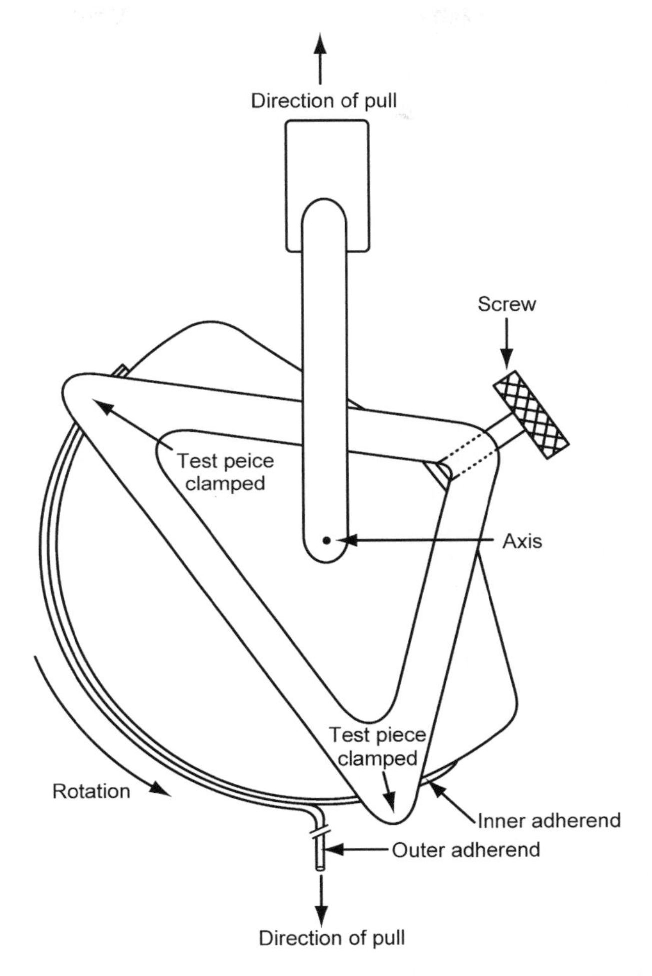

Figure 20: Apparatus for conducting drum peel test for rigid assemblies according to the British Standard BS5350-C14 1979.

more rigid adhesives, the thicker bondline results in lower peel strength. The modulus of the substrates is also a function of the peel strength of an adhesive, as is the thickness of the two substrates being peeled [59]. Therefore, in practical applications, the type of the backing on a pressure-sensitive tape may have a significant effect on the peel strength of the tape.

Peel tests for sealants are usually carried out to determine adhesion and cohesion properties of the material.

In these tests, the sealant is placed between two concrete or aluminium parallel supports. The material is then subjected to an extension/compression amplitude of either 25 or 15% at a rate of 1 mm/min for 100 cycles. The specimen is normally left for 1 h before it is inspected for loss of adhesion/cohesion. Peel tests include EN 1372: 1999, EN 431: 1994, EN 1939: 2003, among others (refer to Table 24).

3.7.4. Impact Testing Adhesive Joints

In common with most polymeric materials, adhesives and sealants are sensitive to the rate of applied force. There are several different types of impact test for adhesives. The most common test (similar in application to the Izod impact test for plastic materials) has been withdrawn without replacement; however, the test was not widely used in a production situation due to the difficulty in achieving reproducible results. A more informative test for impact is the cleavage test in which the test specimen is subject to an impact force through a wedge (refer to Table 24). This can be initiated by a pendulum impact machine with an impact energy range of 50–300 J and an impact speed of 3–5.5 m/s.

The rate at which the impact occurs also affects the impact strength of a viscoelastic material. Often, it is very difficult to achieve very high rates of impact with conventional laboratory testing. In certain applications (military and aerospace) high-speed impact tests have been developed. In order to attain the high speeds required chemical explosives or electromagnetic energy are employed [59].

3.7.5. Environmental Testing Adhesives and Sealants

Short-term tests do not always give an adequate indication of an adhesive's performance during end user applications. Laboratory tests may only be carried out over a thousand hours, and the joint may only be exposed to one type of environmental effect at a time. During its service life, however, the bond is exposed to multiple types of environments. Temperature and humidity may cause an adhesive bond to degrade much faster than when exposed to any single environment. Applied stress accelerates the effect of the environment on an adhesive joint. Data are not always readily available from manufacturers on this phenomenon. Environmental testing should be carried out on the proposed adhesive and adherend prior to its use in service.

Due to the significant use of sealants within the construction industry environmental effects are of importance. Sealants used in the glazing industry will undoubtedly be subject to the effects of ultra-violet light. It is relatively easy to test these effects using an environmental chamber; in this type of test the sample can be subject to the effects of UV light at a certain humidity and a regulated temperature.

Chemical and solvent contact can also inhibit the performance of both adhesives and sealants. While a mild solution of detergent may not affect performance, using the same undiluted may cause failure within hours of contact. Testing for the effects of chemical and solvent contact is relatively straightforward, but should only be carried out under local health and safety regulations; most manufacturers will hold this information.

Heat ageing can reduce the life of a sealant or adhesive, which is generally due to the loss of plasticiser from the material. Elevated temperatures may also increase the amount of crosslinking (post-cure hardening) and oxidisation, both of which may reduce the moveability of the sealant and induce brittle fracture in adhesive joints. In order to understand the effects of heat ageing, thermal testing should be carried out.

3.7.6. The Effect of Glass Transition Temperature on Adhesives and Sealants

Polymers have physical properties that are important in adhesion. The chemical structure of the base polymer of the adhesive or sealant and the various formulating agents used in modification of the polymer control the physical characteristics. The glass transition temperature (Tg) is a physical property measurement that reflects the behaviour of polymers. Tg gives us an understanding of the molecular motion that occurs in a polymer system: the higher the molecular movement the softer the material.

When considering adhesion, cohesion, and other properties of adhesives and sealants the amount of molecular freedom is an important aspect. At low temperatures they exhibit solid characteristics; however, as the temperature increases the material starts to soften and exhibits viscous flow. The temperature at which this occurs is called the glass transition temperature, and signifies a transition of the polymer from a glassy to a rubbery state. Raising the temperature further above its Tg, allows the molecules more freedom of movement until the material melts into liquid flow.

Polymeric materials have transition phases above and below normal room temperature. Many polyurethane elastomers show a Tg ~ -50°C while structural foams show a Tg $\sim$ 83°C.

The transition phase of a polymeric material does not occur at a single unique temperature. Usually, the phase change happens over a short temperature range, the Tg is the approximate mid-point temperature of that range.

The Tg of a polymeric system is not a single, fundamental property, but a complex property that is dependent on the basic characteristics of the polymeric structure and the conditions by which it is measured, these include:

1. molecular weight;
2. molecular weight of polyol components;
3. degree of crosslinking;
4. percentage crystallinity;
5. type of test used to determine Tg;
6. thermal conductivity;
7. rate of heating or cooling.

Generally, the glass transition temperature is measured by observing the variation of some thermodynamic properties, as is shown in Table 23.

For good bond strength and creep resistance at elevated temperatures, the Tg of a cross-linked adhesive or sealant should be above the upper end use temperature. However, when Tg is considerably above the upper end use temperature the peel strength is lowered.

In low temperature applications of a relatively high Tg material the mechanical properties are limited because the material tends to become brittle.

Polymers with a lower Tg show that the molecules are more flexible giving increased impact resistance at a given temperature compared to higher Tg polymers. Low Tg resins (Tg below 0°C) retain their flexibility as the resin remains within its rubbery plateau. However, as the service temperature falls below the Tg, these materials will also become brittle [59]. As the service temperature exceeds the Tg a sharp decrease in tensile strength and creep resistance is evident. This is more pronounced in thermoplastic adhesives and sealants than with thermoset types.

Most materials considered to be pressure-sensitive have a Tg below room temperature. This fulfils two of the major requirements for a pressure-sensitive adhesive. Firstly, that it must undergo plastic flow on contact and secondly, it must

Table 23: Thermal property tests.

Test method	Property
Differential thermal analysis	Thermal expansion coefficient
Dielectric analysis	Dissipation (loss) factor
Differential scanning calorimetry	Heat flow
Dilatometry	Specific volume
Dynamic mechanical analysis	Storage modulus

wet the substrate surface. While this provides the desired characteristic, because they are used above their Tg, these types of adhesives do not achieve high cohesive strengths whilst in this rubbery phase as the molecules are mobile enough to flow under load until the bond fails.

The final Tg of pressure-sensitive adhesive formulation is generally a combination of a low Tg base elastomer and high Tg additives. This ability to adjust the Tg provides the opportunity for the manufacturer to optimise the adhesive for a specific end use.

3.7.7. Durability of Bonded Systems

The long-term performance (durability) of a bonded joint depends on the properties of both the adhesive/sealant system and the materials being joined. As previously stated, adhesives are affected by high temperatures, solvents, and moisture. Durability of the joint is also dependent on the effects of these agents on the materials being joined. The overriding factor which determines the durability of a bonded system is the condition of the joint surfaces when the bond was made [60]. Poor surface preparation generally results in lower initial strength and lower durability.

3.7.8. Single Lap Joint Test

The single lap joint test is the most widely used test method for durability studies. Its popularity lies in its relative simplicity and perceived ease of interpretation. This test method is sensitive to changes in adhesives and surface treatments in joint combinations. Small differences in joint conditions can give rise to variability.

This test method is a measure of the response of the bonded system to the compounded effects of many different mechanisms and changes which occur during the ageing process. Some systems exhibit increase in strength during early weeks of durability testing. This may be due to a number of effects including: plasticisation of the adhesive due to moisture absorption; the elevated temperature of the ageing environment causing post-cure of the system; or residual stress relaxation within the system.

3.7.9. Tensile Butt Joint Test

The tensile butt joint test is less commonly used as an adhesive test method; this may be due to the perception that it is more difficult to prepare the test specimens;

Table 24: BS and EN standards for some adhesive and sealant properties.

Standard	Description
EN 12962:2001	Adhesives – determination of elastic behaviour of liquid adhesives ("elasticity index")
EN 1965-1:2001	Structural adhesives – corrosion – part 1: determination and classification of corrosion to a copper substrate
EN 1965-2:2001	Structural adhesives – corrosion – part 2: determination and classification of corrosion to a brass substrate
EN 12962:2001	Adhesives – determination of elastic behaviour of liquid adhesives ("elasticity index")
EN 12963:2001	Adhesives – determination of free monomer content in adhesives based on synthetic polymers
EN 14173:2002	Structural adhesives – T-peel test for flexible-to-flexible bonded assemblies (ISO 11339:1993 modified)
BS ISO 11600 2002	Building and construction jointing products classification and requirements for sealants
BS EN 29046 1991	Building and construction sealants determination of adhesion/cohesion properties at constant temperature
BS 5350-C14 1979	90° peel test for rigid to rigid assembly
BS 5350-C13:1990	Climbing drum peel test
BS 5131-1.9 1985	Test for footwear and footwear materials. Measurement of green strength of adhesive joints
BS EN 14187 2 2003	Cold applied sealants test method for the determination of tack free time
BS EN 14493 2002	Structural adhesives, determination of dynamic resistance to cleavage of high strength adhesive bonds under impact conditions
EN ISO 10365 1995	Adhesives designation of main failure patterns
BS 5131-1.7 1991	Footwear and footwear materials preparation of test assemblies using hot melt assemblies for creep and peel tests
BS EN 1465 1995	Adhesives determination of tensile lap-shear strength of rigid-to-rigid bonded assemblies
BS EN ISO 14615 1997	Adhesives durability of structural adhesive joints exposure to humidity and temperature under load
BS EN ISO 9664 1995	Adhesives test methods for fatigue properties of structural adhesives in tensile shear
BS EN 26922 1993	Adhesives determination of tensile strength of butt joints

in addition, the adherend materials (typically bar stock) used for bonded engineering structures may not be readily available [60].

Degradation rate in tensile butt joints is more sensitive to the bond area when compared with the lap shear test, therefore, smaller diameter joints allow a more rapid indication of durability. The absence of distortion and the general uniformity of fracture surface provide a good facility for analysis of the locus of failure. This makes this type of test particularly suitable for detailed analytical studies on specific surface treatments.

3.7.10. Perforated Hydrothermal Test

The application of a controlled stress to a bonded joint during environmental exposure can accelerate the degradation process. When compared with other durability tests, the application of different stress levels in different environments requires a greatly extended matrix of test conditions. The selection of appropriate load and environmental combinations requires a well-defined rationale for the application of the test and some prior knowledge of the bond system characteristics [60] (Table 24).

For more information about testing methods, testing standards and testing equipment please refer to the chapters "Technical Characterisation of Adhesives and Sealants in" Volume 1, and "Testing Equipment for Adhesives and Sealants" in this Handbook.

References

[1] Bayer, O., Siefhen, W., Rinke, H., Orther, L., & Schild, H. (1937). Germany Patent No. DRP 728981 (to I.G. Farberindustrie AG.), November 13.

[2] Saunders, J. H., & Slocombe, R. J. (1948). The chemistry of the organic isocyanates. *Chemical Reviews*, **43**, 203–218.

[3] Saunders, J. H., & Frisch, K. C. (1962). *Polyurethanes chemistry and technology*, High Polymers, Chemistry. Interscience Publishers, New York, Vol. XVI, Part I.

[4] Kogelnik, H.-J. (2001). Polyurethanes. *Kunststoffe*, **91**, 10, 339–346.

[5] Szycher, M. (1999). *Szycher's handbook of polyurethanes*. CRC Press, Boca Raton, FL.

[6] Oertel, G., & Abele, L. (Ed.) (1994). *Polyurethane handbook*. Hanser Gardner Publications Cincinnati, OH.

[7] Wirpsza, Z. (1993). *Polyurethanes: chemistry, technology and applications*, Series in Polymer Science and Technology. Barnes & Noble, New York, p. 517.

[8] David, D. J., & Staley, H. B. (1979). *Analytical chemistry of the polyurethanes*, High Polymer Series. Krieger Publishing Company, Melbourne, FL, Vol. 16.

[9] Dombrow, B. A. (1965). *Polyurethanes*. Reinhold Publishing Corporation, New York.

[10] Saunders, J. H., & Frisch, K. C. (1962). *Polyurethanes chemistry and technology*, High Polymers, Technology. Interscience Publishers, New York, Vol. XVI, Part II.

[11] Russel, D., & Smiley, R. A. (1991). *Practical chemistry of polyurethanes and diisocyanates*.

[12] O'Connor, J. M., Grieve, R. L., & McClellan, R. (2002). *Understanding polyurethanes*, Seminar, September 30–October 3, 2001, Columbus, OH.

[13] Richardson, P. N., & Kierstead, R. C. (1969). Organic chemistry for plastics engineers. *SPE Journal*, **25**, 54–62.

[14] Saunders, J. H. (1959). The reactions of isocyanates and isocyanates derivatives at elevated temperatures. *Rubber Chemistry and Technology*, **32**, 337–345.

[15] Kimball, M. E. (1984). Polyurethane adhesives for structural bonding. *Polymer News*, **9**, 198–202.

[16] Bayer Corp., (1997). *The chemistry of polyurethane coatings. A general reference manual*. Bayer Technical Information, Leverkusen, Germany.

[17] Wicks, Z. W. (1975). Blocked isocyanates. *Progress in Organic Coatings*, **3**, 73–79.

[18] Wicks, Z. W. (1981). New developments in the field of blocked isocyanates. *Progress in Organic Coatings*, **9**, 3–28.

[19] Wicks, D. A., & Wicks, Z. W. Jr. (1999). Blocked isocyanates III: Part A. Mechanisms and chemistry. *Progress in Organic Coatings*, **36**, 3, 148–172.

[20] Wicks, D. A., & Wicks, Z. W. (2001). Blocked Isocyanates III Part B: Uses and applications of blocked isocyanates. *Progress in Organic Coatings*, **41**, 1–83.

[21] DOW, (2003) DOW Chemical Company. *DOW MDI based products*, http://www.dow.com/webapps/lit/litorder.asp?objid = 09002f138023d7cb&filepath = /noreg [accessed: June, 2003].

[22] CAS, (2003). Chemical Abstracts Service, American Chemical Society, http://www.cas.org/faq.html [accessed: 21.05.03].

[23] SpecialChem. S. A. (2003). *Raw material database*, http://www.specialchem4adhesives.com/resources/search/additive.aspx [accessed: June, 2003].

[24] Evans, R. M. (1993). *Polyurethane sealants: technology and applications*. Technomic Publishing Co. Inc., Lancaster, USA.

[25] Bandlish, B. K., & Barron, L. R. (1990). *US Patent* No. 4, 916, 199 (to The B. F. Goodrich Company (Akron, OH)), April 10.

[26] Mowrer, N. R., & Rojas, J. L., (1998). *US Patent* No. 5,760,155 (to Ameron International Corporation), June 2.

[27] Florio, J. (2000). Paint & coatings industry. *Troubleshooting metal catalyzed urethane systems*, http://www.pcimag.com/CDA/ArticleInformation/features/BNP__Features__Item/0,1846,11371,00.html [accessed: June, 2003].

[28] Heath, R., & Rungvichaniwat, A. (2002). The examination of the structure property relationships of some water-dispersed polyurethane elastomers. *Progress in Rubber and Recycling Technology*, **18**, 1, 1–48.

[29] Sykes, P. A. (1996). Structure–property relationships in polyurethane elastomers. *Progress in Rubber and Recycling Technology*, **12**, 3, 236–257.

[30] Frisch, K. (2003). *Structure and property relationships of urethanes*, Adhesive and Sealant Council Urethane Short Course. St. Louis Renaissance Grand Hotel, St. Louis, Missouri.

[31] IAL, C. (2001). *Global overview of the polyurethane case markets*. IAL Consultants, London, UK.

[32] Mach, T. (2000). *Sealants find their niche in the North American market*, http://www.adhesivesmag.com

[33] Potter, T. A., Schemelzer, H. G., & Baker, R. D. (1984). High-solids coatings based on polyurethane chemistry. *Progress in Organic Coatings*, **12**, 321–338.

[34] Kimball, M. E. (1984). *Urethane adhesives: an overview*, Adhesives 1984 Conference in 1984. Society of Manufacturing Engineers (SME), Cleveland, Ohio, 84.

[35] Eling, B., & Phanopolous, C. (2003). Huntsman-polyurethanes. *Polyurethane adhesives and binders*, http://www.huntsman.com/pu/Media/Loughborough.pdf (Key word search for adhesives) [accessed: January, 2003].

[36] Randall, D., & Lee, S. (Eds.) (2003). *The polyurethanes book*. Wiley, New York.

[37] Rumack, D. T. (2003). *US Patent* No. 6,613,836 (to National Starch and Chemical Investment Holding Corporation (New Castle, DE)).

[38] Altounian, G. N. (2003). *US Patent* No. 6, 525, 162 (to Eastman Chemical Resins, Inc. (Kingsport, TN)).

[39] Rajalingam, P., & Radhakrishnan, G. (1988). *Polyurethanes adhesives*, *Popular Plastics*, **33** (12), 45–47, December 1988.

[40] Henderson, R. *Waterborne polyurethane coatings: one and two component systems*. Bayer Technical Information, Leverskusen, Germany.

[41] Dieterich, D. (1981). Aqueous emulsions, dispersions and solutions of polyurethanes; synthesis and properties. *Progress in Organic Coatings*, **9**, 281–340.

[42] Rosthauser, J. W., & Nachtkamp, K. (1987). Waterborne polyurethanes. In: K. C. Frisch, & D. Kempler (Eds), *Advances in Urethane Science and Technology*, pp. 121–162.

[43] Stuart, J. T. (2001). *US Patent* No. 6, 265, 517 (to Bostik, Inc.).

[44] Maris Polymers®, (2003). *Polyurethane systems*, http://www.marispolymers.gr [accessed: June, 2003].

[45] Kyung, O. Chemical Co. Ltd. (2003). *Product information*, www.kyung-o.co.kr [accessed: June, 2003].

[46] Huber, L. (2003). Adhesives & sealants industry. *Hot melt polyurethane reactive adhesives targeted for specific applications*, http://www.adhesivesmag.com [accessed: June, 2003].

[47] Hürter, H.-U. (2003). DuPont performance coatings GmbH & Co KG. *Automotive industry agenda: The best of both worlds in automotive adhesives*, http://www.automotivetechnology.net/editorials/Dupont.htm [accessed: July, 2003].

[48] Dow Chemical Company (2003). *INSTA-GRIP in polyurethane systems*, http://www.dow.com/pusystems/product/instagrp.htm [accessed: June, 2003].

[49] Sika Corporation (2003). *Transportation products*, http://www.sikaindustry.com [accessed: June, 2003].

[50] SpecialChem Adhesives and Sealants (2003). *Special feature: technical trends & innovations on adhesives & sealants markets*, http://www.specialchem4adhesives.com [accessed: May, 2003].

[51] British United Shoe Machinery Limited (2003). *PUR adhesives brochure*, http://www.usmgroup.com [accessed: June, 2003].

[52] Jowat Corporation (2003). *Jowat AG brochure* [accessed: June, 2003].

[53] Dunn, D. J. (2004). Engineering and structural adhesives. Report 169. *Rapra Review Reports*, **15**, 1, Chapter 4.

[54] Vantico Ltd. a company of Hunstman Advanced Materials (2002). *Araldite 2000 Range*.

[55] Hardcats (2003). *PT-304 polyurethane construction sealant*, 2002, http://www.hardcast.com [accessed: June, 2003].
[56] Ro-An corporation (2003). *CHEMREX Sonneborn — sealants and caulking*, 2002, http://www.roancorp.com/sonneborn.html [accessed: June, 2003].
[57] OSI® Brands (2003). PL® *Polyurethane sealants*, http://www.stickwithpl.com/brands/osi_brands_index.htm [accessed: July, 2003].
[58] 3M (2003). *3M worldwide product index*, http://www.3m.com [accessed: June, 2003].
[59] Petrie, E. M., & SpecialChem, S. A. (2003). Important characteristics of several common adhesive tests, http://www.specialchem4adhesives.com [accessed: 2003, July].
[60] *Durability of adhesive joints: A best practice guide materials and metrology DTI*: Department of Trade Industry.

Chapter 4

Surface Pretreatment for Structural Bonding[☆]

John Bishopp

It is regrettable that it has needed actual loss of life to secure that general acceptance and the necessary preventative measure

Comment by Norman de Bruyne, founder of Aero Research Limited, on the Air Ministry's reluctance to make surface pretreatment of plywood mandatory prior to bonding.

Abstract. This chapter is concerned with an in-depth examination of the adherend surface pretreatments used prior to structural adhesive bonding. It encompasses the various substrates encountered, particularly but not exclusively, in the aerospace industry. It compares and contrasts mechanical, chemical and electrochemical methods used for substrates comprising aluminium alloys, titanium, stainless steel, thermoplastic and thermoset fibre reinforced composites and non-metallic honeycomb. Scanning and transmission electron microscope techniques are used to analyse and characterise many of the pretreated surfaces so produced.

Keywords: Abrasion; AC anodising; Acid etching; Adherends; Alkaline degreasing/etching; Alocrom; Aluminium; Aluminium–lithium; Anodising; Boric acid/sulphuric acid anodising; Chemical pretreatment; Chromate free anodising; Chromate free treatments; Chromic acid anodising (CAA); Chromic/ sulphuric acid pickling (CSA); Composite; Conversion coating; Corona discharge; Degreasing; Electrochemical pretreatment; Electron microscopy; Fibre reinforced composites; Grit blasting; Honeycomb; Iridite; Laser; P2 etching; Pasa Jell; Peel ply; Phosphoric acid anodising (PAA); Pickling; Plasma; Pretreatment; Silanes; Sodium hydroxide anodising; Sol–gel; Solvent degreasing; Specifications; Stainless steel; Sulphuric acid anodising (SAA); Surface analysis; Thermoplastics; Thermosets; Titanium; Ultramicrotomy; Vapour degreasing

---

[☆] *Photographic Media*: unless otherwise stated, schematic diagrams and photographs are drawn from the Hexcel Composite archives or are the property of the author.

Handbook of Adhesives and Sealants
P. Cognard (Editor)

4.1. Introduction

Although the examples given below are specifically related to the aerospace industry, the techniques, in general, are valid for all industries using such substrates for structural adhesive bonding.

To form a strong, integrally bonded, load-bearing structure, the surface of the adherend should be pretreated before application of the adhesive; this is vital if good environmental or thermal durability is required. Such a procedure ensures that the surface is in as clean a condition as possible; removing weak boundary layers, which could adversely affect the performance of the resultant joint. These weak boundary layers include: oxide shale, weak oxide layers, release agents, any low molecular weight organic species and contamination (particularly airborne) in general. Further, any surface roughness requirement can be optimised and, to maximise its wetting characteristics, the surface free energy of the adherend is raised to as high a level as possible. For many metallic adherends, this latter action also increases the stability of the surface oxides and can maximise potential interactions across the interface.

This general principle is as true for wooden substrates as it is for the "high tech" carbon fibre laminates. Indeed, it was with the bonding of wood in aircraft structures that the requirement for surface pretreatment first came to light.

In November 1937, M. Langley, then the Chief Designer of the British Aircraft Manufacturing Company Limited, noted that when new the plywood (Tego bonded) which his company was using to manufacture airplanes could not be bonded using Aerolite® urea–formaldehyde adhesives. This was attributed to a "case hardening" effect in the outer plies. At Aero Research Limited, the case-hardened plies were treated with glass-paper, sanding "almost down to Tego glue film but leaving an unbroken surface of wood" [1,2]. This solved the problem. This phenomenon was encountered again in 1938 when de Bruyne was repairing a Desoutter monoplane. In this instance the bond between the plywood and the spruce longerons was found to be very weak.

In October 1938 these findings were reported to the Air Ministry but the technique of sanding along the grain prior to bonding was not made mandatory until 1942; unfortunately, "after actual loss of life to secure that general acceptance and the necessary preventative measures" [1].

The principles of surface pretreatment are now well understood and methods have been developed for each type of substrate encountered in aerospace construction: wood, aluminium alloys, titanium alloys, stainless steel, thermoplastic composites, thermosetting composites and non-metallic honeycombs.

4.2. Overview of Surface Pretreatment Techniques

Surfaces can, generally, be pretreated using one of the following procedures; for many substrates, this list is in increasing order of effectiveness:

(1) Degrease only
(2) Degrease, abrade and remove loose particles
(3) Degrease and modify the substrate surface by chemical or electrochemical pretreatment or by techniques such as surface bombardment by activated plasma.

This, of course, does not apply to wooden structures. Here, the same pretreatment is used today as was instigated in 1942, namely sanding along the grain.

Care must be taken to avoid contaminating the surfaces during or after pretreatment. Contamination can be caused by finger marking, by cloths which are not perfectly clean, by contaminated abrasives or by sub-standard degreasing or chemical solutions.

Contamination can also be caused by other work processes taking place in the pretreatment and bonding areas. Particularly to be excluded are oil vapours from machinery, spraying operations (paint, mould release agents, etc.) and procedures involving airborne powdered materials.

Whatever procedure is used, it is a good practice to bond the substrates as soon as possible after completion. It is then that the surfaces are most "active" — i.e. their surface properties are at the optimum level for bonding.

The procedures for cleaning and degreasing and, to some extent, the mechanical abrasion techniques will be the same for virtually all metallic substrates. An overview of these two aspects will, therefore, be followed by a more specific, in-depth review of the pretreatments, both mechanical and chemical, which can be used for the individual substrates encountered in aerospace applications.

4.2.1. Degreasing Methods

The removal of all traces of oil and grease from the adherend is essential for nearly all bonding applications. This can be accomplished using one or other of the following procedures. If plastic adherends are being used then careful consideration of both solvent and temperature must be made.

(1) Solvent washing: three possible approaches
   (a) If safety considerations permit, brush or wipe the adherend surfaces with a clean brush or cloth soaked in clean acetone. For fine work, washing down with solvent applied by aerosol spray may be a more suitable alternative; this technique also ensures that the solvent used is clean.

Allow to stand for about 5 min to permit complete evaporation from the joint surfaces. Good local extraction will have to be employed, which ensures compliance with the requirements of any local or national environmental regulations.

(b) Immerse the substrate successively in two tanks each containing the same solvent. The first tank acts as a wash, the second as a rinse. Currently, trichlorotrifluoroethane is often used, but in view of the pending legislation the use of acetone, in spite of the associated flammability problems, should be considered. When the solvent in the wash tank becomes heavily contaminated, the tank should be emptied, cleaned out and refilled with fresh solvent. This tank is then used for the rinse and the former rinse tank for the wash.

Environmentally more acceptable alternatives to these solvents are under development and include materials based on alcohols, terpenes and water.

(c) Scrub the adherends in a solution of liquid detergent. Wash with clean hot water and allow to dry thoroughly — preferably in a stream of warm air (ca. 40°C), for example in an air-circulating oven or using a domestic forced-air heater.

(2) Vapour degreasing

The adherend is first washed in a suitable solvent (using one of the methods given above) and then is immersed in the vapour of a suitable degreasing agent in a vapour degreasing unit. Currently trichloroethylene is favoured for vapour degreasing but, again, legislation might enforce changes.

(3) Alkaline degreasing

For metallic substrates, and particularly aluminium, the vapour degreasing process can be augmented, or, in certain cases, replaced by immersion of the substrate in a warm, aqueous solution of a suitable alkaline degreasing agent (for example, for aluminium, a 10-min immersion in an aqueous solution of Turco® T 5215 NC-LT at 70°C) followed by a spray-rinse in clean water. If no further treatment is contemplated then the adherend should be dried thoroughly — preferably in a stream of warm air at a maximum temperature of 40°C.

4.2.1.1. Test for a Clean Bonding Surface

Irrespective of the degreasing method used, the water-break test is a simple method to determine whether the surface to be bonded is clean; it is best suited to metals. The surface may be assumed to be free of contamination if either a few drops of distilled water, which have been applied to the adherend, wet and spread or if, on drawing the substrate from out of an aqueous medium, the water film does not break up into droplets. Uniform wetting of the surface by water indicates that it will probably be likewise wetted by the adhesive.

It must be borne in mind that certain plastics, even when clean, may not be wetted by water but will be wetted by the adhesive. Furthermore, satisfactory wetting gives no information as to the potential bond strength. At best, it is a necessary, but not necessarily sufficient, requirement for the achievement of high bond strengths.

4.2.2. Abrasion Methods

For many substrates light abrasion of the surfaces to be bonded can allow the adhesive to key better than when a highly polished adherend is used. Extremely active surfaces, such as those produced immediately following abrasion, particularly where mild steel components are concerned, tend to have a better affinity for the adhesive. This technique, however, is not recommended for many aluminium adherends.

As well as producing an active surface, abrasion pretreatments are intended:

- To remove surface deposits, such as oxide tarnish, rust or mill scale, on metallic substrates; particularly those which are ferrous-based.
- To remove the immediate surface layer of plastics or reinforced plastics to ensure elimination of all traces of release agent, etc. However, if due care is not taken, the release agent can be embedded into the surface which is being pretreated, instead of removing it; this is particularly true if a grit-blasting process is used.

Before the abrasion process starts, irrespective of the method used, the substrate should undergo at least one of the cleaning processes highlighted earlier.

If the substrate to be pretreated (either metallic or plastic) is delicate or if other suitable, more controlled equipment is not available, the surfaces to be bonded can be pretreated using a suitable abrasive cloth (e.g. Scotchbrite®), a hand- or power-operated wire brush or water-proof abrasive paper. In this latter case, the abrasive particles bonded to the paper should, ideally, have a particle-size range of 125–315 μm.

When a more vigorous abrasion of the surface is required, or the component is robust enough to withstand such a technique, the use of air- or water-borne grit-blasting is, generally, the best method of achieving these ends. Here the use of dry, clean compressed air and/or water is essential as is the prevention of contaminated abrading media coming into contact with the surface to be pretreated.

The choice of grit type (fused alumina, silica, quartz, chill-cast iron shot or silicon carbide) will be dependent on the substrate to be abraded. For example, alumina grit would not be used on steel components because of the possibility of galvanic corrosion; chill-cast iron shot would be used.

Selection of grit size will also depend on several factors: the metal to be pretreated, the type of equipment being used, the pressure and angle of blast impact and the blasting time. Grits in the range of 125–315 μm are suitable, but the optimum size for the work in hand can only really be determined by experiment.

When an air-borne abrasion medium is used to pretreat thermoplastic materials, pretreatment times should be kept to a minimum to avoid surface melting.

Water-borne grit-blasting generally gives more control of the depth and severity of the abrasion process and is particularly recommended for plastic and reinforced plastic materials.

Operating under aqueous conditions can also assist in the removal of residual abrading medium and any other contaminants and keeps dust generation to a minimum. If wet techniques are used, then the substrate should be thoroughly dried immediately after pretreatment; this is particularly true for ferrous materials. In this latter case, another approach is to add a rust inhibitor to the water-grit medium.

Any abrasion pretreatment carried out must be followed by a further operation to ensure complete removal of loose and loosely-bound particles to ensure that the substrate is completely free from loose and loosely-bound particles. This can be accomplished by using a soft brush to clean the surface or, better, it can be blown clean using an uncontaminated, dry, filtered, compressed-air blast.

Finally, the abraded substrate should be degreased.

4.3. Pretreatment of Specific Substrates

The surface pretreatments described earlier, i.e. degreasing alone or degreasing followed by abrasion and removal of the loose particles are sufficient to ensure that on some substrates, particularly ferrous adherends and for fibre-reinforced "plastic" laminates, good strong bonds will be formed with the adhesive being used. For many adherends, however, to obtain reproducible bonds having maximum strength and long-term durability, a chemical pretreatment will be required to modify the topography and/or the chemistry of the surface, in such a way as to make it suitable for structural adhesive bonding [3].

For metallic adherends most of these pretreatments either involve acid etching or an acid etch followed by an acidic anodising process. However, individual alloys and the particular surface structures caused by different heat treatments may respond differently to a given pretreatment; for example, aluminium clad aluminium alloys (Alclad®) as opposed to the "bare", unclad alloys.

The effectiveness of one pretreatment over another can be shown only by comparative trials; using both the type of metal and the adhesive specified for the work.

In virtually all cases where chemical pretreatment has been used, the water-break test can be used to confirm the effectiveness of the process.

Surface modification of plastic materials is, nowadays, frequently carried out by exposing the surface to be bonded to a controlled flame, plasma or corona discharge.

4.3.1. Aluminium Alloys

This is, currently, by far the most important substrate encountered in the aerospace industry. It is used in sheet form as well as the basis for lightweight honeycomb.

4.3.1.1. Aluminium Honeycomb

Unless there are obvious signs of contamination, aluminium honeycomb does not require pretreatment prior to bonding.

If, however, any oil or grease contamination be evident, then the affected slice should be immersed in the vapour of a suitable hydrocarbon solvent in a vapour degreasing unit. After immersion, sufficient time should always be allowed for the honeycomb core to drain dry. This is particularly important as liquid solvent held in the corners of the honeycomb cell can be very difficult to detect and must be removed before bonding.

Although honeycomb is not pretreated per se, it is now a standard procedure to pretreat the aluminium foil used in honeycomb manufacture prior to application of the node-bond adhesive (see Fig. 36 in Chapter 5: "Aerospace: A Pioneer in Structural Adhesive Bonding"). This not only gives a high energy surface but also, more importantly and the real driving force behind the introduction of such processes, gives a final structure having a significant degree of environmental resistance [4]. Four examples are given below; Davis [4] gives a comprehensive list.

4.3.1.1.1. Iridite® process

The initial treatment used in Europe was a chromium conversion coating process using a proprietary product called Iridite. This was a two-component system which contained a blend of highly toxic materials such as chromium salts and ferro- and ferricyanides.

The Iridite process used a standard cleaning operation (a hot alkaline etch/degrease) followed by a spray-rinse and then immersion in the mixed Iridite

14-5L solution at ambient temperature. The conversion coated foil was subsequently washed and dried (about 1 min at 70°C) to "set" the coating. The maximum chromium coating thickness was about 1.27 μm.

4.3.1.1.2. Alocrom® process

For various reasons, not least the toxicity of the components of the treatment baths, the use of Iridite was abandoned in Europe, in favour of the Alocrom® process. However, a modified version of Iridite does still exist and appears to be used extensively in the United States of America.

Two relevant Alocrom products are currently available. The first is Alocrom 1200 U which is a three-component system where the liquid and the toner components compromise a mixture of chromic acid (Cr_2O_3) and hydrofluoric acid (HF) and the powder is potassium ferricyanide [$K_3Fe(CN)_6$]; on the face of it, no safer than the Iridite it replaced!

The other system, and that which is currently favoured, is the single component Alocrom 1290. This is an aqueous solution containing about 2% of HF, 18% of fluoroboric acid (HBF_4) and 20% of sodium dichromate ($Na_2Cr_2O_7$); again, highly toxic.

The Alocrom chemically reacts with the aluminium foil so that a relatively thick layer of aluminium oxide is grown on the surface; relatively thick when compared with the oxide film produced by natural exposure to the air. At the same time Cr^{3+} and, to a lesser extent, Cr^{6+} ions are incorporated into this film. In the presence of water or electrolytes, it is the Cr^{6+} ions which migrate to the area under attack and preferentially oxidise preventing oxidation of the aluminium and hence suppressing any potential for corrosion.

The process can be summarised as follows [5]:

- The aluminium foil, which is coated in rolling oils, is first degreased/etched using an aqueous alkaline solution of Ridoline® 72 (sodium hydroxide: NaOH) or, better, Ridoline 305 (potassium hydroxide: KOH). The use of potassium hydroxide allows a lower bath temperature to be used — generally about 55°C.

  Once the surfactants in the Ridoline have enabled the surface oils to be removed, the alkali first removes the weakly bound, air-formed oxide film and then reacts with the aluminium to form a mixed oxide with the liberation of hydrogen thus:

$$Al_2O_3 + 2KOH \rightarrow 2KAlO_2 + H_2O$$

$$2Al + 2KOH + 2H_2O \rightarrow 2KAlO_2 + 3H_2\uparrow$$

- The etched foil is washed in clean cold water and is immediately run through the Alocrom tank.

- The Alocrom solution is maintained at about 35°C. The initial reaction is between the hydrofluoric acid and any oxide film formed, post-alkaline etch. This removes the oxide layer:

$$2Al + 3H_2O \rightarrow Al_2O_3 + 6H^+ + 6e$$

$$Al_2O_3 + HF \rightarrow 2AlF_3 + H_2O$$

- The hydrogen ions liberated reduce Cr^{6+} to Cr^{3+} with the result that a new complex inorganic layer (basic chromium chromate in conjunction with alumina; hence "Alocrom") is precipitated, as a loosely attached gel, onto the surface of the aluminium foil, thus:

$$6H^+ + 2HCrO_4^- + 2Al \rightarrow 2Al^{3+} + {}^{2-}Cr_2O_3{\cdot}3H_2O + H_2O$$

  The initial "coating" comprises numerous distinct "seeds" of the gelatinous Alocrom precipitate on the aluminium surface; second phase particles at the grain boundaries probably act as nucleation sites. The individual seeds then grow laterally until they impinge on their nearest neighbours. Further growth increases the thickness of the new coating.
- The foil is then washed in cold water to remove excess chromic acid and is then dried, initially at about 110°C, to cause crystallisation of the Alocrom surface layer to take place. This leads to a hard, corrosion-resistant coating on the surface of the aluminium foil. To give adequate corrosion protection, whilst still allowing an effective bond to be formed at the nodes, the film coating weights and thicknesses are tightly controlled. Although this will be dependent on the aluminium alloy, they are generally in the region of 0.11–0.38 g/m^2 and 0.5–1.0 μm, respectively.
- It is usual to apply the node-bond adhesive to the foil immediately after drying has taken place.

A typical pretreatment and printing line is shown schematically in Fig. 1.

4.3.1.1.3. CR III process

This is a proprietary pretreatment developed by Hexcel Composites and, as such, is an alternative coating process to the traditional Iridite or Alocrom methods. It has been used in both Europe and the United States of America for several years. In the CR III process, the chromic acid etched foil is further treated to give a hard, fully hydrolysed, polymeric organic–metallic coating. This provides a surface ready for bonding with no further pretreatment necessary.

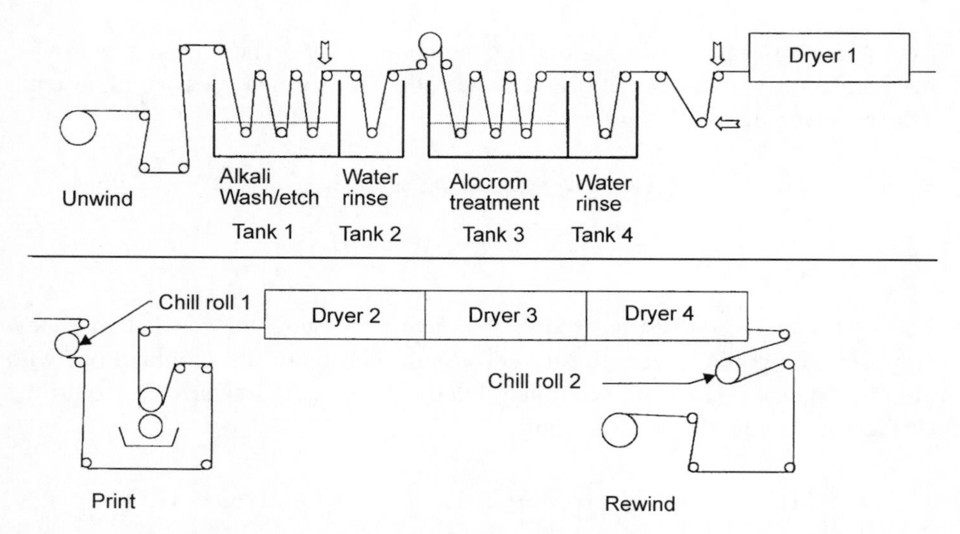

Figure 1: Schematic of an Alocrom pretreatment and adhesive printing plant for honeycomb.

4.3.1.1.4. Phosphoric acid anodising

In this relatively new process, phosphoric acid anodising techniques are used to produce a foil having a high-energy "oxide" surface. Such a surface ensures an enhanced bond with the node-bond adhesive as well as producing an oxide film, which contains bound phosphates that will give improved durability. Generally the anodising process is carried out under alternating rather than the conventional direct current conditions.

One example is the CR-PAA range from Hexcel Composites; a range designed for aerospace structures that are exposed to demanding environmental conditions. CR-PAA core outperforms standard MIL-C-7438 core in salt spray and crack propagation tests.

Here, a single, continuous anodising process allows a thin, uniform oxide film to be grown on both sides of the aluminium foil travelling through the entire process line; the anodiser is proprietary engineered equipment. Whilst this particular technology facilitates economic line speeds, it also minimizes burns, blemishes, and pinholes that can occur when using older methods of anodising. After the substrate passes through the anodiser, it is rinsed with high-purity water, dried, and then passes immediately into the primer applicator; this seals the pretreated surface to protect the oxide film during the remainder of the process.

4.3.1.2. Aluminium Sheet

Techniques range from simple mechanical abrasion to a variety of chemical and electrochemical pretreatments, of which most are in general use. Currently undergoing evaluation trials are techniques using exposure to plasma under atmospheric pressure. All the methodologies encountered can be summarised as follows:

- "Light" Abrasion: wire wool or Scotchbrite
- "Heavy" Abrasion: grit blast with alumina or silica particles
- Chemical: chromic/sulphuric acid pickle
- Chemical: P2 etch
- Electrochemical: chromic acid anodising
- Electrochemical: phosphoric acid anodising
- Electrochemical: sulphuric acid anodising
- Electrochemical: hard anodising
- Electrochemical: boric acid/sulphuric acid anodising
- Chemical: sol–gel procedures
- Activated plasma

4.3.1.2.1. Light abrasion

This approach uses a vapour or solvent degreasing operation in accordance with one or other of several European and American specifications [6–12] followed by an abrasion of the aluminium surface using wire wool or Scotchbrite. This is followed by a second degreasing step. As the aluminium is relatively ductile, the abrasion process tends to cut easily into the surface. Although this produces a macro-rough surface, loosely attached aluminium detritus from the abrasion process will provide points of weakness in the bonded joint. Further, as Fig. 2

Figure 2: SEM micrograph of Scotchbrite particle trapped on an abraded Alclad 2024-T3 surface.

shows, small pieces of abrading material can attach themselves to the pretreated surface; giving good centres for environmental attack [13].

4.3.1.2.2. Heavy abrasion

This technique uses a vapour or solvent degreasing operation in accordance with one of the specifications already indicated. This is followed by an abrasive blasting of the aluminium surface in accordance with one or other of several European and American specifications [14–16]. Many of the available techniques have already been mentioned and include: a "vacu-blast" process, grit blasting utilising pressurised air and, possibly, wet grit-blasting using a combination of water and compressed air. The abrasion operation is followed by a "dusting" or drying procedure using clean air and then by a second degreasing step.

A technique, developed by Professor George Thompson at the Corrosion Protection Centre of the University of Manchester Institute of Science and Technology (UMIST), has been used to examine the surfaces generated by various methods of pretreatment. The specimen to be examined is mounted on a stub and is trimmed using an ultramicrotome set up with a glass knife. Very thin slices (25–200 Å thick) are then cut through the interphasial region of the joint, this time using a diamond-bladed knife in the ultramicrotome. The sections so exposed can then be examined by means of transmission electron microscopy (TEM) [13,17] to show the transition from the underlying metallic substrate, through its boundary layer to the modified surface produced by the pretreatment, through the interphasial region between substrate and adhesive and finally into the bulk of the adhesive itself.

These techniques have been used to evaluate the effect of grit-blasting aluminium substrates. It can be seen that such a pretreatment tends to do considerable damage to the aluminium, giving it a significantly rough surface with loosely bound aluminium particles acting as potential stress raisers and occasionally even promotes some sub-surface stress cracking in the bulk of the substrate. Fig. 3 shows the interface between the grit-blast aluminium and the adhesive; loosely bound detritus is clearly evident. Fig. 4 shows the induced sub-surface stress cracks in the bulk of the aluminium alloy.

Apart from the obvious difficulties with the degree of roughness seen with both forms of abrasion, there is a further complication when film adhesives are being used.

Film adhesives, as their name implies, are solid at ambient temperature. Once the bonded joint is closed, heat is applied to affect cure and pressure is employed to aid the wetting of the pretreated substrate by the adhesive (for further information see Section 5.3.2.3 in Chapter 5: "Aerospace: A Pioneer in Structural Adhesive Bonding"). As the temperature rises, and before gelation begins, the adhesive matrix melts and begins to flow across the adherend to ensure that intimate wetting

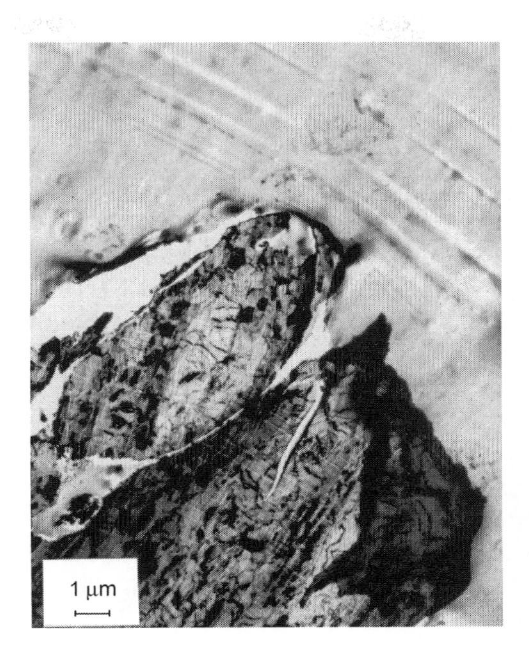

Figure 3: Surface detritus: ultramicrotomed/TEM micrograph through an alumina grit-blasted Alclad 2024-T3 substrate.

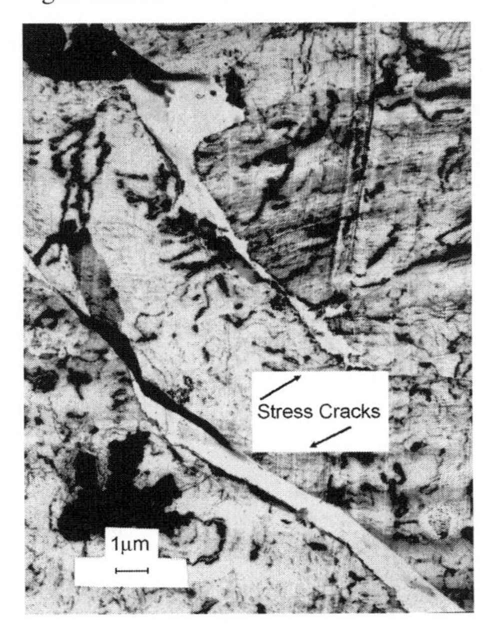

Figure 4: Sub-surface cracking: ultramicrotomed/TEM micrograph through an alumina grit-blasted Alclad 2024-T3 substrate.

occurs in order to maximise the forces of adhesion. As the adhesive flows it displaces the air, which is inevitably trapped between film and substrate. Should the substrate be too rough then air can be trapped under the film as gelation sets in which leads to the so-called "voidy glueline" effect. This is evidenced, on breaking the bonded joint, by the appearance air bubbles in the fracture surface of the cured matrix.

Due to the relatively high ductility of aluminium, neither of the abrasion techniques, therefore, can be readily recommended as a means of surface pretreatment when bonding structural, load-bearing components, particularly should they be subjected to thermal or fatigue loading or exposed to harsh environments. However, there is one abrasion method, which is gaining acceptance. This uses a combination of grit blasting and silane-coating.

4.3.1.2.3. Grit blast and silane treatment

This procedure was developed by the Aeronautic Maritime Research Laboratory (AMRL) in conjunction with the Royal Australian Air Force [18].

Substrates are first degreased using a solvent wash. The surfaces to be bonded are then hand-abraded, cleaned with fresh solvent and then grit blasted using alumina grit with a nominal 50 μm particle size [significantly smaller than that specified for the conventional grit-blasting process: 125–315 μm (a range of 75–420 μm is sometimes used)]. The abraded surface is then swept clean with dry N_2 and the panel is flooded with a pre-hydrolysed silane solution. The most suitable has been found to be a 1–3% solution of γ-glycidoxypropyl trimethoxy silane (Silquest® A-187) in deionised water where the pH has been maintained at about 5 during hydrolysation.

$$H_3C{-}O{-}Si(OCH_3)_2{-}CH_2{-}CH_2{-}CH_2{-}O{-}CH_2{-}\underbrace{CH{-}CH_2}_{O}$$

The area to be bonded is kept "wet" with silane for 10 min and then any excess is removed using a stream of dry nitrogen at about a 210 kPa line pressure. The coated substrate is then baked for 1 h at 104°C and then primed, if so required, or the adhesive is applied.

However, although it is gaining in popularity, this method of aluminium substrate pretreatment is essentially still only used for repair.

4.3.1.2.4. Chemical pretreatments

The oxides associated with aluminium have high surface energies but are relatively unstable; they vary in thickness from a few nanometres to up to 10–20 μ and all have a certain degree of micro-roughness.

For primary bonding in the aerospace industry, a vapour and/or alkaline degrease followed by an acid etch (pickling) and/or by a suitable anodising process is the preferred approach. A controlled film of active, aluminium oxide, highly suitable for structural bonding, is grown on the surface of the aluminium; its thickness being dependent on the chemical process and the alloy used. It should be noted that trace elements and salts can be critical in the efficacy of each of the following processes. Thus aluminium, copper, iron chloride, fluoride and sulphate concentrations are as rigidly controlled as the free acid contents of the relevant baths. For further details, reference should be made to the specifications quoted in the text.

Chromic–sulphuric acid pickling (CSA)

The first stage of the process is a degreasing step, to any of the above specifications, which is possibly augmented with an alkaline degreasing step in accordance with a relevant specification [19–21].

Pickling is carried out in accordance with one or other of several European and American specifications [7,22,23]. It should be noted that the DEF STAN documentation replaces the original DTD 915B (ii) specification of 1956 that was used throughout Europe until recently. The BAC (Boeing) specification and the original Forest Laboratory's Process (FLP etch), used in the United States of America, are similar.

A suitable pickling solution of sodium dichromate in sulphuric acid can be made up as follows:

Water	1.500 l
Concentrated sulphuric acid (Sg: 1.83)	0.750 l
Sodium dichromate ($Na_2Cr_2O_7{\cdot}2H_2O$)	0.375 kg
[or chromium trioxide (CrO_3)]	0.250 kg
Water	make up to 5.0 l

The pickling bath is regulated at 60–65°C and the substrate to be pretreated is then immersed for 30 min. At the end of this time it is removed and immersed in a tank of water at ambient temperature. Any acid remaining on the surface can then be removed using a spray-rinse with cold water. The pretreated components can then be air-dried, preferably in an air-circulating oven whose air temperature is no greater than 45°C. Bonding should take place within 2–8 h of pretreatment.

Figure 5: Ultramicrotomed/TEM micrograph through a CSA pretreated Alclad 2024-T3 substrate.

Following such a pretreatment, a controlled, high-energy film of active, aluminium "oxide/hydroxide", eminently suitable for structural bonding, will have been grown on the surface of the aluminium. The oxide film can be seen in Fig. 5 at the interface between the alloy and the adhesive. It is a needle-like structure growing from the underlying aluminium substrate and has a thickness of about 300 Å. It is clear that the adhesive completely wets and penetrates this surface.

There are many chromic–sulphuric acid pickling specifications which the industry uses and many people treat them all as roughly equivalent. Table 1 shows just how misleading this supposition can be. It is important, therefore, to bear these differences in mind when comparing the bonding properties of various structural adhesives.

Chromic-free pretreatment: P2 etch

An important chromate-free acid etching technique is also available, it is designated the P2 [24] process. Here, as ferrous sulphate replaces the dichromate salts used in the more conventional chromic–sulphuric acid pickling, it is seen as being more friendly to the environment. Although not currently specified as the principle etching process, specifications for pretreatments such as phosphoric acid

Table 1: Comparison of various chromic–sulphuric acid pickling processes.

Process	**$Na_2Cr_2O_7 \cdot 2H_2O$ (%)**	**H_2SO_4 (%)**	**H_2O (%)**	**Temperature (°C)**	**Time (min)**
FPL	2.5	24.3	73.2	68	15–30
DEF STAN 03-2	6.4	23.3	70.3	60–65	30
DIN 53 281	7.5	27.5	65.0	60	30
Alcoa 3	3.2	16.5	80.3	82	5
Optimised FPL	5.0	26.7	68.3	65	10
FPL–RT	6.4	23.4	70.2	22	240

anodising are beginning to offer this as an alternative "deoxidising" treatment prior to anodising.

As for the CSA procedure, the first stage of the process is a conventional solvent degreasing step which can possibly be augmented with an alkaline degreasing step.

A suitable P2 etching solution can be made up as follows:

Water	48% by weight
Concentrated sulphuric acid (Sg: 1.83)	37% by weight
Ferrous sulphate ($FeSO_4$)	15% by weight

The etching bath is regulated at 60–70°C and the substrate to be pretreated is then immersed for between 8 and 15 min. At the end of this time it is removed and immersed in a tank of water at ambient temperature. Any acid remaining on the surface can then be removed using a spray-rinse with cold water. The pretreated components are air-dried, preferably in an air-circulating oven whose air temperature is no greater than 45°C.

Davis and Venables [25] state that it is also possible to run the bath at ambient temperature and produce a suitably pretreated aluminium surface; compare the micrographs of conventional FPL etched aluminium and P2 etched aluminium in Fig. 6.

Researchers report that, not only are the surfaces so produced equivalent to those following a chromic–sulphuric acid pickle and only slightly inferior to PAA pretreated surfaces [26] but they permit the development of strong adhesive bonds [25].

4.3.1.2.5. Electrochemical pretreatments

Chromic acid anodising (CAA)

A thicker, more "robust" oxide film can be grown if the adherend is anodised in chromic acid. The first step is, usually, to produce pickled aluminium substrates

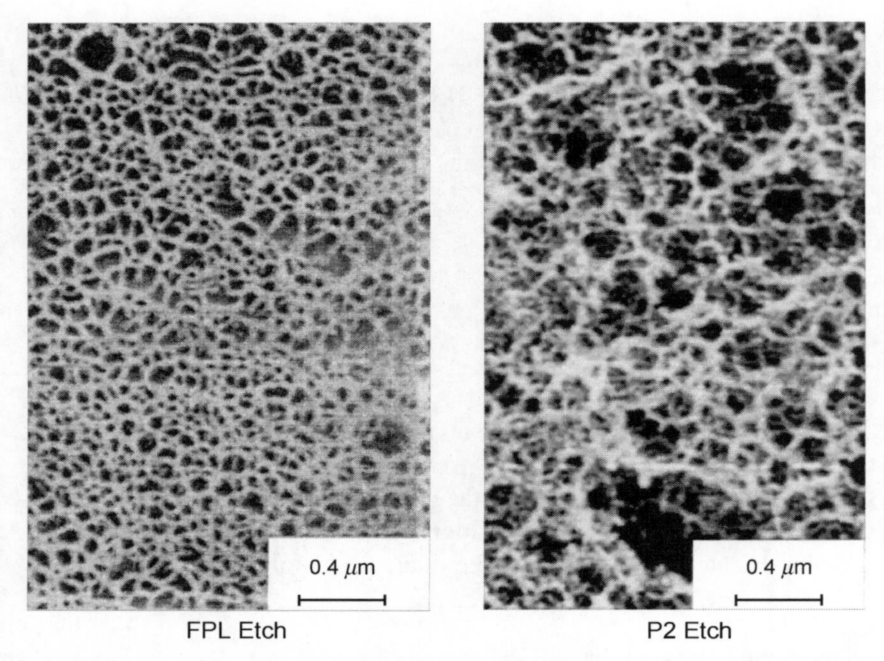

Figure 6: SEM micrographs of FPL etched and P2 etched aluminium alloy [25] (Courtesy: Elsevier Limited).

in accordance with the specifications quoted above. Under certain circumstances it may be sufficient only to degrease in an alkaline medium before anodising. The etched panels are clamped to the anode of a standard anodising bath and are immersed in a solution of chromic acid at 40°C. Anodising then takes place in accordance with one or other of several European specifications [27,28].
A typical bath composition is:

Chromium trioxide (CrO_3)	0.500 kg
Water	10.0 l

It should be noted that the specification given in Ref. [27] is a replacement for the well-known Defence Specification DEF 151 as well as for prEN 3002 (Issue P1, 1990) and BS EN 2101 (Issue 1, 1991).

Typically, the anodising voltage is raised, over a 10-min period, to 40 V, held for 20 min, raised over a 5-min period, to 50 V and held for 5 min. At the end of this cycle the components are removed and immersed in a tank of water at ambient temperature. This is followed by a spray-rinse with cold water. The anodised

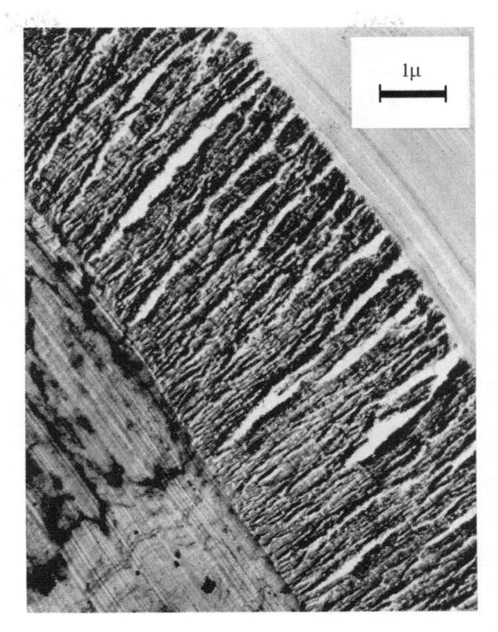

Figure 7: Section through the joint: ultramicrotomed/TEM micrograph through CAA pretreated Alclad 2024-T3 substrates.

components can then be air-dried, preferably in an air-circulating oven whose air temperature is no greater than 45°C. Bonding of the *unsealed* components should take place within 4–6 h.

Again, a controlled, high energy film of active, aluminium oxide/hydroxide will have been grown on the surface of the aluminium. The final film thickness will be about 2.5–4.5 μm. This can be seen in Fig. 7. Fig. 8 shows

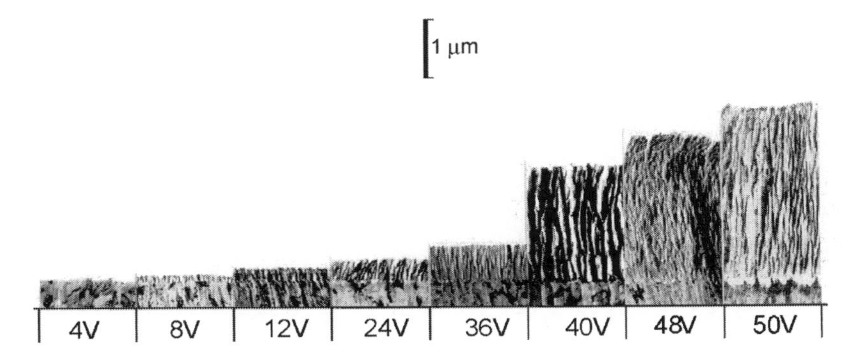

Figure 8: Oxide growth with increasing time and voltage: ultramicrotomed/TEM micrographs through CAA pretreated Alclad 2024-T3 substrates.

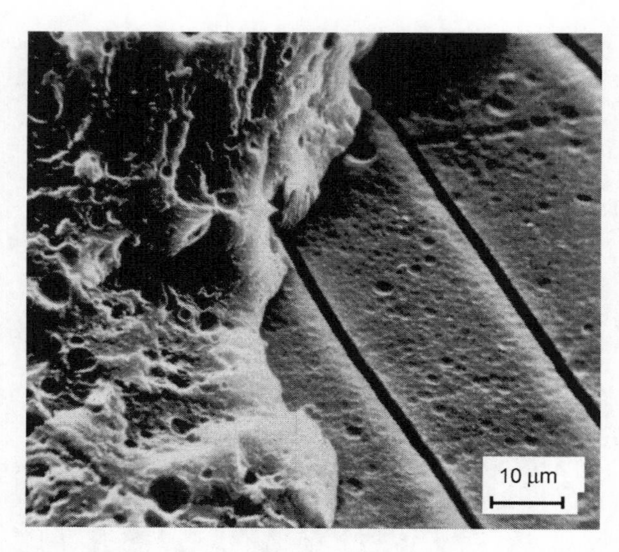

Figure 9: SEM micrograph of a CAA pretreated Alclad 2024-T3 substrate.

the film growth associated with different stages of the anodising cycle; effectively a "plot" of film thickness against time.

Fig. 7 clearly shows that a robust "oxide" film is grown by this electrochemical process. It can also be seen that whilst the surface is well wetted by the adhesive (top right-hand corner of the micrograph), the depth of the oxide film is not penetrated.

Further, the fact that the oxide film is significantly thick means that it can fracture fairly easily. This can be seen in Fig. 9 where cracks through the oxide film can be clearly seen following a standard peel test. The substrate being viewed is the thick adherend which only bends slightly during the test.

Anodising in chromic acid is the electrochemical pretreatment procedure favoured in Europe. European aerospace companies view this as giving a less friable oxide film than the phosphoric acid anodising process favoured in the United States and believe that it imparts much better durability to the bonded joint. This latter statement is somewhat debatable in the light of both experience in the laboratory and the actual chemical structure of the oxide films themselves.

Phosphoric acid anodising (PAA)

To obtain a more open oxide film, i.e. one which is more readily penetrated by the adhesive being used, aluminium adherends can be anodised in phosphoric acid. Certainly, the resultant oxide films are more friable than their CAA counterparts but, in this case, they contain "bound" phosphate ions which will impart some degree of durability to the final adhesive joint; contrary to popular supposition the films produced in chromic oxide do not contain bound chromates in their structure [17].

In this procedure the first step, once again, is to pickle the aluminium substrates in accordance with the specifications quoted above. Under certain circumstances it may be sufficient to use only an alkaline degrease before anodising. The etched panels are then clamped to the anode of a standard anodising bath and are immersed in a solution of phosphoric acid at 25°C. Anodising then takes place in accordance with one or other of several European and American specifications [29,30]. A typical bath composition is:

"Syrupy" orthophosphoric acid (H_3PO_4) (Sg: 1.65)	1.0 l
Water (concentration of phosphoric acid is 75 g/l)	16.6 l

Typically, the anodising voltage is set at 10–15 V (preferably 15 V) and is held for 20–25 min. At the end of this time the adherends are removed and immersed in a bath of water at ambient temperature. This is followed by a spray-rinse with cold water. The anodised adherends can then be air-dried, preferably in an air-circulating oven where the air temperature is no greater than 45°C. Bonding of the unsealed components should take place within 2–4 h.

Following such a pretreatment, a controlled, high-energy film of active, aluminium oxide/hydroxide will have been grown on the surface of the aluminium. During the anodising process, part of this film will have dissolved in the electrolyte. This yields, as already stated, a much more open film with a somewhat "ferny" appearance. Film thicknesses will be in the region of 0.5–0.8 μm; significantly thinner than those produced by chromic acid anodising. This can be seen in Fig. 10.

Fig. 11 examines another PAA pretreated bonded specimen. Here, an ultramicrotomed slice has been taken parallel to, and just above, the boundary layer of the aluminium substrate, i.e. at the bottom of the oxide pore. It is evident that adhesive has, indeed, penetrated to the bottom of the pore; the light-greyish adhesive can be seen in the interstices between the darker coloured oxide pores.

Even without this corroborating evidence, Fig. 9 clearly shows the ferny oxide structure is well wetted and penetrated by the adhesive.

Sulphuric acid anodising (SAA)

Sulphuric acid anodising techniques are more generally used for the production of decorative aluminium sheets to improve the durability of the aluminium (cf. chromic acid anodising when 20 μm, or greater, oxide films are grown and then sealed in hot water/steam) and for the preparation of surfaces prior to priming

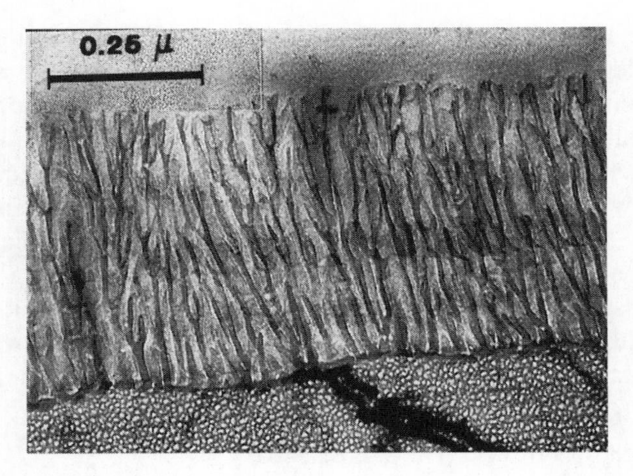

Figure 10: Section through the joint: ultramicrotomed/TEM micrographs through PAA pretreated Alclad 2024-T3 substrates.

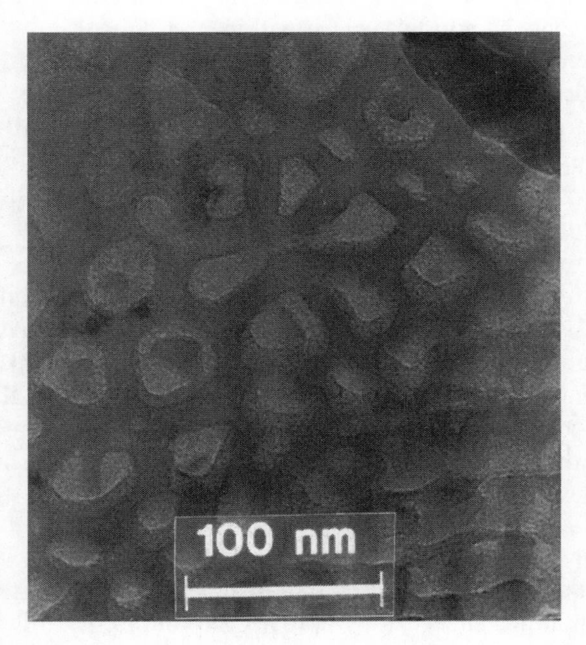

Figure 11: Section parallel to the boundary layer: ultramicrotomed/TEM micrographs through PAA pretreated Alclad 2024-T3 substrates.

and *painting*. However, this technique can also be used to pretreat aluminium and its alloys prior to structural adhesive bonding but, when compared with the performance obtained with bonded structures utilising CSA and/or CAA or PAA pretreatments, significantly lower adhesive strengths and durabilities generally result.

This process produces one of the thickest, most "robust" and least wetable oxide films. The cleaning stage is carried out in accordance with the specifications quoted above. This is then followed by an intermediate treatment comprising an etch in an alkaline solution plus a desmutting treatment [31]; when treating aluminium for all other end uses, these last two stages are often omitted.

Typical bath compositions are:

Etch:	
Sodium hydroxide ($NaOH$)	0.25–0.50 kg
Sodium heptonate [$CH_3(CH_2)_5CO_2Na$]	0.075–0.1 kg
{or sodium gluconate [$HOCH_2[CH(OH)]_4CO_2Na$]}	
Water	Make up to 10.0 l
Desmutter	
Nitric acid (HNO_3) (Sg = 1.42)	30–50% by volume

The etching operation is carried out for between 15 and 60 sec at 60–65°C; rinsing should take place in clean, cold water. The desmutting operation should take place immediately after the components are rinsed. The process is carried out at ambient temperature for about 60 sec, or until the surface is clean. The components are washed in clean, cold water.

The resultant panels are clamped to the anode of a standard anodising bath and are immersed in a solution of sulphuric acid, at 20°C. Anodising then takes place in accordance with a suitable specification [32].

A typical bath composition is:

Sulphuric acid (H_2SO_4)	90–400 g/l

Bonding applications would use sulphuric acid concentrations towards the lower end of the specification whilst decorative applications would use the higher end.

In this instance, the anodising procedure is controlled by maintaining the current density rather than the voltage. Although the anodising parameters will be highly dependent on the alloy being used, a good rule of thumb is that 0.15 μm of oxide will be grown per hour for every amp/m^2 applied across the anode surface. This means, therefore, that using the current-density limits set by DEF STAN 03-15 [32], 100 A/m^2 will give a film growth of about 15 μm/h and 200 A/m^2 will give 30 μm/h. Anodising times will, therefore, be determined by the final oxide thickness required. It is clear, though, that relatively short anodising times can give rise to relatively dense oxide films; DEF STAN 03-15 quotes thicknesses of 8–13 μm as standard with possibilities of reaching 25 μm.

Anodising bath temperatures are generally maintained in the region of 20–25°C. However, changing the anodising bath temperature has little effect on the final product due to the change in the resistance of the electrolyte with temperature — resistance increases as temperature decreases. Thus the voltage–time parameters have to be changed to ensure the correct current density results.

It is possible to control this process by specifying voltage profiles, time and bath temperature rather than current density but these parameters have to be found by experimentation.

At the end of the anodising cycle the components are removed and immersed in a tank of water at ambient temperature. This is followed by a spray-rinse with cold water. The anodised components can then be air-dried, preferably in an air-circulating oven whose air temperature is no greater than 45°C.

Hard anodising

This is a variant of the standard sulphuric acid anodising procedure where the bath temperature, anodising time and current density parameters are somewhat modified to give the densest oxide film growth of all. The hardness of the oxide film increases and the pore size is significantly reduced.

The cleaning, alkaline etch and desmutting stages are carried out as for conventional sulphuric acid anodising.

The resultant panels are clamped to the anode of a standard anodising bath and are immersed in a solution of sulphuric acid at 0–10°C; temperatures as low as −5°C have been used but these are not really practical. The electrolyte concentration in the anodising bath will be dependent on the alloy being used but this will generally be in the region of 10% sulphuric acid (H_2SO_4). Anodising then takes place in accordance with a suitable specification [33].

The anodising process is again controlled by monitoring the current density. However, in hard anodising this can be up to three- or five-times higher than for the conventional SAA process. In conventional engineering applications, current densities up to 600 A/m^2 are used to give oxide thicknesses in the region of 25–75 μm/h; the optimum thickness is believed to be around 50 μm. For adhesive bonding applications, a current density of 400 A/m^2 has been used.

Anodising times will, again, be highly dependent on the alloy being used and the oxide film thickness required; for structural bonding applications, a time of about 20 min is quoted in the literature. As for conventional sulphuric acid anodising, it is possible to control this process by specifying voltage profiles, time and bath temperature rather than current density, but these parameters have to be found by experimentation.

At the end of the anodising cycle the components are removed and immersed in a tank of water at ambient temperature. This is followed by a spray-rinse with cold water. The anodised components can then be air-dried, preferably in an air-circulating oven whose air temperature is no greater than 45°C.

As already indicated, it is generally accepted that the oxide film produced either by conventional sulphuric acid anodising or by hard anodising in sulphuric acid, does not give the optimum surface structure for structural adhesive bonding, which is why these techniques seem to be rarely used in the aerospace industry.

However, Arrowsmith et al. [34] found that the situation could be improved by dipping the clean, anodised substrate in a solution of phosphoric acid to dissolve away some of the anodic oxide layer in order to reveal a more open structure more amenable to adhesive bonding. This can be seen in Fig. 12.

Once again, it does appear that this technique, often referred to as the SAA/PAD process, is rarely encountered in industry in general and the aerospace business in particular.

Boric acid–sulphuric acid anodising (BSAA)

BSAA, now being called a chromate-free anodising process [35], theoretically uses no chromate salts in the cleaning stages or as the electrolyte. A minor exception to this chromate-free concept does, however, occur as some operators use a tri-acid based deoxidiser which does contain some chromic acid together with sulphuric and hydrofluoric acids.

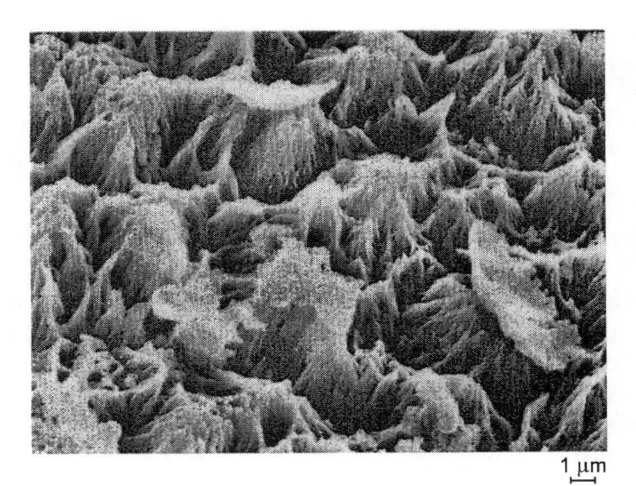

Figure 12: SEM micrograph of an SAA pretreated aluminium substrate, post-dipped in phosphoric acid (Courtesy: IOM Communications Limited).

This procedure has been used for several years to prepare aluminium surfaces for painting. Even in an unsealed state, it was not generally thought that the oxide film produced, which is extremely dense, was suitable for either structural or semi-structural adhesive bonding; although its corrosion resistance was known to be exceptional. It is a modern adaptation of the standard procedure which is now of interest to structural adhesive bonders.

In this modified procedure the first step, once again, is to clean the aluminium substrates in accordance with the specifications quoted earlier. This is followed either by a deoxidising operation in accordance with BAC 5765 [36], a conventional grit-blasting or, better still, an alkaline-etch cleaning stage. The cleaned panels are then clamped to the anode of a standard anodising bath and are immersed in an aqueous solution of boric acid and sulphuric acid, at 25–30°C. Anodising then takes place in accordance with BAC 5632 [37]. A typical bath composition is:

Sulphuric acid (as H_2SO_4)	30.5–52.0 g/l
Boric acid (as H_3BO_3)	5.2–10.7 g/l

To prevent substrate "burning", the anodising voltage is typically raised to 15 V (DC) at a maximum rate of 5 V/min. Once at 15 V, the parts are anodised for 18–20 min. At the end of this time the adherends are removed and immersed in a bath of water at ambient temperature. This is followed by a spray-rinse with cold water. The anodised adherends can then be air-dried, preferably in an air-circulating oven where the air temperature is no greater than 88°C.

Following such a pretreatment, a controlled, dense, high energy film of active aluminium oxide/hydroxide will have been grown on the surface of the aluminium. The film thickness will be in the region of 3–4 μm.

Research by Critchlow and co-workers [35] has shown that these pretreated substrates can be made suitable for structural adhesive bonding if either the anodising temperature is raised to 35°C or the anodised adherends are dipped in dilute phosphoric acid prior to bonding – i.e. similar to the techniques employed by Arrowsmith et al. [34] when using sulphuric acid anodising.

Both procedures appear to open up the pore structure (Fig. 13) and hence make the surface more amenable to wetting by conventional structural adhesives.

Alternating current anodising techniques

For many years, AC anodising of aluminium foil has been a commonplace in the coil coating industry. Since the early 1990s, researchers have evaluated and optimised these techniques to adapt and adopt their use for aluminium foil to be used, for example, in the manufacture of metallic honeycomb (see earlier). Work is now in progress to modify these processes for use with the sheet aluminium

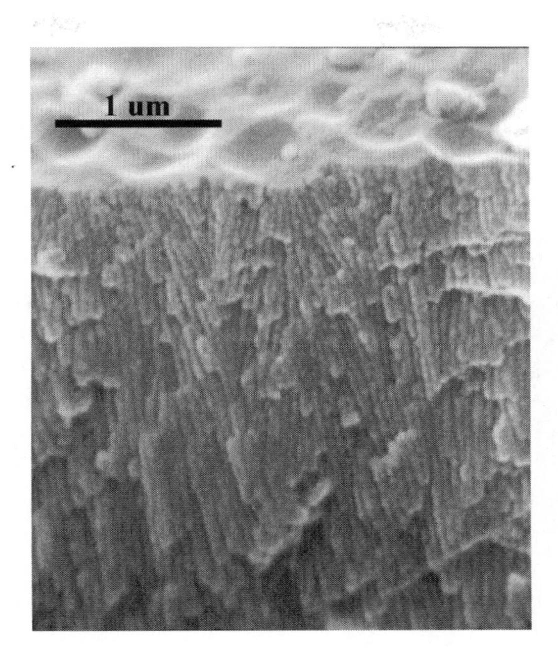

Figure 13: SEM micrograph of a fracture through a BSAA pretreated aluminium substrate (bath temperature: 35°C) (Courtesy: Loughborough University, UK).

substrates typically encountered in the aerospace industry. Although still in the development stage, such an approach could well become important as the need to reduce environmental hazards and increase productivity whilst reducing costs becomes paramount.

Three benefits have been identified. Firstly, all systems so far evaluated do not require the use of hexavalent chromium, which means significantly reduced environmental hazards are associated with these processes. Secondly, the time to grow a suitable thickness of oxide on the substrate surface is reduced by an order of magnitude and thirdly, the hydrogen gas produced on the surface whilst anodising, cleans the substrate prior to oxide growth, removing the need for complex degreasing operations prior to pretreatment.

A typical procedure has been described [38] where either phosphoric acid or sulphuric acid can be used as the electrolyte. Other researchers are evaluating the benefits of various additives in the acidic electrolyte.

Johnsen et al. [38] pretreated aluminium adherends, prior to structural bonding, as follows:

The panels were washed in solvent to remove surface dirt and then immersed in 10% phosphoric acid at 50°C and AC anodised for 30 s using a current density of 4 A/dm^3. Another approach was to wash the panels in solvent to remove surface

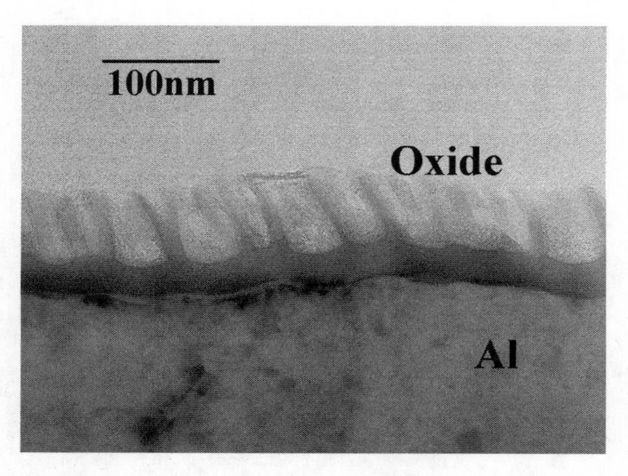

Figure 14: AC anodising in phosphoric acid: ultramicrotomed/TEM micrograph through anodised 6060-T6 alloy substrates [39] (Courtesy: Elsevier Limited).

dirt and then immerse in 15% sulphuric acid at 80°C and AC anodise for 12 s using a current density of 10 A/dm^3.

In both cases the anodised substrates were washed in water and dried at 60°C.

AC anodising of aluminium produces significantly thinner oxide films than by the more conventional DC processes. Johnsen et al. estimated the thicknesses as 0.2 μm for sulphuric acid and 0.1 μm for phosphoric acid.

Some idea as to this significantly reduced thickness can be gained by examining Figs. 14 and 15. Here sections have been taken through 6060-T6 aluminium alloy

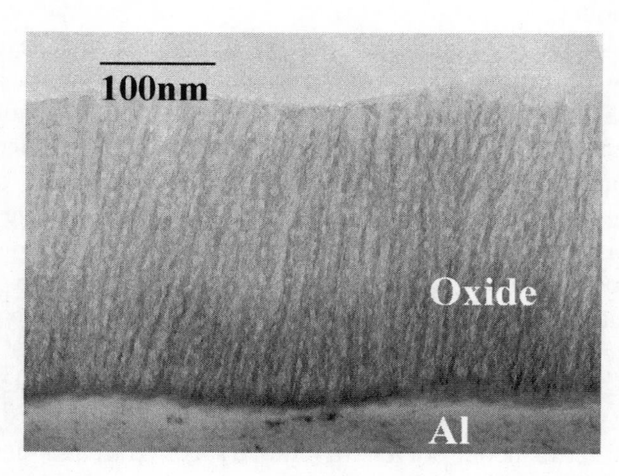

Figure 15: AC anodising in sulphuric acid: ultramicrotomed/TEM micrograph through anodised 6060-T6 alloy substrates [39] (Courtesy: Elsevier Limited).

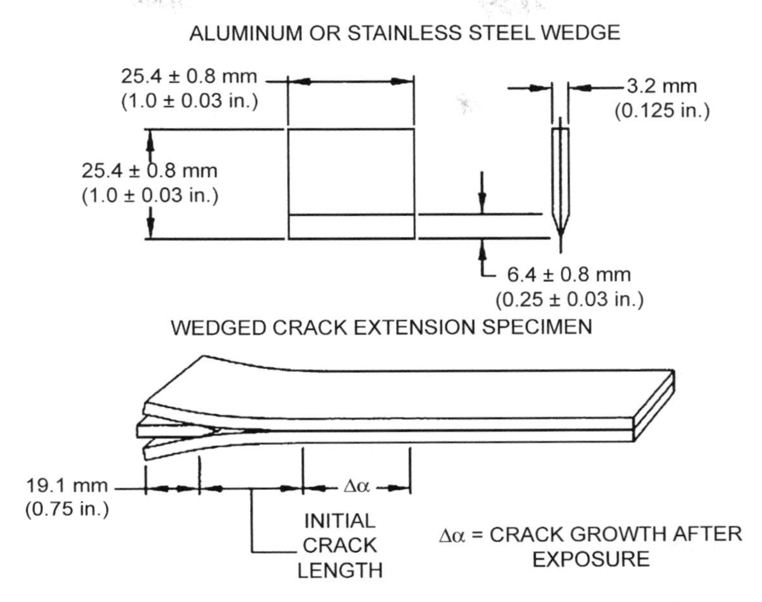

Figure 16: Boeing Wedge Test after ASTM Test Method D3762.

sheets which have been pretreated using an AC anodising procedure in phosphoric and sulphuric acid, respectively [39]; compare the oxide structure produced by DC anodising in phosphoric acid (Fig. 10).

Johnsen et al. found that the durability of the bonded joints, assessed by the Wedge Test [40] (Fig. 16), almost irrespective of electrolyte used, to be nearly as good as obtained using conventionally pretreated adherends, which had been CSA pickled and then DC anodised in phosphoric acid [38]. Bjørgum et al., however, found that an AC PAA pretreatment yielded joints which failed cohesively whereas the AC SAA joints tended towards adhesion failure [39].

4.3.1.2.6. Sol–gel procedures

Work on this novel method of surface pretreatment is still in its relative infancy but initial results are very promising and are currently generating considerable interest.

The sol–gel procedure is different from all other chemical pretreatment processes in that it relies on the formation of chemical bonds at the interface to aid in the production of a strong durable bond, rather than a modification of the surface structure of the substrate [25].

The process uses a combination of hydrolysis and condensation reactions which lead to the formation of purely inorganic or hybrid inorganic–organic polymer networks. It is the hydrolysis reactions, which lead to the formation of the hydroxide groups on the metallic substrate that are absolutely essential to the process as a whole; this hydrolysed oxide surface promotes the condensation reactions.

Alkoxides based, for example, on aluminium, titanium, silicon or zirconium will all hydrolyse in aqueous media but it is the silicon alkoxides which are favoured in Europe due to their slower overall reaction rates. Although silicon based alkoxides will readily hydrolyse, condensation reactions leading to structure growth are relatively latent when compared with zirconium, titanium and aluminium. This gives the end user a significant degree of control.

A good example, which is based on earlier work using alkoxide primers [41,42], is the Boeing sol–gel system (originally designated Boegel [43]). This is a dilute aqueous solution of tetra-n-propoxy zirconium (TPOZ) with a silane coupling agent [44–46]. In this instance, the actual silane is chosen to give optimum compatibility as well as having the ability to form strong bonds and to enhance the final surface durability. When epoxy based adhesives are to be used, the choice is often γ-glycidoxypropyl trimethoxy silane (GTMS or Silquest A-187). Finally acetic acid is added to the solution to control stability and rate of reaction.

This system has now been commercialised by Advanced Chemistry & Technology as AC®-130 [47].

Fig. 17 gives a schematic representation of a sol–gel structure which could be produced using the Boeing approach.

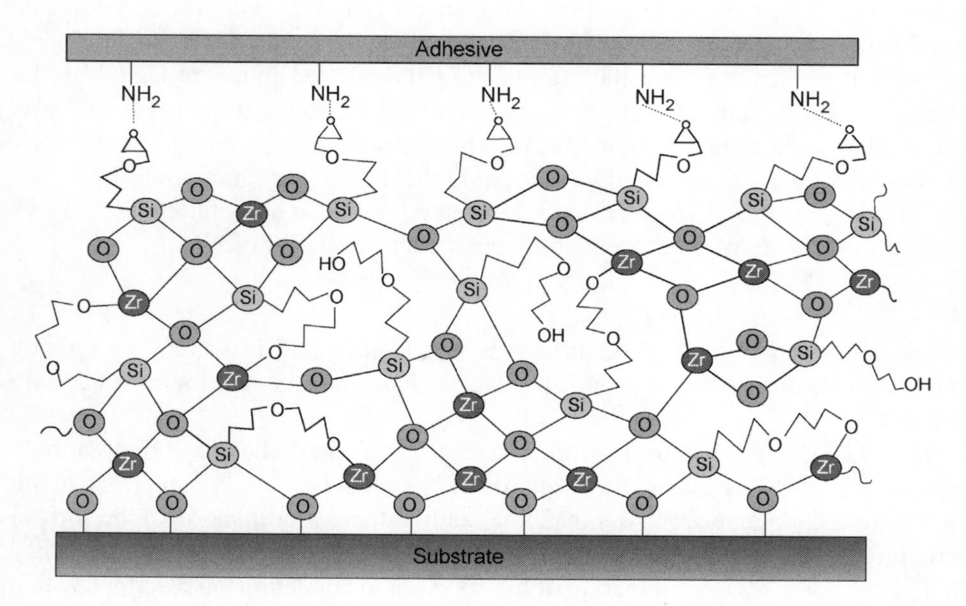

Figure 17: Schematic representation of a typical sol–gel structure [25].

It should be noted, however, that although the ability to form strong bonds is paramount, the need for solution compatibility is not. Dependent on the actual formulation used, some degree of incompatibility is acceptable; this widens the choice of silane. Thus, researchers in Europe have found that other silanes, such as γ-methacryloxy propyl trimethoxy silane (Silquest A-174) have proved to be equally, if not more, efficacious when used with various adhesive chemistries, including epoxy.

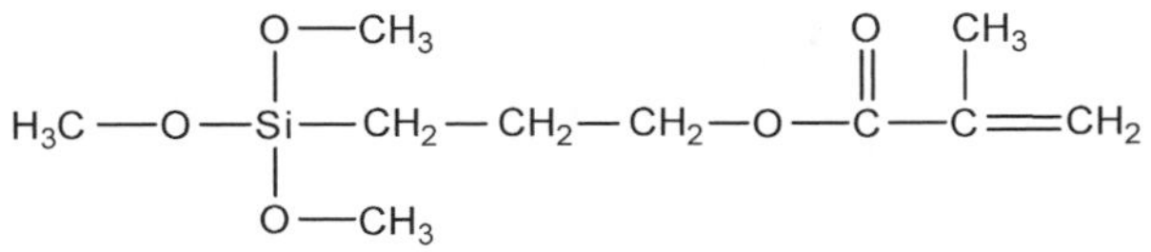

$$H_3C{-}O{-}Si(OCH_3)_2{-}CH_2{-}CH_2{-}CH_2{-}O{-}C({=}O){-}C(CH_3){=}CH_2$$

Further, as well as the use of silicon-based alkoxides to control reaction rates, it is possible to prepare the sol–gel system in alcohol rather than water; water is then added to the system only when reaction (hydrolysis) is required.

Immersion, spraying, curtain-coating or painting procedures can be used to coat the sol–gel system onto the substrates to be bonded; there is no need to wash off the solution after pretreatment.

Films grown by the sol–gel method can be up to 1–2 μm thick; using the Boeing procedure typical thicknesses of 50–200 nm are quoted. Film thicknesses in the region of 50 nm would be ideal for surface pretreatment/priming. Under these conditions, it is believed that some form of gradient exists from the inorganic metallic substrate to the organic primer or adhesive. Thus, in the Boeing case on aluminium substrates, the interface between the aluminium and the coating could well be zirconium rich with Zr–O bonds possibly being formed across this interface [45] and the interface between the coating and the adhesive being organosilane rich.

This is likely at coating thicknesses of about 50 nm as the alkoxy group (at about 10 nm) can easily stand proud of the surface. This inorganic-to-organic gradient is extremely unlikely to occur at higher film thickness as some degree of inhomogeneity would be observable in the coating and no such reports are to be seen in the literature.

The importance of this technique can be seen by the fact that several approaches have already been patented by both Boeing [48] in the United States of America and by TWI [49,50] in the UK.

4.3.1.2.7. Activated plasma

It is certainly possible to modify the surface of aluminium alloys utilising various activated plasma pretreatments. It is reported that excellent joint strengths are possible utilising such an approach. However, due to the overall size and complexity of most aerospace structures, this is not envisaged as a viable method for the foreseeable future.

4.3.1.3. Aluminium–Lithium Alloys

In the early 1980s, alloys of aluminium and lithium (typically BA 8090C) were introduced as possibly replacements for, especially, carbon fibre composites in aircraft structures. They had a significantly lower density than the conventional alloys — 2024 and 7075 — (Sg of 2.54 as opposed to 2.78) which could make a considerable contribution to weight saving. Aluminium–lithium alloys also had an increased stiffness over other aluminium alloys of similar strength. Conversely, they appear to be more ductile than the conventional alloys, as evidenced by the load-extension curves for simple lap shear joints [51].

Bishopp et al. [51] have shown that these alloys can be pretreated, prior to structural bonding, in a similar manner to those methods already outlined for the conventional aluminium alloys. In view, though, of their increased ductility, the recommendation not to use abrasion pretreatments is even more relevant.

However, it should be remembered that during the heat treatment of the alloy, both lithium and lithium intermetallics are known to migrate to the surface. Fig. 18 not only shows the expected PAA oxide film at the interface between metal and adhesive but also what appears to be a nodule on the substrate surface. This can be explained by the fact that surface lithium has been dissolved during acid pretreatment leaving a localised, heavily pitted surface; or what appears to be a nodule.

A further phenomenon, as yet unexplained, can also be seen following the same PAA pretreatment of the alloy. Fig. 19 shows the unexpectedly high distribution of copper-rich intermetallics at the aluminium boundary layer — the black inclusions at the base of the oxide film. Their chemical composition — essentially copper with some lithium — has been confirmed by both EDX (energy dispersive X-ray) and EELS (electron energy loss spectroscopy) analysis [51].

Both these phenomena need to be borne in mind when pretreating aluminium–lithium alloys for structural adhesive bonding.

4.3.1.4. Pre-anodised Aluminium

Decorative anodised aluminium or aluminium alloys are, as such, not suitable for adhesive bonding as they have been sealed; these types of substrate require stripping prior to use. Stripping is sometimes accomplished by abrasive blasting but this sort of treatment is not recommended. The anodic oxide film is best removed by immersion in the chromic–sulphuric acid solution given above [22,23].

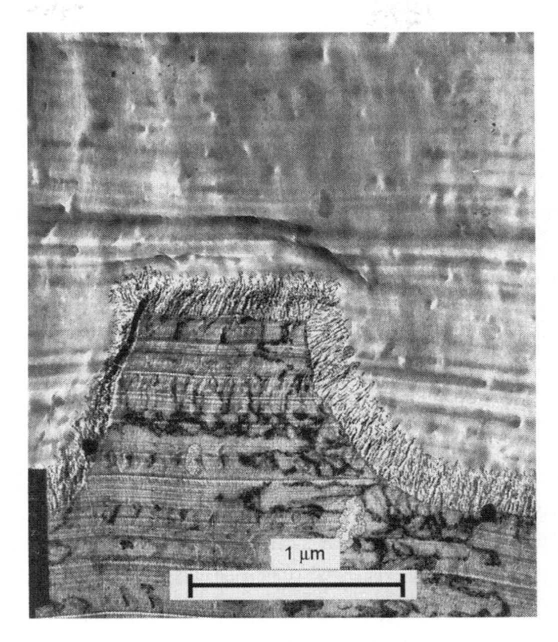

Figure 18: Dissolution of lithium: ultramicrotomed/TEM micrograph through PAA pretreated BA 8090C substrates.

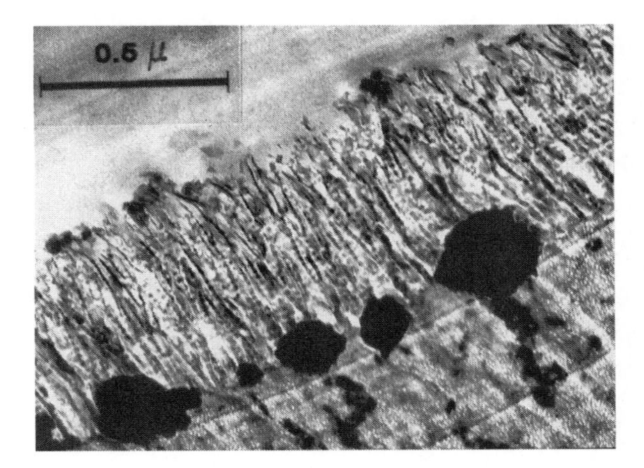

Figure 19: Copper concentration at the boundary layer: ultramicrotomed/TEM micrograph through PAA pretreated BA 8090C substrates.

In some cases a thin anodic oxide film is grown on the aluminium surface using what the industry calls "flash-anodising"; this is usually accomplished in a conventional chromic acid bath. This oxide film protects the initial cleaning of the aluminium surface until the component is ready for further processing and structural bonding. In this instance the oxide film is removed by vacu-blasting using a fused alumina grit of controlled particle size. It is known that at least one company uses glass-bead peening to remove the oxide layer; this is not to be recommended due to the stresses induced into the substrate prior to bonding.

Once the anodic oxide layer has been removed, one of the conventional pretreatments for aluminium can be used.

4.3.2. Titanium and Titanium Alloys

The oxides associated with titanium are significantly thinner than those grown on aluminium; all have a certain degree of micro-roughness. If high temperature adhesives are to be used (350–400°C cure) very thin oxide layers are required otherwise the oxygen from the oxide layer will dissolve in the base metal giving rise to voids and microcracks.

In a similar manner to the alloys of aluminium, titanium and its alloys can be pretreated for adhesive bonding by both mechanical and chemical methods. When chemical pretreatments are used, the oxide films so formed are not only thinner than those produced on aluminium but are significantly more stable to both heat and humidity [25].

Whichever route is being followed, the first stage is that of degreasing, by one of the methods already discussed, to clean the surfaces to be bonded. This can be followed or, occasionally, even replaced by a suitable alkaline etch or acidic emulsion cleaning [52].

All the methodologies encountered can be summarised as follows:

- Heavy Abrasion: grit-blasting
- Chemical: alkaline peroxide etching
- Chemical: acid etch plus phosphate–fluoride treatment
- Chemical: Pasa Jell® treatment
- Electrochemical: chromic acid anodizing
- Electrochemical: sulphuric acid anodizing
- Electrochemical: sodium hydroxide anodizing
- Chemical: chromate free: alkaline perborate treatment
- Chemical: sol–gel procedures

4.3.2.1. Abrasive Blasting

After the substrates have been cleaned, the surface to be bonded is abraded using a dry grit-blasting process [53] which uses alumina grit with a particle size range of about 40–50 μm and a line pressure of about 140 kPa. Alternatively, a wet process [54] can be employed utilising a 35%, by volume, slurry of alumina grit in water; the grit having a nominal particle size of 70 μm. In this latter case, significantly higher line pressures (about 275–425 kPa) are used. After abrasion the surfaces are cleaned using a clean air jet followed by a solvent wash. Substrates should be bonded, or further pretreated, within 1–2 h of blasting.

4.3.2.2. Acid and Alkaline Etching

Unlike aluminium, alkaline or acid etching techniques, apart from two known exceptions, are not, on their own, ideal as pretreatments prior to structural bonding; they are, though, an invaluable intermediate stage prior to anodising or a phosphate–fluoride treatment [55,56]. When this intermediate etching stage is used, standard cleaning techniques are followed by immersion in sulphuric acid, nitric acid/hydrofluoric acid or hydrochloric acid–orthophosphoric acid mixtures.

In the phosphate–fluoride process, the adherend is etched in 3% HF. The etched titanium is then immersed in an aqueous solution of the following composition for 2 min:

Trisodium phosphate (Na_3PO_4)	5% by weight
Potassium fluoride (KF)	2% by weight
Hydrofluoric acid (HF)	2.6% by weight
Water	Make up to 100% by weight

The anatase oxide coating produced, which has a thickness in the range of 150–295 nm, consists of titanium, oxygen, phosphorus and fluorine.

The two exceptions, already mentioned, are the so-called Pasa Jell treatment [54,55], and alkaline peroxide etching [25,57]. The Pasa Jell process, which gives an oxide thickness of about 20 nm, can be summarised as follows:

- Standard surface cleaning and degreasing
- Wet-abrasion and/or alkaline etching (Turco alkaline rust remover at 93–100°C) — optional
- 15–20 min immersion in the acidic Pasa Jell solution at ≤38°C.

A typical bath composition is:

Concentrated etch (Pasa Jell 107-CV)	10% by volume
Nitric acid (as HNO_3)	21% by volume (180–217 g/l)
Chromic acid (as CrO_3)	30 g/l (26.2–31.4 g/l)
Deionised water	Remainder

(Although a proprietary material, it is believed that Pasa Jell comprises a mixture of nitric acid, fluorides, chromic acid, coupling agents and water)

Once etching is complete the parts are removed, washed in deionised water and then dried in an air-circulating oven at 38–65°C; bonding or priming should take place within 4 h.

It should be noted, however, that the resultant bonds' durability performance is not as good as for those produced by chromic acid anodising or the novel sol–gel pretreatment. This is perhaps not surprising when the resultant surface topography is examined (Fig. 20).

The alkaline peroxide etch, dependent on bath conditions, gives an oxide thickness of 60–200 nm. The process comprises a 3–5 min alkaline degrease in Kelite® 19 at 60–71°C. The washed components are then immersed for 23–27 min, at 55–65°C, in a bath containing 0.5 M sodium hydroxide and 0.4 M hydrogen peroxide (H_2O_2). The etched substrates are then washed and air-dried at 65°C.

The resultant surface topography is very reminiscent of that produced by submitting aluminium alloys to a chromic–sulphuric etch (Fig. 21).

The main drawback with this process is the lack of stability of the hydrogen peroxide; the bath has to be renewed every 30 min.

4.3.2.3. Chromic Acid Anodising

As for aluminium, titanium substrates can be pretreated by anodising in chromic acid [55,57]. The cleaning stage is carried out in accordance with the specifications quoted above. This is then followed by an acidic pickling. The typical bath composition is:

Concentrated nitric acid (Sg: 1.42) (HNO_3)	4.5 l
Hydrofluoric acid (Sg: 1.17) (HF)	0.450 l
Water	10.0 l

The adherends are pickled at ambient temperature for about 10–20 min. The etched specimens are removed from the bath and then washed under clean, cold, running water; any black deposits should be brushed off with a clean, stiff-bristle, nylon brush.

The substrates are then attached to the anode of a standard anodising bath (anode to cathode ratio of about 3:1) and are immersed, at 40°C, in chromic acid of the following composition:

Chromium trioxide (CrO_3)	0.700 kg
Water	10.0 l

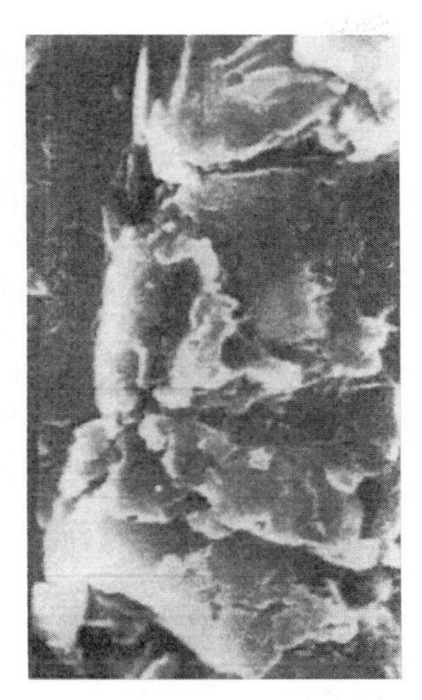

Figure 20: SEM micrographs of Pasa Jell etched titanium alloy [25] (Courtesy: Elsevier Limited).

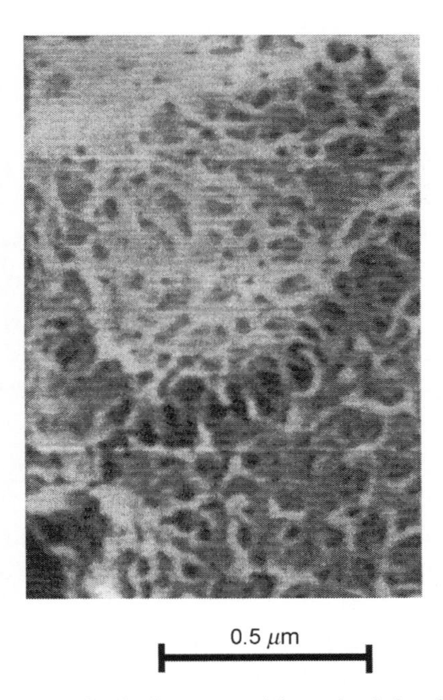

Figure 21: SEM micrographs of alkaline peroxide etched titanium alloy [25] (Courtesy: Elsevier Limited).

The voltage is raised to 20 V over a 5-min period and, dependent on alloy type, is held for 5–30 min; by then, the titanium should have developed a distinctive *blue* colouration. Lower voltages, as low as 5 V, can be used if less brittle oxide coatings are required.

The anodised sheets are then removed from the bath, spray-rinsed with cold, clean water and then air dried, preferably in an air-circulating oven whose air temperature does not exceed 45°C. Bonding of the components should take place within 4–6 h.

US Patent 3959091 [59] gives an alternative anodising method. In this case titanium components are anodised at ambient temperature for 20 min at 10 V. The electrolyte should contain 50 g/l of CrO_3 and sufficient ammonium bifluoride to maintain the current density at 22 A/m^2.

4.3.2.4. Sulphuric Acid Anodising

Unlike aluminium, DC sulphuric acid anodising is an acceptable pretreatment prior to bonding titanium components.

The first stage of the process is a grit blasting and/or degreasing step, to any of the above specifications. This is usually augmented with an alkaline degreasing step in accordance with a relevant specification.

Pickling and anodising is then carried out in accordance with one or other of several European and American specifications [60,61].

A suitable pickling solution of nitric and hydrofluoric acid can be made up as follows:

Nitric acid (HNO_3)	180–380 g/l
Hydrofluoric acid (HF)	13–21 g/l

Generally, a 20–30 s pickle at 15–30°C suffices to remove any abrasive particles and/or weak oxide layers. The titanium sheet should then be spray-rinsed in cold, clean water.

The resultant panels are clamped to the anode of a standard anodising bath and are immersed in a solution of sulphuric acid, at 21°C. Anodising then takes place in accordance with the above specifications. A suitable aqueous anodising solution comprises sulphuric acid (H_2SO_4) at a concentration of 50–400 g/l.

The voltage is raised to 15–20 V and, dependent on alloy type, is held until the titanium has developed a distinctive bluish-violet colouration; the actual time is pre-determined experimentally. The anodised sheet should then be spray-rinsed in cold, clean water and dried in a stream of clean air.

Figure 22: SEM micrograph of alkaline anodised titanium alloy [62] (Courtesy: Elsevier Limited).

4.3.2.5. Sodium Hydroxide Anodising

In 1983, researchers found [62] that by anodising titanium under alkaline conditions, not only could the unstable alkaline hydrogen peroxide etch be effectively replaced but also bonded joints having good durability could be produced. They evaluated various electrolyte molarities from 0.4 to 5.0 M, various applied voltages from 5 to 15 V and various bath temperatures from 16 to 30 °C. Their optimised procedure, to obtain the best durability results, was as follows:

Titanium substrates were first alkaline degreased using conventional commercially available cleaners. They were then anodised in 5 M sodium hydroxide for 30 min at 20°C using an applied voltage of 10 V. The performance of the anodised surfaces appeared to be independent of the anode: cathode ratio used. A scanning electron micrograph of a typical titanium surface pretreated using this procedure is given in Fig. 22; the resultant open, nodular oxide structure is clearly evident.

4.3.2.6. Chromate-free Pretreatment

In the late 1990s, researchers at NASA/LaRC developed a chemical pretreatment for titanium alloys which did not require the use of chromic acid or chromate salts [63]. The resultant surface is reputed to have excellent compatibility with structural adhesives of high thermal-oxidative resistance; in other words, suitable for bonding titanium components in supersonic aircraft.

The adherends are initially degreased using standard procedures. This process then makes use of an acid etch (to replace grit blasting) followed by an alkaline oxidation (to replace a peroxide oxidation). The details are as follows:

The cleaned substrates are dipped in aqueous 9 M sulphuric acid solutions for about 10 min at ambient temperature. The specimens are rinsed in cold, clean water and then dipped, again for about 10 min at ambient, in an aqueous, alkaline perborate solution — a mixture of 0.5 M sodium borate ($NaBO_3$) and 1 M sodium hydroxide (NaOH). The pretreated substrates are immediately washed in clean, cold water.

During the research programme, immersion times and bath temperatures were varied; longer times and higher immersion temperatures resulted in thicker and less stable oxide films. The optimum immersion for about 10 min at ambient temperature yielded oxide films in the region of 2 μm thick.

4.3.2.7. Sol–gel Procedures

This novel method of pretreatment has also been shown to be suitable for titanium. As well as the TPOZ/GTMS system, already discussed under the section dealing with the pretreatment of aluminium substrates, other aromatic [*p*- and *m*-aminophenyl triethoxy silane (a) and (b), respectively] and aliphatic [γ-aminopropyl triethoxy silane: Silquest A-1100: (c)] silanes have been used.

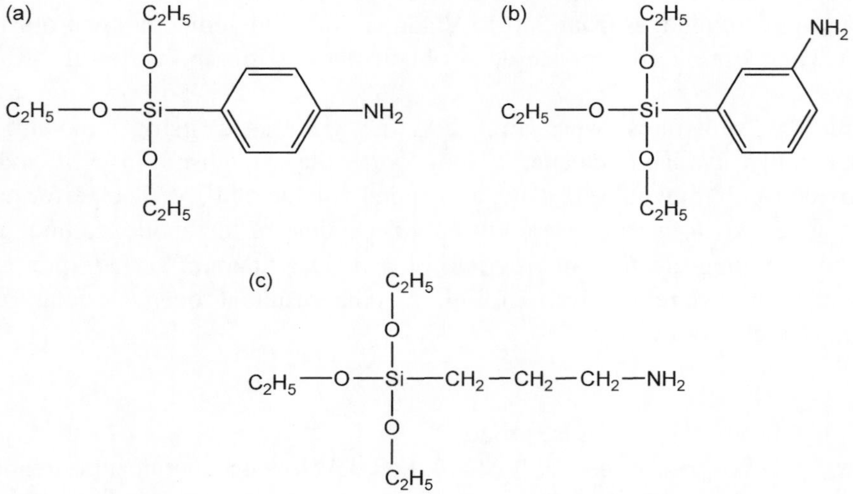

(a)
C_2H_5
O
C_2H_5—O—Si—[benzene ring]—NH_2
O
C_2H_5

(b)
C_2H_5
NH_2
O
C_2H_5—O—Si—[benzene ring]
O
C_2H_5

(c)
C_2H_5
O
C_2H_5—O—Si—CH_2—CH_2—CH_2—NH_2
O
C_2H_5

Another novel approach has been to use polyimide–silica hybrids, for example pendent phenylethynyl imide oligomeric bis-silanes with tetraethoxy silane [46,64,65].

4.3.3. Stainless Steel

Stainless steel is well known as being difficult to bond, especially when long-term environmental resistance is required. The degree of difficulty increases with the increasing chromium content of the steel. The oxides associated with low carbon steels are very stable but are thin and very smooth; they have none of the micro-roughness which is typical for other substrates.

The correct pretreatment, therefore, is vital and, amongst other considerations, it will be dependent on the grade of stainless steel being used as well as the minimum specified tensile strength of the substrate and the projected end use of the bonded component.

It is generally to be recommended that, before attempting to bond stainless steel components, bonding trials should take place to determine the optimum method and conditions needed to obtain the best bond strengths with the particular stainless steel being used. Such trials will also take into account end usage and, particularly, the durability requirements.

The possible methodologies can be summarised as follows:

- Heavy Abrasion: grit-blast with chill-cast iron shot, glass or alumina
- Chemical: acid etching
- Laser Treatment: CLP (Ciba Laser Pretreatment)
- Chemical: sol–gel procedures
- Surface Bombardment: plasma spray coating

Several pretreatments are recommended by the British Standards Institute [66]. In essence, these methods cover solvent and/or alkaline degreasing followed by surface abrasion or by the use of a chemical etchant. Grit blasting, using chill-cast iron shot, glass or alumina, is the ideal abrasion technique. Etchants based on sulphuric, hydrochloric or phosphoric acid are recommended; etching conditions are 5–30 min at temperatures from ambient to 65°C.

In many cases, the chemically pretreated substrates will require desmutting after etching and washing. This can be accomplished by immersion in the standard CSA pickling solution (see above) for 5–20 min at 60–65°C. Once such a bath has been used for desmutting stainless steel the pickle cannot be used again for the pretreatment of aluminium.

It has also been shown that adequate bond strengths can be obtained on fresh specimens following a combination of grit blasting and sulphuric–oxalic acid or hydrochloric acid–formalin–hydrogen peroxide etching. The latter has been shown to give better bath stability [58]. Other researchers [67] have shown that a simple sulphuric acid/oxalic acid etch is sufficient. Here, the stainless steel adherends are immersed in the aqueous acid mixture for up to 30 min at 90–95°C. A typical etchant formulation is:

Oxalic acid ($HO_2C{\cdot}CO_2H$)	10 pbw
Concentrated sulphuric acid (H_2SO_4)	10 pbw
Distilled water	80 pbw

Some stainless steels will "smut" under these conditions and others will not. If a black smut does form then this can, in this particular case, be removed by brushing, under running cold water. The adherends are then washed in distilled water and dried at 100–120°C; bonding should follow immediately.

There is mention in the literature of another pretreatment which works well on stainless steel. This is a proprietary, patented, experimental pretreatment developed by Ciba AG (now Huntsman Advanced Materials) and, as such, is an alternative to the more conventional "wet" processes. It is designated CLP (Ciba Laser Pretreatment).

CLP was developed for typical substrates used in the automotive industry but as it is equally efficacious in pretreating aluminium, titanium and stainless steel, it could prove of considerable use to the aerospace industry.

The process is a two-step one where the initial stage is to prime and then dry/stove the substrate to be bonded. Although the literature does not give examples of the primer used, it is fairly clear that a suitable surface protection or corrosion protection primer could be used; by inference, this could be extended to a silane-based primer.

The area to be bonded, and only the area to be bonded, is then exposed to the beam of a suitable laser; laser type, power and speed of treatment have to be optimised for each type of substrate.

No further treatment is needed and Huntsman Advanced Materials quote excellent joint durability using this approach [68].

Finally, there are two further experimental surface pretreatment methods which are worthy of note.

Firstly, work has started to evaluate the use of sol–gel procedures on stainless steel. Early results are encouraging. Secondly the use of plasma spraying techniques has been tried on low carbon steels [69]. Researchers report that, although temperature control is both difficult and critical, the initial results obtained are good.

4.3.4. Fibre-reinforced Laminates

There are, essentially, two types of fibre-reinforced laminates and both — those with thermoplastic and those with thermosetting matrices — can be pretreated utilising somewhat similar methodologies; each type of matrix, though, has its own special considerations.

The optimum pretreatment for the preparation of composite substrates for adhesive bonding can be defined as one which removes all the surface matrix system, to expose the reinforcing fibres, without doing any damage to the fibres themselves.

Obviously, at present, it is not feasible to achieve such a surface. However, the best possible procedures which can be employed are given below.

4.3.4.1. Thermoplastic Matrices

Currently the most commonly encountered substrates are those produced using polyetheretherketone (PEEK) or polyethersulphone matrices, especially reinforced with carbon fibre.

The earliest approaches to achieve suitable surface pretreatments on fibre-reinforced thermoplastic composites used some form of surface cleaning/degreasing followed either by a controlled hand-abrasion or semi-automated wet grit-blasting process. However, later researchers showed that this was not an ideal technique. Kinloch et al. [70] compared typical abrasion techniques with corona discharge processes and found that the adhesive fracture energy (G_c) values for bonded double cantilever beam specimens varied from 0.2 to 0.3 kJ/m^2 when abrasion techniques were used to 1.5–3.75 kJ/m^2 when an optimised corona discharge treatment had been used. They also found that the locus of failure for the abraded joints was at the composite/adhesive interface whereas for the joints pretreated by corona discharge it was generally cohesive in nature.

When the substrate was reinforced with carbon fibre, the plateau results obtained depended heavily on the type of adhesive being used: Cytec-Fiberite film adhesive FM®-73M (curing at 120°C) and room-temperature curing Loctite Aerospace paste adhesive EA® 9309. The composite matrix chemistry exerted significantly less influence.

This improvement in performance in the bonded joint was attributed to the polarity of the composite surface. The intensity of corona treatment required to achieve the necessary degree of polarity was found to be dependent on the cure characteristics of the adhesive used and, to a certain extent, on the chemistry of the matrix. Thus, for the EA 9309, carbon/PEEK composites needed >10 J/m^2 and carbon–polyamide composites required a corona energy of >3 J/mm^2. When FM-73M was used, the figures were >20 J/mm^2 and >6 J/mm^2, respectively — i.e. double the energy requirement.

Blackman et al. [71] compared the use of conventional abrasion pretreatments with corona discharge and oxygen plasma treatments. Again, the need to optimise the surface polarity was confirmed. As in the previous work, both adhesive and matrix chemistry influenced the optimum plasma conditions. For carbon/PEEK composites plasma treatment times of 3 min were required to optimise the adhesive fracture energy, at about 2.75 kJ/m^2, when using the FM-73M, whereas treatment times of >12 min were required for the EA 9309. However, in the latter case the adhesive fracture energy plateaued at about 3.7 kJ/m^2. For carbon/PPS composites the optimum fracture energy was independent of adhesive chemistry. The ideal treatment time was found to be >6 min to give a fracture energy value of 1.5 kJ/m^2.

Analysis of the surfaces generated by these different pretreatment processes [71] shows the various surface topographies which are attained (Fig. 23); it is clear how much more efficacious the plasma treatment is.

Finally, Buchman et al. [72] have evaluated the use of excimer laser treatment of carbon fibre reinforced PEEK composites. Once again, the final performance of the bond depended on the intensity of the pretreatment but these workers quoted that the resultant adhesive strengths were nearly double those for conventionally pretreated (hand-abrasion) substrates.

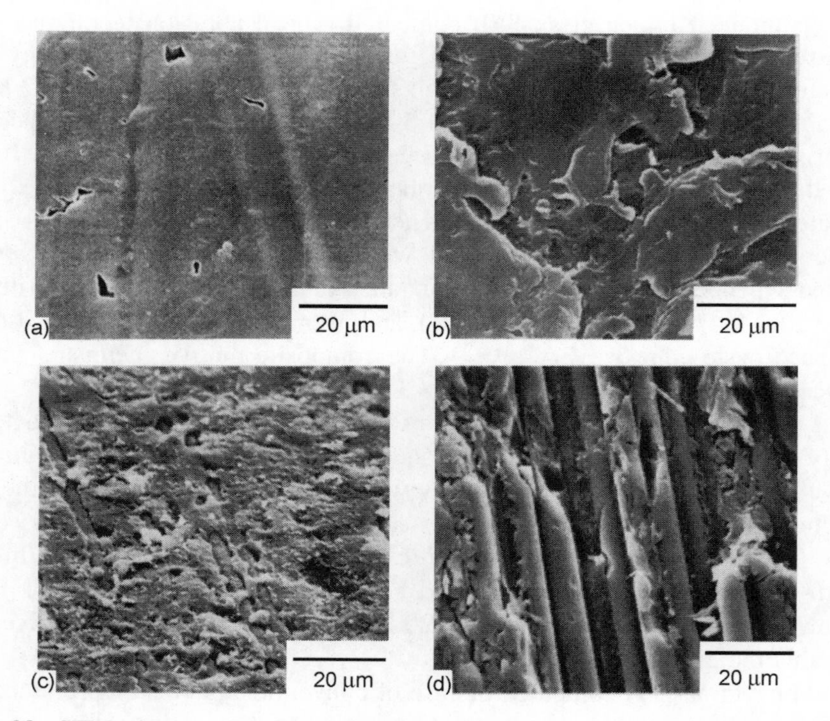

Figure 23: SEM micrographs of carbon/PEEK composites: a) as moulded; b) abraded and solvent cleaned; c) corona treated (20 J/mm^2); d) oxygen plasma treated (10 min) (Courtesy: Imperial College London).

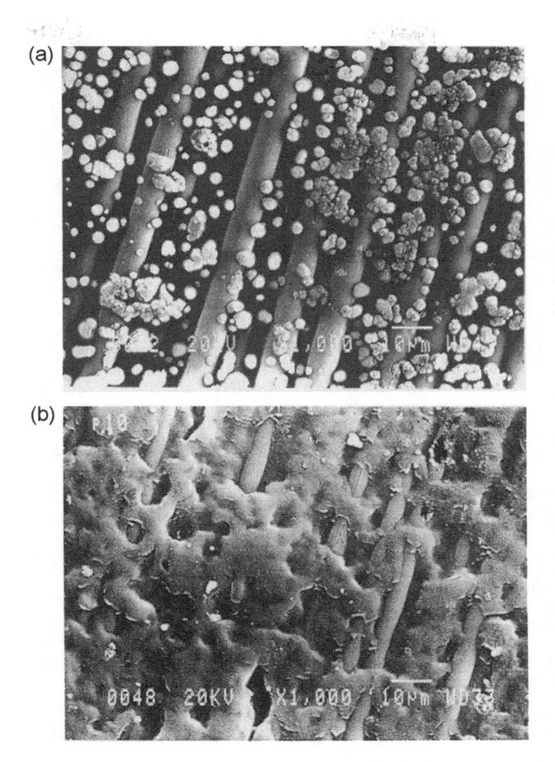

Figure 24: SEM micrographs of Excimer-Laser ablated carbon/PEEK composites: a) low energy (0.18 J/P cm^2, 100 pulses); b) high energy (1 J/P cm^2, 10 pulses) (Courtesy: VSP Publishers).

Microscopic analysis of the pretreated substrates does reveal that considerable amounts of the surface matrix resin can be removed, thus exposing the reinforcing fibres. However, Fig. 24 shows how dependent the efficacy of matrix removal is on the energy of the laser beam (cf. Fig. 23d for plasma treatment).

It could be argued that, in spite of the excellent results seen when using corona discharge, plasma treatment and laser ablation surface pretreatment techniques, it is unlikely that they will be cost effective and will be difficult to introduce for the treatment of large components. However, in view of the poor performances now often seen by using conventional abrasion techniques, it is probable that ways round these difficulties will have to be found.

4.3.4.2. Thermosetting Matrices

The following procedures are all possible and, indeed, are used in the industry to a greater or lesser extent.

- *No surface pretreatment*: such an approach will leave a surface which is possibly contaminated with mould release agents and airborne dirt, oils, etc. It will also have a relatively thick layer of cured matrix resin at the proposed adhesive interface between composite and adhesive. Not only will the contaminants ensure that a weak boundary layer is produced at this interface but experience shows that, even under scrupulously clean surface conditions, the most durable joints are produced when the resin fraction at the composite surface is kept to a minimum.
- *Solvent clean*: this technique will remove some, but not necessarily all, surface contamination. The procedure is seen as little better than no pretreatment at all.
- *Peel ply/tear ply*: with RTM or RIM components or laminates prepared from prepregs, it may be possible to design the laminating process so that a layer of a suitable perforated fabric ("peel ply": coated with a release agent and "tear ply": uncoated) is placed on the surface to be bonded as part of the lay-up procedure. Thus, when the laminate is cured this peel ply becomes part of the laminate itself. Just prior to bonding, the peel ply is peeled off. As it is perforated, matrix resin will flow through the holes during cure and thus, when the peel ply is stripped off, some of the excess cured matrix material is removed, which exposes a fresh, clean surface for bonding. This is shown, schematically, in Fig. 25.

However, even with the improved peel plies now available, such a technique is seen as very much a "second best" approach. Certainly contamination from mould release agents and airborne substances is kept to a minimum, as the surface only needs be exposed just before bonding, and some of the resin-rich

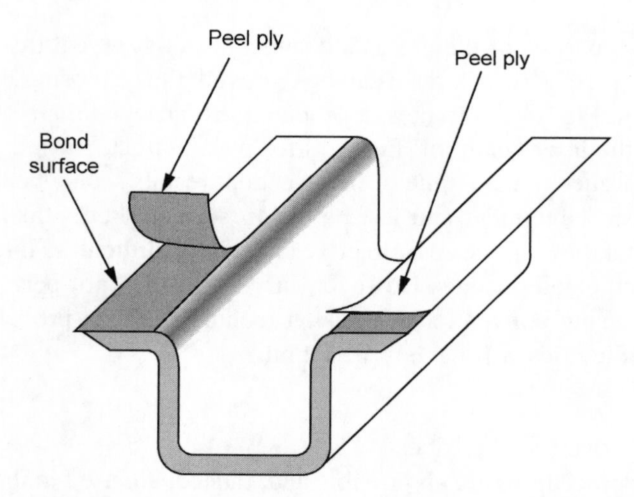

Figure 25: Schematic of a typical composite structure utilising peel plies.

surface layer can be removed. However, the degree of resin removal is not uniform across the laminate surface and a relatively resin-rich surface layer still remains. Both aspects can lead to a reduction in the actual level of bond strengths achievable. Further, some transfer of the peel-ply polymer system and in some cases of the peel ply (or more likely, the tear-ply) itself, to the bonding surface, cannot be ruled out. Experience shows that the bond is generally improved but not optimised. The same is probably true for the durability performance of the bonded structure.

- *Hand abrasion techniques*: pretreatment is usually accomplished using wet-and-dry paper and it often augments the peel ply approach. Although this can lead to an improvement in bond strengths, the techniques are highly operator-dependent as well as being arduous and slow. However, with a good operator, it is probably a more than adequate method to prepare the surface for bonding but it will not be reproducible from operator to operator. Fibre damage is a distinct possibility and post-treatment cleaning will be critical and difficult. In many cases, if inadequately carried out, this will lead to poor quality bonds and even worse durability. The situation can be improved by automating the process, which makes the operation more controllable and more conducive to routine production. However, the post-treatment cleaning problems still apply.
- *Grit-blasting techniques*: these procedures are much more controllable than the above as such parameters as grit chemistry and hardness, grit size, air-line pressure, distance from the blasting head to the job, rate and frequency of pass are all readily programmable.

  When dry grit-blasting is used then there are concerns as to the cleanliness of the grit being used and the fact that the process itself might "drive" surface contamination into the component rather than removing it. Most of these concerns can be overcome if wet grit-blasting is used. Here the abrasive grit, usually alumina or silica, is mixed into a slurry with water and this blend is used to abrade the composite surface. The water not only acts as a cushion, which means that a more gentle abrasion results, but it is also constantly washing down the surface which means that there is less chance for residual contamination to remain on the bonding surface.

  Care, though, still needs to be exercised over the final cleaning and drying operations. Fig. 26 shows a schematic of a typical wet grit-blaster.
- *Surface modification techniques*: these can include such approaches as the laser ablation of the surface using, for example, excimer lasers. Other possibilities are to use plasma, corona and flame treatments to modify the laminate surface, chemically, to make it more amenable to adhesive bonding. As for thermoplastic matrices, it should be possible to introduce highly polar groups onto the surface to improve bonding across the interface.

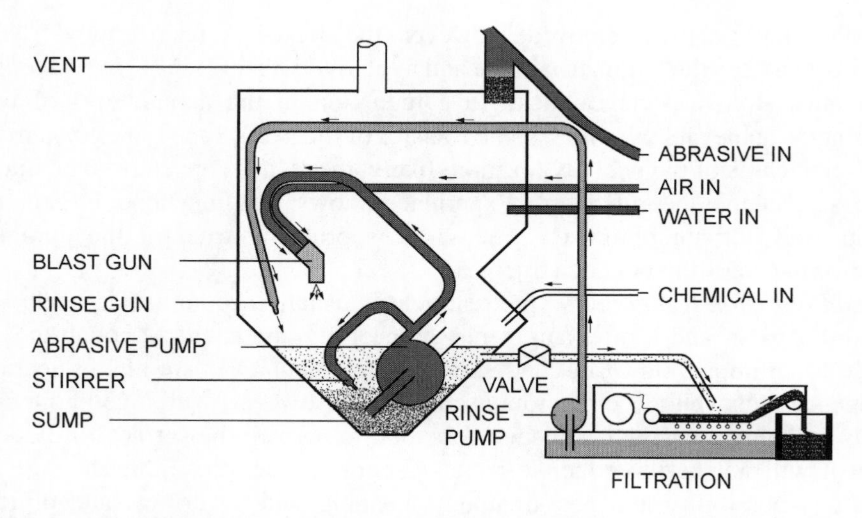

Figure 26: Schematic of a typical wet grit-blaster (Courtesy: Vapormatt Limited).

Although much has been written on the use of these techniques to pretreat polymeric substrates prior to bonding, there is very little in the literature related to fibre-reinforced composites; most is related to thermoplastic composite substrates.

Should such approaches prove to be beneficial it is unlikely that they will be cost effective and will be difficult to introduce for the treatment of large components. In the light of the good performances which are possible using peel plies and abrasion techniques, unlike the situation which exists for thermoplastic composite substrates, these techniques are only likely to find applications where small, delicate components are to be bonded.

Irrespective of pretreatment procedure, optimum bond strengths can be attained if the laminates are dried before bonding; to remove any moisture absorbed from the atmosphere. This should be carried out, immediately before joint assembly, in an air-circulating oven at a temperature at which no thermal damage will be imparted to the laminate.

4.3.5. Non-metallic Honeycombs

Honeycombs produced from such materials as aramid paper (Nomex®) require no pretreatment unless contaminated. However, higher bond strengths are obtained if the honeycomb is dried for 1 h at 120°C just before bonding.

References

[1] de Bruyne, N. A. (1996). *My life*. Midsummer Books, Cambridge.
[2] Aero Research Technical Notes, *Difficulties in Gluing Plywood*, Bulletin No. 21, September 1944.
[3] Snogren, R. C. (1974). *Handbook of surface preparation*. Palmerton Publishing Company Inc., New York.
[4] Davis, J. R. (1999). *Corrosion of aluminum and aluminum alloys*. ASM International, Materials Park, OH, pp. 191–214.
[5] Henkel Chemicals Limited, Product Data: Alocrom 1290–(S821-1290), (1991).
[6] UK Defence Standardization, DEF STAN 03-2 (1995). *Cleaning and Preparation of Metal Surfaces*, Issue 3 (Process A: Method A1 or A2).
[7] British Standards Institute, BS EN 2334 (1997). *Chromic–Sulphuric Acid Pickle of Aluminium and Aluminium Alloys*, Issue 1 (Treatments A, B and C).
[8] Boeing Specifications, BAC 5408 (1988). *Vapor Degreasing*, Issue M.
[9] Boeing Specifications, BAC 5744 (2002). *Manual Cleaning: Cold Alkaline and Solvent Emulsion*, Issue G.
[10] Boeing Specifications, BAC 5750 (2001). *Solvent Cleaning*, Issue L.
[11] Boeing Specifications, BAC 5763 (1995). *Emulsion Cleaning and Aqueous Degreasing*, Issue E.
[12] SAE Standard, ARP1842 (1984). *Surface Preparation for Structural Adhesive Bonding, Aluminum Alloy and Low Alloy Steel Parts*.
[13] Bishopp, J. A., Sim, E. K., Thompson, G. E., & Wood, G. C. (1988). The adhesively bonded aluminium joint: the effect of pretreatment on durability. *Journal of Adhesion*, **26**, 237–263.
[14] UK Defence Standardization, DEF STAN 03-3 (2002). *Protection of Aluminium Alloys by Sprayed Metal Coatings*, Issue 3 (Section 2).
[15] British Standards Institute, BS EN 2334 (1997). *Chromic–Sulphuric Acid Pickle of Aluminium and Aluminium Alloys*, Issue 1 (Treatments B and C).
[16] Boeing Specifications, BAC 5748 (2000). *Abrasive Cleaning, Deburring and Finishing*, Issue N.
[17] Bishopp, J. A., & Thompson, G. E. (1993). The role of electron microscopy in the study of adhesion to aluminium substrates. *Surface and Interface Analysis*, **20**, 485–494.
[18] Baker, A. A. (1994). Bonded composite repair of metallic aircraft components — overview of Australian activities, *AGARD [Advisory Group for Aerospace Research and Development] Conference on Composite Repair of Military Aircraft Structures,* 3–5 October 1994, Seville.
[19] UK Defence Standardization, DEF STAN 03-2 (1995). *Cleaning and Preparation of Metal Surfaces*, Issue 3 (Method B 2(ii)).
[20] British Standards Institute, BS EN 2334 (1997). *Chromic–Sulphuric Acid Pickle of Aluminium and Aluminium Alloys*, Issue 1 (Treatment A).
[21] Boeing Specifications, BAC 5749 (2002). *Alkaline Cleaning*, Issue T.
[22] UK Defence Standardization, DEF STAN 03-2 (1995). *Cleaning and Preparation of Metal Surfaces*, Issue 3 (Process O).
[23] Boeing Specifications, BAC 5555 (2001). *Phosphoric Acid Anodizing of Aluminum*, Issue N (Deoxidising in accordance with paragraphs 8.2.3 and 9.3).
[24] Rodgers, N. L. (1981). *Proceedings of 13th National SAMPE Technical Conference, SAMPE,* Azusa, CA, 1981, 640 p.

[25] Davis, G. D., & Venables, J. D. (2002). Surface treatment of metal adherends. In: M. Chaudhury, & A. V. Pocius (Eds), *Adhesion science and engineering — volume 2: surfaces, chemistry and applications* (pp. 947–1008). Elsevier, Amsterdam.
[26] Desai, A., Ahearn, J. S., & McNamara, D. K. (1985). *Cleanliness of external tank surfaces*, Martin Marietta Laboratories, Baltimore, MD, Unpublished Technical Report MML TR 85-65.
[27] UK Defence Standardization, DEF STAN 03-24 (1997). *Chromic Acid Anodizing of Aluminium and Aluminium Alloys*, Issue 3 (unsealed process).
[28] British Standards Institute, prEN 2101, Aerospace Series (1998). *Chromic Acid Anodising of Aluminium and Wrought Aluminium Alloys*, Draft Issue 2 (Type A).
[29] SAAB Specification, STD 1991 (1999). Issue 6.
[30] Boeing Specifications, BAC 5555 (2001). *Phosphoric Acid Anodizing of Aluminum*, Issue N.
[31] UK Defence Standardization, DEF STAN 03-2 (1995). *Cleaning and Preparation of Metal Surfaces*, Issue 3 (Process P).
[32] UK Defence Standardization, DEF STAN 03-15 (1997). *Sulphuric Acid Anodizing of Aluminium and Aluminium Alloys*, Issue 3.
[33] British Standards Institute, BS 5599 (1993). *Specification for Hard Anodic Oxidation Coatings on Aluminium and Its Alloys for Engineering Purposes*.
[34] Arrowsmith, D. J., Clifford, A. W., Davies, R. J., & Moth, D. A. (1984). *Durability of adhesively bonded aluminium pretreated by hard anodizing followed by a phosphoric acid dip*, Extended abstracts for the international adhesive conference. PRI, London.
[35] Yendall, K. A., Critchlow, G. W., Andrews, F. R., & Bahrani, D. (2002). *An evaluation of chromate-free anodising processes for aerospace applications*, Extended abstracts for Euradh 2002. IOM Communications, London.
[36] Boeing Specifications, BAC 5765 (1995). *Cleaning and Deoxidizing Aluminum Alloys*, Issue T.
[37] Boeing Specifications, BAC 5632 (1999). *Boric-Acid Sulfuric Acid Anodizing*, Issue C.
[38] Johnsen, B. B., Lapique, F., & Bjørgum, A. (2004). The durability of bonded aluminium joints: a comparison of AC and DC anodising pretreatments. *International Journal of Adhesion and Adhesives*, **24**, 2, 153–161.
[39] Bjørgum, A., Lapique, F., Walmsley, J., & Redford, K. (2003). Anodising as pre-treatment for structural bonding. *International Journal of Adhesion and Adhesives*, **23**, 5, 401–412.
[40] ASTM D 3762. (1981). *Standard test method for adhesive-bonded surface durability of aluminum (wedge test), Part 22: wood; adhesives*. American Society for Testing and Materials, West Conshohoken, PA.
[41] Pike, R. A. (1985). *International Journal of Adhesion and Adhesives*, **5**, 1, 3–6.
[42] Pike, R. A. (1986). *International Journal of Adhesion and Adhesives*, **6**, 1, 21–24.
[43] Boeing, *Advanced Coating Materials for Air Force Aircraft*, A&M Environmental Tech Notes, 7(2) May 2002.
[44] Blohowiak, K. Y., Osborne, J. H., Krienke, K. A., & Sekitis, D. F. (1996). *Proceedings of 28th International SAMPE Technical Conference,* Corvina, CA, 1996, 440 p.
[45] Blohowiak, K. Y., Osborne, J. H., Krienke, K. A., & Sekitis, D. F. (1996). *Environmentally benign sol–gel surface treatment for aluminium bonding applications,* Boeing Defence and Space Group, Seattle, Final Report.

[46] Blohowiak, K. Y., Krienke, K. A., Osborne, J. H., & Greegor, R. B. (1998). *Proceedings of Workshop on Advanced Metal Finishing Techniques for Aerospace Applications,* Keystone, CO.

[47] Fiebig, J. W., Mazza, J. J., & McKray, D. B. (2002). An ALC consideration of simple sol–gel preparations for improved durability of field and depot-level bonded repairs, *Proceedings of the Aging Aircraft Conference, 2002,* Galaxy Scientific.

[48] Chung, Y. J., Jeanjaquet, S. L., & Kendig, M. W. US Patent US 6,579,472, Corrosion Inhibiting Sol/Gel Coatings for Metal Alloys, June 2003.

[49] Taylor, A. World Intellectual Property Organization WO 01/25343, Coating Materials, April 2001.

[50] Taylor, A. World Intellectual Property Organization WO 02/24824, Coating Compositions, March 2002.

[51] Bishopp, J. A., Jobling, D., & Thompson, G. E. (1990). The surface pretreatment of aluminium–lithium alloys for structural bonding. *International Journal of Adhesion and Adhesives*, **10**, 3, 153–160.

[52] UK Defence Standardization, DEF STAN 03-2 (1995). *Cleaning and Preparation of Metal Surfaces*, Issue 3 (Process B: Method B1 or B2 or Process C).

[53] British Standards Institute, BS EN 2497 (1990). *Specification for Dry Abrasive Blasting of Titanium and Titanium Alloys.*

[54] SAE Standard, ARP 1843A (1991). *Surface Preparation for Structural Adhesive Bonding, Titanium Alloy Parts*, Revision A.

[55] Tiwari, R. K. (2002). The thermal stability of anodic oxide coatings — strength and durability of adhesively bonded Ti–6Al–4V alloy, Dissertation submitted to the Faculty of the Virginia Polytechnic Institute and State University.

[56] Miller, P. D., & Jefferys, R. A. (1958). US Patent US 2,864,732, Method of Coating Titanium Articles and Products Thereof.

[57] Cotter, J. L., & Mahoon, A. (1982). Development of new surface pretreatments, based on alkaline hydrogen peroxide solutions, for adhesive bonding of titanium. *International Journal of Adhesion and Adhesives*, **2**, 1, 47–52.

[58] Redux Bonding Technology, Hexcel Composites Limited, 2000.

[59] Moji, Y., & Marceau, J. A. (1976). Method of Anodising Titanium to Promote Adhesion, US Patent 3959091.

[60] British Standards Institute, BS EN 2808 (1997). *Anodizing of Titanium and Titanium Alloys.*

[61] International Standard, ISO 8080 (1985). *Specification for Anodic Coating of Titanium and Titanium Alloys by the Sulphuric Acid Process.*

[62] Kennedy, A. C., Kohler, R., & Poole, P. (1983). *International Journal of Adhesion and Adhesives*, **3**, 3, 133–139.

[63] Lowther, S. E., Park, C., & St. Clair, T. L. (1999). A novel surface treatment for titanium alloys, *22nd Annual Meeting of the Adhesion Society,* Florida.

[64] Park, C., Lowther, S. E., Smith, J. G. Jr., Connell, J. G., Hergenrother, P. M., & St Clair, T. L. (2000). *International Journal of Adhesion and Adhesives*, **20**, 457.

[65] Hergenrother, P. M. (2000). *SAMPE Journal*, **36**, 30.

[66] British Standards Institute, BS 7773 (1995). *Code of Practice for Cleaning and Preparation of Metal Surfaces.*

[67] Polymerics GmbH, Berlin, Surface Pretreatment Methods, http://www.polymerics.de/products/pretreatments.html.

[68] Broad, R., French, J., & Sauer, J. (1999). New effective, ecological surface pretreatment for highly durable adhesively bonded metal joints. *International Journal of Adhesion and Adhesives*, **19**, 2–3.
[69] Davis, G. D., Groff, G. B., Biegert, L. L., & Heaton, H. (1995). *Journal of Adhesion*, **54**, 47.
[70] Kinloch, A. J., Kodokian, G. K. A., & Watts, J. F. (1992). The adhesion of thermoplastic fibre composites. *Philospohical Transactions of the Royal Society of London, Series A*, **338**, 83–112.
[71] Blackman, B. R. K., Kinloch, A. J., & Watts, J. F. (1994). The plasma treatment of thermoplastic fibre composites for adhesive bonding. *Composites*, **25**, 5, 332–341.
[72] Buchman, A., Dodiuk, H., Rotel, M., & Zahavi, J. (1996). Laser-induced adhesion enhancement of polymer composites and metal alloys. In: K. L. Mittal (Ed.), *Polymer surface modification: relevance to adhesion* (pp. 119–212). VSP, The Netherlands.

Chapter 5

Aerospace: A Pioneer in Structural Adhesive Bonding[☆]

John Bishopp

My biggest mistake was to agree to the riveting of the window surrounds in the Comet 1 rather than insisting that they were "Reduxed" as planned

Statement, just before his death, by RE Bishop, Designer of the de Havilland Aircraft Company's DH106 Comet 1.

John Bishopp is a Fellow of the Institute of Materials, Minerals and Mining and a Member of the Royal Society of Chemistry.

He has a deep knowledge of the science of adhesion and a long–nearly forty years–track record in the development of structural adhesives: both as an individual bench chemist and then as the leader of Hexcel Composites' worldwide Adhesive R and D Group.

Since retirement, he has acted as a consultant: working with academe, industry and research establishments in the UK and the USA.

He has presented numerous papers to international conferences and scientific journals. He was the founder chairman of the Society for Adhesion and Adhesives.

Abstract. This chapter falls, quite logically, into four distinct sections. The first deals with wooden aircraft from the first heavier-than-air machines through to the mid-1940s. Here animal-based adhesives were a natural choice for bonding wood; furniture makers had been using them for centuries. This section ends with the success story of the de Havilland Mosquito; a wooden airplane held together with a urea–formaldehyde (U/F) glue – one of the first truly synthetic adhesives.

The second section traces the use of structural adhesives, first, in all-metal aircraft and then in aeroplanes having complex structures manufactured from both metal and composite components; indicating why the industry persevered with adhesive bonding when, perhaps, mechanical joining appeared to be more logical.

[☆]*Photographic Media*: Where not stated, schematic diagrams and photographs are drawn from the Hexcel Composite archives or are the property of the author.

Handbook of Adhesives and Sealants
P. Cognard (Editor)

The third section devotes itself to an in-depth examination of the structural bonded joint: the substrates, primers and the structural adhesives themselves. In this last area, a full appraisal is made of a typical range of commercially available structural adhesives. This examines their role in the bonded structure, the formats in which they are supplied, the basic chemistries employed with their relative cure cycles and generic formulations. This is augmented with key properties of selected adhesives from this range and a typical qualification package generated for one adhesive to meet typical aerospace specifications. The section concludes with brief outlines as to how the adhesives are made, how they are applied to the substrates to be bonded and the methods by which they can be cured.

The fourth and final section is, in essence, a series of case studies showing how various sections of the industry use structural adhesives in modern aircrafts. This deals with each of the relevant adhesive chemistries in turn and covers both military and commercial engines and aeroplanes, helicopters, bonded components within the airframe, satellites and missiles.

Keywords: 3Ms; Acrylic; Aero engines; Aileron; Airbus; Aluminium; Aluminium–lithium; Animal glue; ARALL; Aramid and aramid fibre; Autoclave; BAE Systems; Beam-Shear; Bismaleimide or BMI; Bis-silane; Boeing; Bryte Technologies; Carbon; Carbon fibre; Carrier; Casein; Collagen; Comet; Composite; Control surface; Corrosion inhibition; Creep; Cure Temperature; Curing; Cyanate ester; Cytec-Fiberite; De Bruyne; De Havilland; Doubler; Epoxy; Epoxy-phenolic; Fatigue; Fibre–metal laminates; Fillet and Filleting; Film and film adhesive; Floating roller peel; Foaming adhesive; Foaming film; Fokker; Fuselage; Glare; Glass fibre; Helicopters; Henkel Corporation; Hexcel; Honeycomb; Honeycomb flatwise tensile; Honeycomb peel; Huntsman Advanced Materials; Kevlar®; Laminate; Lap shear; Leading edge; Lord Corporation; Missiles; Nacelle; Nomex; Novolac; Paste adhesive; Permabond; Phenol-formaldehyde or P/F; Phenolic; Polyimide; Polyurethane; Primer; Primer: Corrosion and corrosion-inhibiting; Primer: Surface protection; Radome; Redux; Resole; Resorcinol-formaldehyde or R/F; Rib; Rivet and riveting; Rohacell; Rotor blade; Rudder; Sandwich panel; Satellites; Sea Hornet; Service temperature; Silane; Spar; Specification; Spoiler; Statistical experimental design; Steel and Stainless Steel; Stiffener; Stringers; Surface pretreatment; Syntactic adhesives; Temperature resistance; Test fluid immersion; Tests; Thixotropic; Titanium; Urea-formaldehyde or U/F; Water-based; Water resistance; Westland; Wings; Wood and plywood; Wright Brothers

5.1. The Wright Brothers

On 17 December 1903, when Wilbur and Orville Wright flew a heavier-than-air machine the distance of 120 feet at Kitty Hawk in North Carolina, a new industry was born (Fig. 1). An industry which, in 100 years, has caused the "size" of the world to be cut so that no two places are more than a few hours apart, has sent men to the moon and has sent unmanned spacecraft to other planets as well as deep into space. Thus, today, the aerospace industry stands at the "cutting edge" of science and technology, with a history of innovation, of radical designs and the adoption of novel advances in technology.

One of these has been the use, from the very early days, of structural adhesives which have enabled the construction of lighter, stronger and more durable airframes and aircraft.

Perhaps it was fortuitous that the first aircraft were built of wood. The use of natural glues in producing wooden furniture and musical instruments has a history going back several hundreds of years. So it cannot be just by chance that such a means of joining structures was so readily adopted from the very first.

A comparison with the automobile industry is illuminating. Although, some forms of steam-powered road transport were designed and built from the 1770s, it was the 1880s that saw the real birth of the automobile industry. Although this makes it slightly older than the aerospace industry, it was not until the last 20–30 years that automobile designers have seriously considered incorporating structural adhesive bonding into their construction techniques.

The Wright Brothers' plane was constructed of spruce, ash and muslin and contained two 8-ft propellers, which were carved from laminated spruce; the lamination being achieved using a typical protein-based wood-working glue.

Today's aircraft, on the other hand, are constructed of aluminium, titanium, stainless steel and fibre-reinforced composites and are structurally bonded with

Figure 1: The Wright Brothers 1903 flyer. First flight (SI 2002-16646). (Copyright of: National Air and Space Museum, Smithsonian Institution.)

synthetic adhesives based on such diverse chemistries as: phenolic, epoxy, bismaleimide, polyimide and cyanate ester.

5.2. Structural Bonding in Wooden Aircraft [1]

The aircraft industry was only 11 years old when the First World War broke out. These 5 years of conflict revolutionized aircraft design and fully established the use of bonding or, more accurately in those days, gluing, as a means for manufacturing and joining structural aircraft components.

5.2.1. Animal Glues

Early aircraft were constructed using spars of solid wood and laminated wooden structures. The latter being used where constructions containing simple or complex curvatures were required. Initially ash was used but this was later replaced, to a great extent, by spruce. The glues used at this time for bonding wood were based on proteins extracted from animal products such as bone, hide, fish skins, blood or milk.

These adhesives were split into two broad classes:

- Assembly (or laminating) glues which were used to bond specific assemblies/components within the airframe structure – spars, ribs, etc.
- Glues for plywood manufacture

Plywood is a generic term for bonded wooden panels of thicknesses in the region of 1.5–12 mm or thicker. Plywood is made by gluing together thin wood veneers, between 0.5 and 4.0 mm thick, which are laid-up with grain directions at 0°/90° (only rarely is a $\pm 45°$ configuration used), to achieve the required thickness and properties. Such panels were used, at the lowest thicknesses, for wing and fuselage skins and the thicker sections for reinforcement at the attachment points for undercarriage and engine-mounting bulkheads.

The more solid structures, such as propellers and curved frame members, were prepared by laminating solid pieces of timber, between 3.0 and 15.0 mm thick, with all grain directions parallel to each other. Such structures are defined as "layers" [2] or "glued laminated structures" [3].

The applications using assembly glues could tolerate adhesives with short working lives, whereas those used for plywood manufacture needed much longer set-up windows in order to accommodate the time needed to lay up the packs of veneers before pressing took place.

The first assembly glues would have used collagen-based systems [4], which are all traditionally called "animal" glues; these would have later been replaced by casein-based glues.

Glues based on casein, blood albumen and blends of these two natural products were used for plywood manufacture during World War I. Plywood panels would have first been assembled from glue-coated veneers and the adhesive would then have been "cured" in heated hydraulic presses. Whilst blood albumen glues require hot "curing", casein can set at temperatures as low as 5°C. Cold-setting plywood, glued with casein adhesives, was manufactured up to the mid 1920s, but was then seen as too poor quality for aircraft use. The casein adhesives were replaced with blood albumen glues or those using casein/blood albumen blends, until the advent of the phenolic-based Tego® film adhesive (q.v.) in the early 1930s.

5.2.1.1. Collagen Glues

These so-called "animal glues" are hydrocolloids derived from collagen (*colla* being the Greek for 'glue' and *gen* meaning 'creator') [5]; they are also known as "hot" glues.

Collagen is isolated by water/steam extraction from either animal hide or bones or, occasionally, fish skins. For the best quality glues, these extracts are then chemically treated to produce gelatine solutions; in the case of bone glues by a demineralisation process and by the addition of lime to the hide-based systems. These gelatine solutions are then concentrated and dried to produce the jelly-like gelatine product whose major constituent is an amino acid/polypeptide referred to as gelatin; the model structure is given in Fig. 2.

The glue was produced by dissolving the jelly in hot water. This hot solution would have been applied to the substrates, the joint would have been closed and, on cooling, the glue solidified to produce the final bonded structure.

The main disadvantage associated with these glues was that the "working time" could be very short. A few degrees drop in temperature could be sufficient to set the adhesive; hence their unsuitability as adhesives for plywood applications. Further they were very susceptible to attack by moisture.

Fig. 3 shows the boss of a typical wooden propeller construction; the bonded laminated structure can be clearly seen.

5.2.1.2. Blood Glues

These glues are based on the globular protein, albumen, which would have been extracted from blood; essentially as a by-product of the massive meat processing

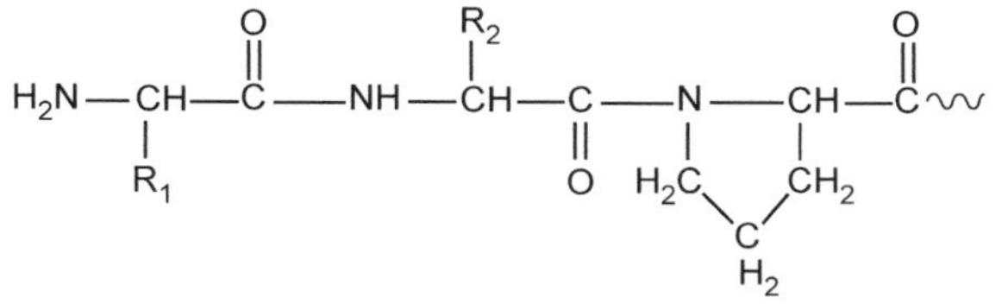

$$H_2N{-}CH(R_1){-}C(=O){-}NH{-}CH(R_2){-}C(=O){-}N^{1}{-}CH^{1}(CH_2{-}CH_2{-}CH_2){-}C(=O){\sim}$$

Figure 2: Structure attributed to gelatin.

Figure 3: A 110-Lerhone bonded laminated propeller from a Nieuport scout of 1916–1920. (colour version of this figure appears on p. xiv).

industry. Albumen has a molecular weight in the region of 20,000–70,000 and, unlike other globulins, is soluble both in water and in dilute salts. It is important that the adhesive is supplied in an uncoagulated form or water solubility is lost. Many of the glues also contain an addition of alkali (particularly sodium hydroxide) to improve their adhesive characteristics. Although the biggest suppliers and users appear to have been in Finland and the United States of America, there are many instances of their use throughout Europe.

Blood glues are another example of hot glues. In this case, however, they are applied at ambient temperature and are converted to an infusible solid (actually an insoluble protein structure) by the application of heat.

This requirement for "hot curing" was their major disadvantage as many structures were too delicate to withstand the thermal cycles. However, they were ideal for plywood manufacture, the major advantage being their hot water resistance and their ability to withstand "limited periods" of boiling water immersion.

The literature [5] indicates that there were many adhesives based on blood albumen or combinations of blood albumen with phenol-aldehydes, casein, soybean meal and other protein-based products. However, in aircraft structures blood-based glues would have been almost exclusively used in plywood [6,7]. With the aircraft industry's reluctance to use casein adhesives, due to their poor water resistance, blood albumen glues and/or blood albumen/casein or occasionally blood albumen/soya bean extract became the primary adhesive systems for the preparation of plywood until ousted by the Tego phenol-formaldehyde (P/F) system in the early to mid-1930s.

Norman de Bruyne (q.v.) commented [8] in the late 1980s that in the early 1930s, blood glues were state-of-the-art for plywood applications and quoted their use in preparing laminated wood structures for the Snark, his first aircraft designed and built in 1932–1934; he also used casein-based assembly glues in the plane's construction.

5.2.1.3. Casein Glues

During the First World War, due to lack of improvement in engine power, the weight of aircraft had to be reduced to improve their speed and performance; adhesive bonding was key in being able to achieve this The aerofoil section became a truly bonded structure:

- The solid wooden spars were replaced by considerably lighter, hollow box spars, generally made by bonding strips of spruce together to form the required box section.
- The ribs were bonded into place; their position being held, until the glue set, by applying a rivet at either end.
- The curved aerofoil surfaces were thin, pre-glued plywood sections which were then bonded into place; any other curved structures were also produced from bonded plywood.

In addition, the longerons, initially of ash and spruce, were screwed together and wrapped in a cotton tape, which was then impregnated with the adhesive solution to form an integrally bonded structure. When it became difficult to obtain wood of suitable length to produce the longerons, two shorter pieces were first glued together before being submitted to the treatment just described.

Amongst many others, the Sopwith airplanes are typical of this technique of reducing the weight of the airframe whist still retaining its integral strength; the Sopwith Triplane in Fig. 4 is a typical example.

Casein had been the basis of general-purpose wood glues, particularly in Switzerland and Germany, since the early 1800s but now their use was adopted, across the board, by the aircraft industry as assembly adhesives and glues for plywood. Fig. 5 shows a post-war application in the construction of a fully bonded rib; this would have mirrored the military approach of a few years earlier. The

Figure 4: Sopwith Triplane of 1916. (Copyright: John M. Dibbs at www.planepicture.com.)

Figure 5: Casein bonded rib of a 1923 ANEC II.

two close-ups show the bonded box spars at the leading and trailing edge areas and the reinforcing cross-members which are bonded in place; being initially held in position, by a simple rivet at either end, until the glue set.

Casein, like gelatin and blood albumen, is essentially a protein. It is obtained as a precipitate from skimmed milk, which has been treated with sulfuric, hydrochloric or lactic acid. The glue is prepared by blending the precipitated casein (which also contains several milk-product impurities) and alkaline salts (generally of calcium, sodium or boron) in an aqueous medium. To "set" – i.e. to achieve a solid state – the adhesive solution pH is adjusted to between 9 and 13.

Casein has a micelle structure, which consists of sub-micelles linked through $Ca_9(PO_4)_6$ moieties as is shown in Fig. 6.

The advantage of these glues over their gelatin counterparts was that they could be applied "cold", i.e. at ambient temperatures. Further, the polypeptide chains in the casein contained bound phosphorous in the form of mono- and di-esters of phosphoric acid as well as pyrophosphates. The di-esters and the pyrophosphates caused a degree of cross-linking to take place within the adhesive during the setting process.

Steam forming is used to produce wooden structures with tight radii of curvatures. Casein glues are capable of bonding these structures, which will have high moisture contents. They are not, however, recommended in the manufacture of very thin plywood as here the relatively high moisture content cannot be tolerated and can lead to the "blowing" of the skins on de-pressing.

Their main disadvantages were that the pressing times were long but this situation could be alleviated by submitting them to a hot cure. Further, the gluelines in the bonded structures were readily attacked by acidic-, alkaline- or enzymatic-hydration.

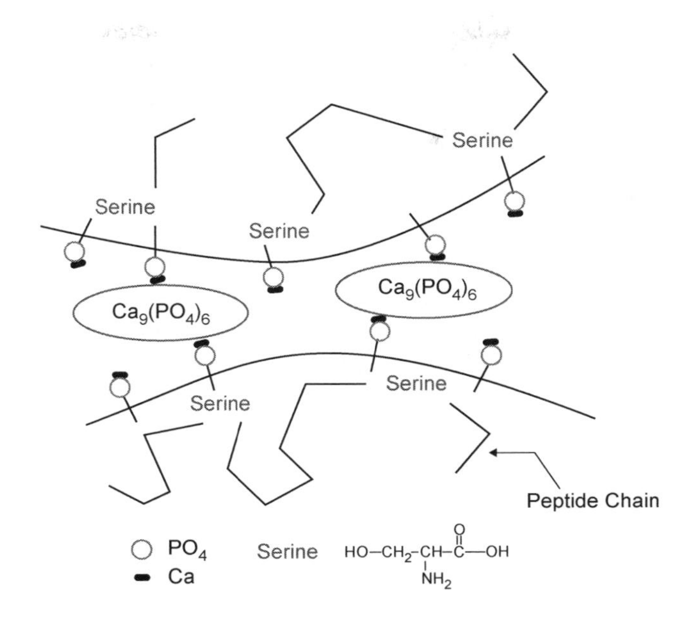

Figure 6: Structure attributed to casein micelles.

This latter occurrence led to many observations of fungi spores being seen in early aircraft and to the story that if the pilots could detect the smell of sour milk, the aircraft structure was about to fail!

In spite of this, the use of casein-based glues to bond aircraft structures continued until the middle of the Second World War; their use in general plywood construction was phased out in the 1930s with the advent of P/F adhesives and, to a certain extent, by the U/F systems.

5.2.2. Synthetic Adhesives

Although the use of glues based on animal products continued well after the successful introduction of synthetic adhesives, the use of the latter products in aeroplane construction became almost universal in the early 1940s.

5.2.2.1. Urea–Formaldehyde Adhesives

The most significant revolution in bonding wooden aircraft structures occurred with the introduction of adhesives based on U/F resins. Although the basic resin had been patented in 1920, it was not until 1937 that adhesives based on this chemistry were produced by BASF in Germany and Aero Research Limited at Duxford in the UK (Aerolite® range). Such was their impact that they were in significant commercial production before the outbreak of the Second World War.

However, due to the considerable commercial pressure by the aluminium companies, many aeroplanes, by that time, were essentially "wooden aircraft built out of metal". The use of U/F glues, therefore, was initially confined to marine applications. With the outbreak of the Second World War and the potential shortages envisaged for aluminium, the industry turned back to the use of wood and to U/F adhesives.

U/F adhesives, as their name implies are made by the reaction of urea with formaldehyde (Fig. 7). They are, essentially, two-component systems. In the initial commercial products the U/F, resin was applied to one substrate and the hardener to the other. A relatively rapid cure resulted when the joint was closed; this was often accelerated by heating. Hardeners were either acids or compounds, which would react with the U/F resin (usually the free formaldehyde) to release acidic species. The resultant lowering of the overall pH of the system initiated the cross-linking reaction and produced the hard infusible solid, which was the cured adhesive.

The cured adhesive had excellent resistance to cold water and relatively good resistance to high humidity. It was, though, susceptible to hydrolysis under acidic conditions; such an attack was significantly accelerated under the influence of heat.

The other problem, and this was a serious one, was that the adhesive had no gap-filling properties. Any thick film of adhesive was extremely brittle and "crazed" very easily. The first attempt to address this situation used the addition of fillers. This slowed the crazing process down but did not eliminate it. However, when C Rayner of

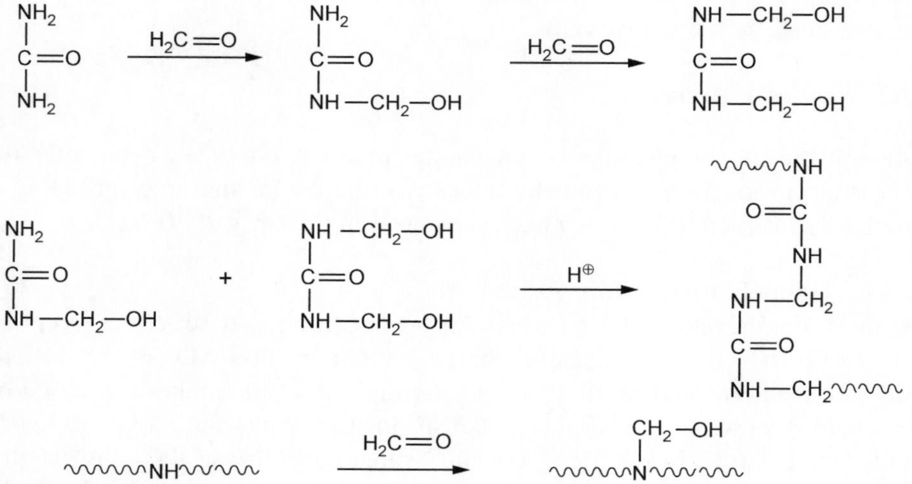

$$NH_2-C(=O)-NH_2 \xrightarrow{H_2C=O} NH_2-C(=O)-NH-CH_2-OH \xrightarrow{H_2C=O} HO-CH_2-NH-C(=O)-NH-CH_2-OH$$

$$NH_2-C(=O)-NH-CH_2-OH + HO-CH_2-NH-C(=O)-NH-CH_2-OH \xrightarrow{H^{\oplus}} \sim NH-C(=O)-NH-CH_2-NH-C(=O)-NH-CH_2\sim$$

$$\sim NH \sim \xrightarrow{H_2C=O} \sim N(CH_2-OH) \sim$$

Figure 7: Reaction schematic for a typical urea–formaldehyde resin.

Aero Research Limited discovered the efficacy of formic acid as a hardener [9], the crazing problem was solved; the resultant adhesive was designated Aerolite 300.

Another approach to solve this problem, which in addition gave good gap-filling properties and improved the long-term durability, was to add cellulosic fillers and an aliphatic or alicyclic alcohol to the resin and cure with thiocyanic acid released from the ammonium thiocyanate hardener.

In the UK, the considerable knowledge on using such systems in the preparation of laminated wooden hulls and general bonding applications in marine craft was available, in 1939, for the construction of wooden gliders.

The design of the Airspeed Horsa glider [10] is a good example of this. The fuselage comprised six bonded "barrels" – plywood skins bonded with Aerolite 300 to solid wood frames (Fig. 8). The bulkheads were wooden sandwich structures bonded with Aerolite and the wooden stringers were bonded along the length of the fuselage. The same adhesive was used to bond the wing and tail assemblies and to build up the laminated main spar.

One of the most famous examples of powered aircraft construction is that of the de Havilland Mosquito which first flew in 1940 and was in service by mid-1941 (Fig. 9). The DH98 Mosquito was developed from the DH91 Albatross of 1932, which was an all-wood airliner and mail plane; the Albatross contained structures bonded using casein glues.

The Mosquito was constructed from plywood skins, laminated wooden spars of birch and spruce and balsa wood [10]. The plywood was bonded with the Tego phenolic film adhesive and the overall assembly was initially glued together with a casein adhesive. The better properties, particularly the superior resistance to biodegradation, and the shorter curing times seen with the U/F adhesives (an estimated reduction factor of 60 times) soon saw Aerolite, amongst others, replacing the casein. The construction proved to be an ideal exemplar for the concept of reducing weight by bonding but not sacrificing strength as it easily outperformed the then current marques of Spitfire.

Figure 8: Construction of the airspeed Horsa Glider.

Figure 9: de Havilland DH98 Mosquito (courtesy of: Colin Davies at www.users.bigpond.com).

The fuselage comprised an inner 3-ply plywood skin bonded to a laminated wood frame. The balsa wood, acting as a sandwich material, was then glued in place to fill the spaces between the skins. The outer 3-ply skin was then bonded on top to complete the sandwich.

Bonded box spars were used in the wings together with laminated spruce booms. To build up the rear spars, sections were joined using scarf joints and adhesive.

After the war the use of U/F adhesives persisted, particularly with gliders [13] where similar construction techniques were used, but these adhesives also entered the jet age with their extensive use in the de Havilland Vampire [10] and limited use in the Gloster Meteor. The Vampire not only used a similar fuselage construction to the Mosquito but the production of the air intakes, which had an incredibly complicated multi-curvature design, was only possible through the ability to be able to produce bonded wooden laminates. The Meteor used bonded wood spacers in the trailing edges and control surfaces of the wing.

It is interesting to note that since it was first approved in 1937, Aerolite is still the only repair adhesive specified by the Royal Air Force for wooden aircraft [9] and is extensively used for both renovation of early aircraft and the construction of replicas; as typified by the work of the Shuttleworth Collection at Old Warden in the UK.

This is, in part, due to its ability to repair successfully structures that were previously bonded with casein glues. The residual casein in the joint to be repaired gave a substrate, which was highly alkaline. The acidic hardener of the U/F adhesive could first be used to neutralise the interphasial region and then further applications could be used to cure the U/F resin.

5.2.2.2. Phenol–Formaldehyde Adhesives

The first patents for these resins appeared in 1904. The first P/F adhesive appeared in 1929 and comprised cellulosic film impregnated with a spirit-soluble resole resin. This was the hot-curing (typically 5 min at 150°C under a bonding pressure of 17.5 MPa) Tego film adhesive of *Th*eodor *Go*ldschmidt that was used in the Focke-Wulf TA 154 Moskito (Fig. 10); the German attempt at matching the DH98 Mosquito.

Tego film was an important innovation as far as plywood adhesives were concerned. It eventually replaced casein as the universal adhesive for preparing water-proof plywood structures in aircraft construction and plywood in general.

In the UK, Tego was declared to be a strategic material in 1936 and Micanite and Insulators Limited, the UK agents for Theodor Goldschmidt AG of Essen, undertook manufacture of this product for the home market [14]. To do this, they launched a new company – British Tego Gluefilm Limited – which was part-owned by Goldschmidt!

As in Germany, two grades were offered: "Commercial" and "Aero" (equivalent to the "Standard" and "Flugzeug" ("FZ") grades in Germany). As its name implies, the Aero version was used for plywood intended primarily for the aircraft industry. The Aero grade gave stronger bonds than its Commercial counterpart and cured more quickly. This would indicate that the phenolic resole used in this version was more advanced and/or more highly catalysed. This conclusion is born out by the fact that the shelf-life of the Aero film was quoted as being "rather less" than the 3 years given for the Commercial grade; probably about 1 year at ambient temperatures.

The Aero grade was made for two different end-users; the water-proof plywood manufacturers and those companies making densified wood — laminated wood where, due to the laminating pressure used and the actual construction of the plywood, the final product had a density significantly above that of the timber species used in its construction.

As has already been indicated, the water-proof plywood was used extensively for the construction of aircraft and gliders and the densified wood was used in propeller manufacture for such aircraft as the Avro Lancaster, the Handley Page Halifax, the Supermarine Spitfire and the Hawker Typhoon.

Figure 10: Focke-Wulf TA 154 *Moskito* (photographer: unknown).

During the early days of Tego film manufacture in England, it was found that the key to good performance and reproducibility was the quality, and hence the method of manufacture, of the cellulosic tissue which acted as the adhesive support/carrier.

Once this problem was solved then manufacture of the film proceeded apace with rolls up to 84 in. wide (2.1 m), but more usually about 50 in. (1.25 m), being produced. Production peaked in 1943 with 10,000,000 ft^2 (about 1,000,000 m^2) being manufactured per week. This type of formulation was so important and successful that it remains the preferred choice for water-proof plywood applications up to the current day.

In Germany, as already mentioned, Tego film was the adhesive of choice for the bonded plywood used in the prototypes of the Moskito. The story of this aeroplane's development is an extreme example of the problems that can occur on choosing the wrong adhesive [15]. The original bonded airframe performed well but a design fault with the undercarriage caused numerous crashes, which delayed any major build. However, when the Goldschmidt factory, which made the Tego film, was destroyed in a RAF bombing raid, a new adhesive was introduced by Dynamit AG of Leverkusen.

Not only was this new adhesive weaker but it also actually degraded the wooden substrates. This, fairly rapidly, led to the termination of the programme following a spectacular crash where the wings disintegrated in flight.

Although few references exist, it is known that the Dynamit adhesive was also a P/F system. However, it is believed that this must have been a cold-curing adhesive; probably similar to a system developed by BIP in the UK. This would have certainly utilised an acidic compound (probably *p*-toluene sulphonic acid) of very low pH to accelerate the cure; it has since been shown that this type of adhesive does, indeed, lead to a rapid degradation of wooden adherends.

The effect of acid catalysis/curing on the integrity of the wooden substrate was not the only problem with these ambient-curing systems; the water boil test also had to be altered to accommodate the presence of acidic species.

The original test for bonded plywood was 72 h in boiling water but for the acid-cure systems (essentially assembly glues) the immersion time had to be dropped to 6 h as significant acid loss (for example *p*-toluene sulphonic acid) occurred if the bonded joints were left any longer.

5.2.2.3. Resorcinol–Formaldehyde Adhesives

Although they were patented in 1930, resorcinol–formaldehyde (R/F) adhesives, based on resorcinol novolac resins, started to appear in the mid- to late 1940s, following the discovery of how to control their reactivity by ensuring that their pH was close to, or just slightly above, neutral. However, their formulation is somewhat complex and the pure R/F systems have, essentially, only been used in marine applications. However, there is an exception as these systems do still have one application in the aircraft industry. They are used in wooden propeller manufacture where the laminates are bonded together with Aerodux® R/F adhesives [16].

To gain better control over assembly times these initial products were reformulated to contain a blend of a pure R/F novolac and a P/F resole of high formaldehyde content. As an unexpected benefit, the resole acted as a diluent for the system, which allowed longer assembly times, making the system easier and more tolerant to use.

It is these hybrid systems (for example Aerodux 500), which are based on phenol and resorcinol that have found extensive usage, up to the current day, in wooden aircraft applications both for assembly and repair and for the manufacture of plywood.

5.3. Early Structural Adhesive Bonding in Metallic Aircraft Structures

Both towards the end of and immediately following the Second World War, wooden structures in military, passenger and freighter aircraft were slowly replaced by metals; more recently composites have begun to replace the metallic structures. This meant that individual components and structures were required to carry higher loads, due to more demanding aircraft performances, and this, in its turn, demanded significantly improved adhesive systems.

The first of this new generation of structural adhesives, developed in the UK by Norman de Bruyne and George Newell of Aero Research Limited, Duxford, Cambridge, was Redux® (*Re*search at *Dux*ford), which was based on a formulated P/F resin and was the first synthetic adhesive for bonding structural components constructed from composite and metallic materials.

5.3.1. Redux and the Redux Process

It was the search for a suitable adhesive to bond Gordon Aerolite®, a structural composite developed in 1936 and comprising a phenolic resin matrix reinforced with flax fibres, which led to the development of Redux and the Redux Process.

Gordon Aerolite was the result of work on phenolic moulding materials for the manufacture of variable pitch propellers for de Havilland. The use of continuous flax fibres enabled de Bruyne to produce unidirectional prepregs instead of producing reinforced mouldings.

These prepregs were laid up in sheets – either in the 0° direction or cross-plied at 0°/90° – and cured to give a structural composite having, at an Sg of 1.36, nearly half the density of the Duralumin then in use. It exhibited an ultimate tensile strength of nearly 500 MPa and a Young's modulus of nearly 50,000 MPa.

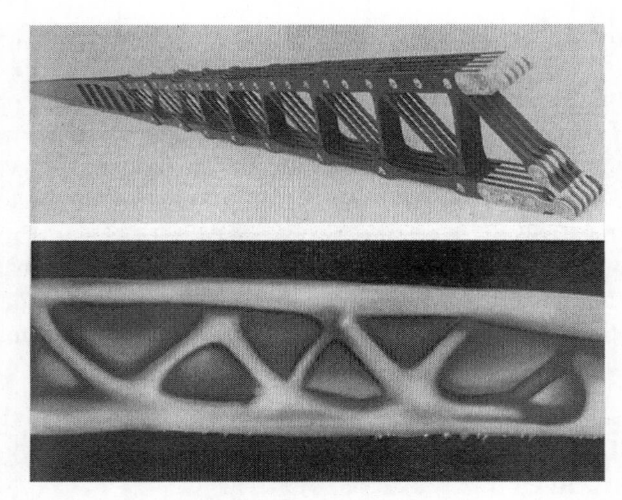

Figure 11: Gordon Aerolite Blenheim Spar modelled on the Wing of the Vulture.

This novel concept of using synthetic structural composites was immediately taken up by the Air Ministry who placed an initial order for a prototype 30-ft wing spar for the Bristol Blenheim fighter-bomber; the design based on the metacarpal bone of the vulture (Fig. 11).

With the outbreak of war, work was immediately commenced on producing a prototype Supermarine Spitfire cockpit and fuselage out of Gordon Aerolite [17]

Figure 12: The Gordon aerolite spitfire prototype.

(Fig. 12). In the event, although a replacement for aluminium was not required for the Spitfire, some 30 Miles Magister tailplanes were produced out of Gordon Aerolite.

From the start, de Bruyne and Newell realised that use of the available adhesives (or glues) would not be suitable for joining non-porous substrates such as Gordon Aerolite. As Gordon Aerolite was based on phenolic resole chemistry, it was towards these systems that de Bruyne turned for his adhesive. Several years of non-continuous research led to the discovery, on 24 February 1942 of an adhesive system which could not only successfully bond Gordon Aerolite but also, for the first time, gave a shear strength on aluminium substrates of over 2000 lb/in^2 (>13.8 MPa) (Fig. 13).

The liquid component of the adhesive (Liquid E) was a phenolic resole having a controlled phenol to formaldehyde ratio of 1:1.57 and a relatively high alkaline

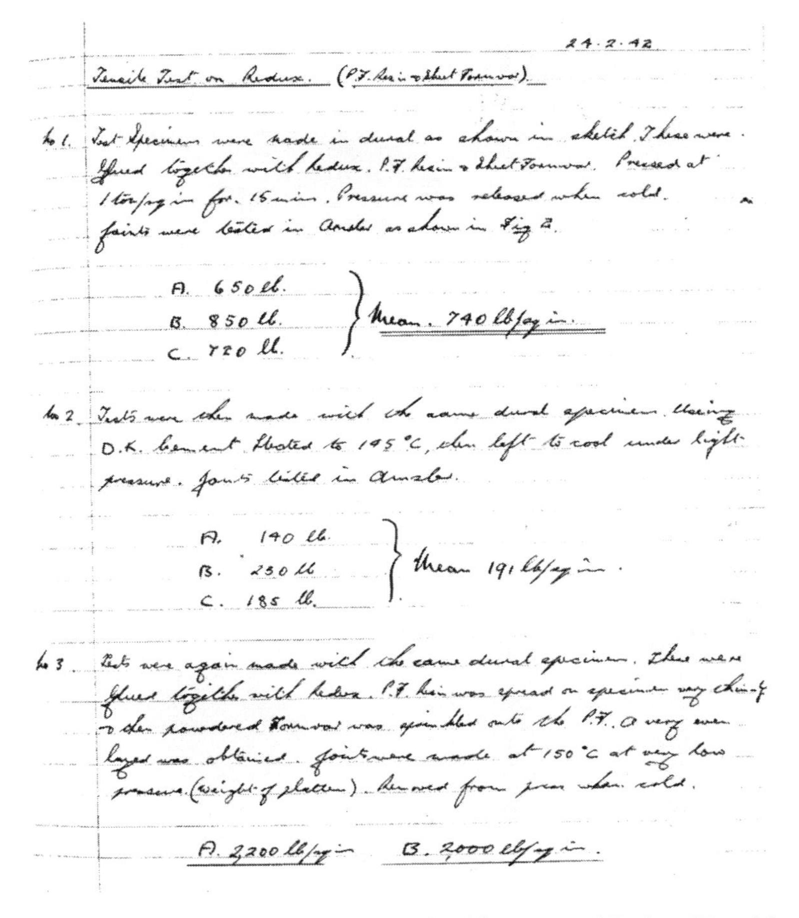

24.2.42

Tensile Test on Redux. (P.F. Resin & Sheet Formvar).

No 1. Test Specimens were made in dural as shown in sketch. These were glued together with Redux. P.F. Resin & Sheet Formvar. Pressed at 1 ton/sq in for 15 mins. Pressure was released when cold. Joints were tested in Amsler as shown in Fig 2.

A. 650 lb.
B. 850 lb.
C. 720 lb.
Mean. 740 lb/sq in.

No 2. Tests were then made with the same dural specimens using D.K. Cement heated to 145 °C, then left to cool under light pressure. Joints tested in Amsler.

A. 140 lb.
B. 250 lb
C. 185 lb.
Mean 191 lb/sq in.

No 3. Tests were again made with the same dural specimens. These were glued together with Redux. P.F. Resin was spread on specimens very thinly & then powdered Formvar was sprinkled onto the P.F. A very even layer was obtained. Joints were made at 150 °C at very low pressure (weight of platten). Removed from press when cold.

A. 2200 lb/sq in B. 2000 lb/sq in.

Figure 13: George Newell's notebook — The Discovery of Redux (Exp. No. 4.3).

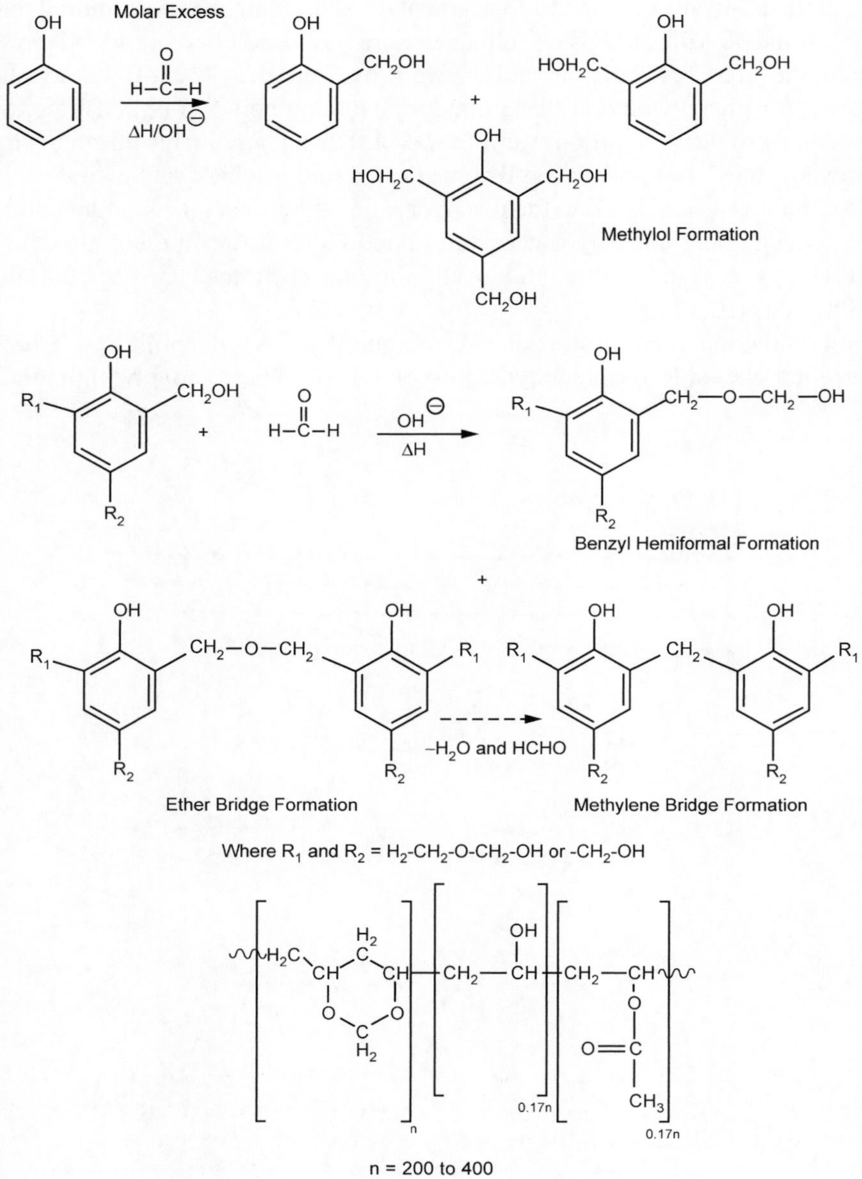

Figure 14: Chemical structure of a typical phenolic resole and of polyvinyl formal resin.

Figure 15: Application of Redux liquid and powder adhesive to an airframe component.

content. It was painted onto the substrate to be bonded and was then covered with polyvinyl acetal, a high molecular weight polymer in powder form. Fig. 14 shows the chemical structure of the adhesive and Fig. 15 illustrates the typical application process for Redux Liquid and Powder, which is still valid today.

It is no small example of the trust that the aircraft manufacturers had in adhesive bonding that bonded structural components using Redux were in operational aircraft within 2 years of the product's development.

Oddly, the first use of Redux was not in a metal aircraft or to bond Gordon Aerolite components but in the essentially all-wood, long-range versions of the Mosquito: the de Havilland DH103 Hornet and Sea Hornet; the latter was the version intended for naval carrier use [11,18,19].

The design of both aircraft was such that the wings were expected to have to carry very high tensile stresses in flight. Further, in the naval version, the wings were folded when stowing the aircraft below decks, putting even greater stresses on the "hinge" and the wing roots. To ensure that the wing was strong enough to withstand these loads without increasing the overall weight of the aircraft, aluminium (duralumin) stiffeners were Redux bonded to the wooden components in the wing ribs, spar booms and stringers as is shown in Fig. 16.

It is interesting to note that British Tego film had also been evaluated for bonding aluminium-skinned sandwiches but this approach had never progressed beyond the initial laboratory trials.

Stiffeners were also bonded into the fuselage to strengthen areas where internal attachments were to be made and this culminated in the first structural metal-to-metal bonding when integrally bonded metal reinforcements were built up on the rear bulkhead at the points where the arrestor hook and the tail wheel were located.

From these beginnings have come the commercial ranges of structural adhesives based, particularly, on phenolic, epoxy, polyimide and now cyanate

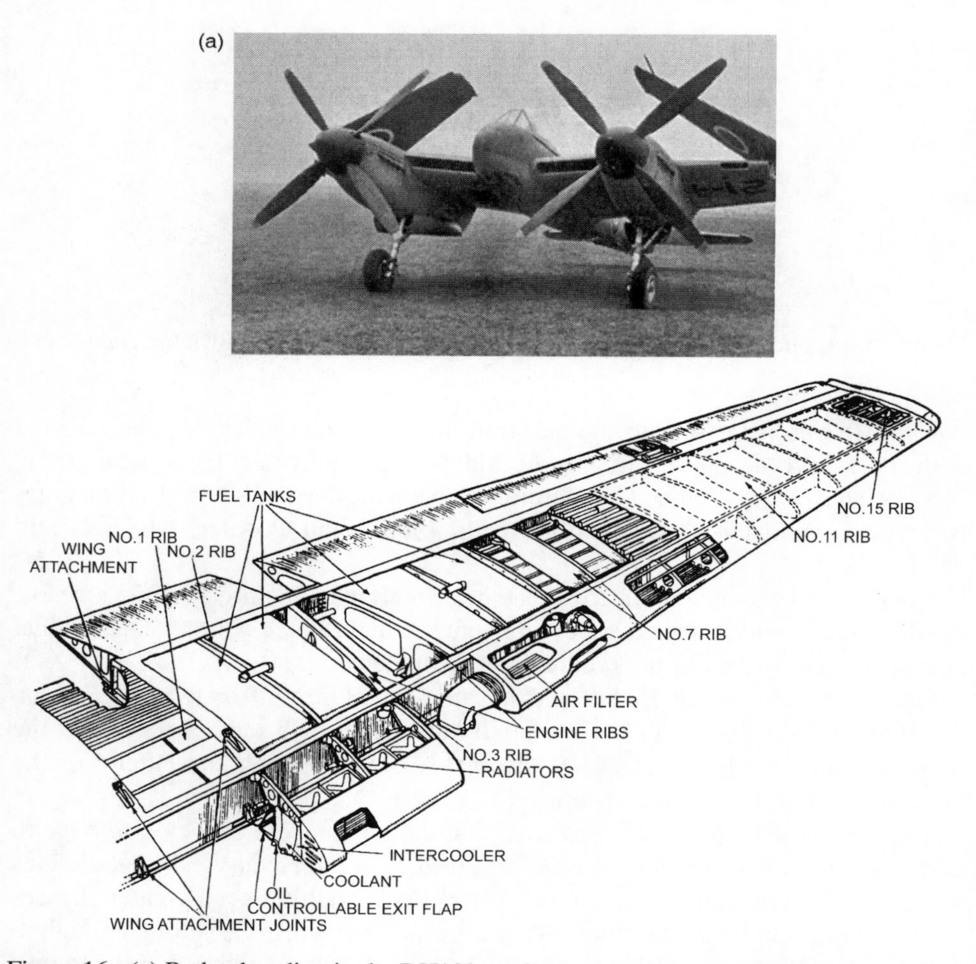

Figure 16: (a) Redux bonding in the DH103 sea hornet – structure of the wing. (b) Redux bonding in the DH103 sea hornet – reinforced spars. (c) Redux bonding in the DH103 sea hornet – reinforced stringer.

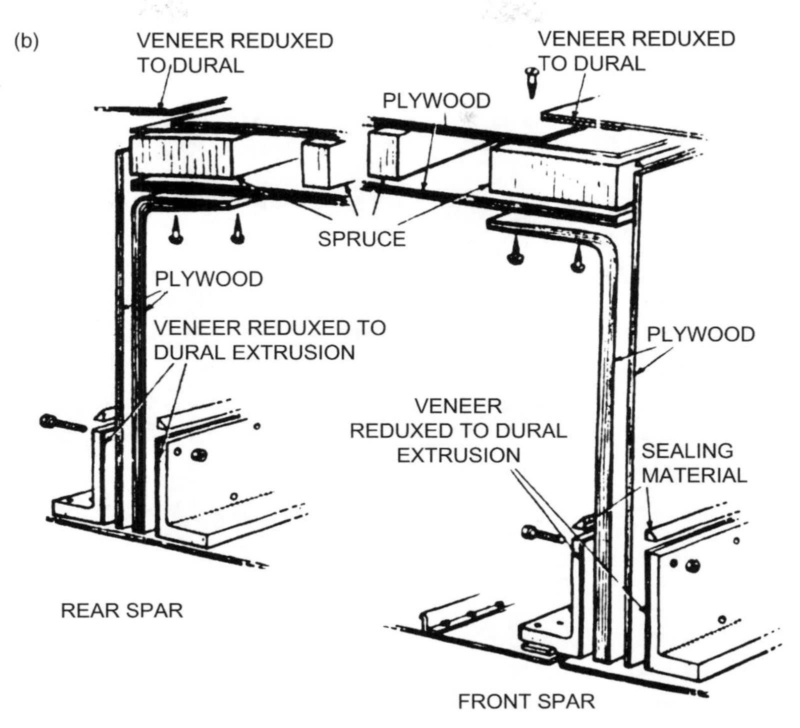

Figure 16: Continued.

ester chemistries which have been tailor-made for the various applications within the aircraft industry and, in the last few years, for space applications. These adhesives impart, as required, a combination of toughness, thermal resistance and durability to the bonded joint.

5.4. The Acceptance of Structural Bonding by the Post-war Aerospace Industry

It is clear from the foregoing that the aircraft industry embraced the concept of adhesive bonding from the very start in 1903; the step from gluing furniture to gluing wooden structures in aeroplanes obviously was not seen as a major one.

However, following the end of the Second World War, structural bonding in the burgeoning aerospace industry was no longer simply an extension of techniques used for centuries in the furniture industry. To all intents and purposes, with the advent of the all metal constructions, for example, the de Havilland DH104 Dove (Fig. 17) (the first all-metal passenger aircraft; initial flight in 1945), the day of the wooden commercial and military aircraft had begun to come to an end.

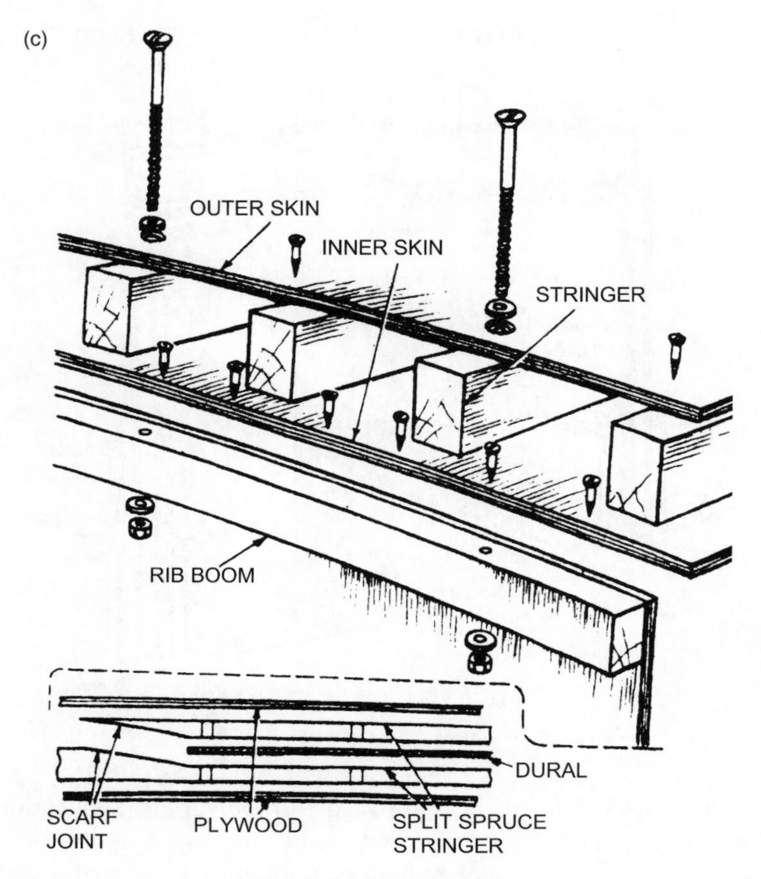

Figure 16: Continued.

Figure 17: de Havilland DH104 Dove.

Figure 18: Fokker F-27 Friendship.

The industry, therefore, needed to reassess its commitment to adhesive bonding. The extension of this acceptance to bond structures, which used permeable substrates, to the concept of bonding non-permeable metal components is, at first sight, not so readily understandable. There are probably three major reasons why it happened.

Firstly, the immediate post-war industry was emerging from a major conflict, which had driven the acceptance of innovation at a greater pace than in peacetime. Secondly, the strong support of de Havilland was clear for all to see and thirdly there were, in Europe, two passionate advocates for this approach. One was Bob Schliekelmann who was head of the Production Research Department at Fokker (then called N.V. Koninklijke Nederlandse Vliegtuigenfabriek Fokker) who, from the very first, used Redux bonding in Fokker's aeroplanes. This commenced in the late 1940s with the Fokker F-27 Friendship (Fig. 18) and continued to the present day with the F-28, F-50, F-70 and F-100.

The other was, not unnaturally, Norman de Bruyne who was effectively the "father" of modern day structural adhesives and structural adhesive bonding. De Bruyne had 11 compelling reasons why structural adhesive bonding was the better option to bolting and/or riveting and, hence, why it would be beneficial for the post-1945 aircraft industries [8]. These concepts were disseminated to aircraft designers and engineers at series of "Summer Schools" which were run by de Bruyne in the early 1950s. The reasoning is as valid today as it was in 1950.

5.4.1. Why Bond?

De Bruyne's 11 reasons can be summarised as follows:

Reduction in weight: As for timbered structures, the weight of the final metallic or composite structure can be significantly reduced by, for example, using essentially "hollow", preformed stringers which are bonded into place, or honeycomb sandwich structures (see below) instead of a solid constructions. Further weight reductions can be achieved by the use of thinner gauge metal; reinforcements can then be bonded into place only in areas of high load or stress.

Increase in fatigue life and improved sonic damping: Rivet and bolt holes act as stress concentrators which can readily lead to failure – for example, the Comet disaster where catastrophic fatigue crack growth occurred following unacceptably high stresses around the rivets in the window construction.

As Fig. 19 shows, a bonded joint gives no such areas of high stress concentration; a uniform stress distribution across the bonded joint is achieved. The glueline also absorbs some of the acoustic energy associated with the structure which leads to reduced high frequency noise propagation.

In Fig. 20, the fatigue curves for both riveted and Redux-bonded joints — plotting cyclic load against number of cycles to failure — are shown. It is clear that all of the bonded specimens failed outside the bonded area. Further, it can be seen that the bonded joint can support about 6–7 times the fatigue load for any given "cycles to failure" value or, to look at it in another way, at any given cyclic load, the bonded joint can be seen to withstand many orders of magnitude more cycles.

Simplification in design: In general, the designs associated with bonded structures are simpler than those using rivets. For example, only simple design criteria were used to produce the Redux bonded wing sections of the Fokker F-27 Friendship and the RJ 80 (Figs. 21 and 22, respectively).

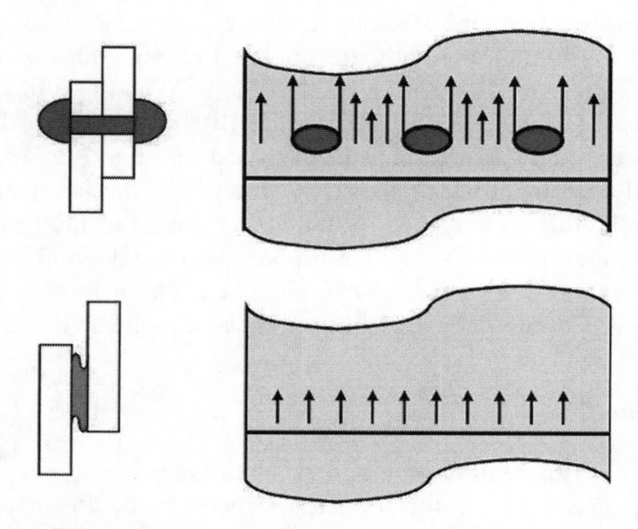

Figure 19: Stress concentrations in riveted and bonded joints.

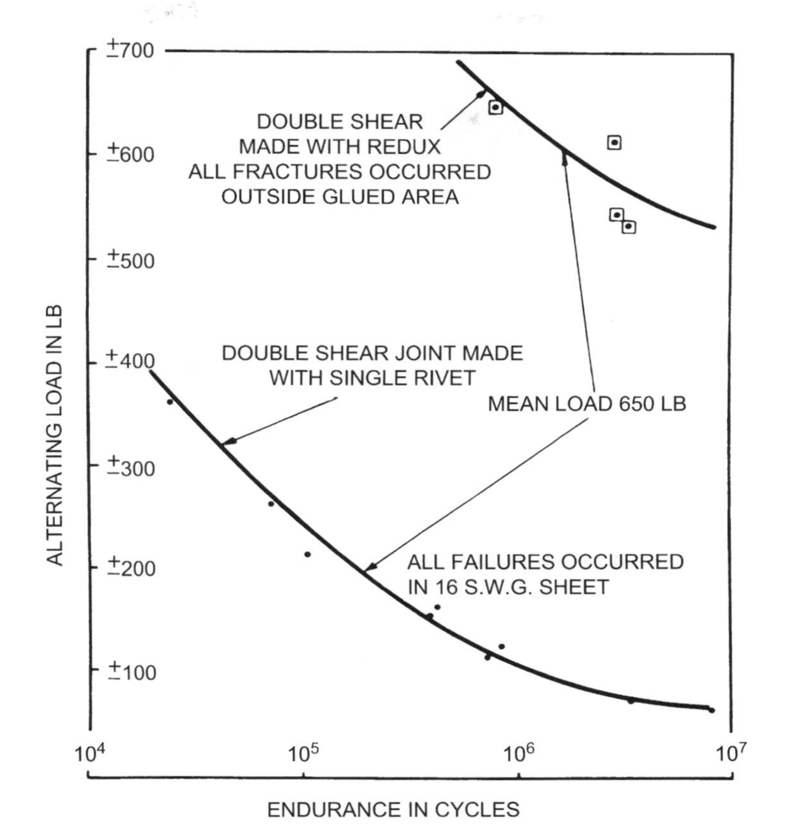

Figure 20: Fatigue curves for riveted and bonded joints.

Figure 21: Redux bonded wing assembly in the Fokker F-27.

Figure 22: Redux bonded wing assembly in the BAE SYSTEM's RJ 80 (formerly the BAe 146).

Smooth external finish: When attachments are riveted to external skins, the closing of the rivet tends to pull in the skin and cause local distortion, which can lead to some disruption in the aerodynamic properties of the surface. If the same area is bonded, distortion is reduced to an acceptable level and, as can be seen in Fig. 23, bonding also improves the overall stiffness of the final structure when compared with a riveted component of the same pitch between points of joining.

The ability to join thin substrates: If thin pieces of, for example, aluminium are joined using rivets, the final component is susceptible to a tearing failure through the substrate which emanates at the rivet and its associated hole in the substrate. Any tensile load across the riveted joint (as seen in Fig. 19) is maximised at the point where the rivet enters the skin. Thus, when the substrate is thin, this can readily lead to crack and/or tear initiation and failure of the structure. Using bonding technology within these components eliminates these bearing stresses and minimises the chance of premature failure.

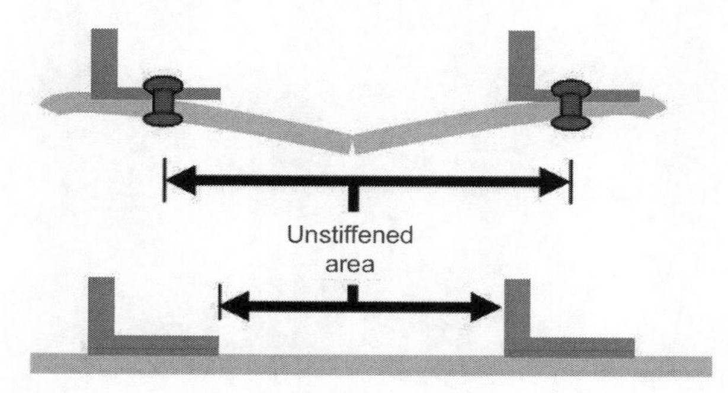

Figure 23: Schematic of the effect of riveting and bonding on the aluminium skin.

Reduction in production costs and time: When adhesive bonding is used, as opposed to producing a riveted structure, an entire component can be assembled and bonded in one operation. Further, the jigging required is often simpler, which means that assembly is quicker. The thinner gauge metal that can be used reduces costs. As soon as the bonded component reaches a certain size, irrespective of all the other benefits, bonding becomes progressively cheaper than automatic riveting. This latter point can readily be seen from de Bruyne's calculations carried out in the early 1950s (Fig. 24). He evaluated the labour and monetary commitments and showed that, at stringer lengths of over about 8 m, bonding was more economic both in terms of time and overall costs.

Crack growth arrest: A crack growing in an external skin is neither stopped nor is its rate of growth retarded in a riveted structure; the crack simply grows round the rivet or through the rivet hole. Any crack meeting, for example, a bonded stringer such as those represented schematically above, is turned back on itself and cannot progress.

As an example, Fig. 25 shows the slow rate of crack growth in two samples of Glare® (q.v.) — a bonded laminate — compared with the catastrophic crack growth associated with aluminium alloy sheet. For an explanation of the Glare designations see Section 5.5.1.6.

Joint sealing: As can be judged from Fig. 23, the integral bonding of components in the main aircraft frame and, particularly in the fuel tanks, makes internal pressurisation a much simpler process.

Increased strength: The strength of a riveted structure is essentially that of the bearing strength of the rivet on the skin to which it is fixed. One simple experiment showed that a riveted structure having an area of about 7000 mm^2 failed at a shear load of about 3.5 kN. The same structure, when adhesively bonded, failed at a load in the region of 55 kN.

New types of structure utilised: As far as the aerospace industry is concerned, one of the most important, novel structures since the early 1950s has been the honeycomb sandwich panel (Fig. 26). It could be argued that the wood-skinned balsa sandwich panels used in the Mosquito or the metal skinned version used by Chance Vought (Metalite® — q.v.) could have been joined by some other method than adhesive bonding. However, it is nearly impossible to argue such a case for the structures using aluminium, aramid, glass or carbon honeycomb. It is the filleting effect of the adhesive (q.v.) which enables the skins to be securely fixed to the honeycomb core.

The ability to join dissimilar materials: De Bruyne originally had in mind the use of adhesives to join combinations of metal, wood and Gordon Aerolite. However, it is just as valid today when considering the structural bonding of the range of fibre reinforced composites that are now used in

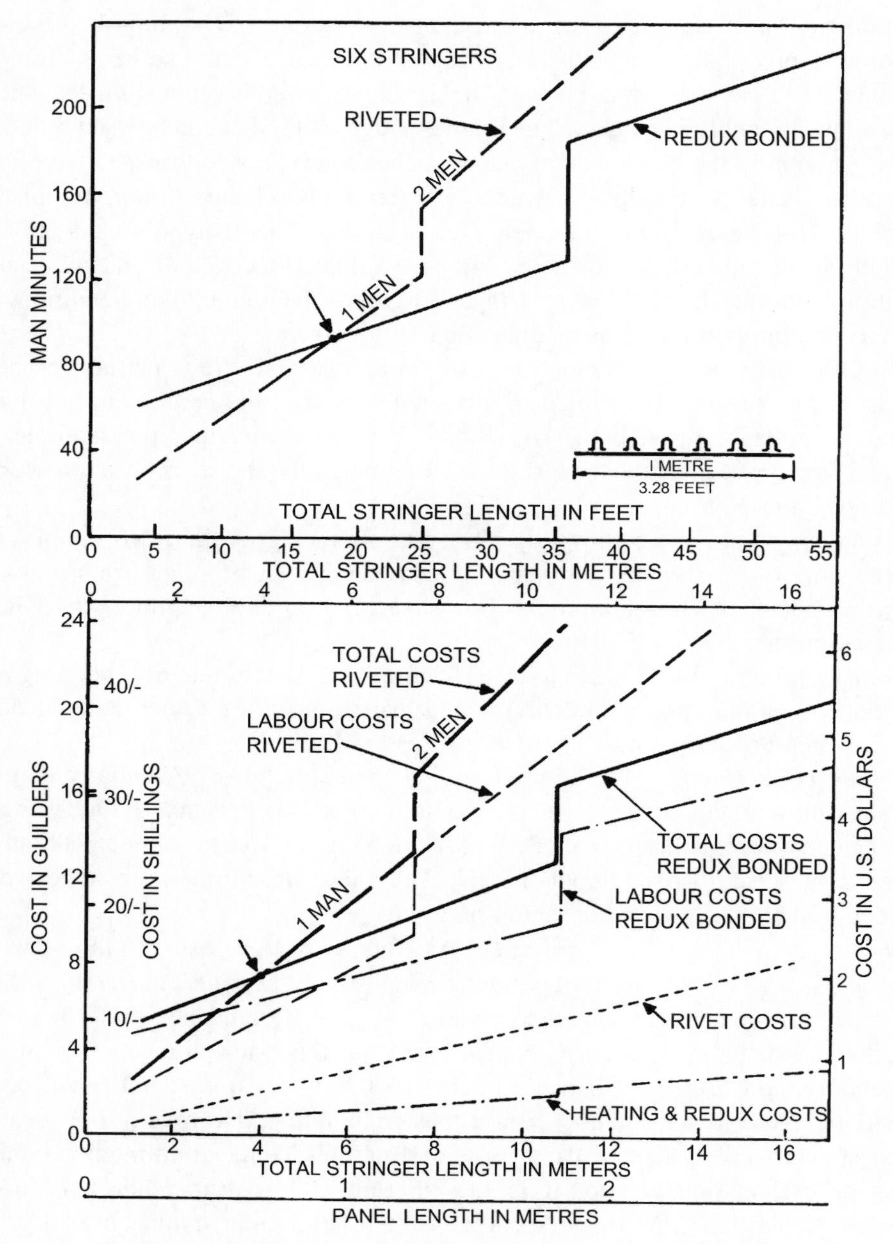

Figure 24: The economics of bonding versus riveting (early 1950s).

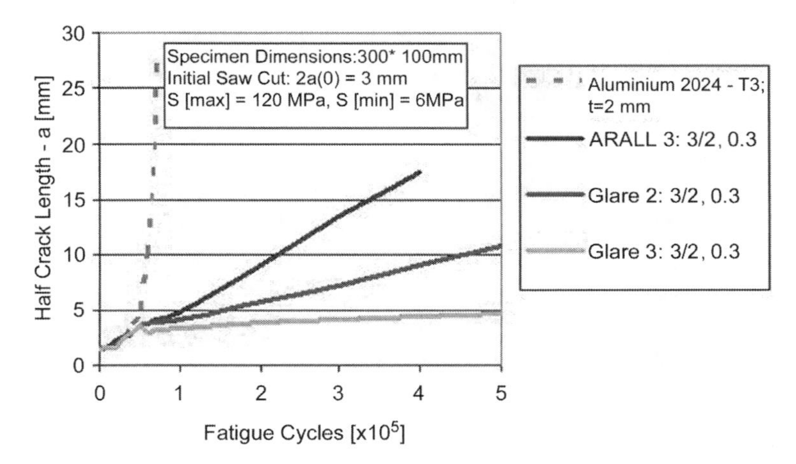

Figure 25: Crack growth in fibre–metal laminates compared with aluminium alloy (courtesy of: Fibre Metal Laminates, Delft).

the aerospace industry: particularly glass, carbon and aramid fibre-reinforced epoxy, phenolic, cyanate ester and polyimide matrices.

Having spent a significant amount of effort in both optimising the chemistry and the structure of the matrix and prepreg and the orientation of the prepreg in the final laminate to produce optimum properties, the last thing that many designers want to do is to drill holes through it so that other components can be attached. What is needed, therefore, is a range of adhesives which will join components without affecting their overall properties.

Fig. 27 shows helicopter rotor blade sections where adhesive bonding is used to produce components from carbon and glass composites, aluminium and Nomex® honeycombs and simple “plastic” foams.

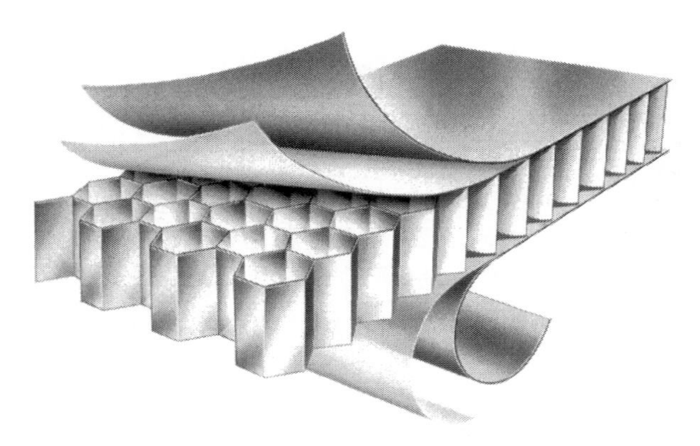

Figure 26: Schematic of a bonded honeycomb sandwich.

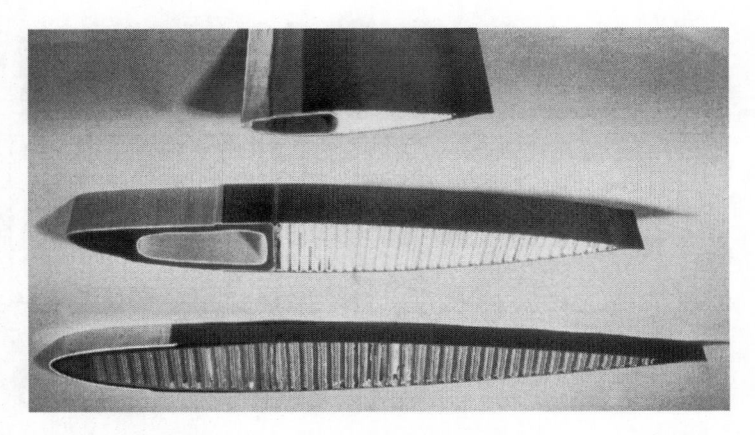

Figure 27: Bonded Westland helicopter rotor blade sections. (colour version of this figure appears on p. xv).

5.5. The Adhesive Joint

Any adhesive joint comprises two substrates, which may or may not be of the same material, which are joined by means of an adhesive.

5.5.1. Substrates

Although aircraft structures comprise thousands of components produced from a myriad of basic materials, the most common substrates for structural adhesive bonding are wood, aluminium, titanium, stainless steel and the composite materials such as bonded sandwich structures, fibre reinforced plastic (FRP) laminates and fibre–metal laminates (FML) where the metal is usually aluminium.

5.5.1.1. Wood

In post-war applications, wood is nearly always used as bonded plywood.

5.5.1.2. Aluminium

The aluminium alloys most frequently encountered are the 2000, 7000 and, occasionally, the 6000 ranges with 2024 and 7075 grades being the most common aluminium–lithium alloys such as 8090C are also used in special applications.

5.5.1.3. Titanium

Occasionally, aerospace applications utilise commercially pure titanium but the most commonly encountered substrate is an alloy of titanium which is designated Ti 6Al4V that contains 6% aluminium and 4% vanadium.

5.5.1.4. Stainless Steel

Stainless steels are alloys of iron with a low carbon content (usually $\ll 2\%$) and a minimum of 10.5% chromium. Dependant on the grade, stainless steels can also contain nickel, manganese, molybdenum, titanium, copper and nitrogen. These other alloying elements are used to enhance properties such as formability, strength and cryogenic toughness. The main requirement for stainless steels is that they should be corrosion resistant for a specified application or environment.

Typical stainless steels specified for aerospace use (many are covered by the BS 500 specifications) contain $\leq 0.15\%$ carbon, 11–20% chromium, $\leq 2.0\%$ manganese, 4–10% nickel and 0–2% molybdenum. Other trace elements, including titanium and niobium, can also be present as stabilisers.

A few structural applications, particularly in missile construction, utilise the so-called carbon steels (mild steels); these are of high carbon and low chromium content. These will be covered under the relevant application section.

5.5.1.5. Fibre Reinforced Plastics

FRPs are usually constructed of unidirectional or woven fibres imbedded in a specifically formulated resin matrix. The fibres are usually of glass and/or carbon although some specific applications call for aromatic polyamide (for example, Kevlar®), quartz or even boron fibres.

The resin matrix is usually a formulated thermoset system (i.e. a reactive matrix, which on the application of heat and pressure, chemically reacts to form an infusible reinforced laminate). The thermosetting matrices are most often based on epoxy chemistries although there are plenty of examples of phenolic, bismaleimide and polyimide matrices (for example, the HexPly® range from Hexcel Composites) and a few where the resin is based on cyanate esters. Thermoplastic matrices are also encountered (i.e. matrices that can change from a solidus to a liquidus form by the application of heat and pressure and then revert to the solid state on cooling) which are usually, but not exclusively, based on polysulphone, polyethersulphone or polyether ether ketone chemistries.

In all cases, the substrates are prepared by the careful orientation of the individual prepreg layers (to obtain the properties required) which, after being cured under pressure at temperatures between 120 and 175°C, form the composite laminates that become the adherends in the structure.

5.5.1.6. Fibre–Metal Laminate

FML materials are the most recent substrates to be used in aircraft construction. The first products, initially designated Arall® (aramid fibre reinforced aluminium laminates), Glare (glass fibre reinforced aluminium laminates) and

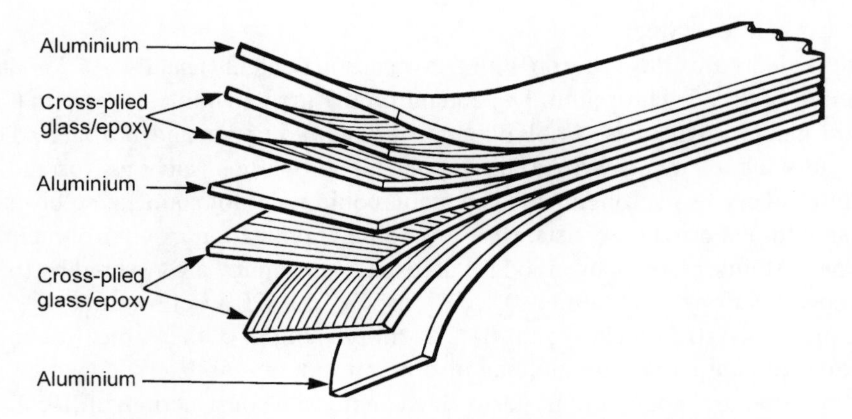

Figure 28: Schematic of a typical glare (FML) structure.

Care® (carbon fibre reinforced aluminium laminates), were invented by Prof. Boud Vogelesang of the Aerospace Faculty of the Technical University of Delft. Development to full commercial product status took place from 1978 to 1989 for Arall and to 2001 for Glare.

The laminates consist of alternate layers aluminium (0.2–0.5 mm thick], which has been anodised and then primed with Cytec-Fiberite's BR-127 corrosion-inhibiting primer, bonded together with fibre-rich, epoxy-based "adhesive-prepregs" (about 0.125 mm thick]. The reinforcing fibre is usually unidirectional in nature, although woven fabric is occasionally used. Generally two or three sheets of adhesive separate each sheet of aluminium; this can be in the 0° direction or laid up in a cross-plied orientation.

In all cases, a fatigue-resistant metal laminate is produced, which not only resembles aluminium but can also be worked, machined and bonded in a similar manner.

A typical Glare structure (glass fibre reinforcement) is given in Fig. 28 and Glare product designations are given in Table 1; the designations for ARALL are very similar.

The product designation, for example Glare 3-4/3-0.4, indicates, therefore, that the laminate is produced from 0.4 mm thick 2024-T3 aluminium alloy and comprises 4 layers of aluminium and 3 fibre layers; each fibre layer being two sheets of adhesive prepreg orientated in the 0°/90° direction.

5.5.1.7. Sandwich Panels[1]

Bonded sandwich structures are vital in aircraft design as they enable significant weight reduction without the loss of stiffness and strength. A sandwich panel

[1] More information will be given in the specific chapter "Sandwich panels" in a volume to be issued later.

Table 1: Glare product designations.

Grade	**Sub-grade**	**Metal sheet thickness in mm, alloy**	**Prepreg orientation in each fibre layer**
Glare 1		0.3–0.4, 7475-T761	0/0
Glare 2	2A	0.2–0.5, 2024-T3	0/0
	2B	0.2–0.5, 2024-T3	90/90
Glare 3		0.2–0.5, 2024-T3	0/90
Glare 4	4A	0.2–0.5, 2024-T3	0/90/0
	4B	0.2–0.5, 2024-T3	90/0/90
Glare 5		0.2–0.5, 2024-T3	0/90/90/0
Glare 6	6A	0.2–0.5, 2024-T3	+45/−45
	6B	0.2–0.5, 2024-T3	−45/+45

comprises metallic or synthetic surface skins integrally bonded to a central core. In some instances, this core is balsa wood or, because of objections to using "natural" products in structural components [20], a foamed "plastic". Typical examples of the latter are foams of polyvinyl chloride, phenolic and polyurethane.

It was the former, however, which, initially, proved to be of great interest to the aircraft industry. The essence of an idea of de Bruyne was worked on by Chance Vought, in the United States of America. They used Redux 775 Liquid and Powder adhesive to bond thin aluminium skins to "end-grain" balsa wood sheets in order to produce sandwich panels. These panels were called Metalite (Fig. 29).

This was first used in skinning the fuselage and wings of the relatively unsuccessful F-6 – Pirate and the F7 – Cutlass. It is reported that Metalite was used

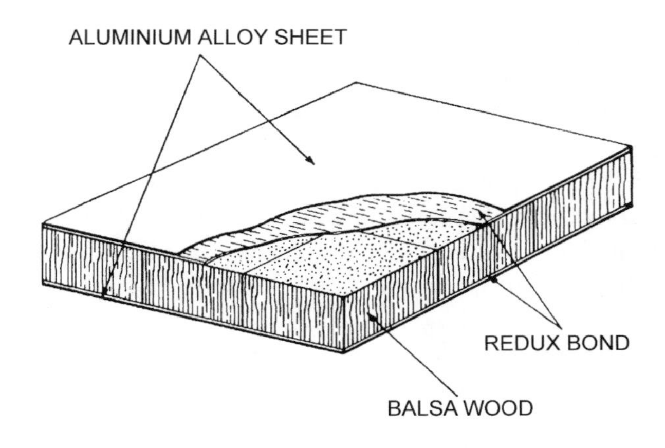

Figure 29: Schematic of metalite-bonded sandwich structure.

Figure 30: Chance Vought Regulus guided missile (image courtesy of: Lockheed Martin missiles and fire control).

extensively in skinning the Chance Vought F4-Corsair [18], but no corroboration can be found for this statement. However, Metalite did come into its own when the skins of the wings and tail planes of the Regulus guided missile (Fig. 30) were constructed from it, leading to significant weight reduction whilst retaining the required stiffness.

However, the majority of sandwich panels now utilise a honeycomb core rather than either balsa wood or plastic foam; the adhesive can be based on either thermosetting or thermoplastic chemistries. The three basic components used in honeycomb sandwich construction are discussed below.

5.5.1.7.1. Honeycomb core[2]

In 1938, de Bruyne was contemplating the use of a metallic "reinforcement" of the Miles Magister tailplane, which was to be constructed of either Gordon Aerolite or plywood. This reinforcement was to be a "honeycomb of hexagons" and would be inserted between the upper and lower skins of the tailplane. Like the Blenheim spar mentioned above, the idea for honeycomb had its roots in the natural world. This time it was the lightweight cellular structure seen in the human skull (Fig. 31). This concept is shown in his laboratory notebook (Fig. 32). A potential method of manufacture was also patented in 1938 [21].

In 1938, when de Bruyne patented honeycomb, there were no suitable adhesives to join the outer skins to the inner core, so he dropped the idea. Eventually it was

[2] For more information about honeycomb and honeycomb sandwich panels, refer to the chapter "Sandwich panels" in this Handbook.

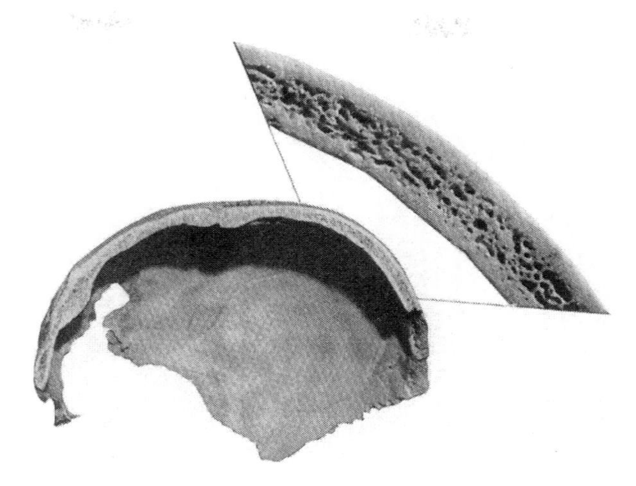

Figure 31: The structure of a human skull cellular bone between two solid outer surfaces.

Hexcel, in the United States, which started honeycomb production in 1946 supplying glass honeycomb sandwich panels for the construction of the wings of the B 36 in 1949. In Europe, it was not until the early 1950s, with the introduction of Redux 775 Film that de Bruyne's original concept came to fruition with

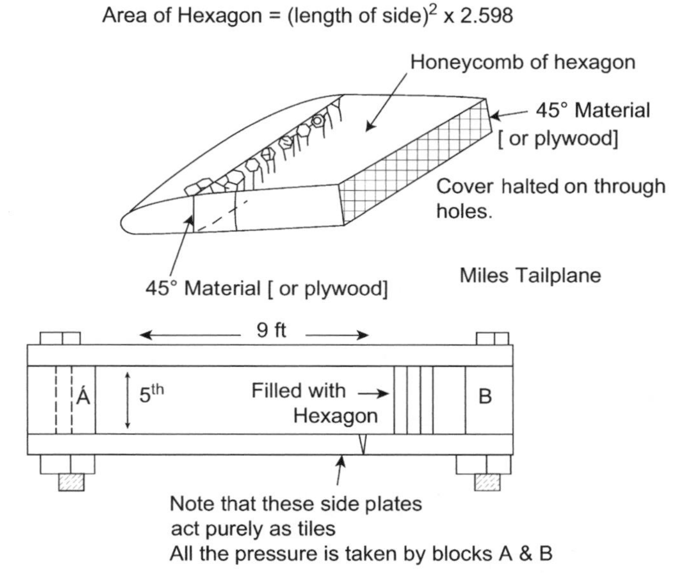

Figure 32: De Bruyne's notebook for August 1938 — The use of honeycomb "Reinforcement".

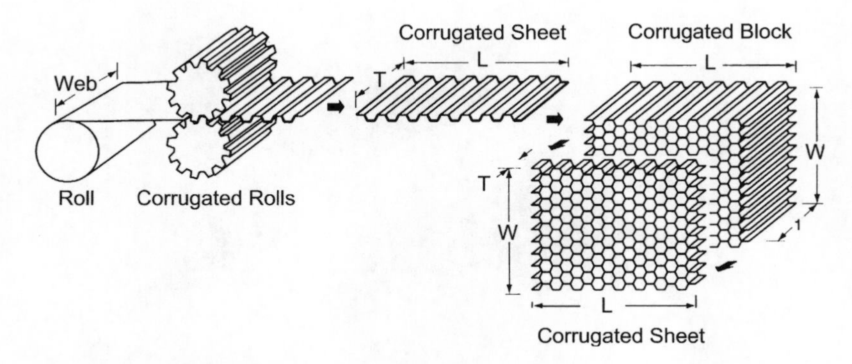

Figure 33: Schematic showing the manufacture of corrugated honeycomb.

the production of the so-called "corrugated" honeycomb which was sold as Aeroweb®. Here thin foils (up to about 0.1 mm thick) were corrugated to give a sheet of half-hexagons (Fig. 33). An adhesive, initially an epoxy-based Araldite® formulation, was applied to the flat surface of the half-hexagon. Layers of corrugated foils were stacked on top of each other until the required thickness was achieved. Glass rods were inserted into the hexagonal apertures and the whole assembly was clamped together, to apply pressure to the gluelines, and then heated in an oven to cure the adhesive.

The resultant honeycomb had a perfect hexagonal structure which has never been improved even by modern advances in manufacturing technology.

As can be seen in Fig. 34, the flat areas of foil which are coated with adhesive are called the "nodes" of the resultant honeycomb cell and hence the adhesives used are often referred to as node-bonding adhesives.

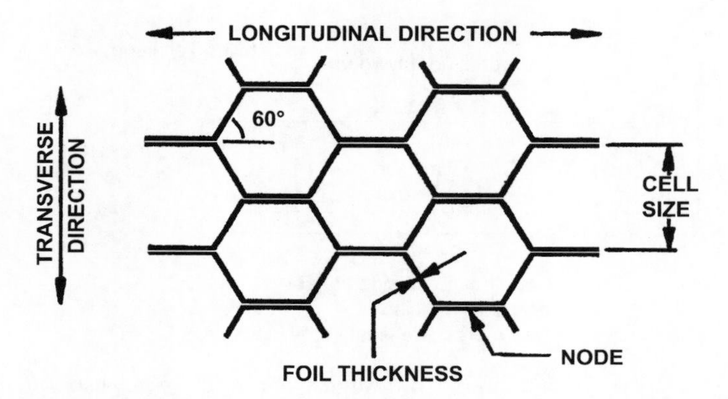

Figure 34: Schematic showing the lay up of corrugated honeycomb.

Although of very high quality, manufacture of the corrugated honeycomb was both time consuming and labour intensive. The major advance, to improve the situation, was the production of "expanded" honeycomb (Fig. 35). In this case, stripes of adhesive (usually a formulated phenolic resole or epoxy) are printed across the aluminium foil. The next step is the accurate cutting of the roll of printed foil into sheets, which are then stacked so that the stripes of adhesive are off-set by exactly half a cell's width from one sheet to the next. The laid-up block is put into a heated press (usually at about 150–160°C) to cure the adhesive. "Hobes" are then cut from the block, dimensions dependant on the requirements, and these are then expanded to give the resultant honeycomb slice. In some instances, see below, the whole block, rather than the smaller hobes, is expanded.

Either process lends itself to the use of a significant variety of aluminium foils. For example, the HexWeb® (originally Aeroweb) range of honeycombs uses 3003, 5052 and 5056 alloys – the aerospace grades having a corrosion-resistant treatment (for example, HexWeb CR-PAA/CRIII 5052). The expanded process is also particularly suited to a variety of non-metallic foil materials.

For these non-metallic honeycomb cores, the manufacturing procedure, as given above, is augmented with one or two further steps. Once the basic block or hobe has been expanded into the conventional honeycomb configuration, it has to be pinned as it has as yet no integral strength. Then, for materials such as Nomex, the structure is heat set by taking its temperature above its glass transition point. The final step is to convert the so-called "green honeycomb" structure into a product having the required density and mechanical performance properties. To achieve this, aramid and paper cores are generally dipped, often several times, in suitable baths containing simple or formulated phenolic resole or polyimide systems. Once dried, the phenolic coating can be fully cured to give the final honeycomb structure its required density and its well-known engineering properties. Typical products are: Nomex® (for example, HexWeb HRH-10), Kevlar® (for example, HexWeb HRH-49), the hybrid Korex® and cellulosic paper (for example, HexWeb HRH-86).

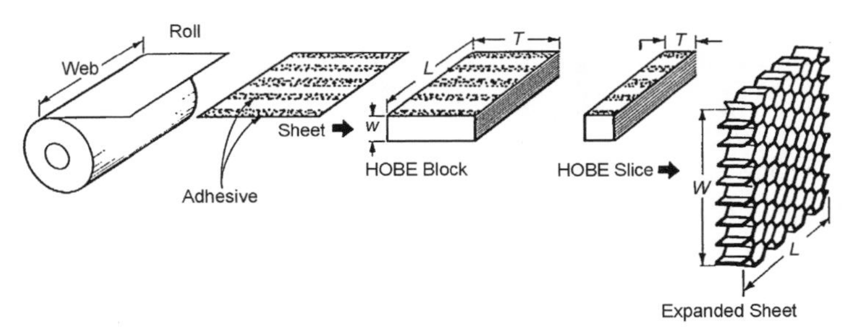

Figure 35: Schematic showing the manufacture of expanded honeycomb.

A similar procedure is carried out for fabric-based honeycombs. Here the pinned honeycomb structure is immersed in an impregnating solution, which is then dried and cured. Typical chemistries used are peroxide cured polyesters, for glass fabrics, and phenolic resole formulations or polyimide systems for glass and carbon. Typical products are: glass (for example, HexWeb HRH-327) and carbon (for example, HexWeb HFT-G).

For particularly, but not exclusively, aluminium core, the conventional hexagonal configuration can be varied with a range of other shapes:

"OX Core" where the core has been over-expanded in the "W" direction.
"Flex Core", which enables exceptional formability of the core to take place with tight radii of curvature being possible.
"Double Flex Core", which has a large cell size and possesses high compression properties.
"Reinforced Hexagonal Core", which provides a honeycomb of higher density for use at, for example, attachment points.

Fig. 36 shows these configurations, in comparison with the conventional hexagonal core.

If the adhesive used to form the integrally bonded sandwich panel is of such a chemistry that volatile matter is evolved during cure, then the potential pressure built up inside the individual honeycomb cells can be alleviated by using perforated honeycomb core; this is particularly relevant to aluminium honeycombs. During processing, prior to applying the node bond adhesive stripes, the aluminium foil is passed over a perforating roll, which punches small holes in the foil.

Finally, honeycombs made from titanium are also available (for example, from Darchem Engineering). These are manufactured by a corrugation technique, the nodes being laser spot-welded rather than adhesively bonded.

5.5.1.7.2. Skins

The potential skinning material for honeycomb sandwich panels can cover all the other substrates mentioned above – in flat sheet form – and can be augmented with decorative laminates, based on melamine–formaldehyde, P/F laminates or any other suitable plastic. The latter are to be seen in aircraft interiors.

5.5.1.7.3. Adhesive

The adhesive can be of any chemistry and/or format; the latter includes liquid, paste or film adhesives. It must rigidly attach the skins to the core and, generally, should be of high modulus when cured. It is accepted that brittle adhesives, i.e. adhesives showing relatively low peel strength – should not be used in very light sandwich structures which are likely to be submitted to a significant degree of abuse in use.

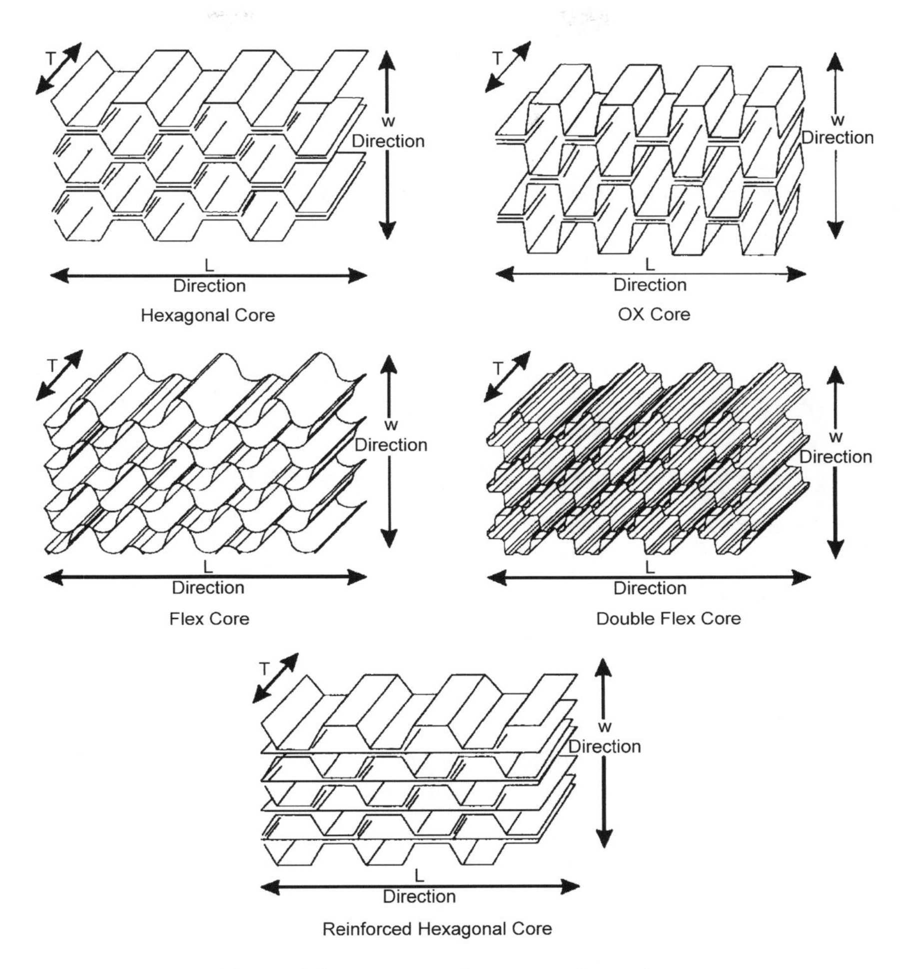

Figure 36: Honeycomb core configurations.

It is of paramount importance that the adhesive should be capable of forming the so-called "fillets" during cure. This means that during its cure cycle, the adhesive will melt and flow away from the skin lying at the centre of the honeycomb, first to the cell edges and then down the sides of the cell. This leaves a conical "fillet" of adhesive on the cell walls and only a very thin coating on the skin. The adhesive between the edge of the cell and the skin should form a butt joint between the two materials.

It is common with many structural adhesives that the size of the fillet on the bottom of the panel is larger than at the top. This is attributed to preferential flow from the top fillet to the bottom one.

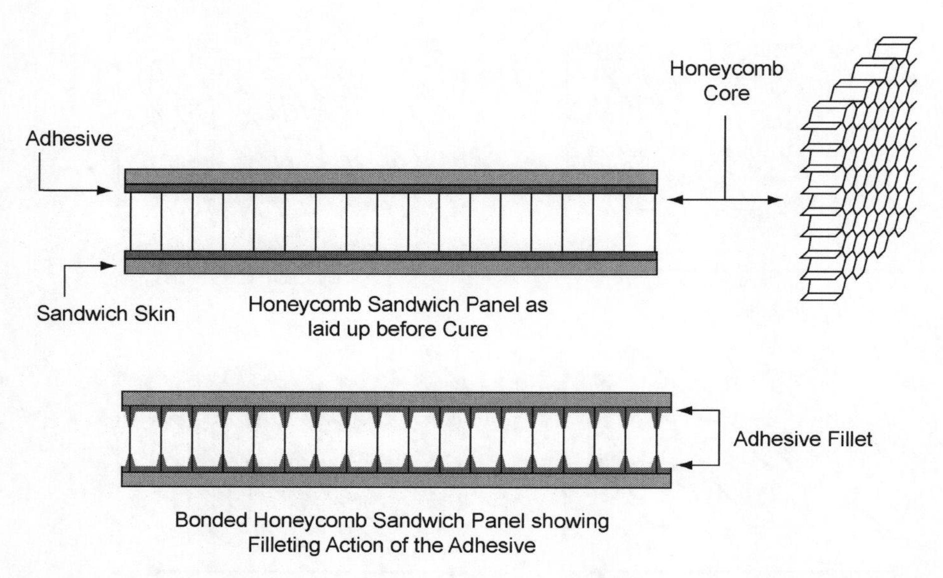

Figure 37: Schematic of the adhesive filleting action in a bonded honeycomb sandwich panel.

Fig. 37 illustrates the filleting phenomenon; the size of the fillet has been deliberately exaggerated.

The combination of core (aluminium, aramid, paper, carbon, glass or balsa), skins (aluminium, steel, titanium, FRP, melamine or wood) and adhesive (film or paste, epoxy, phenolic, PUR, PI, cyanate ester, acrylic or thermoplastic) gives a sandwich panel which is an integrally bonded, load-bearing structure.

Using bonded sandwich beams, of different core thicknesses (t and $3t$), in place of all metal components (a bonded metal beam of overall thickness t), it is possible to increase the stiffness and strength, with minimum weight penalty. This can be seen in Table 2.

If, however, weight saving is the important criterion, rather than an increase in strength or stiffness, then Table 3 shows what is possible by indicating the beam thickness required to maintain a deflection of 1 mm under a load of 500 kg over a span of 1 m.

5.5.2. Substrate Surface Pretreatment

To form a strong, integrally bonded, load-bearing structure, the surface of the adherend should be pretreated before application of the adhesive; this is vital if good environmental or thermal durability is required. Such a procedure ensures that the surface is in as clean a condition as possible, removing weak boundary layers which could adversely affect the performance of the resultant joint.

Table 2: Comparison of weight, stiffness and strength for bonded beams.

Bonded beam thickness	Honeycomb thickness	Relative		
		Stiffness	Strength	Weight
t	0	1.0	1.0	1.0
$2t$	t	7.0	3.5	1.03
$4t$	$3t$	39.0	9.25	1.06

This important topic is dealt with in the chapter entitled "Surface Pretreatment for Structural Bonding", in this volume.

5.5.3. Primers and Priming

In aerospace applications, once metallic adherends have been pretreated, it can be several weeks before they are bonded. In view of the fact that most of the oxide or "fresh metal" surfaces are stable for considerably less than 24 h, these surfaces have to be protected during the time that the component is stored prior to bonding. This is invariably achieved by priming the dried adherend immediately after pretreatment; such techniques rarely, if ever, apply to fibre-reinforced composite components.

The perceived drawback with primers, however, is that both the surface protection and, particularly, the corrosion inhibiting systems can be very sensitive to coating thickness. It is possible almost to halve the peel performance of some adhesives when going from primer coating thicknesses in the region of 2–5 μm to thicknesses above 8–10 μm.

Currently, the structural adhesive market essentially relies on two classes of primer: surface protection primers and corrosion inhibition primers. There is a further class of materials which are said to act as both surface pretreatment agents in their own right as well as surface protection and, it is claimed, bond enhancing

Table 3: Bonded sandwich panels — potential weight savings over metal plate.

Beam material	Beam thickness (mm)	Beam areal weight (kg/m^2)
Steel plate	12.5	98
Aluminium alloy plate	18.3	49
Aluminium honeycomb sandwich panel	33.0	9

primers: these are the organosilanes. By far the greatest numbers of commercial products are now corrosion-inhibiting primers.

5.5.3.1. Surface Protection Primers

These are usually solvent-based systems of low solids content (ca. 10%). Their chemistries should match those of the structural adhesives to be used in the bonding application. To this end, in a commercial range of adhesives, each primer offered is often a simple solution of its equivalent adhesive, where any insoluble fillers, etc., have been removed.

This means, therefore, that the chemistries associated with the primers could be as varied as the adhesives used with them. In aerospace applications, however, the predominant surface protection primer chemistries, for structural bonding, are those based on epoxy; phenolic-, polyimide- and polyurethane-based primers are also encountered.

These primers are applied to the abraded or chemically pretreated substrate by brush, roller or spray-gun. The wet coating is dried, usually in an extracted, air-circulating oven at about 60–80°C, to remove the solvent(s).

There is, however, no hard and fast rule about the next step in the process. In some applications, the primed substrates are then stored in a clean, dry area until needed for component assembly and bonding. However, other applications call for the primer to be cured, by oven-stoving, at the requisite temperature. These components would then be similarly stored until required.

In the first case, the primer co-cures with the adhesive and in the second, the adhesives wet and bond to the already cured primer coating. Irrespective of which route is followed, the primer protects the integrity of the pretreated surface prior to bonding ensuring that as high a quality bond as possible is formed.

It should, however, be remembered that surface protection primers essentially donate nothing to the bonded joint apart from the protection of the pretreatment applied to the substrate and acting as an adequate “key” between adhesive and adherend. Enhancement of the mechanical performance of the adhesive being used is rarely seen. Indeed, in a few cases a reduction in bonding strength occurs.

5.5.3.2. Corrosion Inhibiting Primers

These do the same job as the surface protection primers but also impart a degree of corrosion resistance into the bonded joint protecting it against the attack of moisture and electrolytic corrosion. As for the surface protection systems, the primer can be dried or dried and cured following application.

Until recently the classic example of this type of primer has been Cytec-Fiberite’s BR-127. This comprises an epoxy–phenolic system dissolved in a blend of solvents. Corrosion inhibition is achieved through the addition of a significant loading of strontium chromate to the resin solution; this is present as a fairly unstable suspension.

The mechanism of corrosion inhibition, with these heavy metal chromates, hinges on the fact that they can passivate aluminium [22–24]. When such a corrosion-inhibited bonded joint is attacked, a mixture of hydrated aluminium oxide and chromic oxide (Cr_2O_3) is formed (cf. the Alocrom process) This not only seals the "oxide" film, repairing the damage caused by the ingress of the electrolyte, but the presence of the stable chromic oxide also reduces the rate of dissolution of the aluminium oxide. The longevity of such a protection is due to the low solubility (~ 1.2 g/l at 15°C) of the chromate in water [25], which means that the chromate remains "active" for a considerable period of time.

The possibility of using other inorganic compounds such as the oxides, hydroxides, phosphates and borates of calcium, magnesium and zinc, to replace the potentially carcinogenic chromates, has recently been investigated by many companies including Hexcel Composites and 3Ms.

Another approach has been the evaluation of suitable "ion-exchanged silicas". In the presence of electrolytes, such materials give protection by releasing passivating ions that can interact not only with the matrix but also with the electrolytic species themselves and the substrate. This type of inhibitor is characterised by the fact that it is completely insoluble in water and only works on demand, i.e. only when invading electrolytic species are present. Fletcher [24] has given a schematic mechanism for this method of corrosion inhibition, which is summarised in Fig. 38.

Research is now strongly centred on producing primers with low volatile organic contents (VOCs) by using water as the primary solvent or dispersion medium or actual water-based primers where the organic solvent content is zero. Further, as indicated above, the move to chromate-free systems is also underway.

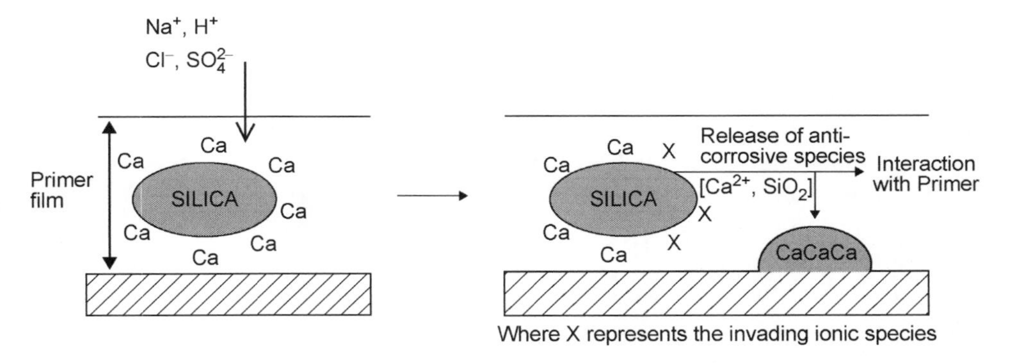

Figure 38: Schematic of the corrosion inhibition reaction mechanism for ion-exchanged silicas.

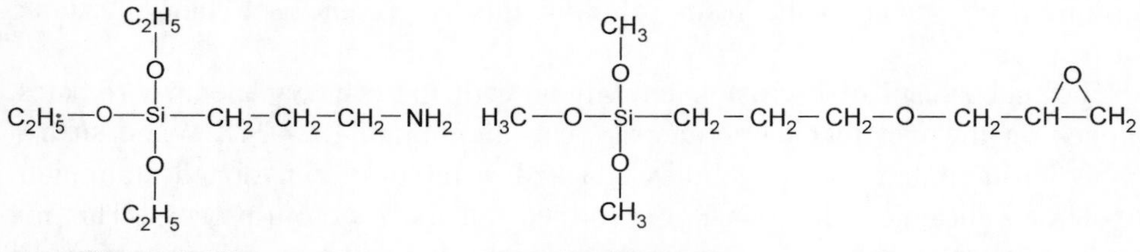

$$C_2H_5{-}O{-}\overset{\overset{\displaystyle C_2H_5}{|}}{\underset{\underset{\displaystyle C_2H_5}{|}}{\overset{O}{\underset{O}{Si}}}}{-}CH_2{-}CH_2{-}CH_2{-}NH_2$$

Silquest A-1100

$$H_3C{-}O{-}\overset{\overset{\displaystyle CH_3}{|}}{\underset{\underset{\displaystyle CH_3}{|}}{\overset{O}{\underset{O}{Si}}}}{-}CH_2{-}CH_2{-}CH_2{-}O{-}CH_2{-}\overset{O}{CH{-}CH_2}$$

Silquest A-187

Figure 39: Typical monofunctional silanes.

This has culminated in the 2003 qualification by Boeing, to BMS 5-42, of a water-based (low rather than zero VOC) phenolic primer from SIA Adhesives Inc. (Aqualock® 2000); a polyacrylate and amorphous silica are used as toughener and corrosion inhibitor, respectively.

5.5.3.3. Organosilanes*

Researchers in industry, governmental institutes and academe, have been evaluating the efficacy of organosilanes for two to three decades. Pluedemann, who is seen as the "father" of this class of compounds, has advocated their use [26] both as methods of surface pretreatment as well as adhesive primers and adhesion promoters in structural bonding applications.

The generic chemistry associated with most of the silanes used in these applications can be represented as:

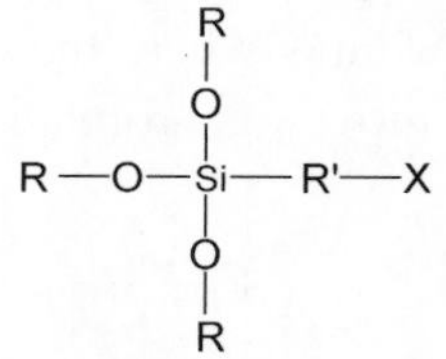

$$R{-}O{-}\overset{\overset{\displaystyle R}{|}}{\underset{\underset{\displaystyle R}{|}}{\overset{O}{\underset{O}{Si}}}}{-}R'{-}X$$

Where R represents a simple alkyl group, which is generally methyl or ethyl, R′ is an alkyl chain which is often propyl, X is the end group which can be H but is more usually a functional group such as: amino, epoxy, mercapto, ureido, isocyanato, vinyl and methacryloxy.

Although some of the silanes encountered can be very complex, the two most commonly employed (Fig. 39) are γ-aminopropyl triethoxy silane (for example, Silquest® A-1100) and γ-glycidoxypropyl trimethoxy silane (for example, Silquest A-187).

---

* Editor's note: For information on the silanes and their chemistry, refer to "Silicone Adhesives and Sealants" in Volume 3.

The theory behind their use is that the alkoxysilane groups will hydrolyse to form silanol groups, which can then react with the metal oxide/hydroxide surface. Further hydrolysation will form an organic–inorganic hybrid at the metal surface that can lead to either a continuous or a randomly discontinuous film, as shown in Fig. 40. This is very much akin to one particular aspect of the sol–gel pretreatment approach (see Chapter 4).

The key to the use of silanes, whether as sole pretreatment systems or adhesion promoting primers, is how they are prepared for application. Researchers have found the following parameters to be critical:

Carrying medium (water or alcohol) silane concentration, solution pH and temperature

Current indications are that low concentrations of silanes in water at a pH of about 5 yield optimum performance.

However, although their structures have, over the years, multiplied in number and complexity and although a considerable amount of work has been carried out in this field, the use of silanes as an accepted pretreatment or primer, prior to structural bonding in the aerospace industry is still relatively limited; 3Ms do, though, market two silane bonding primers: EC 2333 and EC 3903. Certainly, they are used as formulating ingredients in epoxy-based primers, paste adhesives and film adhesives. However, many formulators admit, though, that they are not included for any particular scientific reason; simply because "they don't appear to detract from the properties and might enhance them"!

More complex bis-silanes are now being considered not only as potential replacements for conventional metal pretreatment but also as chromate replacements in the formulation of corrosion inhibiting primers for metals in

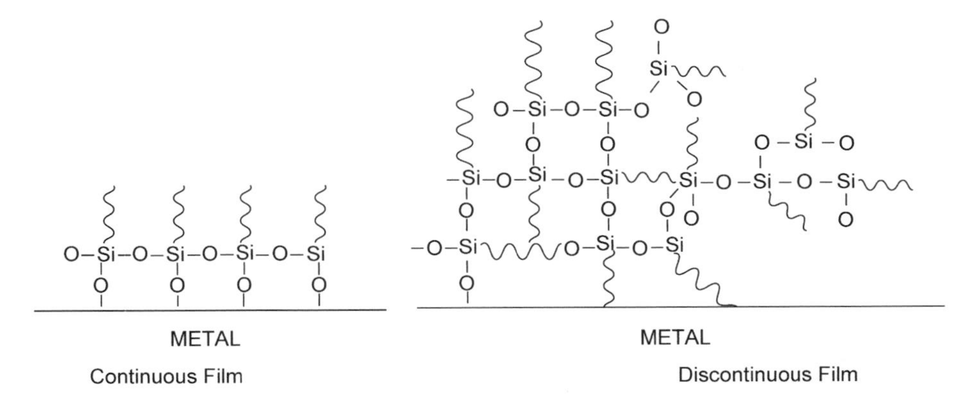

Figure 40: Films of monofunctional silanes on metal oxide/hydroxide surfaces.

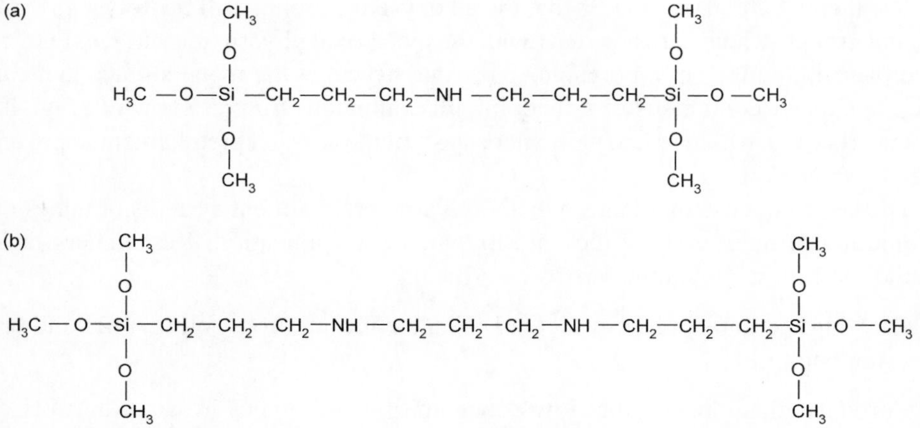

(a)

$$H_3C-O-Si(OCH_3)_2-CH_2-CH_2-CH_2-NH-CH_2-CH_2-CH_2-Si(OCH_3)_2-O-CH_3$$

(b)

$$H_3C-O-Si(OCH_3)_2-CH_2-CH_2-CH_2-NH-CH_2-CH_2-CH_2-NH-CH_2-CH_2-CH_2-Si(OCH_3)_2-O-CH_3$$

Figure 41: (a) *Bis*-(trimethoxysilylpropyl) amine. (b) *Bis*-(trimethoxysilylpropyl) ethylenediamine.

many varied application areas [27]. van Ooij advocates, amongst others, the use of both mono- and di-amino based *bis*-silanes (Fig. 41).

These *bis*-silanes can readily form continuous networks on the metal surface (Fig. 42) potentially giving excellent protection. Their post hydrolysis film thickness has been measured in the region of 400–500 nm.

In several instances, van Ooij et al. have found that mixtures of mono and *bis*-silanes give the best corrosion protection.

As with their mono functional counterparts, these materials are not yet in common use.

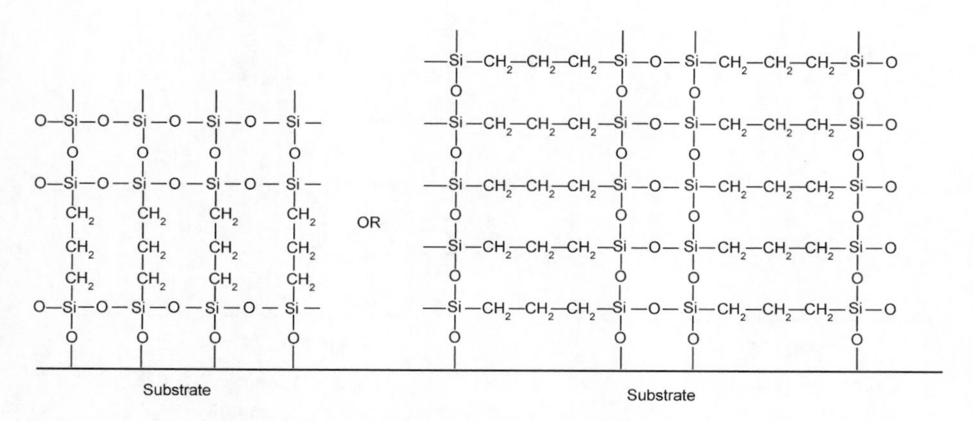

Figure 42: Continuous network formations of "*Bis*-Silanes" on metal surfaces.

5.5.4. Adhesives

The heart of any bond is, of course, the adhesive. Most bonding in the aerospace industry relies on the so-called structural adhesives, i.e. adhesives based upon monomer compositions which polymerise, or cure, to give high modulus, high strength bonds between relatively rigid adherends, such as those discussed above, so that a load-bearing structure is produced.

In order to achieve this state, many of these adhesives are reactive, i.e. multi-component systems which, after application to the adherend, change their physical format from liquidus to infusible solid by chemical reaction. These polymerisation and cross-linking reactions, often thermally induced, are achieved using one or more of the following reaction mechanisms:

- Condensation,
- Addition,
- Rearrangement,
- Polymerisation through the double bond.

5.5.4.1. Adhesive Ranges for Aerospace Applications

There are no "universal adhesives" for the diverse applications seen in the aerospace business. Adhesives not only have to operate in space applications from about −100 to +120°C, in civil and military aircraft under "hot/wet" climatic conditions, where relative humidities can approach 100% at temperatures up to 60–80°C, and close to aero engine assemblies where temperatures can reach in excess of 220°C but also have to resist the numerous fluids which are present in most modern aircraft: de-icing fluid (alcohols and phosphate esters), fuels, bilge fluid, oils and lubricants, water and water-based electrolytes.

To meet these varied requirements a "family" of adhesives generally has to be offered to the end-user. These adhesives will vary in the actual role they have to play, in their format, the chemistry used and the cure temperature employed to achieve the required property levels. The end-users, drawing on advice from the relevant adhesive producers, may, and probably will, use several adhesives from this range to meet their design requirements.

5.5.4.1.1. Adhesive "Role"

Generally any structural adhesive range will include the primary adhesive itself plus two different types of supporting products which have structural adhesive properties in their own right; the chemistries and formulations of these supporting products are often strongly related to that of the primary adhesive which they accompany.

The first of these supporting adhesives, the surface protection primers and/or the corrosion inhibiting primers, have been dealt with in Section 5.5.3.

The second range of products contains the foaming and syntactic adhesives. These are the materials, which can be used to splice sheets of honeycomb together, to join honeycomb to metallic channelling or simply to seal the edges round honeycomb sandwich structures. These adhesives are either formulations which, during the thermal cure cycle, generate controlled gaseous products (usually, but not exclusively, nitrogen) to yield an infusible structural lightweight foam or they are lightweight pastes.

The lightweight syntactic paste adhesives are conventional, usually two-part, adhesive formulations which contain a considerable loading of a lightweight filler to reduce the cured Sg to well below 1; Sgs of 0.55–0.75 are typical. The lightweight fillers are invariably the so-called "microballoons". These are thin-walled, generally, ceramic hollow spheres; in the case of glass microballoons, the original solid glass density of 2.48 is reduced to a value in the region of 0.3.

In the case of the foaming adhesives, the formulation very often mirrors that of the primary adhesive being used. Invariably anti-slump (thixotropic) fillers have been added as well as calculated quantities of suitable foaming (or blowing) agents which break down as shown in Fig. 43.

The choice of blowing agent will depend on the cure temperature being used; the curative will also often have an unexpected influence. Thus, AZDN, although

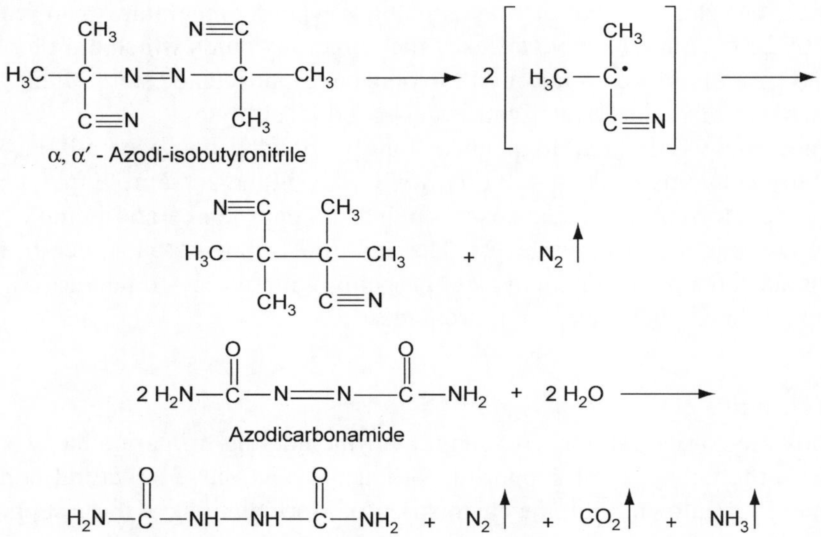

$$H_3C-C(CH_3)(C{\equiv}N)-N{=}N-C(N{\equiv}C)(CH_3)-CH_3 \longrightarrow 2\left[H_3C-\dot{C}(CH_3)(C{\equiv}N)\right] \longrightarrow H_3C-C(N{\equiv}C)(CH_3)-C(CH_3)(C{\equiv}N)-CH_3 + N_2\uparrow$$

α, α′ - Azodi-isobutyronitrile

$$2\,H_2N-C(=O)-N{=}N-C(=O)-NH_2 + 2\,H_2O \longrightarrow H_2N-C(=O)-NH-NH-C(=O)-NH_2 + N_2\uparrow + CO_2\uparrow + NH_3\uparrow$$

Azodicarbonamide

Figure 43: Reaction schematic for two typical blowing agents.

its theoretical breakdown temperature is much higher, can be used in epoxy-based foaming adhesives which are designed to cure at 120°C. The key to achieving this is the hardener. Curatives such as dicyandiamide have been found to act as powerful initiators/accelerators for the thermal decomposition reaction. For the same reasons, the azodicarbonamide, which breaks down at >200°C, can be used under 175°C curing conditions.

5.5.4.1.2. Adhesive formats

Essentially, all structural adhesives are offered in one of three formats:

- Solvent-based liquids: These are usually single-component systems, which are dissolved and/or dispersed in a liquid medium which can be aqueous.

  Although many non-structural adhesives can be solvent or water based, in aerospace applications, this format is usually the preserve of the surface protection and corrosion inhibiting primers. However, as indicated, there are exceptions and these are often those adhesives which are based on phenolic resoles (for example, Redux 775 Liquid) or on polyimides (for example, PMR®-2).
- Pastes: These can be single-component or two-component systems which contain no solvent and whose physical state can vary from very low viscosity liquids, for example, adhesives which can be injected into the glueline, to relatively high viscosity, pasty materials which are often thixotropic. Primary adhesives, lightweight syntactic and foaming paste adhesives fall into this category.

  In the case of the two-component systems, one component is usually a formulated resin system and the other a formulated hardener. The two-component systems are often supplied in twin cartridges, which pump out the two parts of the adhesive through an efficient mixing head. Thus the adhesive can be easily mixed and applied to the substrate in one operation.
- Film adhesives: Film adhesives are supplied as foils and can be cut to the shape and size of the area to be bonded. The matrices, therefore, are solid at room temperature but, critically, will liquefy but do not cross-link on moderate heating. Thus, on heating from ambient to cure temperature they will first melt, then flow and displace the air in the bond and, in so doing, wet the substrates to be bonded and finally cure to an infusible solid. This can be readily seen by examining the dynamic viscosity traces for three typical film adhesives: Redux 312, Redux 319 and Redux 322 (Fig. 44).

It can be clearly seen that all three systems yield low viscosity resinous matrices at temperatures well below their cure temperature (120, 175 and 175°C, respectively) and that there is an adequate window for flow of the adhesive to displace any trapped air and hence ensure that an intimate wetting of the substrate occurs prior to gelation taking place. The latter is indicated by the asymptotic rise in matrix viscosity.

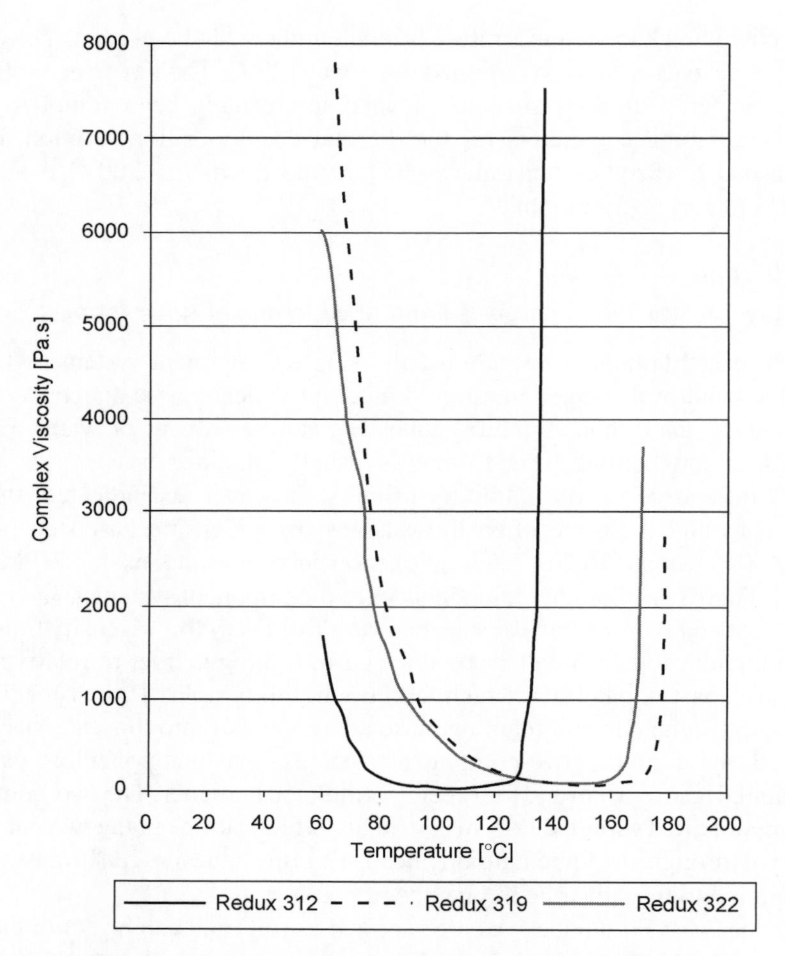

Figure 44: Dynamic viscosity traces for three film adhesives.

Further, due to their ability to be easily cut, these adhesives can be supplied in any desired width from over 1 m down to about 5 mm. Their very format means that contact between operator and adhesive can be kept to a minimum. This ensures that the adhesive can be easily maintained in a contamination-free condition and, more importantly, virtually eliminates any chances of skin irritation as far as the operator is concerned.

Primary adhesives and foaming film adhesives fall into this category.

5.5.4.1.3. Adhesive chemistries

There are many formulations, based on a large variety of backbone chemistries, which can be used as adhesives. As indicated above, apart from a few thermoplastic adhesives such as polyethersulphone:

$$HO-\left[C_6H_4-SO_2-C_6H_4-O\right]_n-H$$

and polyetheretherketone (Victrex® PEEK):

$$-\left[C_6H_4-O-C_6H_4-CO-C_6H_4-O-C_6H_4-CO-C_6H_5\right]_n$$

which are used in specialist applications, most aerospace adhesives utilise "reactive" chemistries. These can be summarised as:

- Acrylic: anaerobics, conventional acrylics, cyanoacrylates, the so-called "second generation", toughened acrylics and adhesives having UV-activateable or UV-curing capabilities
- Amino-formaldehyde
- Ceramic
- Epoxy
- Cyanate ester
- Phenolic: novolacs and resoles
- Polyimide: bismaleimides (BMI) and polyimides (PI — e.g. PMR-2)
- Polyurethane (PUR)
- Silicone

Of these, apart from the amino-formaldehyde systems used on wooden structures, the aerospace industry currently concentrates on epoxies, phenolics and polyimides for the majority of its structural bonding applications; it is the epoxies that are the real "workhorse" for the industry. Acrylic and polyurethane chemistries are used, but in very few applications of a structural nature. The relatively novel chemistry based on cyanate esters, for example Arocy B 10 (dicyanate ester of Bisphenol A):

$$N{\equiv}C-O-C_6H_4-C(CH_3)_2-C_6H_4-O-C{\equiv}N$$

is, though, now gaining a foothold, especially for space applications, due to its proven low moisture uptake in service.

Of the polyimide resins available, it is the specific family based on maleic anhydride that is most commonly encountered; these are the bismaleimide-based adhesives:

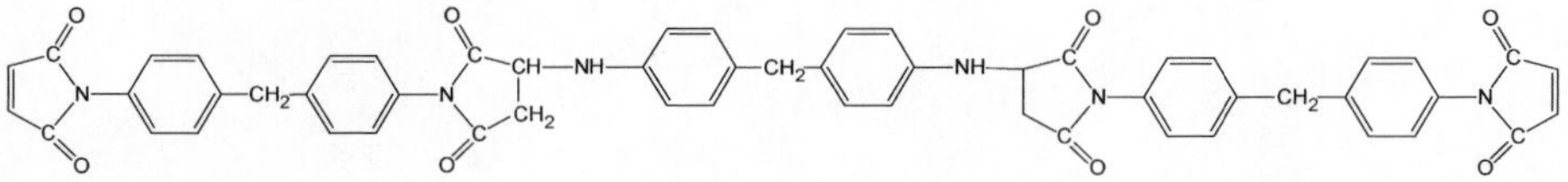

It is of interest to note that before the advent of the bismaleimides, and the polyimides in general, high service temperatures were achieved using combinations of solid Bisphenol A-based epoxies with conventional phenolic resoles; the interaction between phenolic and epoxy often being promoted by the incorporation of dicyandiamide. The first "toughener" for high temperature matrices, namely finely divided aluminium powder, was also incorporated into these formulations.[3]

Thermosetting polyimides are encountered for specialist, usually space, applications. These are the adhesive and composite PMR systems which are based on nadic anhydride and benzophenone tetracarboxylic dianhydride.

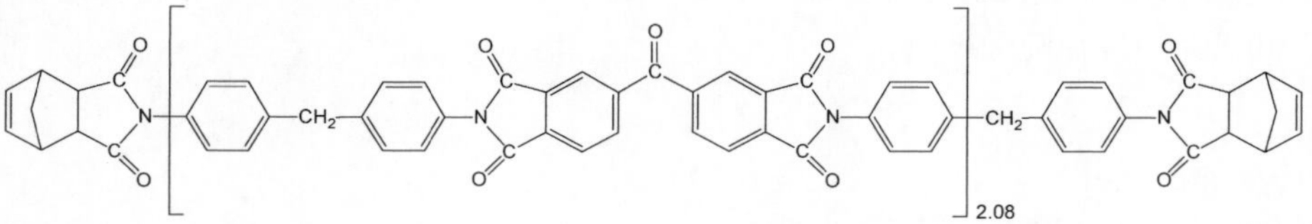

5.5.4.1.4. Adhesive cure temperatures

With such a plethora of chemistries available, the adhesive formulator can generally tailor the adhesive to meet any required cure temperature — usually in the range of ambient (i.e. about 22°C) to as high as 230°C. Cure times can range from several seconds, particularly with polyurethane-based systems and to a certain extent with acrylic systems, to several hours, as is the case with bismaleimide and polyimide adhesives. In this latter instance, cure, or more usually post cure, temperatures as high as 300°C often have to be used to ensure that the final rearrangement reactions go to completion.

5.5.4.1.5. Adhesive formulations

Similarly, by careful formulation using this significant range of resins and the large number of curative chemistries as well as compatible polymeric modifiers which are available, novel adhesives can also be tailored to meet the physico-chemical

[3] A very large study of heat stable adhesive is provided in Volume 2 in the chapter "Heat stable adhesives" by Guy Rabilloud.

and mechanical requirements of any individual bonding application. The following demands can readily be encompassed: adhesive format (liquid, paste or film and, in the latter case whether it is supported or unsupported), shelf life, gap-filling properties, volatility and out-gassing characteristics during cure, shrinkage, service temperature (from sub zero to about 200 to 220°C), strength requirements, environmental resistance and toughness.

Possible adhesive formulations, therefore, are legion, so it is not possible to give examples to cover all potential adhesive ranges. However, as the workhorse of the industry is the epoxy adhesive, it is valid to give the breakdown of a generic formulation.

Thus a typical, epoxy-based structural film adhesive would comprise[4]:

- Liquid epoxy resin(s): the backbone of the adhesive giving the basic properties required.
- Solid epoxy resin(s): primarily as a film former but also to modify the end properties of the adhesive.
- Polymeric modifier(s): primarily as a toughener and in many cases to improve the thermal resistance of the adhesive, it can also aid in film forming.
- Hardener: usually solid and of low solubility to give a degree of latency to the adhesive and hence improve the shelf life.
- Co-hardener/accelerator: to adjust, if necessary, the cure temperature to that which is specified.
- Formulation additive(s): this/these could include such components as flame retardants for low fire, smoke and toxicity products, coupling agents and/or surfactants to improve the wetting and bonding with the substrate, fillers – thixotropic or otherwise – to control Sg, viscosity, flow and slump, and blowing agents for foaming applications.
- Pigments/dyestuffs: essential for the manufacturer to aid in mixing and for the operator to identify that adhesive had actually been applied to the substrate
- Support carrier: to impart better handleability and to give good control of the final glueline thickness. Support carriers come in many formats and chemistries. The most commonly encountered are made from glass, polyamide (usually Nylon 6 or Nylon 6, 6), or polyethylene terephthalate (PET). Their format can range from random mats to highly structured woven or knitted fabrics. Typical examples of the latter are shown in Fig. 45.

5.5.4.1.6. Typical commercial range of structural adhesives

The tables and text below show the make up of a typical range of commercially available structural adhesives (Redux from Hexcel Composites [28]) and their

[4] Detailed information on epoxy adhesives and their formulation will be provided by the chapter "Epoxy adhesives" in Volume 5 of this Handbook.

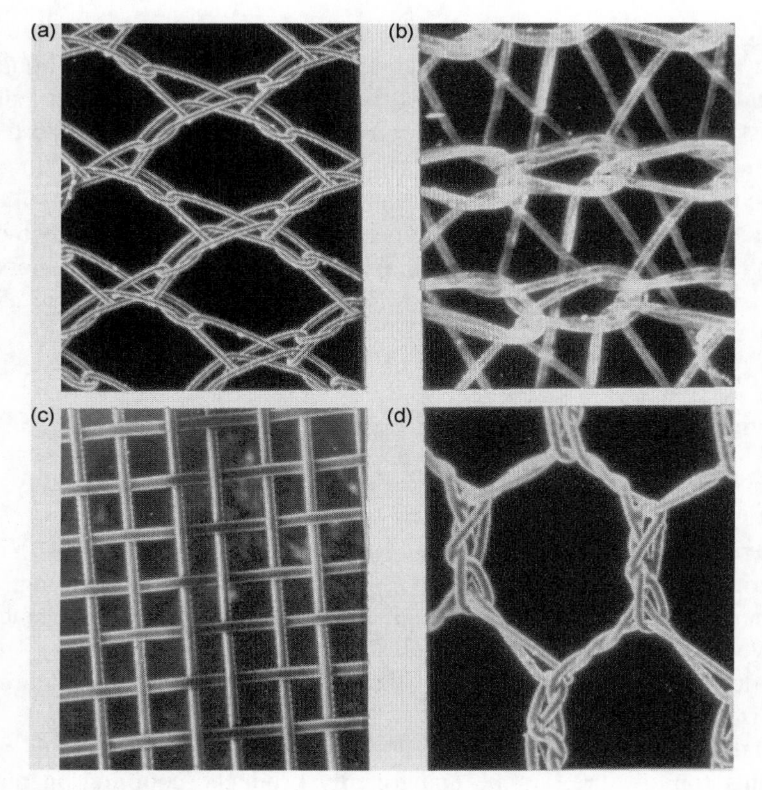

Figure 45: Typical woven and knitted carrier structures: (c) conventional weaving; (a), (b) and (d) various knitting styles.

supporting primers, syntactic pastes and foaming adhesives The whole range is considerably larger – to meet the demands of other industries such as road and marine transport, electrical and electronics etc. – but here only the key products for the aerospace market are shown. Table 4 shows the chemistries employed and Table 5 gives the different formats in which this range is supplied.

Tables 6 and 7 indicate the different cure cycles which need to be applied to obtain optimum properties for each of the adhesives shown above.

It can be seen that several of the range have the ability to be cured under two, or more, completely different regimens which gives them the so-called "dual cure" capability. This is particularly important with room-temperature curing paste adhesives where the extended cure at ambient cannot always be contemplated by the end user. Dual cure capability with foaming adhesives is also important as this allows the bonding shop to use only one foaming adhesive with a large range of structural adhesives.

Table 4: Chemistries employed in the Redux range of structural adhesives.

Role	**Chemistry**			
	Phenolic	**Epoxy**	**Cyanate ester**	**Bismaleimide**
Primer	Redux 101	Redux 112, Redux 119, Redux 122		Redux HP655P
Foam		Redux 206, Redux 212, Redux 219, Redux 260, Redux 840		
Light-weight syntactic		Redux 830		
Primary adhesive	Redux 775	Redux 312, Redux 319, Redux 322, Redux 330, Redux 340SP, Redux 810, Redux 850	Redux A54	Redux AP 655

Table 5: Adhesive formats employed in the Redux range of structural adhesives.

Format	**Chemistry**			
	Phenolic	**Epoxy**	**Cyanate ester**	**Bismaleimide**
Solution	Redux 101	Redux 112, Redux 119, Redux 122		Redux HP655P
One-part paste		Redux 840		
Two-part paste		Redux 810, Redux 830, Redux 850		
Film	Redux 775	Redux 206, Redux 212, Redux 219, Redux 260, Redux 312, Redux 319, Redux 322, Redux 330, Redux 340SP	Redux A54	Redux HP655

Curing pressures range from minimal contact pressure (about 70 kPa) to about 350 kPa for the epoxies and the cyanate esters, about 415 kPa (at 130°C) for the bismaleimides and about 700 kPa for the phenolic adhesives to counterbalance the condensation products evolved during cure.*

* Editor's note: For detailed information about the chemistry and formulation of epoxy adhesives, please refer to the chapter "Epoxy Adhesives" in Volume 5 of this handbook.

Table 6: Typical cure cycles for the Redux range of primers and structural film and paste adhesives.

Chemistry	Products	Cure cycles	Service temperature (°C)
Phenolic	Redux 101	15 min at 45°C (to dry) + 30 min at 150°C or co-cure with the adhesive	
	Redux 775	30 min at 150°C	75
Epoxy	Redux 810	5 days at ambient, 1 h at 70°C, <30 min at 100°C or <10 min at 120°C	100
	Redux 850	16 h at ambient + 1 h post cure at 120°C, 16 h at 60°C or 1 h at 120°C	100
	Redux 112	20 min at 70°C (to dry) + 30 min at 120°C or co-cure with the adhesive	
	Redux 312	30 min at 120°C	100
	Redux 119 and Redux 122	30 min at 70°C (to dry) + 60 min at 175°C or co-cure with the adhesive	
	Redux 330, Redux 319 Redux 322 and Redux 340SP	60 min at 175°C	135/150/175/175, respectively
Cyanate ester	Redux A54	120 min at 175°C	160
Bismaleimide	Redux HP655P	30 min at 70°C (to dry) + co-cure with the adhesive	
	Redux HP655	4 h at 190°C + 16 h free-standing post-cure at 230°C	230

For the structural paste and film adhesives (Table 6) some indication as to the maximum service temperatures obtainable, under these cure cycles, are given. In the case of the syntactics and foams (Table 7), the cured Sg or expansion ratio is given.

A more detailed examination of several products within this range not only reaffirms why a range, rather than a "universal" adhesive, is needed but also reinforces the versatility of structural adhesives and their "tailoring" to specific end applications.

Redux 312: A cure temperature of 120°C gives the ability to bond more delicate components whilst, at the same time, being more energy efficient. The service

Table 7: Typical cure cycles for the Redux range of syntactic and foaming adhesives.

Chemistry	Products	Cure cycles	Foaming ratio/(cured Sg)
Epoxy	Redux 830	5 days at ambient or 5 h at 50°C	(0.66–0.68)
	Redux 840	30–60 min at 120°C, 30 min at 150°C or 20–60 min at 175°C generally co-curing with the adhesive	1:1.5–1:2.1 dependant on cure cycle
	Redux 206	30–60 min at 120°C co-curing with the adhesive	1:3–1:4
	Redux 212	30–60 min at 120°C co-curing with the adhesive	1:1.5–1:2
	Redux 260	60 min at 120°C or 60 min at 175°C co-curing with the adhesive	1:2.4
	Redux 219	60 min at 175°C co-curing with the adhesive	1:1.9–1:2.0

temperature of 100°C is sufficient for most civil aircraft applications and its performance in honeycomb structures is excellent (Fig. 46). It is, therefore, a good general purpose adhesive for structural bonding in many airframe applications.
Redux 319 and Redux 322: A cure temperature of 175°C ensures a good elevated temperature performance for these two adhesives. The two adhesives are complementary with excellent toughness, as evidenced by good peel strength, being seen with Redux 319. Optimum high temperature performance is shown by Redux 322, which has a service temperature of 175°C with thermal excursion up to about 200°C being possible.

Applications in military aircraft and in areas closer to engine structures are, therefore, possible with these two systems; the choice being whether toughness or very high temperature performance is required by the design. The comparative shear performances can be seen in Fig. 47. Fig. 48 shows the peel properties.
Redux HP655: This adhesive utilises bismaleimide chemistry and, therefore, a more demanding cure cycle is required to extend the structural adhesive's service temperature to well above 200°C. Apart from the use of polyimides based on benzophenone tetracarboxylic dianhydride, Nadic anhydride and diamino diphenyl methane, for example PMR-2 (see above), the formulated bismaleimide adhesives offer the best resistance to extreme thermal oxidative conditions.
Redux 810: Redux 810 is a good example of two-component, high strength, high peel, general purpose paste adhesives used in structural applications, particularly in the aerospace industry. It is capable of bonding a variety of metallic and non-metallic substrates utilising cure cycles, which can range from days at room

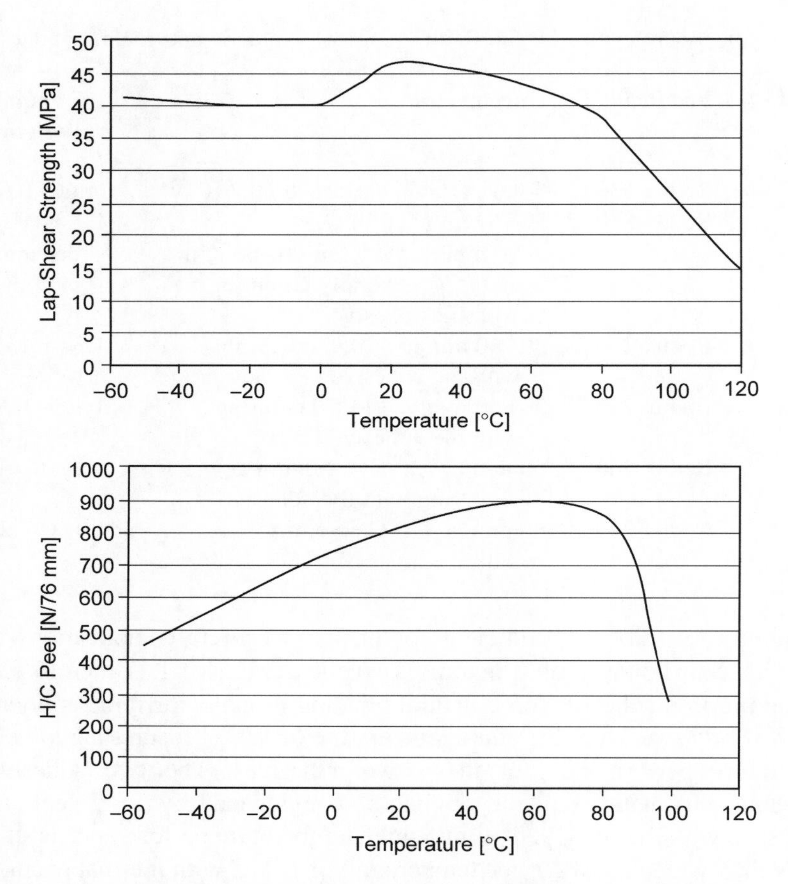

Figure 46: Lap-shear and honeycomb peel performance of Redux 312.

temperature to minutes at elevated temperatures up to 120°C. The system rheology is carefully controlled to ensure easy mixing and, on application, a thixotropic (non slump) behaviour. These rheological properties are also important in allowing the adhesive to be supplied in cartridges enabling the adhesive to be metered, mixed by a static mixer in the dispensing head and dispensed directly into the glueline.

Excellent mechanical properties are augmented by a corrosion-inhibiting performance and the incorporation of solid glass beads having a specific particle size distribution, builds in an automatic glueline thickness control.

Redux 850: This two-component, essentially hot-curing, paste adhesive is a good example of the formulator co-operating with the end-user.

One of the research teams within the Airbus consortium (TANGO – Technology Acquisition of Near-term Goals and Objectives) was examining the possibility of producing the wing sections for the Airbus A.380 airliner out of carbon fibre

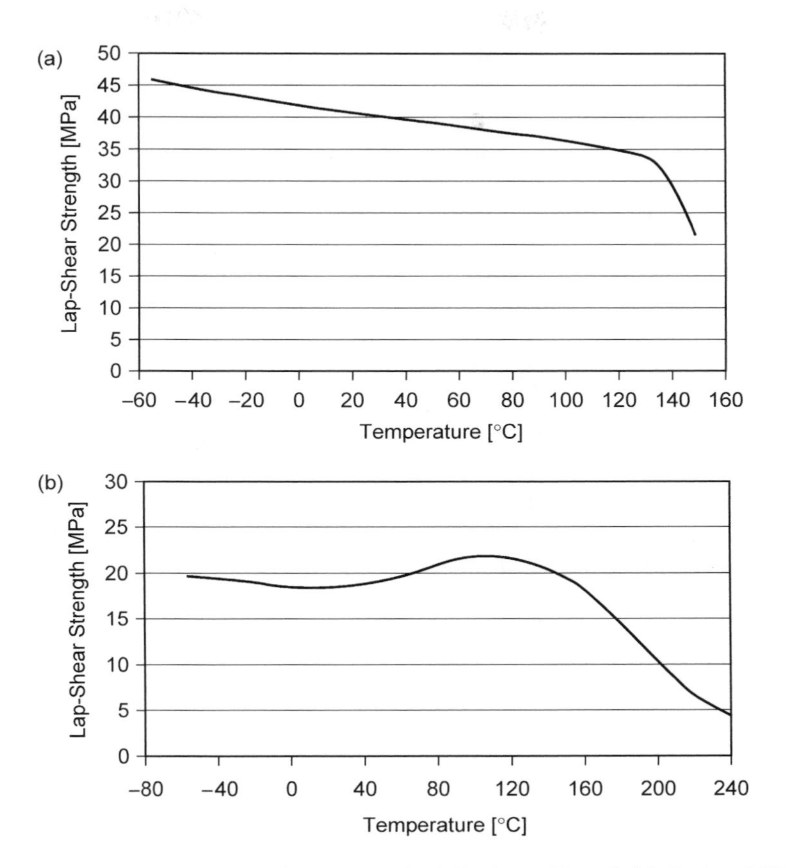

Figure 47: Lap-shear performance of (a) Redux 319 and (b) Redux 322.

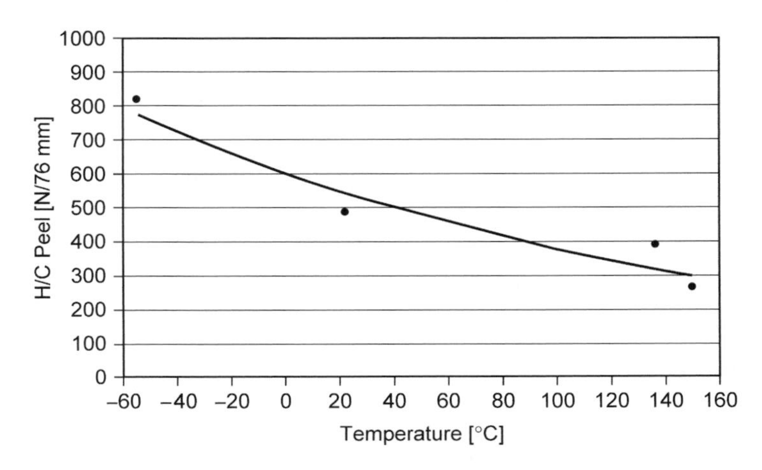

Figure 48: Honeycomb peel performance of Redux 319.

composite. Of the various means of joining such a structure, adhesive bonding was clearly a significant possibility. With the design engineers from Airbus working very closely with the adhesive chemists from Hexcel Composites, a novel composite-bonding and shimming paste adhesive was tailor-made for the application giving the desired rheological and handling characteristics, pot life, cure cycles and performance at elevated temperatures on both metallic and composite adherends.

Finally, to gauge the complexity of gaining acceptance for any new structural adhesive by the aerospace market, all the data that have to be generated to meet relevant specifications in both Europe and the United States of America are shown for Redux 319 and Redux 319A (supported version).

The data sets are split into: standard metal-to-metal tests, honeycomb sandwich tests, effect of multiple cures, fatigue and creep, thermal oxidative testing and finally immersion testing. Where necessary, an explanation of the tests being used is given with each individual table (Tables 8–13). Until recently, few physico-chemical tests were carried out for qualification purposes. Thus, for Redux 319, the only tests performed were: Flow (53.7%) and Volatile Content (0.99%).

Table 8: Specification data for Redux 319: physical properties and standard metal-to-metal properties.

	Test temperature (°C)	**Redux 319 (175°C cure)**		**Redux 319A (175°C cure)**	
		Mean	**Range**	**Mean**	**Range**
Physical properties					
Flow (%)				53.700	
Volatiles (%)				0.99	
Standard metal-to metal properties					
Lap shear (MPa)	22	40.5	37.2–47.0	35.7	29.2–39.3
	135	32.8		19.2	17.8–22.9
	150	20.7	17.2–25.0	16.4	13.4–23.6
Blister detection (MPa)	−55			31.3	
	22			28.8	
	135			22.1	
Floating roller peel (N/25 mm)	22	178	133–223	178	
BS 5577 peel (N/25 mm)	22	75		157	
MMM-A-132 T-Peel (N/25 mm)	22	80		61	

Table 9: Specification data for Redux 319: standard honeycomb sandwich properties.

	Test temperature (°C)	Redux 319 (175°C cure)		Redux 319A (175°C cure)	
		Mean	Range	Mean	Range
Honeycomb peel (N/76 mm)	−55			820	666–974
	22	594	320–790	487	
	135	546		390	320–460
	150			264	
Beam shear (kN)	−55			10.7	
	22			10.5	
	135			8.8	
Honeycomb flatwise tensile (MPa)	−55	10.7		10.0	
	22	8.2	4.9–11.8	9.0	8.1–10.2
	135	6.5		3.6	4.3–2.9
	150	5.5	3.3–9.0	2.9	

Table 10: Specification data for Redux 319: effect of multiple cures on metal-to-metal and honeycomb sandwich panels.

	Test temperature (°C)	175°C cure (mean)	
		Redux 319	Redux 319A
Lap shear (MPa)			
Control	22	44.0	37.3
	135		22.9
	150	25.0	
Second cure	22	46.0	36.7
	135		19.4
	150	22.5	
Third cure	22	45.5	40.9

(Continued)

Table 10: Continued.

	Test temperature (°C)	175°C cure (mean)	
		Redux 319	Redux 319A
	135		21.6
	150	24.2	
Fourth cure	22		36.7
	135		17.1
Fifth cure	22		34.7
	135		16.4
DTD 5577 peel (N/25 mm)			
Control	22	242	
Second cure	22	240	
Third cure	22	234	
Honeycomb peel (N/76 mm)			
Control	22		481
	135		460
Second cure	22		378
	135		396
Third cure	22		392
	135		371
Fourth cure	22		182
	135		331
Fifth cure	22		369
	135		416
Flatwise tensile (MPa)			
Control	22		8.8
	135		2.9
Second cure	22		7.7
	135		3.2
Third cure	22		7.1
	135		3.5
Fourth cure	22		8.0
	135		3.4
Fifth cure	22		7.5
	150		3.9

Table 11: Specification data for Redux 319: fatigue and creep.

	Test temperature (°C)	Redux 319A (175°C cure)
		Mean
Fatigue		
Lap shear at 50 Hz ($s_{max} = 5.2$, $s_{min} = 0.5$ MPa)		$< 10^6$ cycles
Creep		
Lap shear: 8 days at 22°C under 11 MPa (mm)	22	0.013
Lap shear: 8 days at 22°C under 5.5 MPa (mm)	135	0.090
Beam shear: 8 days at 22°C under 4.4 kN (mm)	22	0.14
Beam shear: 8 days at 135°C under 3.5 kN (mm)	135	1.14

Table 12: Specification data for Redux 319: thermal oxidative ageing.

	Test temperature (°C)	Redux 319 (175°C cure)		Redux 319A (175°C cure)
		Mean	Range	Mean
Lap shear (MPa)				
42 days at 135°C	135	31.8		23.7
90 days at 150°C	22			28.0
183 days at 150°C				25.6
42 days at 150°C	150	20.8	15.1–25.7	
90 days at 150°C				18.5
183 days at 150°C				15.8
Honeycomb peel (N/76 mm)				
8 days at 150°C	22			246
21 days at 150°C				264
42 days at 150°C				192
Beam shear (kN)				
8 days at 135°C	135			8.9

Table 13: Specification data for Redux 319: immersion resistances to fluids.

	Test temperature (°C)	Redux 319 175°C cure		Redux 319A 175°C cure	
		Mean	**Range**	**Mean**	**Range**
Lap shear (MPa)					
JP4 at 22°C					
7 days				33.1	
30 days		41.4			
90 days	22	39.0			
365 days		38.3			
30 days		16.2			
90 days	150	17.6		35.0	
365 days		14.8		11.0	
Kerosene fuel at 22°C					
30 days		36.6			
42 days		43.3			
90 days	22	39.7			
365 days		38.3		34.0	
30 days		16.9			
90 days	150	13.1			
365 days		13.4		13.0	
Silcodyne "H" at 22°C					
30 days		37.2			
42 days		41.6	39.9–43.4		
90 days	22	38.6			
365 days		39.7		34.0	
30 days		16.2			
90 days	150	17.9			
365 days		14.5		12.0	
Hydraulic oil at 22°C					
7 days				35.7	
30 days		37.9			
42 days	22	42.0			
90 days		26.9			
365 days		36.9		35.0	

(*Continued*)

Table 13: Continued.

	Test temperature (°C)	Redux 319 175°C cure		Redux 319A 175°C cure	
		Mean	Range	Mean	Range
30 days		18.6			
90 days	150	17.6			
365 days		17.9		13.0	
Standard test fluids at 22°C					
7 days				36.5	
30 days		42.5	39.0–46.1		
90 days	22	38.6			
365 days		40.0			
30 days		24.4	20.3–28.5		
90 days	150	18.3			
365 days		14.1			
De-icing fluids at 22°C					
7 days		36.2			
30 days		39.3			
42 days	22	41.5	38.7–44.3		
90 days		39.0			
365 days		39.7		36.0	
30 days		14.1			
90 days	150	14.8			
365 days		15.2		13.0	
Synthetic ester-based lubricant at 22°C					
30 days		37.9			
42 days	22	41.3	39–43.7		
90 days		39.3			
365 days		40.0		35.0	
30 days		16.9			
90 days	150	16.6			
365 days		15.9		12.0	
Skydrol 500A at 22°C					
30 days		34.8			
42 days	22	38.4	35.9–42.2		
90 days		39.7			
365 days		38.6		39.0	

(Continued)

Table 13: Continued.

	Test temperature (°C)	Redux 319 175°C cure		Redux 319A 175°C cure	
		Mean	**Range**	**Mean**	**Range**
30 days		18.6			
90 days	150	14.8			
365 days				13.0	
Skydrol 500A at 70°C					
30 days				29.0	
90 days	22			33.8	
365 days				28.3	
30 days				16.2	
90 days	150			15.5	
365 days				11.7	
Water/Methanol at 22°C					
30 days		33.8		35.2	
42 days	22	40.6	38.2–43.1		
90 days		16.6		35.2	
365 days		10.3		33.8	
30 days		12.4		12.1	
90 days	150	5.9		10.3	
365 days		0.0		5.9	
Distilled water at 22°C					
30 days		34.1		34.1	
42 days	22	39.5	37.2–41.5		
90 days		32.1		34.8	
365 days		32.4		31.0	
30 days		14.8		12.4	
90 days	150	9.0		10.3	
365 days		3.4		7.6	
Tap water at 22°C					
30 days	22	36.6		34.2	
90 days		27.6		32.4	
365 days		31.0		32.8	
30 days	150	19.7		9.7	
90 days		13.4		10.3	
365 days		3.4		7.6	

(*Continued*)

Table 13: Continued.

	Test temperature (°C)	Redux 319 175°C cure		Redux 319A 175°C cure	
		Mean	Range	Mean	Range
Salt spray at 35°C					
30 days	22	33.4		32.8	31.8–33.8
90 days		30.7		33.8	
365 days		0.0		33.8	
30 days	150	12.8		12.7	
90 days		9.7		11.0	
365 days		0.0		2.1	
100% relative humidity at 49°C					
30 days	22	32.5	29.3–35.8	31.6	30.1–33.1
90 days		30.0		30.3	
365 days		32.8		31.7	
30 days	150	14.6	12.8–16.4	11.7	
90 days		4.8		6.2	
365 days		1.0		4.8	
98% relative humidity at 70°C					
30 days	22	29.7		26.5	
42 days					
90 days		21.4		27.6	
365 days		9.3		18.6	
30 days	150	8.3		3.8	
90 days		2.8		2.4	
365 days		1.4		2.8	
Honeycomb peel (N/76 mm)					
100% relative humidity at 49°C					
8 days	22			420	
21 days				480	
42 days				252	

Lap Shear: Conventional single overlap specimens.

Blister Detection: Large area metal-to-metal bonding to ascertain adhesive flow characteristics during bonding. Conventional lap shear specimens are machined from the bonded sheet and are tested. The specimens are examined for evidence of air entrapment ("blistering").

Floating Roller and BS 5577 Peel: Fig. 49 shows a typical peeling rig for the floating roller peel test. The differences between this test and the BS 5577 specification are the peeling angle, which is about 70° for the floating roller or Bell peel test and 90° for the BS 5577 test, and the aluminium alloy used; BS 5577 uses a much softer alloy.

T-Peel: No jig is used for this test. Both adherends are the same thickness and the peeling angle is 180°.

Honeycomb Peel: A conventional thin-skinned honeycomb sandwich panel is tested in peel using the so-called climbing drum peel rig (Fig. 50). Applying a measurable torque to the drum forces it to "climb" the specimen and, hence, peel off the bonded skin; enough specimens are tested to be able to assess the adhesive bond to both top and bottom skins in the sandwich.

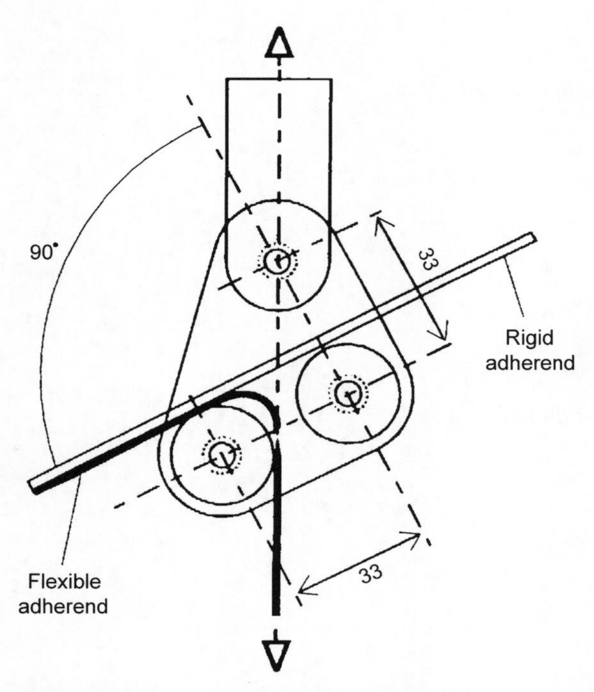

Figure 49: Schematic of the floating roller peel test.

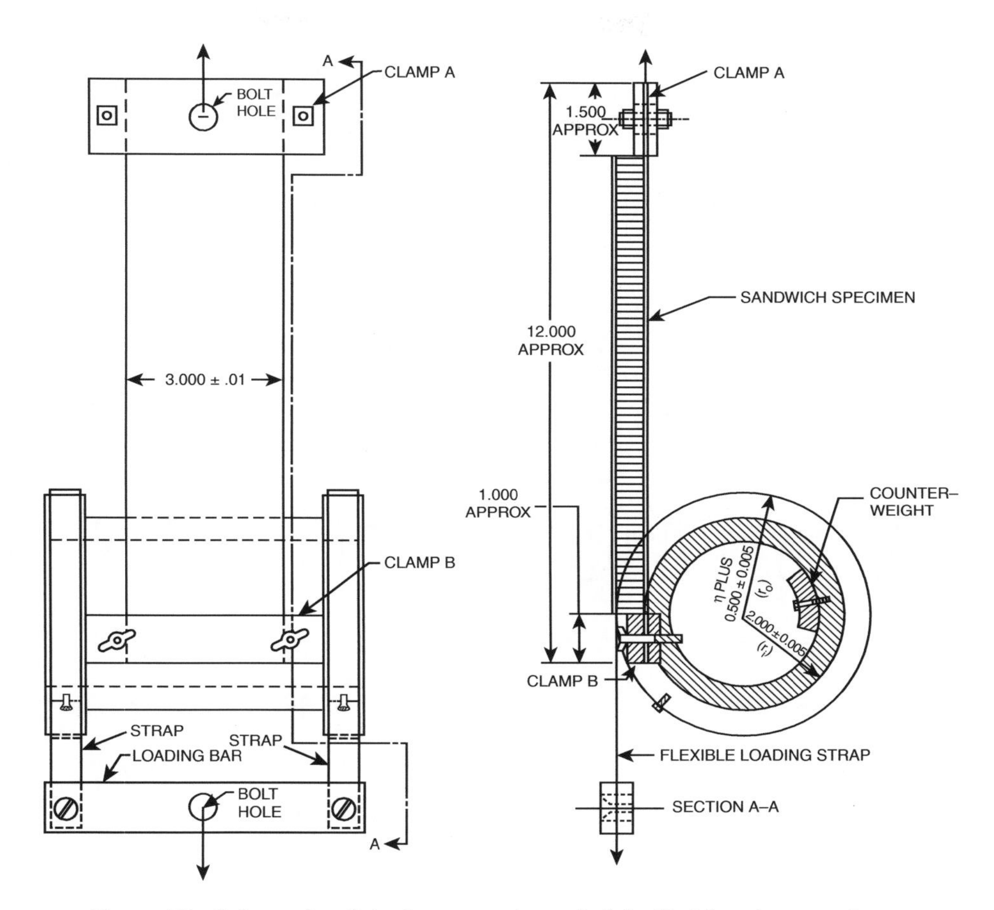

Figure 50: Schematic of the honeycomb sandwich climbing drum peel test.

Beam Shear: Here the bondline in a thick-skinned honeycomb sandwich panel is subjected to shear stresses utilising 3- or 4-point loading (Fig. 51).

Honeycomb Flatwise Tensile: 50 × 50 mm specimens are cut from similar sandwich panels as are prepared for the beam shear test. These are then bonded to rigid blocks, generally using a paste adhesive having a cure temperature at ambient or, at least, lower than that used to cure the adhesive under test. The whole is then loaded in such a manner as to subject the bond between honeycomb and skin to tensile stresses (Fig. 52).

Fatigue and Creep Tests: Conventional single overlap lap-shear and beam shear specimens are used.

Heat Ageing: Conventional single overlap lap-shear, honeycomb climbing drum peel and beam shear specimens are exposed to the requisite temperature in an air-circulating oven.

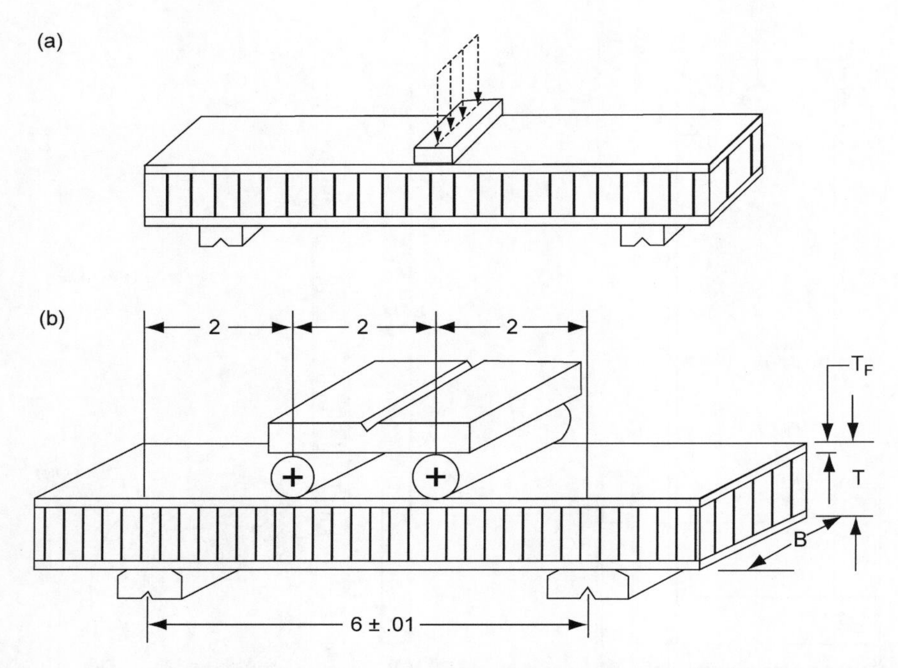

Figure 51: Schematic of the beam shear test. (a) 3-Point and (b) 4-Point loading.

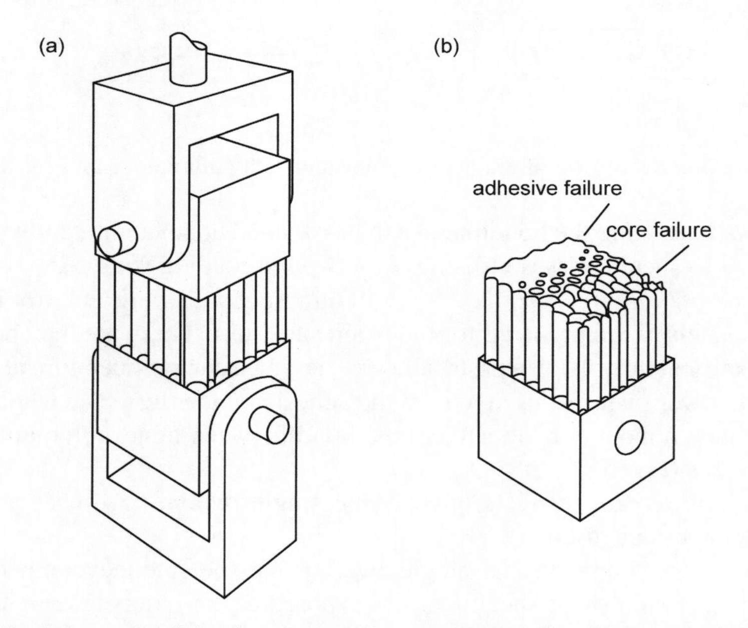

Figure 52: Schematic of the flatwise tensile test. (a) Test jig. (b) Ruptured specimen.

Immersion Testing: Conventional single overlap lap-shear and honeycomb climbing drum peel specimens are immersed in the requisite liquid or vapour for the requisite time.

5.5.4.2. Adhesive Manufacture

5.5.4.2.1. Primers

Simple stainless steel, glass-lined or even mild steel vessels, fitted with a suitable stirrer, are more than adequate for the production of solvent-based primers. These vessels will often be fitted with internal or external heating elements and water-cooled condensers so that for example, polymeric ingredients can be more readily dissolved at temperatures close to the reflux temperature of the solvent or solvent blend. This procedure is also valid for solvent-based structural adhesives.

If water-based primers are being manufactured then the situation is slightly more involved. The use of water means that only glass-lined or stainless steel vessels can be used for the final stage of mixing. In many instances, solid raw materials will not be supplied as a water-based solution or dispersion and, hence, the first stage of any manufacturing process requires suitable solution/dispersions to be made. This is often achieved using a conventional bead mill (Fig. 53) where bead size, bead volume, temperature, pump rate, solids content, pH, etc. all have to be optimised.

This is best carried out using statistical experimental design techniques [29,30], optimising the process conditions against the critical parameter of resultant particle size distribution. Once the variables to be studied have been identified, an experimental design can be created. This is usually a standard quadratic model, which can fit non-linear data. The experimental designs produced allow several variables to be studied at once, which enables a wide area of experimental space to

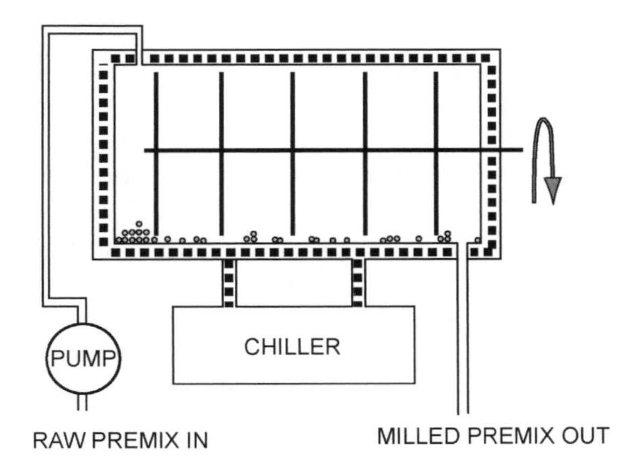

Figure 53: Schematic of a typical bead mill.

be mapped. This enables interactions to be identified, and areas of optimum performance to be found.

The resultant matrix of experiments comprises a series of trials with each chosen variable set at high, low or intermediate values; replication of some trials is used to assess error; other parameters are held at a constant value.

Data analysis then fits a polynomial equation to the collected data. The magnitudes of the coefficient estimates in the equation indicate the importance of the variables. This equation can be simply viewed as a multidimensional French curve to illustrate the relationship between variables and responses. Those coefficient estimates with statistical significance are highlighted, and are used to select the axes for contour plots.

Bishopp et al. [31] show a contour plot, or response curve, for the variation in bead size and pump rate against the $D_{0.9}$ value for the particle size ($D_{0.9}$ is the particle size, in μm, below which 90% of the particles fall); constants are bead volume at 75% and mill speed at 4000 rpm.

Contour plots allow the relationship between significant variables and responses to be visualised. These plots resemble topographical maps in that contour lines are drawn on a two dimensional plane to represent the surface of a response variable. This allows a highly visual, easily interpreted 'picture' to be used to understand the process or system being studied. Thus, in the example given in Reference 31, it is very clear that the lowest particle size is achieved using intermediate bead dimensions and that it is essentially independent of pump rate.

5.5.4.2.2. *Paste adhesives*

Although the general method of manufacture of paste adhesives is fairly straightforward, the viscosity and chemistry of the matrix will determine the actual procedures to be followed.

Generally, both one- and two-component paste adhesives can be made in the simple mixing vessels outlined above without the need for condensers and usually without the need for heating.

Corrosive ingredients will require stainless steel or glass-lined vessels but the remainder can use simple mild steel mixing chambers.

Low viscosity pastes can usually be manufactured using simple stirrers to disperse and/or dissolve the raw materials. However, as the viscosity increases so does the need to use high-shear stirrers such as toothed-bladed stirrers, planetary mixers and/or Z-blade mixers.

Insoluble powders, which can include blowing agents for foaming adhesives, curatives, fillers etc., can be pre-dispersed in any liquid polymer or resin present, thoroughly wetted out using a conventional two-roll paint mill and then added as an intimate dispersion to the main mix.

Other techniques such as mixing under vacuum or under a nitrogen blanket might be necessary should incorporated air prove a problem or, as is the case with isocyanate resins and some aliphatic amines where the raw materials are reactive with moisture in the air.

5.5.4.2.3. Film adhesives

In essence there are two methods of manufacturing film adhesives: from solution or from a melt In the first case, the film has to be cast and the solvent removed in a continuous operation. In the second case, as there is no solvent to be removed, the matrix has to be melted and then cast into film; this is the so-called "hot melt" film technique.

The initial stage, in either case, is to produce the fully formulated matrix; the solvent-based system can use similar equipment as is used to manufacture primers and the hot-melt route would use the same sort of plant as is used for the high-viscosity pastes except that the capability of heating and then cooling the matrix during the mixing cycle will be required.

Film casting also gives rise to, essentially, two different procedures. When preparing films from solvent-based matrices, the actual equipment used can be dependant on the final areal film weight. Low areal weights, for example 50–100 g/m^2, can utilise conventional reverse roll coating of a continuous film onto a backing "paper". This would pass straight into an air-circulating oven whose various zones would be set to drive off the residual solvent without leading to skinning or blistering of the final film. For areal weights above 100 g/m^2, lower weight films can be laminated together through nip rollers or a suitable metering device, for example a doctor knife-over table or knife-over-roll technique can produce film at the correct weight, which can then be dried as before.

When 100% solid matrices are employed, manufacture of the film is simplified due to the fact that no solvents have to be removed but, nevertheless, highly specialised equipment has to be used. Fig. 54 shows a schematic for a "batch" process where the mixed formulation is metered, at a suitable elevated temperature, under a doctor knife to produce the final film.

Using the hot melt approach, it is possible to mix and cast the adhesive film as a continuous operation. Here the individual raw materials are fed into a conventional screw extruder, are mixed under controlled temperature conditions and are then pumped out into a reverse-roll coating machine, which can produce film adhesives from as low as 50 g/m^2 up to about 1800 g/mm^2. The schematic of such a process is shown in Fig. 55.

Mention has already been made of Redux 775, the first structural adhesive film. Its manufacturing process has not changed since 1954. A film of phenolic resole is cast under a doctor knife, a considerable excess of the PVF powder is

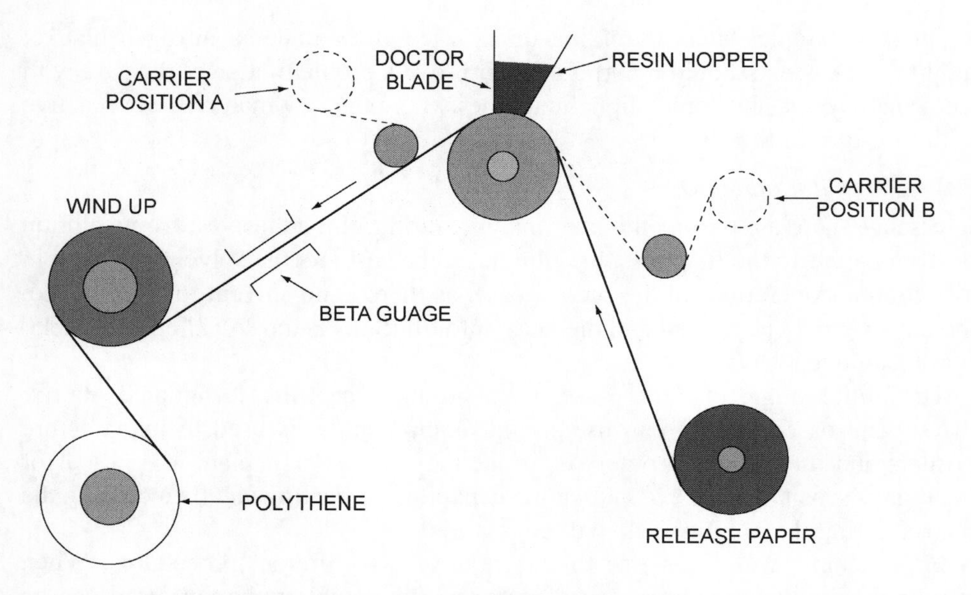

Figure 54: Schematic of a Knife-over-Roll "Hot Melt" film casting procedure.

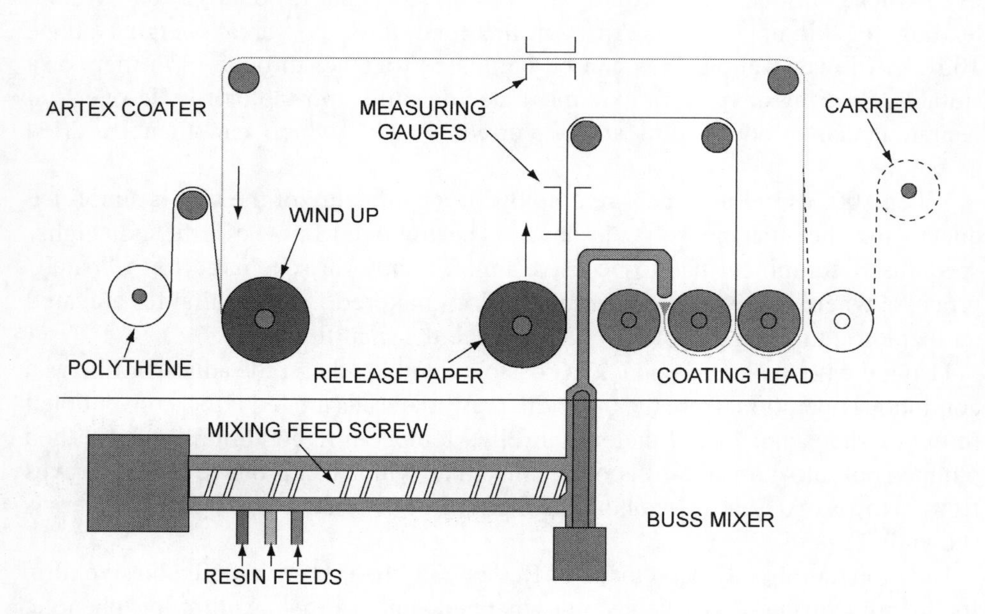

Figure 55: Schematic of the continuous manufacture of a "Hot Melt" adhesive film.

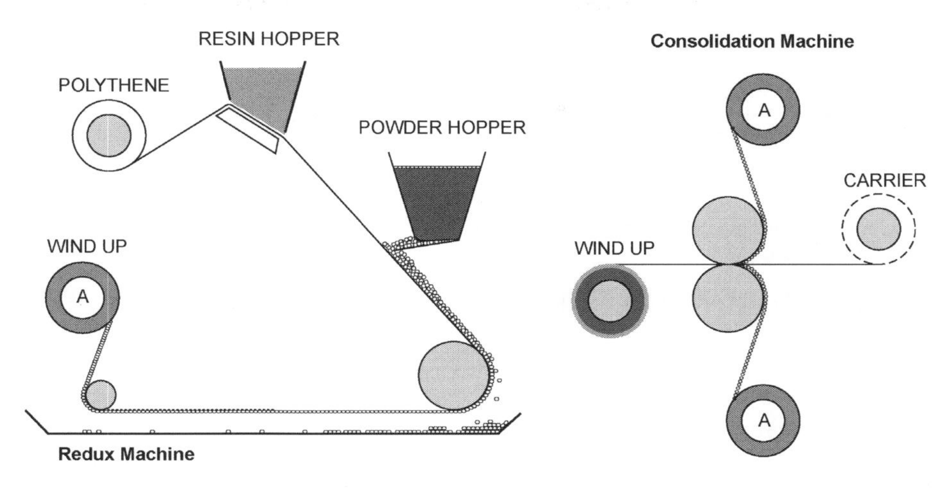

Figure 56: Schematic of the Redux 775 manufacturing process.

Figure 57: The original Redux film machine. (colour version of this figure appears on p. xv).

curtain-coated onto the phenolic film and the excess is then removed. Two of these "half-webs" are then laminated together to produce the final film adhesive. This is shown schematically in Fig. 56 and Fig. 57 shows the slightly modified original, 1954 film-making equipment.

5.5.4.3. Methods of Adhesive Application[5]

5.5.4.3.1. Primers

For small areas, these can be brushed on and slightly larger areas can be covered using a conventional roller-coater. For the large areas generally associated with the aerospace industry, spray coating is recommended.

Spray coating generally uses either a high volume/low pressure or a low volume/high pressure technique often using robotic applicators.

The primed surfaces are then allowed to dry. This can be accomplished by forced drying (at elevated temperatures in an air-circulating oven) or, traditionally, over night at ambient temperature. Depending on the primer and the application, the primed surfaces can be stored in the dried state until required or they can be oven-cured prior to later use.

5.5.4.3.2. Paste adhesives

A variety of methods can be used to apply these adhesives The actual method chosen will often depend on the component to be bonded – its size and the complexity of its shape – and the viscosity of the adhesive.

As many, but not all, of these systems are two-component, they first have to be mixed together. The oldest method is simply to weigh out the components and then mix by hand; where syntactic adhesives, which contain delicate glass microballoons, are concerned, hand mixing is often the safest and ideal method. Other possibilities encompass simple mechanical mixing using a bladed stirrer or pumping the two components, either by hand using a cartridge gun or using some hydraulic method (a typical set-up is given in Fig. 58), through a mixing head, which contains either static or dynamic mixing tubes, to an applicator.

Independent of final adhesive viscosity, hand application, by spatula, palette knife or trowel, is often the favoured method. However, once mixed, low viscosity adhesives are frequently pumped to an applicator gun, which is often robotically controlled. Higher viscosity paste adhesives can make use of hand-held cartridge dispensers which not only mix the adhesive, as indicated above, but also can apply a bead or stripe of adhesive to the substrate, dependant on the design of the applicator head.

[5] For detailed explanation about the application equipment, readers may refer to the chapter "Application equipment for adhesives" in Volume 2.

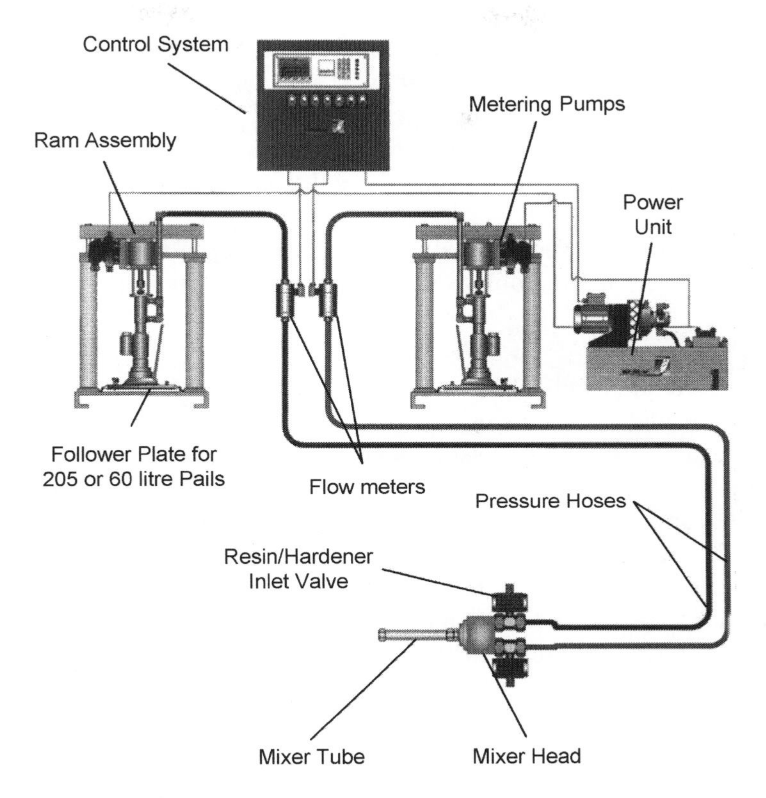

Figure 58: Schematic of a typical paste adhesive mixing and delivery unit (courtesy of: 2KM (UK) Limited).

For the highest viscosity adhesives reverse roll applicators can be used. Here the adhesive is pumped into the nip of the applicator rollers and the adherend to be coated is passed between a pressure roller and the metering roller. This allows an accurate coating of adhesive to be applied to the substrate.

Other methods, particularly for large areas can make use of comb and coating bar devices.

5.5.4.3.3. Film adhesives

Film adhesives offer the simplest method of adhesive application to the substrate to be bonded. This is particularly so when the tack of the film is highly controlled: enough tack to ensure the film stays where it is put but not so high as to prevent repositioning.

For the simplest structures, for example stringers, the adhesive film can be cut into tapes of the correct width by the supplier and can then be laid down onto the stringer by hand or by a tape laying machine.

Tape laying machines can be used for larger bond areas or the film can be pre-cut or die-stamped to the correct geometry and then put into position by hand or by robot.

When using adhesives such as foaming films, the easiest method of application is often by hand; cutting the film to size and dropping it into place.

5.5.4.4. Adhesive Curing

There are, essentially, only three methods of curing structural adhesive joints: press, autoclave and oven.

To some extent, the method used will be dependant on the adhesive but more often than not it is dependant on the complexity of the structure and the overall policy of the company producing the bonded component.

Simple, flat components are best cured in a press; Fig. 59 shows a typical press shop. Generally, the lay up is faster and the total time to produce the cured component is less than by either of the other methods. However, this is, of course, limited by the size of the component over a certain size and the press will be too small and other methods have to be contemplated.

More complex components, which can be accommodated in a simple, flat tool, for example the EH101 rotor blade (Fig. 75) can also be press cured.

Large, complex shapes, where high bonding pressures (> 100 kPa) are required, have to be either clamped up (Fig. 60) or jigged up on an appropriate tool, utilising vacuum-bag techniques (Fig. 61) and this assembly is then cured in an autoclave. A typical autoclave bonding shop is shown in Fig. 62.

Figure 59: Typical press bonding shop. (colour version of this figure appears on p. xvi).

Figure 60: Dornier helicopter rotor blade clamped before autoclave curing.

Figure 61: Rolls Royce engine nacelle component – Vac-Bag assembly before autoclave curing.

Figure 62: Autoclave bonding shop; railway system to take the tools to the autoclave (courtesy of: BAE SYSTEMS, Broughton).

A good example of oven-curing is during the first stage of the process to bond the propellant to the motor lining in rocket motor production. Here pressure bags, rather than vacuum bags, are used to apply $\gg$100 kPa bonding pressures to the surface of the adhesive film (Fig. 63), which can then be cured in a simple air-circulating oven.

5.6. Applications for Structural Adhesives Within the Aerospace Industry

5.6.1. Phenolic Adhesives

Many of the design concepts that were used in the Hornet and Sea Hornet are shown in Fig. 16. Sheets of aluminium reinforcement were "Redux"-bonded to

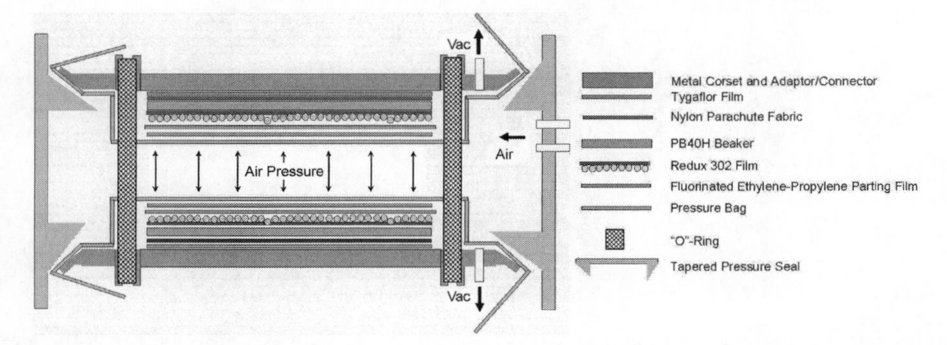

Figure 63: Schematic of a pressurised oven curing rig used in rocket motor applications.

the existing wooden components, using Redux Liquid E and Redux Powder, to strengthen stringers, ribs and spars. Metal to metal joints were also formed to strengthen areas of attachment such as the tail wheel and the arrestor hook.

In the mid-1940s, in order to improve its stability and performance, the resole was changed to one of lower alkalinity and a phenol to formaldehyde ratio of 1: 1.43 (Redux K6); this has remained unaltered until the present day. As well as the expected improvement in stability, there was also an unexpected benefit. This was the marked improvement in the ambient temperature lap-shear performance on aluminium; an increase from an average value of about 15 MPa to bond strengths which were consistently above 30 MPa.

However, in spite of this innovation, the use of Redux Liquid E was retained by de Havilland until 1963 (well into the Comet programme) and by Chance Vought in the United States of America until the termination of the F-7 Cutlass programme.

On a point of terminology, in 1955, Redux was qualified to a new military specification – DTD 775. Following that qualification all Redux products took the specification number into their official designation and became Redux 775.

The bonding in the de Havilland Comet gives a good example of the types of airframe structures which were being bonded and their overall design.

Fig. 64 shows the areas in the Comet where structural adhesive bonding took place. Figs. 65 and 66 give a schematic representation of the bonded stringers and the window surrounds and Fig. 67 a beam assembly which has been coated with Redux 775 Liquid and Powder prior to bonding. All these design concepts come together in the construction of the fuselage. This can be seen in Fig. 68.

These designs and applications are typical of the airframes being constructed in the late 1940s and 1950s many of which were discussed at a 1957 conference in Cambridge, UK, chaired by Norman de Bruyne [32]. Thus, as well as in the Comet range, such concepts were used in the de Havilland Dove and Heron, the English Electric Canberra, the Fokker F-27 and F-28, the Saab J 29, J 32 and J 35 fighter aircraft, the Short Brothers Sperrin, SB 5, Sherpa and Seamew, the Vickers Vanguard and the Chance Vought F4 and F7 fighter aircraft and, to a certain extent, in their Regulus guided missile.

In 1954, a film adhesive was added to the Redux range. Here two half-webs of the powder-coated resole film (Figs. 56 and 57) were consolidated together – powder face to powder face – to produce a handleable film having the same adhesive properties as the original liquid/powder system. Although Fokker and Short Brothers did not make the change from Redux Liquid and Powder until the 1990s, other companies readily adopted this "easier to use" format; utilising the same design concepts as for the liquid and powder version. Thus, not only was this film adhesive used in the later de Havilland Comets but was also extensively in the Armstrong Whitworth Argosy, the BAC 1-11, the Bristol Britannia, the Hawker

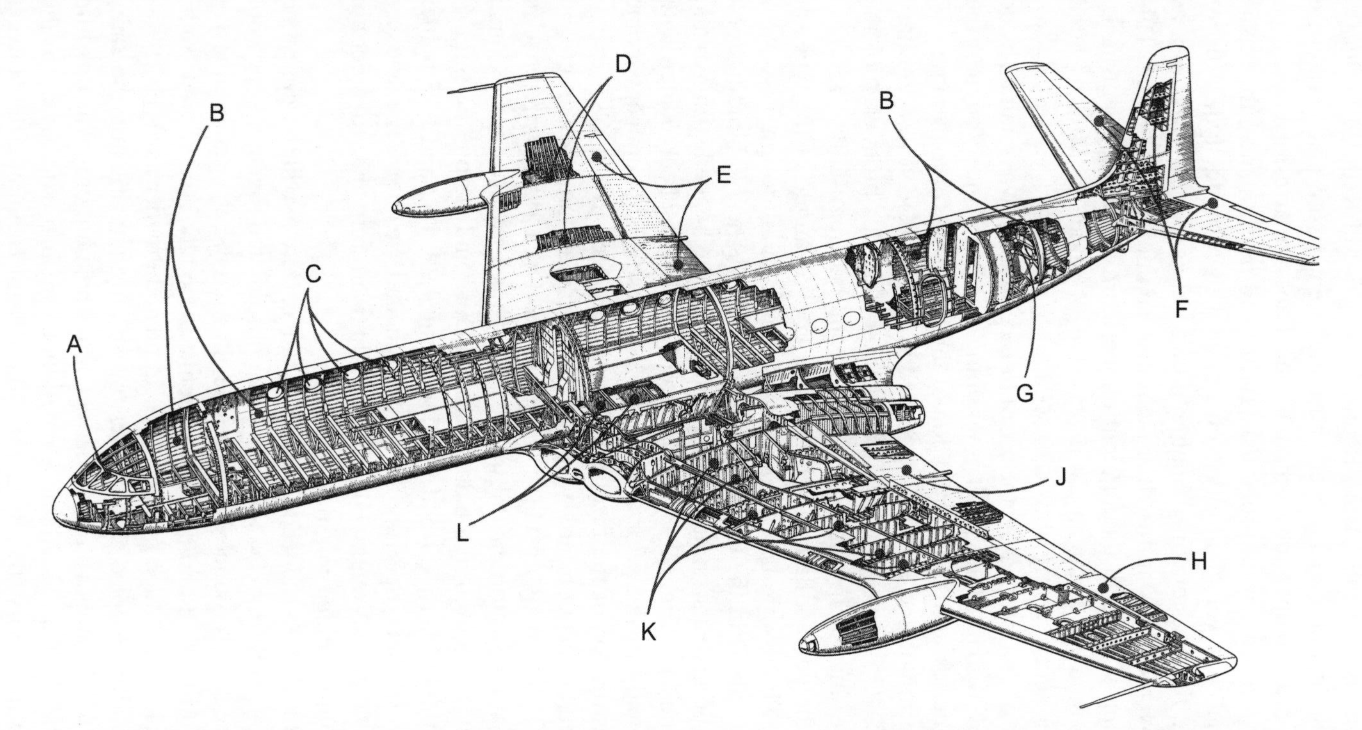

Figure 64: Diagrammatic view of the de Havilland Comet: areas of structural adhesive bonding, (A) Canopy structure, (B) All longitudinal stringers of fuselage shell (Figure 65), (C) Local doublers and reinforcements around window frames (Figure 66), (D) Spanwise stringers in wings, (E) In ailerons and flap structures, (F) Wall stiffeners in fin and elevators, (G) Seals of pressure dome, (H) In aileron structures, (J) In flap structures, (K) Vertical stiffener flanges and doublers in wings, (L) All stiffeners in pressure floor "Redux bonded" to floor shell.

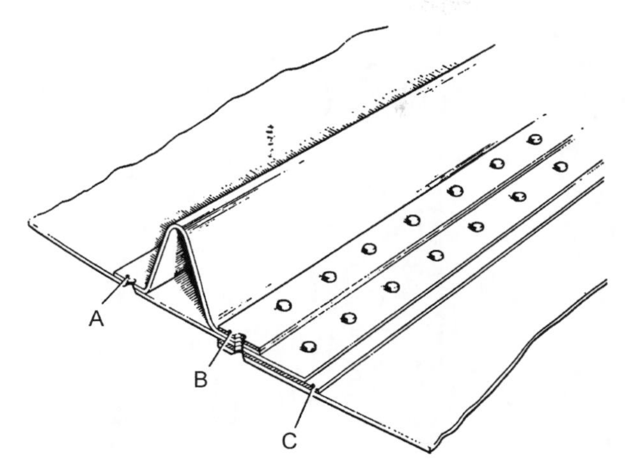

Figure 65: Schematic of a bonded stringer in the comet, (A) Stringer bonded to skin, (B) Reinforcement bonded to stringer, (C) Edge reinforcement bonded to skin.

Siddley HS-121 Trident and HS-125 Jet Dragon, the Handley Page Herald, the SAAB Lansen and Draken, and the Westlands/Saunders Roe Hovercraft.

The outstanding properties of this adhesive can be illustrated by two examples.

The first batch of Redux Film was produced on 1 May 1954 and a small sample has been stored at ambient temperature ever since. Lap-shear tests have since been

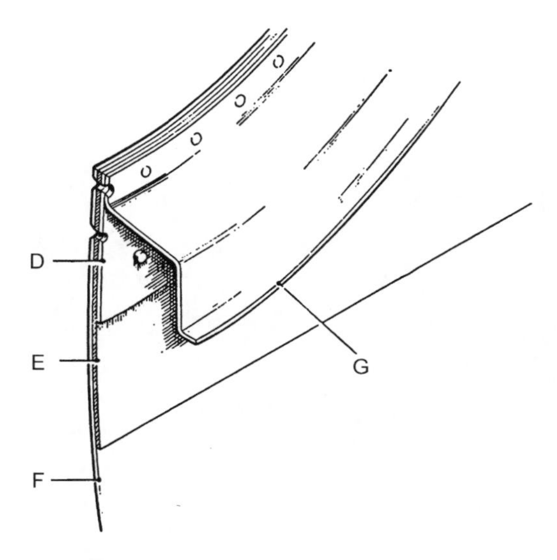

Figure 66: Schematic of the bonded window surround in the comet, (D) Doubler bonded to skin, (E) Reinforcement bonded to skin, (F) Heavy outer skin, (G) Window frame.

Figure 67: Redux liquid and powder applied to a beam assembly in the Comet.

carried out on an intermittent basis. The last test in the series was carried out on 7 June 2004 (50 years). Fig. 69 shows the results over this time scale and emphasises the remarkable consistency of this type of adhesive.

The other example concerns the long-range maritime patrol aircraft, the de Havilland Nimrod. This aircraft was based on the successful Comet IV design utilising a significant extent of structural metal-to-metal and metal-to-honeycomb bonding with Redux 775 Film adhesive. In 1998, this airframe was relifed for a further 25 years, not least in view of the proven long-term durability of the adhesive.

Figure 68: Fuselage interior of the de Havilland Comet.

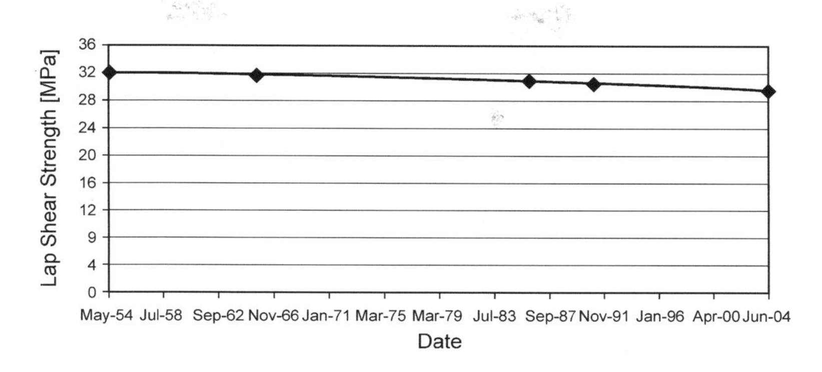

Figure 69: Ambient temperature aging data for Redux 775 film adhesive.

The confidence given to aircraft constructors by such a reliability has led to the extensive use of bonded components in the construction of more recent aircraft such as the BAe 146 (RJ-80), the Fokker F-50, F-70 and F-100 and the Raytheon Aircraft HS-125-800XP (originally the Hawker Siddley HS-125); the latter is still in production. The bonded components, such as fuselage and wing skins, control surfaces, stringers and longerons, all use the essential design concepts of 40–50 years earlier.

Phenolic adhesives were also used in the late 1950s for helicopter rotor-blade bonding. Redux 775 (PVF-phenolic) was used in Sud Aviation's Alouette in the late 1950s and Scotch-Weld® adhesives AF-6, etc. (nitrile-phenolics) are currently in use in Westlands' Sea King helicopter (q.v.).

Apart from the examples quoted above, there is one specialised application for phenolic/PVF films, which has been in operation since the 1950s. This is the use of the so-called half-web Redux 775 film (designated Redux 302: a simple film of phenolic resole covered with a significant excess of PVF powder) for bonding propellant to the walls of numerous rocket motors used in missile construction.

Although, techniques vary from motor to motor the generic method can be outlined as follows:

- The film adhesive is first laid up against the lining of the rocket motor; phenolic face towards the substrate.
- The film is cured to the wall of the motor.
- The propellant is cast into the motor and cured against the PVF-rich surface of the adhesive; the solubility parameters of the PVF and the propellant formulations are similar enough to permit a degree of swelling of the PVF particles which allows an interpenetrating network to be set up between propellant and adhesive which "locks" the propellant in place.

It is the strength of the resole bond to the motor lining and the efficacy of the interpenetrating network, which ensures an efficient burn when the motor is fired.

5.6.1.1. Availability
Phenolic structural adhesives for the aerospace industry are available [33] from 3Ms, Hexcel Composites and Lord Corporation.

5.6.2. Phenolic-epoxy Adhesives

Phenolic-epoxy adhesives were developed in the mid-1950s by Ciba (ARL) Limited (now Hexcel Composites) and American Cyanamid (now Cytec-Fiberite). These phenolic-epoxy formulations were initially developed for bonding components exposed to a much higher service temperature than the 70°C limit for Redux 775.

Initial applications included bonded control surfaces of the Raytheon Hawk (Homing All-The-Way Killer) surface to air missile and components in the Rolls-Royce RB162 aero engine. The RB162 was the lightweight, so called "plastic" engine, which was used as the auxiliary in the tail of the Hawker Siddely HS-121 Trident. Here the metallic retaining strap was bonded round the engine assembly using Redux 307; a version of Hidux® 1197A using a special support carrier.

Although their usage was, to a great extent, limited to aero engines and particularly missile applications, these adhesives were critical for solving other design problems in, for example, the Blackburn Buccaneer (Fig. 70).

The design of the Buccaneer was such that the tail brakes opened into the jet stream. This meant that they not only were exposed to high temperatures but also suffered from sonic fatigue in the e-flux. The problem was solved by bonding the components with Hidux 1197A and C.

Figure 70: Blackburn Buccaneer. (Copyright: Andrew Brooks at www.avcollect.com.)

Although the Hidux range has long since been withdrawn the Cytec-Fiberite, HT®-424, is still on the market. A brief examination of some of the current areas of application shows how advanced the phenolic-epoxy adhesives were for their time. These include the flight control surfaces on the B 52 bomber and the Sidewinder missile, components within the navigation base of the Space Shuttle, applications within the Motorola Mk 45 missile and the wings on the Teledyne Ryan Firebrand missile.

5.6.2.1. Availability
Phenolic-epoxy structural adhesives for the aerospace industry are available [33] from Cytec-Fiberite.

5.6.3. From Phenolic-epoxy to Epoxy Adhesives

The phenolic-epoxy adhesives were, and still are, quoted as having short-duration service temperatures of over 500°C; as good, if not better than many of the polyimide adhesives of today. Their major drawback was, not surprisingly, the condensation products associated with phenolic resoles during cure.

Researchers in the early 1960s tried, with considerable success, to replace the phenolic resoles in these adhesives by the then novel tetraglycidyl derivative of diamino diphenyl methane (Araldite MY 9512). This led to Hidux 1033, which was then further optimised to give Hidux 1233. This latter product was specifically designed for use on such super-sonic aircraft as the Anglo-French Concorde. Although replaced, in the final designs, by other high-temperature epoxy-based adhesives, for example Redux 322 of Hexcel Composites and AF® 191 of 3Ms, Hidux 1233 was used in the ill-fated BAC TSR-2 (Fig. 71).

Figure 71: BAC TSR-2 (Courtesy of: Andrew Bates).

Figure 72: Handley Page Herald (courtesy of: Paul Middleton at Oldprops).

In spite of these apparent setbacks, Hidux 1033 was used in a commercial airframe; it was key to solving a design problem in the Handley Page Herald (Fig. 72). In the case of the Herald, the design had the undercarriage doors opening into the turbo-prop e-flux. This, like in the case of the Buccaneer, led to both sonic and thermal fatigue. The answer was to replace the original adhesive and bond using Hidux 1033M.

5.6.4. Epoxy Adhesives*

The aerospace industry uses more adhesives based on epoxy resin chemistry than those based on any other. It is, therefore, worthwhile to examine several individual sectors within this industry, to illustrate how and where these adhesives are used.

5.6.4.1. Bonding in Helicopter Structures

Here, examples are drawn from Westland Helicopters Limited (an Agusta Westlands Company), one of the most important companies designing and manufacturing helicopters. This is a company that fully believes in the use of structural adhesive bonding [34]. This is evidenced by the fact that no "chicken" rivets (riveting through the bonded glueline in the belief that this will improve safety and eliminate the effects of peeling loads generated through "end lift") are used in any of their structures apart from the end caps on the tail drive shafts where rivets are utilised to change the stress patterns within the joint so that the torque loads are carried throughout the bonded area rather than just at the edges.

* Editor's note: For more information about the chemistry of epoxy adhesives, please refer to the chapter "Epoxy Adhesives" in Volume 5 of this handbook.

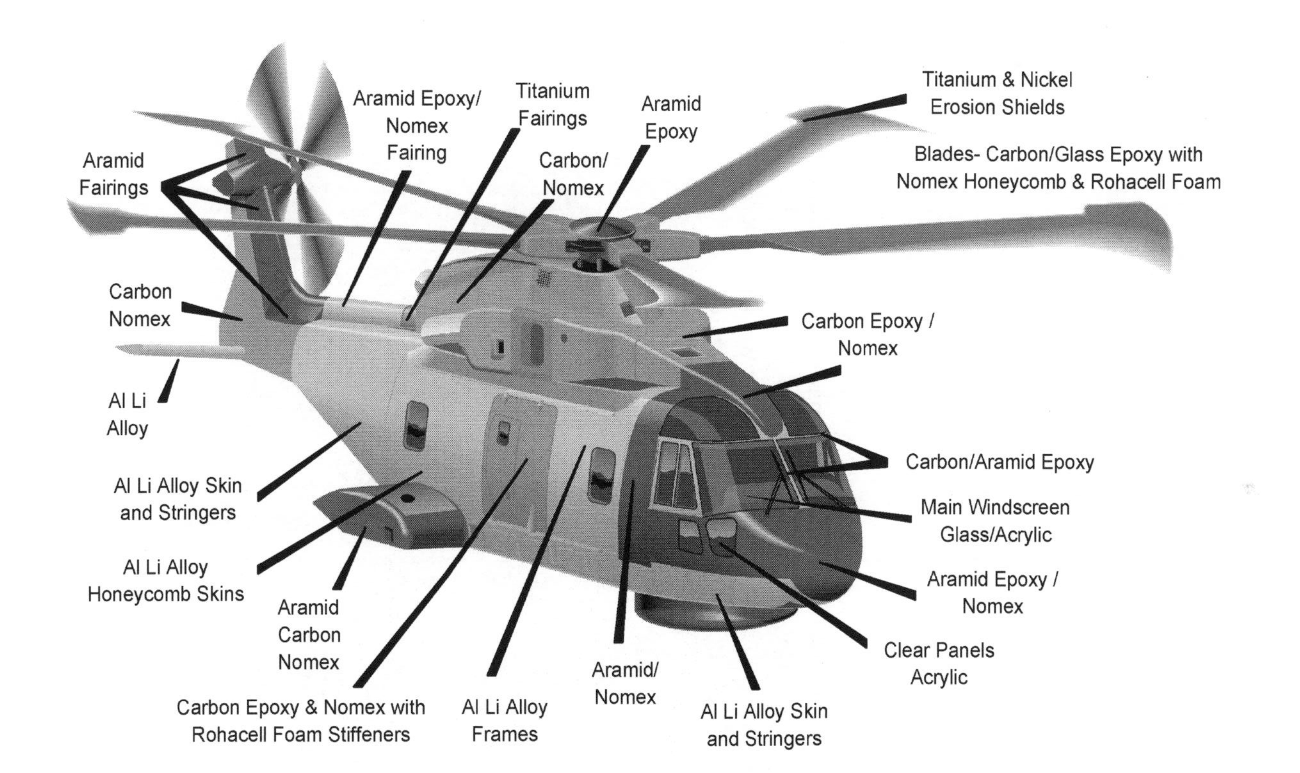

Figure 73: Schematic structure of Westland Helicopter's EH101 Merlin Helicopter (courtesy of: Westland Helicopters Limited). (colour version of this figure appears on p. xvii).

As the performance of structural adhesives has improved, so has the ability for the helicopter engineers to operate under tighter design tolerances. This enables an enhanced performance to be obtained from the re-optimised structures. One example is the EH101 helicopter, which has a significantly improved lifting capacity, over the older Sea King, without the need for an increased rotor diameter or number of blades. The overall construction of the EH101 Merlin helicopter can be seen in the following schematic (Fig. 73).

Apart from the rotor blades themselves, one of the most important areas as far as designing with adhesives is concerned, it is clear that most of the body construction contains elements which are suitable for structural adhesive bonding.

The engine platforms and the drive shafts from the gearbox to the tail rotor are other parts of the structure that rely heavily on structural adhesives.

To meet all the different requirements for these varied applications, this industry has to make use of all the adhesive formats previously discussed, carefully selecting those adhesives which are suitable for each application; of necessity, these adhesives (solvent-based primers, paste and film adhesives) are drawn from several manufacturers.

Rotor Blade Construction: Fig. 74 shows the general design of a typical main rotor blade; tail rotor blades are, typically, very similar. Fig. 75 shows the typical bonding tool for the main rotor blade.

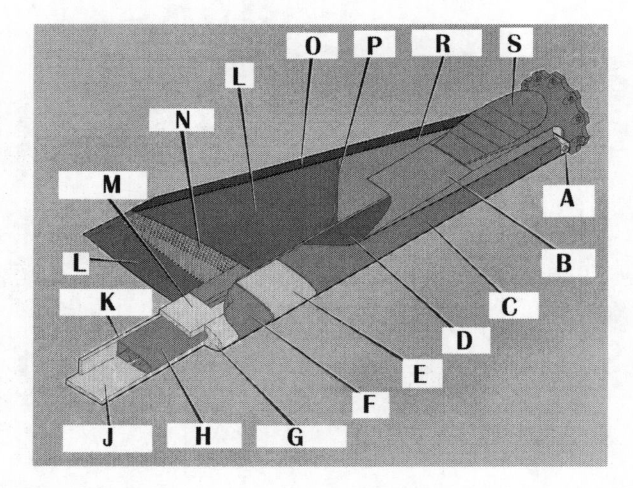

Figure 74: Schematic of the main rotor blade of Westlands' Sea King Helicopter (courtesy of: Westland Helicopters Limited), (A) BIM manifold, (B) Root doublers, (C) Erosion shield, (D) Packing piece, (E) Heater mat, (F) Outer wraps, (G) Uni-directional nose moulding, (H) Inner wraps, (J) Uni-directional sidewall slab, (K) Uni-directional backwall slab, (L) Skin, (M) Uni-directional sidewall slab, (N) Honeycomb, (O) Closing Channel, (P) Caulk, (R) Dummy skin, (S) Cuff. (colour version of this figure appears on p. xviii).

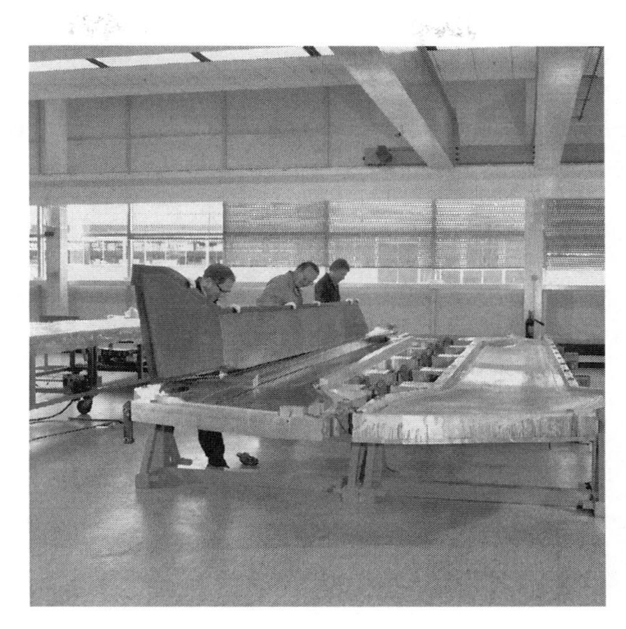

Figure 75: Typical bonding tool for the EH101 main rotor blade (courtesy of: Westland Helicopters Limited).

As can be seen, the design is complex and utilises varied components. It is a good example of the use of different adhesive formats from different companies, all ideally suited to the job they have to perform.

Most, but not all, of the skins and spars on the rotor blades are now of carbon-epoxy composite manufacture. The outer ply of the construction generally being of glass-epoxy so that the wet-abrasion pretreatment, prior to bonding, can create the necessary fresh, high-energy surface without doing any damage to the carbon fibres.

The centre of the spar can either be left empty or filled with Rohacell® foam. In the latter case, a structural epoxy adhesive film is used to splice layers of foam together; an instance of a structural adhesive being used for an essentially non-structural application. The unfilled spar is taken as being 1970s–early 1980s technology (Sea King) whereas the foam filled spar is the technology of the mid 1980s–1990s (EH101 Merlin and Lynx).

The trailing edge is filled with profiled Nomex honeycomb slices, which have been spliced together using a foaming film adhesive. The honeycomb profile is joined to the carbon spar using another foaming film adhesive.

The erosion shield is generally titanium, which has been pretreated using a conventional etch process followed by application of an adhesive primer that matches the 120°C-curing adhesive used for bonding both the shield to the spar and the outer skin of the blade to the inner structure. Thus, Redux 112

Primer will be used with Redux 312 and 3M's EC® 3960 primer is used with Scotch-Weld AF 163-2.

The rotor tips are usually clad in electroformed nickel sheet; this is chemically pretreated (usually a nitric acid etch [35]), primed and then bonded with a film adhesive curing at 120–130°C. In some cases, due to the increased service temperature requirements caused by underlying heater mats, an adhesive is used, which has enhanced temperature performance.

In the case of the swept tips on the Lynx rotor blades, the outer skin is made from polyetheretherketone (PEEK). Here, the surfaces are abraded before bonding; the adhesive is an epoxy paste adhesive curing at ambient temperatures.

In, for example, the EH101, apart from the main rotor structure, many sub-assemblies are adhesively bonded; this gives both structural and non-structural components. Thus:

- Room-temperature epoxy paste adhesives are used to bond weight pots (for balancing), lashing points, lightning-strike connecting strops and foam ducting.
- Syntactic paste adhesives are used for edge-filling and damage-repair of foam components.
- Two-component acrylic adhesives were used for all rubber-to-composite bonding but this is now being replaced by rubber-toughened epoxy paste adhesives, which cure at ambient.
- The trim tabs are bonded in place by a combination of rubber-based contact adhesives (for initial "grab") and an epoxy film adhesive for the ultimate bond.
- Polysulphide-based conductive caulking sealants are used along specified lengths of the rotor blade.
- Polyurethane-based adhesives are used to bond polyurethane rubber to the blade.

As has been indicated, the production of the tail rotors generally follows a similar pattern. The exception is the Sea King. Here there is a difference between the 6-hub and the 5-hub versions. In the former case, the outer skin is of anodised aluminium, which has been adhesively primed and then bonded to the inner aluminium spar and honeycomb structure using 3M's Scotch-Weld AF-6 and AF-10 nitrile-phenolic 170°C curing film adhesives, and the erosion shield is of stainless steel which has been chemically etched and primed before bonding with AF-30 (another 170°C curing nitrile-phenolic film adhesive). The 5-hub version, however, is more conventional. This uses an epoxy film adhesive to bond a co-moulded composite spar and skin structure to a titanium erosion shield.

The engine platforms are constructed from titanium pillars and titanium sheeting. The metal is pretreated with an etchant acid mixture and then bonded at 175°C with Redux 319, to obtain the required higher service temperature.

Much of the body is of a bonded sandwich construction. In the EH101, epoxy film adhesives are used to bond aluminium–lithium skins to aluminium honeycomb and composite skins to Nomex honeycomb; the composites are generally carbon, but aramid (for example Kevlar®) is used over the leading edges.

5.6.4.2. Bonding in Military Aero Engine Structures

Military aero engines, for example those manufactured by Rolls Royce, call for little usage of adhesives, as they tend to be turbojets without bypasses. This means that the temperature build up is sudden and reaches levels not suitable for structural adhesives.

No Grade I structural bonding, therefore, takes place within the main engine structure, although some Grade II bonded components are manufactured. In this case honeycomb sandwich structures are produced, using epoxy-based paste adhesives, as part of the engine compressor assembly.[6]

Most of these sandwich structures utilise either titanium or stainless steel skins bonded to aluminium honeycomb; a low volume of aluminium skinned structures are also made.

The pretreatment for titanium and stainless steel skins, prior to bonding, is a conventional grit abrasion; the aluminium can either be chemically etched or grit blasted. In all cases, no surface protection primers are used.

The few remaining bonding applications generally use structural paste adhesives to produce, essentially, non-structural bonds (Grade III rating). Apart from the BMI systems, mentioned below, by far the greatest proportion of these adhesives utilise epoxy resin chemistry.

Applications include the bonding of brackets, bosses, sleeves and seals within the engine structure, usually accomplished with epoxy paste adhesives.

It is worth noting here some low volume, non-epoxy usage. The bonding of some brackets, etc., makes use of a two-part acrylic system.

Further, anaerobic acrylic systems are used for conventional thread-locking and retaining applications and cyanoacrylates are used as manufacturing aids.

Finally, conventional sealants, based on silicone rubber formulations, are used to bond in silicone rubber sealant strips.

5.6.4.3. Bonding in Civil Aero Engine Structures

Examples can be drawn here from Rolls Royce who make the engine units themselves and Hurel-Hispano who manufacture subsidiary components such as thrust reversers and air intakes for various engines.

[6] Grade I = life depending primary structures, Grade II = secondary structures; and Grade III = non-load-bearing structures/bonding.

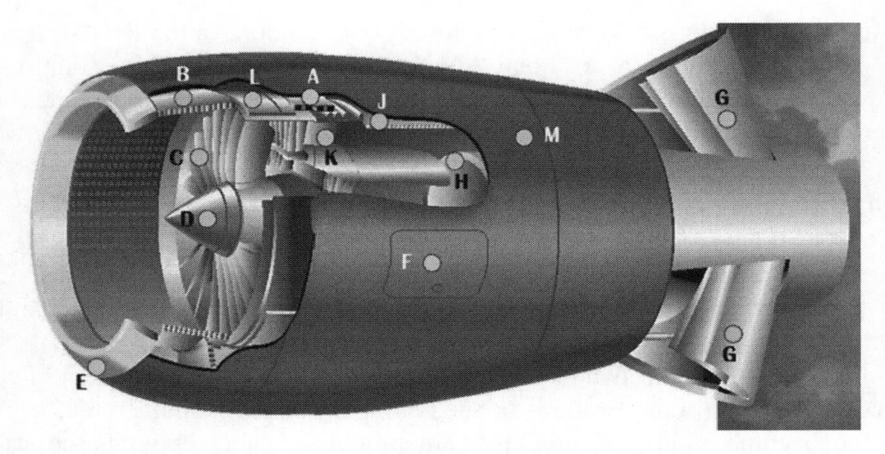

Figure 76: Cutaway of a generic commercial aero engine. (A) Electric control unit casing, (B) Acoustic lining panels, (C) Fan blades, (D) Nose cone, (E) Nose cowl, (F) Engine access doors, (G) Thrust reverser buckets, (H) Compressor fairing, (J) Bypass duct, (K) Guide vanes, (L) Fan containment ring, (M) Nacelle cowling.

Considering first the engine unit, Rolls Royce use a significant range of substrates in their bonded components. These include large quantities of sheet aluminium and much smaller quantities of titanium and stainless steel, large quantities of carbon-epoxy laminates and small quantities of fibre-reinforced bismaleimide laminates and aluminium and Nomex honeycomb.

One specific example concerns the sound-deadening, so-called "acoustic panels". These are fitted to the front (position "B" in Fig. 76) and the rear of the engine and comprise a structurally bonded sandwich panel where a perforated aluminium, titanium or stainless steel top skin is bonded to the honeycomb core.

The perforated skins are usually pretreated by grit abrasion and then, occasionally, primed with a surface protection primer such as Redux 122 or with an adhesion-promotion primer. The adhesive is epoxy based and can be in film or in paste form. The acoustic panel so formed is either bonded directly to the engine casing or is first closed with a glass-reinforced epoxy laminate to promote stiffness and this GRP skin is then bonded directly to the casing.

A second example is the fan track liner (above the fan blades below position "L" in Fig. 76; it can also be seen in the same relative position in Fig. 77). This assembly is a combination of aluminium honeycomb, Nomex honeycomb and a GRP laminate. The aluminium honeycomb is supplied pre-coated with a corrosion resistant epoxy primer ("Blue Dip") and this is bonded directly to the engine casing using a paste or a film adhesive. A GRP septum layer is then bonded on as a top skin using the same adhesive. Finally, a small-cell Nomex honeycomb,

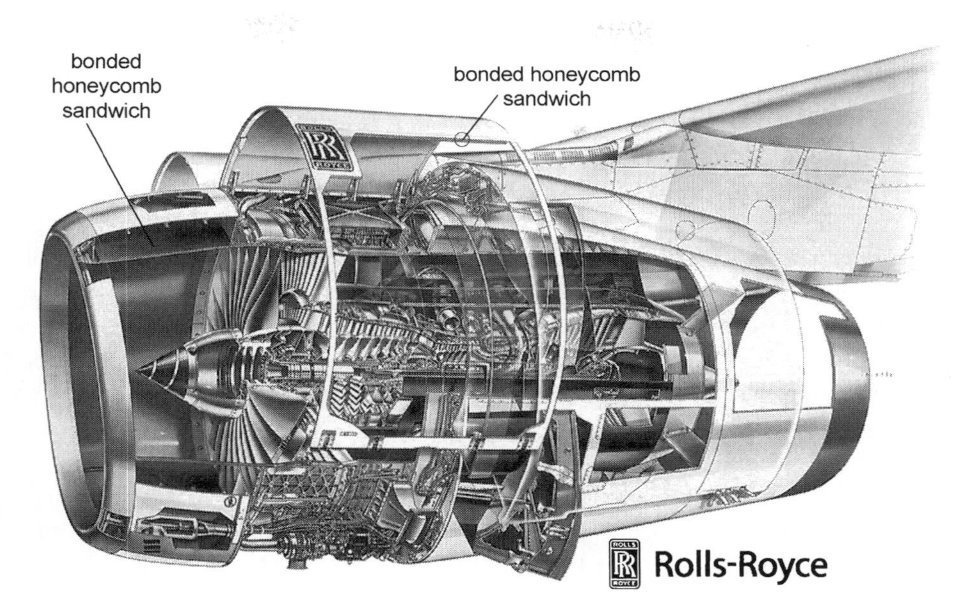

Figure 77: Cut-away schematic of the Trent 700 aero engine (courtesy of: Rolls Royce). (colour version of this figure appears on p. xviii).

pre-filled with an epoxy-based syntactic adhesive, is bonded on top of the septum layer; a film adhesive curing at 120 to 130°C is used.

It is the Nomex honeycomb layer which is critical to the efficacy of the whole assembly. Should the fan blade "stray" slightly from its course it can erode the filled Nomex honeycomb without doing any damage to the engine casing.

Many other structures within the engine assembly, which do not see high service temperatures, are bonded as this helps to reduce the weight of the final assembly. These include access panels (position "F" in Fig. 76) and thrust reversers and their blocker doors (position "G" in Fig. 76). Paste adhesives, having at least 20 MPa lap shear strength at room temperature, are also used to bond in fasteners, bushes, etc.

The various thrust reversers produced by Hurel–Hispano are good examples of how structural adhesives can facilitate the design and manufacture of these types of components as virtually all are structurally bonded. Examples can be drawn from the following engines:

- RB211-524: G and H variants
- CFM 56 (for the Airbus A.319, A.320 and A.321 series)
- Trent 700 (for the Airbus A.330 series)
- Trent 900 (for the Airbus A.380)
- AS 907 (for the Bombardier Aerospace BD 100 Continental)

Air intakes for the AE 3007 engines on the Embraer 145 are also built using integrally bonded components.

One of the major components of the thrust reversers are the blocker doors which, on landing, are deployed to block off or divert the air stream passing through the engine. The blocker doors, therefore, effectively act as brakes. As the majority of these components only see appreciable service temperature on landing, structural adhesives with temperature resistance up to about 150°C can be used for these bonding applications. However, there are some designs, for example the Trent 700, where the service temperature of these components is in the region of 150°C, i.e. the bonded structure experiences temperatures around this figure for most of the time that the engine is operating. In these instances, epoxy adhesives having service temperatures between 150 and 180°C have to be used. For certain parts of these structures, BMI adhesives are also employed. This latter case is covered in the section on bismaleimide adhesives.

There are three main designs of thrust reversers: Fig. 78 shows the cascade reverser (as in the RB211). The pivoting door or fanflow reverser (for example in the Trent 700 and the CFM 56) is shown in Fig. 79; both figures show the blocker doors in the deployed position, diverting and effectively reversing the direction of thrust. There is also the planar exit rear target (PERT) reverser (as in the AS 907).

Although the blocker doors are fairly complex in construction, in the final analysis they are effectively heavy duty honeycomb-filled sandwich structures.

The substrates used in the construction of the bonded components are exclusively aluminium and carbon fibre reinforced composites. Where sandwich panels are used aluminium and aramid (Nomex) honeycombs are used.

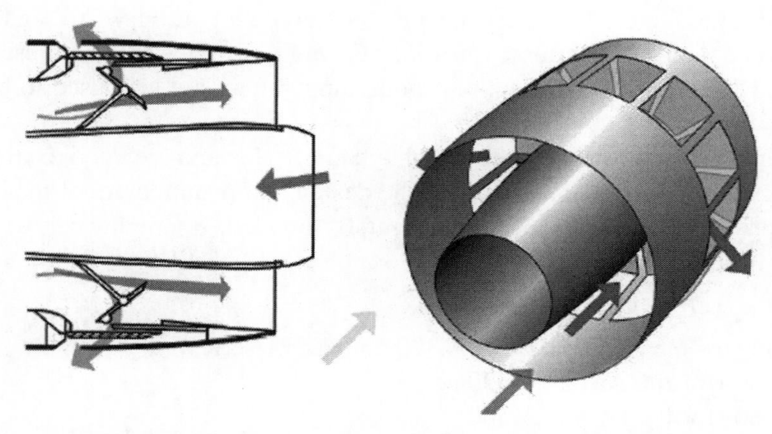

Figure 78: Schematic of the Cascade Reverser in operation (courtesy of: Hurel Hispano).

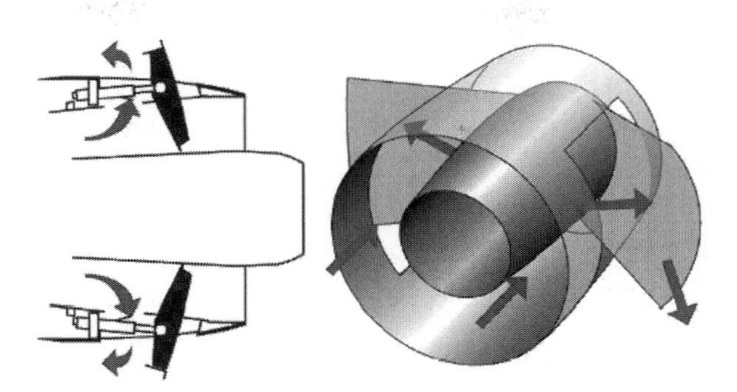

Figure 79: Schematic of the Fanflow Reverser in operation (courtesy of: Hurel Hispano).

In all cases, the aluminium substrates are pretreated using conventional chromic acid anodising techniques, often this is augmented with the application of a pretreatment protection primer. The carbon composites use peel ply techniques to prepare the surfaces for bonding.

It is possible to analyse many of these application in more detail, as follows:

RB-211: The hoop plates within the cascade reverser are Class I bonded sandwich structures where Redux 119 primed aluminium substrates are bonded to aluminium honeycomb, using Redux 319 film adhesive

Trent 700: The inner fixed structure (IFS) utilises BMI chemistries but the outer skins are more conventional comprising carbon fibre composite laminates bonded with Redux 322 (175°C curing) to Nomex honeycomb. The one unusual feature in this design is that a layer of Tedlar® film is bonded to the outer composite ply to reduce moisture ingress.

Trent 900: In the construction of the blocker doors for the Trent 900, Redux 322 is first reticulated onto the pre-cured and perforated carbon skin. This is then bonded directly to the honeycomb. Finally, uncured carbon prepreg skins are co-bonded onto the honeycomb to complete the sandwich panel. (Note: It is accepted that reticulation of the adhesive film onto the perforated skin gives better acoustic deadening properties than if the adhesive were to be reticulated onto the core. However, both methods are used, depending on the engine.)

To avoid crushing, fasteners are supported within these structures by the use of a single-component syntactic paste adhesive: Scotch-Weld Void-Filling Compound 3439.

CFM 56: Here both aluminium and carbon reinforced composite sandwich panel constructions are used. An aluminium bondment forms the inner structure of the acoustic panel. Perforated, Redux 119 primed skins are bonded to aluminium honeycomb, Redux 319 film adhesive having first been reticulated onto the core. The outer structure is a conventional bondment of a carbon reinforced composite skin bonded to the core.

Embraer 145: The air intakes comprise conventionally constructed acoustic sandwich panels where pretreated and primed perforated aluminium skins are bonded to aluminium honeycomb, Redux 319 film adhesive having first been reticulated onto the core.

AS 907: The structural bonding that takes place in this engine construction is a good example of the end-user modifying a conventional adhesive to be "fit for use" for their particular application.

Henkel Corporation's EA 934 NA is a commercial room-temperature curing, liquid shimming adhesive, i.e. a high compressive strength paste adhesive which can be used to fill gaps between two materials in a component. Such adhesives are an effective and efficient way to fill these gaps and by doing so transfer load effectively across the interface to minimize fatigue-induced cracking.

This technique is used extensively in many nacelle structures to shim gaps between aluminium, steel and carbon composites. However, Hurel–Hispano modify the as-received adhesive by the addition of 15% aluminium powder, which then makes it suitable for their applications.

5.6.4.4. Bonding in Military Aircraft Structures

As has already been seen for military aero engines, the amount of structural bonding taking place in military aircraft is considerably smaller than that seen for the civil industry.

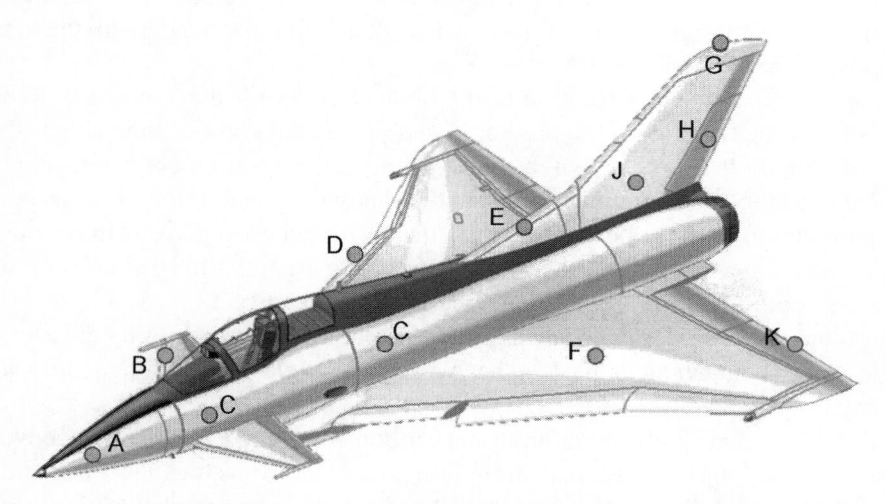

Figure 80: Generic military strike aircraft: (A) radome, (B) foreplane canard wings, (C) fuselage panel sections, (D) leading edge devices, (E) fin fairings, (F) wing skins and ribs, (G) fin tip, (H) rudder, (J) fin, (K) flying control surfaces. (Colour version of this figure appears on p. xix.)

However, there are examples which can be taken from the bonding of airframe structures, of specific components in military transport aircraft and from the design and construction of radomes.

Fig. 80 is a drawing of a generic military strike aircraft showing the relevant areas where structural composite, honeycomb and structural adhesives are used. Specific examples refer back to these particular areas.

5.6.4.4.1. Bonding of Airframe Structures

Here, specific examples can be taken from the last three generations of strike and multi-role combat aircraft: the Anglo-French Sepecat Jaguar, the Panavia Tornado and the Eurofighter Typhoon (Fig. 81). Although all three aircraft have been or are being built by European consortia, the following examples are courtesy of BAE SYSTEMS.

Jaguar: In-board and out-board flaps, trailing edges, tips and slats are constructed from BR-227 primed, chromic acid anodised aluminium bonded with FM-61, a 175°C curing modified epoxy film adhesive.

Tornado: Access doors are constructed from BR-227 primed, chromic acid anodised aluminium sandwich panels bonded with FM-96, a 175°C curing modified epoxy film adhesive. The same primer/adhesive combination is used to produce structurally bonded tips, rudders and heat exchanger panels.

Typhoon: Applications in this aircraft show the use of structural adhesives to bond carbon composite adherends together. The composite adherends use a peel ply method of pretreatment prior to bonding, with the actual areas to be bonded then being abraded (by hand). This is followed by a solvent wash.

Two 175°C curing film adhesives are then used to prepare the bonded components. Redux 322 is used to prepare composite/Nomex honeycomb sandwich structures and FM-300 for the composite-to-composite bonds, for example in the construction of the rib assembly.

Further examples can be taken from the Jaguar, Tornado and the BAE Hawk where unprimed, chromic acid anodised aluminium is bonded with another 175°C curing film adhesive – Redux 308A NA – to produce the sandwich structures needed for the control surfaces. All three aircraft also use EA 934 as a shimming adhesive.

5.6.4.4.2. Bonded structures in military transport aircraft

The first example concerns titanium substrates, which are pretreated using the Pasa Jell technique (Chapter 4) and are then bonded with FM-300 to produce the strakes in the C-17 Globemaster military transport aircraft.

Mention has been made above of fibre-metal laminates. These are not only bonded structural substrates in their own right – the layers of aluminium and unidirectional fibres are joined together using a true adhesive matrix – but in

Figure 81: Three generations of modern military aircraft. (a) Jaguar (Crown Copyright/MoD: Reproduced with the permission of the Controller of HMSO), (b) Tornado (Copyright: Roger Hadlow), (c) Typhoon (courtesy: Eurofighter at www.eurofighter.com). (colour version of this figure appears on p. xx).

component manufacture they must be joined together by some means. This will often be by recourse to structural adhesive bonding.

It is of value, therefore, to examine briefly examples of their use. The earlier of the two main types of structure is ARALL. Here layers of aramid fibres are bonded to the BR-127 primed aluminium sheets using a modified epoxy film adhesive, AF 163-2.

Such structures not only give significant weight reduction (Sg of ARALL is 2.16 whereas that for aluminium alloy is 2.78) but also phenomenal fatigue resistance (cf. Fig. 25). Both of these aspects were utilised in the design of the cargo bay door of the McDonnell Douglas C-17 Globemaster (Fig. 82). A combination of ARALL 3-3/2-0.3 and ARALL 3-4/3-0.3 was used which gave a mass reduction of 26% over an all-aluminium design (see Section 5.5.1.6 and Table 1 for an explanation of the ARALL designations).

To date, after more than 5 years in service, there have been no problems with fatigue, impact, or corrosion.

5.6.4.4.3. Bonding in radome structures

Radomes can be described as composite shields which cover/protect aircraft communications equipment. Their design has to fit in with the duties which the equipment has to perform and also where it is required on the aircraft. Although

Figure 82: ARALL Cargo Bay Door of the McDonnell Douglas C-17 Globemaster (Courtesy of: Fibre Metal Laminates, Delft).

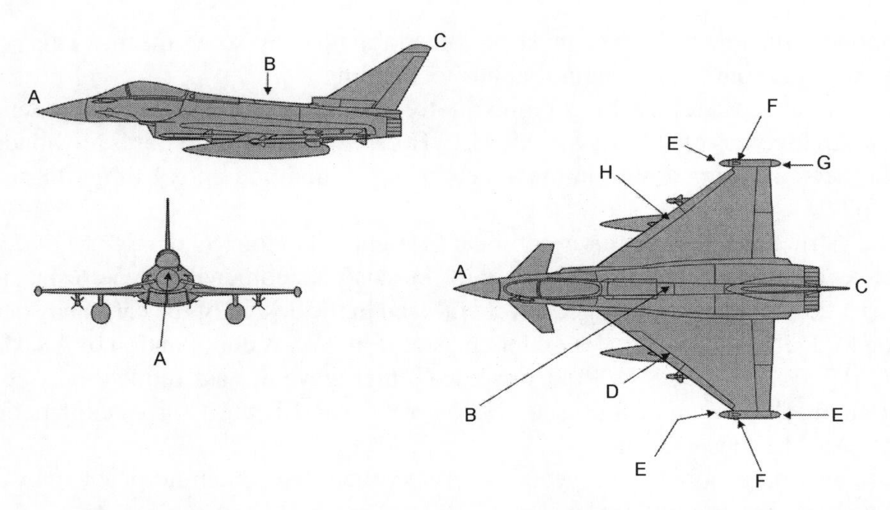

Figure 83: Position of Radomes in the Eurofighter Typhoon (courtesy of: Chelton Radomes), (A) Captor radome, (B) CRPA, (C) Fin cap, (D) EFA maw radome, port, (E) Radome general, (F) Radome blister general, (G) Radome starboard rear, (H) EFA maw radome starboard.

radomes are used on both civil and military aircraft it is in the military aircraft sector that the performance requirements are most demanding.

Examples of such structures can be drawn from Chelton Radomes (part of Cobham plc) who manufacture a variety of radomes which range from cones of about 300 mm in height, for the Sea Wolf missile (Fig. 100), to naval radomes which are about 6 m in height. Between these two extremes lie the radomes for many military aircraft and helicopters; the Eurofighter Typhoon (Fig. 83) being typical of the extensive use of radomes in these aircraft.

Civil and low-performance military radomes are usually Class II structures, which mean that they do not have to carry high structural loads; it is their electrical properties which are paramount. However, fast-jet nose radomes are Class I structures whereby major failure would be critical. This being so, both their design and the choice and treatment of the materials used in their construction, are paramount.

In an ideal situation, 100% of generated electromagnetic radiation should pass through the wall of the radome with no adsorption by the structure. In practice, however, this does not happen due to the dielectric constant of the composite skins, which range between 3.3 for quartz/cyanate ester and 4.5 for glass/epoxy. Once a design has been agreed then the overall dielectric constant has to be rigidly controlled. The finished radome components are required to be interchangeable and as such the dimensional tolerances allowed are very tight to ensure the same performance is obtained.

To obtain the best balance between wall thickness and electrical performance many radomes currently utilise simple honeycomb sandwich structures in their construction. The composite skins are quartz, glass or aramid (generally Kevlar) reinforced epoxy mouldings; carbon reinforcement is not used within the "window area" of a radome because of the poor dielectric constants associated with these structures. Epoxy-based film adhesives are used to bond these skins to the Nomex honeycomb core and/or semi-structural foam.

In view of the critical nature of the electrical properties of these structures, component pretreatment is vital. This is particularly true with respect to the honeycomb or the semi-structural foam (for example, Rohacell) core.

Conventional abrasion processes are used to pretreat the reinforced composite components. Dependant on the finish required either a standard peel-ply approach or a wet-and-dry paper abrasion followed by a solvent wash is used. In some instances, the second skin is co-cured to a fully bonded single-skinned core; here, of course, no further pretreatment is necessary.

In the case of the honeycomb core, this is first degreased, then oven dried and finally vacuum treated to remove all potential sources which could interfere with the final electrical properties, a process unique to radome manufacture. The Rohacell foam, on the other hand, is heat treated at 125°C for 3 h to prevent any further shrinkage during the cure cycle of the sandwich panel.

Conventional sandwich panel lay up is used to prepare the structure for bonding. Here choice of film adhesive is important, particularly as far as the areal weight is concerned. To reduce the overall panel thickness as much as possible films are used with areal weights generally in the region of 150–175 g/m^2. The adhesive chemistry is currently exclusively based on epoxy but the situation is beginning to change with a movement to cyanate ester (q.v.). Even though thickness/weight is critical, if a benefit is seen, the adhesive film can be supported using a glass, nylon or polyester carrier.

Typical of the adhesives being used are: Hexcel Composites' Redux 319L and Redux 335K, Cytec-Fiberite's FM 73 and FM 123-2, 3Ms' AF 163-2 and most recently Henkel Corporation's EA 9696. Conventional foaming films and/or foaming paste adhesives are used for general core splicing.

From the above range of film adhesives, it is clear that this application can tolerate cures from 120°C to as high as 175°C and this is generally achieved using vacuum-bag lay ups and air-circulating oven cures; autoclave processes are only used as and when absolutely necessary.

This procedure gives rise to radomes, such as the nose and tail structures on the BAE SYSTEMS Hawk 200 (Fig. 84), which appear to be relatively simple components. However, the complexity of design, materials and manufacture are not immediately evident.

Figure 84: BAE SYSTEMS Hawk 200 (courtesy of: Chelton Radomes).

Once the basic structure has been produced, and generally before painting, secondary bonding operations take place. These utilise two-component epoxy paste adhesive and syntactic paste adhesives to bond in a range of fasteners, to attach the all-important lightning strike conducting strips and to edge-fill the sandwich panels. The adhesive will be chosen to suit the nature of the components and the actual application; adhesives from Huntsman Advanced Materials (originally Vantico Limited) and 3Ms are typical.

A good example of this secondary bonding work is the Astor Satcom radome for the Bombardier Global Express. Here as well as the conventional lightning strike strips and the edge filling, a significant number of fasteners are let into the top rim of the radome for attachment to the front fuselage. This radome can be clearly seen in Fig. 85 as can the radome attached to the belly of the fuselage. This latter

Figure 85: Bombardier Global Express with Satcom and under-belly radomes (courtesy of: Chelton Radomes).

construction is manufactured in a similar manner to the one just described but not by Chelton Radomes.

Another good example, using similar sandwich panel and paste bonding techniques, is the construction of the "canoe-shaped" fin-tip radomes for the Nimrod.

The construction of radomes is one where constant innovation is taking place. The ongoing search for improved materials is leading to the use of new material combinations such as the current under-belly radome for AgustaWestland Helicopter's EH101 Merlin (Fig. 73) which still uses the conventional sandwich bonding approach but now uses Kevlar/cyanate ester skins.

The majority of fast-jet Class I nose radomes are manufactured using a solid half-wavelength construction, resulting in the limited use of bonding adhesives. This can be illustrated by examining the processes used for the Tornado and, particularly, the Typhoon nose radomes. Here the wall structure is built up using layers of glass or quartz "socks" which are then impregnated with a specially formulated polyester system using a resin transfer moulding technique. The aluminium base-rings are mechanically fastened to the radome and thus the radome is produced with only recourse to adhesive bonding for the application of lightning strike diverter strips.

5.6.4.5. Bonding in Civil Aircraft Structures

Examples in this area are legion with structural components in the fuselage, control surfaces in wing flaps and tips, empennage components, composite fairings, cargo liners, radomes, landing gear doors and floors all being adhesively bonded. Further, semi-structural bonding takes place in the manufacture of the wing and body fairings, the seats, the galleys, the toilets and the luggage bins. This is illustrated in Fig. 86 – a generic passenger aircraft. It indicates the relevant areas where structural composite, honeycomb and structural adhesives are used.

However, as many of the types of bonded structures are very similar to those already discussed under the section on the application of phenolic adhesives, examples are limited to a representative cross-section, which cover bonding of airframe structures, floors and specific components for civil aircraft.

5.6.4.5.1. Bonding of airframe structures

Here examples are taken from both Airbus Industries and the Boeing Aircraft Company

Fig. 87 shows the areas in which bonding takes place within the latest Airbus design: the A.380.

Although it is the A.380 which is illustrated, the following comments, unless otherwise indicated, refer to the Airbus family as a whole.

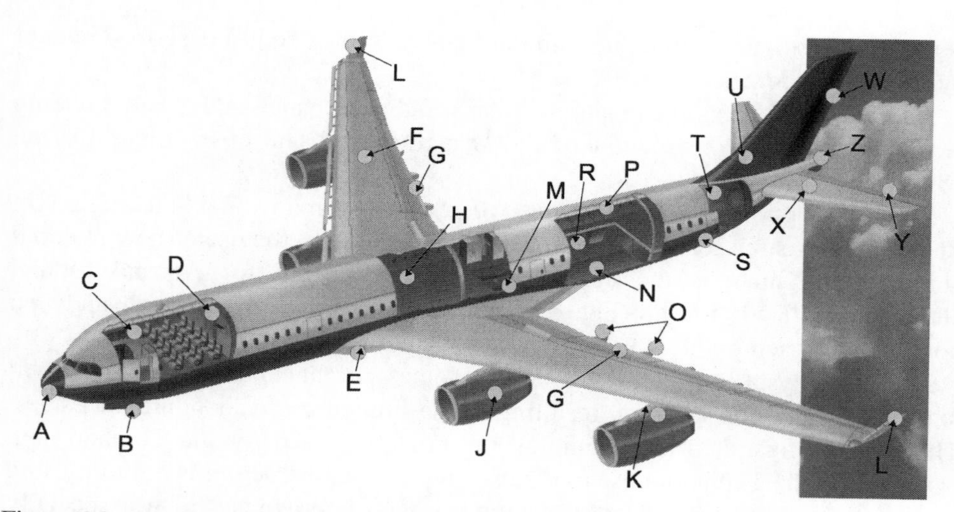

Figure 86: Generic commercial passenger aircraft: (A) radome, (B) landing gear doors and leg fairings, (C) galley, wardrobes, toilets, (D) partitions, (E) wing to body fairing. (F) wing assembly, (G) flying control surfaces: ailerons, spoilers, vanes, flaps and slats, (H) passenger flooring, (J) engine nacelles and thrust reversers, (K) pylon fairings, (L) winglets, (M) keel beam, (N) cargo flooring, (O) flaptrack fairings, (P) overhead stowage bins, (R) ceiling and sidewall panels, (S) airstairs, (T) pressure bulkhead, (U) vertical stabiliser, (W) rudder, (X) horizontal stabiliser, (Y) elevator, (Z) tail cone. (colour version of this figure appears on p. xxi).

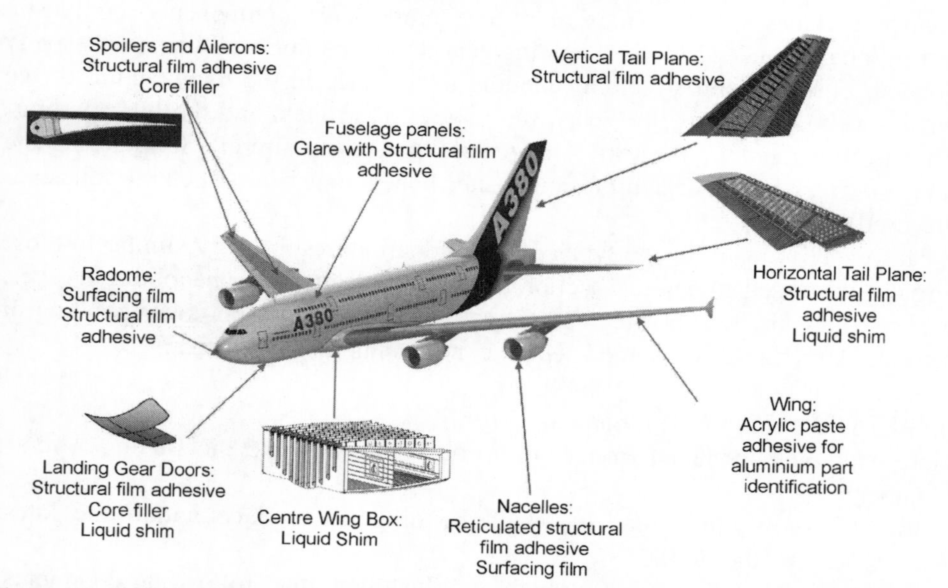

Figure 87: Schematic of the Airbus A.380 showing areas of adhesive bonding. (courtesy of: Airbus UK).

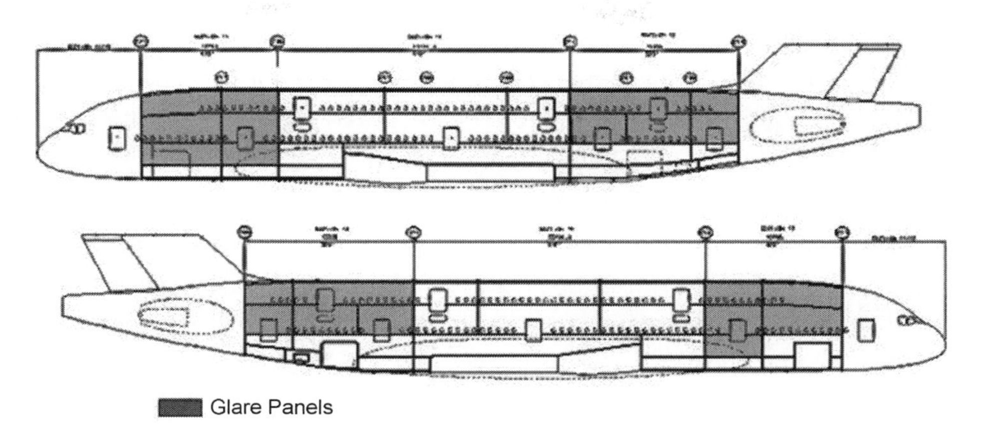

Figure 88: Schematic of the Airbus A.380 showing glare panel usage. (Courtesy of: Airbus.)

In the various components where aluminium is used, conventional chromic acid anodising pretreatment is used prior to bonding. This is preferred over any other pretreatment as it is believed that the oxide layer so formed gives corrosion protection away from the glueline. In most cases, the pretreated substrates are then primed. A corrosion inhibiting primer – BR-127 – is used where structural adhesives curing at 120°C or below are to be used. Where 175°C curing film adhesives are to be used, a conventional surface protection primer is used; generally Redux 119. It is believed that the adhesives and primers used at the higher cure cycle temperatures are intrinsically more environmentally resistant that their lower-curing counterparts.

Where composite materials are encountered, conventional peel-ply techniques are used to prepare the surfaces for bonding.

If an adhesive or primer is qualified for use on Airbus, it can be used in any of the aircraft designs.

Typical film adhesives used for metal–metal bonding are: AF 163-2 (3Ms), effectively a dual-cure adhesive curing at either 120 or 175°C, and FM-73 and FM-74 (Cytec-Fiberite) curing at 120°C and FM-300 (Cytec-Fiberite) curing at 175°C.

Film adhesives for composite bonding are: AF 163-2, PL-795 and PL-780 (SIA Adhesives Inc.), both curing at 175 to 180°C, EA 9695 (Henkel Corporation), a dual-cure adhesive curing at either 120 or 175°C, AF 191 (3Ms), curing at 175°C and FM-300.

Paste adhesives are: Araldite 1590-3 (Huntsman Advanced Materials), curing at ambient or moderate temperature.

Liquid shimming adhesives are: EA 9394 (Henkel Corporation), curing at ambient or moderate temperature and Epibond 1590, curing at ambient or temperatures up to 110°C.

Qualified repair adhesives are: EA 9394 and EA 9695 (Henkel Corporation) and Araldite 1590-3.

Dealing with each section of the aircraft in turn it is possible to make some general comments.

Wings There are no primary bonded structures in the Airbus wings but there are areas, within the outer wing box and the wing tips, where secondary bonded structures are used. In the A.380 the composite wing ribs are structurally bonded.

Film adhesive bonding does take place within the overwing and the underwing panels; the A.380 makes much use of composite materials to replace aluminium structures, otherwise the design concept for the Airbus family is, essentially, identical. Fig. 89 shows the fully bonded (Redux 319 film adhesive) aluminium box and beam structure used in the A.340.

Centre Wing Box: The centre wing box in the A.380 is a monolithic composite structure which utilises liquid shim adhesives in its construction; for earlier

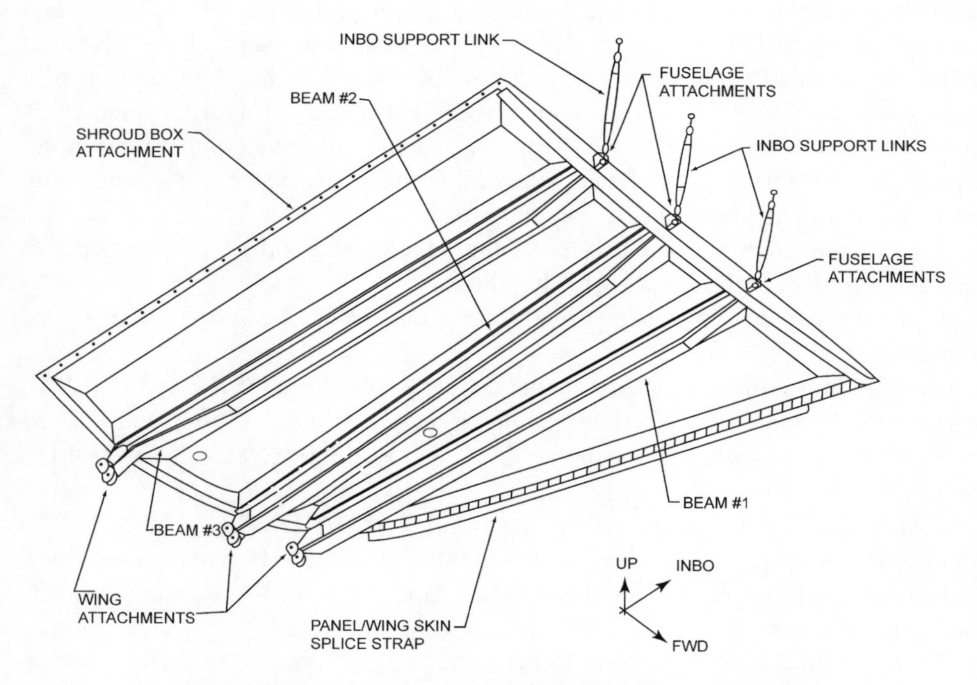

Figure 89: Schematic of the Airbus A.340 overwing panel.

aircraft these tended to be manufactured from machined metal components which were then bolted together.

Spoilers and Ailerons: Typical bonded aluminium skinned aluminium honeycomb sandwich structures are produced using film adhesives such as FM-94 to bond the skins to the core and foaming film adhesives for core splicing; the A.380 tends to use Nomex rather than aluminium honeycomb in these applications. More information on the design of control surfaces is given under the section on Boeing aircraft (Fig. 92).

Radomes: This has already been described in the section on radomes, above.

Landing Gear Doors: These structures are very much as described for the spoilers and ailerons. However, these components also contain composite materials, which are also bonded using film adhesives. There is a certain amount of composite used in the construction of the landing gear legs; these also utilise film adhesives.

Nacelles: This has already been described in the examples given for the civil aero engine components supplied by Hurel–Hispano.

Vertical and Horizontal Tail Planes: These are typical structures again making use of film adhesives and conventional paste adhesives where composite structures are to be joined. This composite design structure is generic to most of the Airbus family.

Fuselage: The early Airbus aircraft essentially used riveted components based solely on aluminium in the construction of the fuselage. The biggest step change of component design is seen in the A.380 and the aircraft, which will follow it. Here, the fuselage panels are constructed from composite materials, Glare and aluminium–lithium sheets as well as conventional aluminium alloys. Metal stiffeners are bonded to the composite skins and there is a certain amount of bonding in the areas where aluminium–lithium panels are used.

The largest use of adhesive systems is in the crown of the fuselage where, immediately aft of the cockpit and immediately forward of the tail, a significant quantity of Glare material is to be used. These areas are shown in Fig. 88.

The principal behind the Glare concept has been described, in detail, above. Essentially it is a laminate of BR-127 primed aluminium and unidirectional glass fibres which is bonded together with Cytec-Fiberite's FM-94 adhesive matrix.

The Airbus A.380 is the first large commercial use of this novel construction material. It is planned that future A.380 variants will use considerably more Glare as confidence grows in the material; the fuselage crown of the military freighter version will possibly be almost entirely constructed of Glare.

Access Doors/Panels: The Airbus aircraft contain, as indeed do most civil and military aircraft, numerous access doors or panels in the wing and body structures.

Many of these are simple aluminium skinned, bonded honeycomb sandwich structures.

Pate [36] has examined, in depth, the adhesive design concepts used by Boeing over the years, particularly for sandwich panel construction. Some of these and other constructions can be briefly summarised as follows:

Metal–Metal Bonded Doublers: These are generally used as reinforcements in fuselage skins around cutouts for doors, windows, access hatches, etc., increasing the cross-sectional thickness of the skin immediately around the cutout thereby compensating for any loss in strength incurred without the need to increase the overall monolithic skin thickness. This has the additional benefit of increasing the fatigue resistance in these areas as the bonded joint can blunt the tip of the fatigue crack as well as slowing down crack-growth rate.

This ability to control crack initiation and crack growth is utilised in fuselage construction to resist catastrophic decompression, which could result from a crack initiated by the cyclic pressurisation loads, which are applied during flight. The bonded doublers have the ability to turn the tip of any crack which might have been formed hence preventing catastrophic crack growth which can and, in the case of the de Havilland Comet 1, has led to uncontrolled decompression.

This bonded design has been and is still being successfully used on much of Boeing's fleet of aircraft.

Stiffened Panels: These designs refer to exterior fairing panels, which are attached on several sides but, essentially, only carry air loads. The initial riveted design was heavy and suffered badly from sonic fatigue. The design of the first bonded replacement used beaded double panels but these proved to be difficult to bond accurately into place and, perversely, they reduced the overall stiffness of the panel.

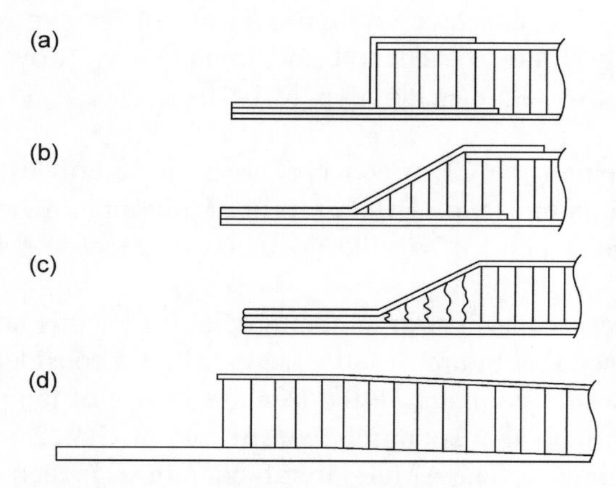

Figure 90: Schematics of stiffened panel designs. (Copyright: Boeing.)

The second approach, and the one still being employed, was to use bonded honeycomb sandwich structures, which not only virtually eliminated the effect of sonic fatigue but also enabled very thin skins to be used. This latter led to a beneficial weight saving.

Various designs are used; schematics of some of these are shown in Fig. 90. The tapered design (b) is typical of control surface applications and was often favoured over the enclosed square edge (a) as it was easier to produce a consistently leak-free structure; though, even here it proved to be difficult to obtain a good fit between core and the inner skin. This was eventually achieved by using the crushed-core design (c).

As both the durability of the structural adhesives and of the design itself improved, it was possible to move to the open square-edge panel (d). Using these designs it has proved possible to form and fold the sandwich panel to produce structures such as that seen in Fig. 91.

Open square edge panels are used extensively in the aft, unpressurised portion of the fuselage of such aircraft as the Boeing 737, 757 and 767.

Trailing Edge Wedges: These make up the rear section of the control surfaces: leading edge slats, flaps, ailerons, spoilers, etc. (Fig. 92). Early bonded designs proved both complex and costly to manufacture and were replaced, from the early 1980s in the Boeing 737, 757 and 767 aircraft, with the so-called sparless trailing edge wedge (Fig. 93); two bonded sandwich structures were dovetailed and bonded together using a foaming epoxy adhesive.

Figure 91: Typical Boeing leading edge bonded empennage structure. (Copyright: Boeing.)

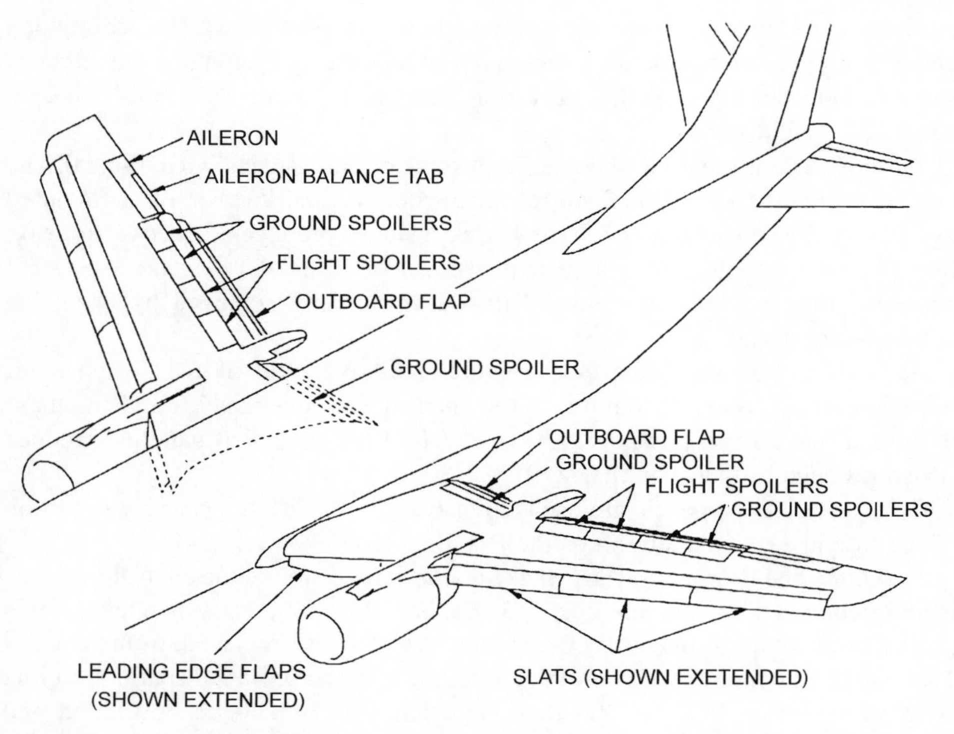

Figure 92: Typical bonded control surfaces on Boeing aircraft. (Copyright: Boeing.)

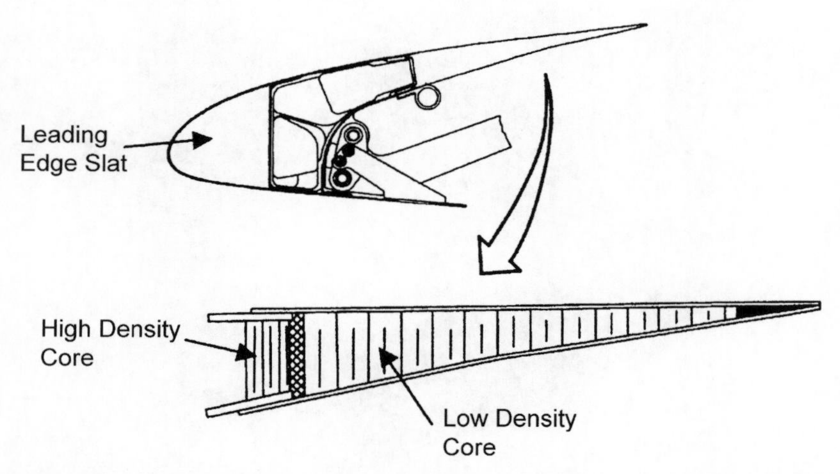

Figure 93: Schematic of the Boeing bonded sparless trailing edge wedge. (Copyright: Boeing.)

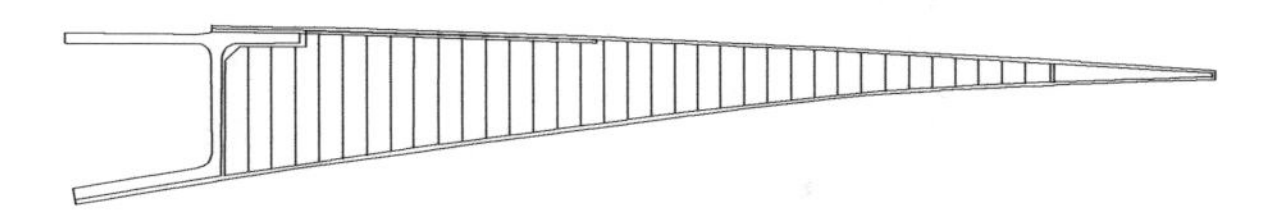

Figure 94: Boeing replacement for the bonded sparless trailing edge wedge (Copyright: Boeing).

However, the sparless trailing edge wedge also gave problems on manufacture and was difficult to maintain. Thus, the design shown in Fig. 94 has now been adopted, the forward honeycomb sandwich being replaced by a simple extrusion which is bonded into place.

Much has been said about the use of bonded honeycomb sandwich panels in spoilers, ailerons, etc. However, the Gulf Stream G5 does not use this technique. Hurel–Hispano manufactures the spoilers for this aircraft from a machined aluminium block (rather than a honeycomb in-fill) to which Redux 119 primed aluminium skins are bonded directly using Redux 319 film adhesive.

5.6.4.5.2. Bonding in aircraft floors

Aircraft floors, and other associated components, are one of the largest consumers of adhesively bonded structural sandwich panels. Examples as to their design and usage can be taken from the Fibrelam® product range of Hexcel Composites.

Early aircraft floors were constructed from bonded plywood or, where weight was a significant factor, from metal sheets which were reinforced with top hat stringers, I-beams, channels, etc.; these stringers would have been riveted to the skins.

In the early 1950s, floor design started to change with the introduction of adhesive bonding to produce aluminium-skinned sandwich panels where the core was balsa wood. The adhesive was generally based on phenolic resole chemistry; many applications being satisfied by using Redux 775. However, to reduce weight the adhesive was eventually replaced by Redux 603L. This was a so-called "half-web" film: a film of phenolic resole which was coated, on one side only, by polyvinyl formal powder having a particle size of < 150 μm.

Excellent performances were seen through the latter innovation where specially orientated balsa wood sheets were used. 50×50 mm^2 beams of balsa wood were bonded together, using an R/F adhesive (Aerodux 185) to form a block. Panels were then sawn off by cutting at right angles to the grain to produce what became known as end-grain balsa wood panels. In Fig. 95, the gluelines between the original beams can clearly be seen.

In the early 1960s, corrosion problems were being experienced by both Boeing (in the 707) and Vickers (in the VC-10). It became apparent that due to the natural variation in the quality of the balsa wood, its compressive strength was not always

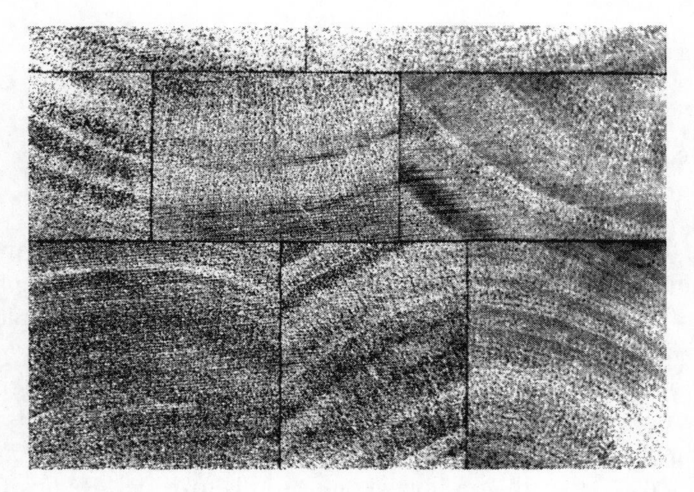

Figure 95: The structure of end-grain balsa sheets for aircraft floors.

high enough to withstand the fatigue loads generated by constant flexing (loading and unloading) of the floor. The balsa wood was seen to fail just under the aluminium skin. Not only did this significantly reduce the strength of the floors but it also allowed the ingress of water, which started to swell the balsa wood and lead to complete breakdown of the floor; this was particularly severe in the region of the toilets. It should be noted, however, that when perfect, the 9 lb balsa-based floors performed as well as the Nomex sandwich panels which eventually replaced them.

These problems led to significant research and development programmes being initiated in the UK and in the United States. Many new designs were offered to solve the problem. These included glass fibre reinforced skins to replace the aluminium and epoxy paste adhesives to replace the phenolics and PVC foams and aluminium honeycomb to replace the balsa wood. The Lockheed Corporation developed their own floors where epoxy paste adhesives were used to bond fibre-reinforced epoxy skins to the balsa wood core; phenolic matrices were used for skins in known "wet" areas.

It soon became clear that the use of aluminium honeycomb gave the compression strengths and the fatigue resistance which were required. Early development work and trials concentrated on the use of carbon fibre reinforced skins with Nomex honeycomb; the combination of aluminium and carbon was not liked because of potential galvanic corrosion problems.

However, in the 1970s, with the arrival of the wide-bodied aircraft, the situation started to change. At that time, aluminium skinned balsa wood panels were still the norm for most passenger aircraft, although it was well known

Figure 96: Lay-up of a typical Fibrelam panel. (colour version of this figure appears on p. xxi).

that aluminium honeycomb sandwich panels could give significant improvement. It was the driving force of weight reduction and, in the case of the Boeing 747, the poor fatigue performance of their aluminium skinned low density PVC foam floors that finally enabled the introduction of honeycomb sandwich panels. This eventually led to the introduction of the "all-synthetic" product, Fibrelam [37].

The first Fibrelam product was manufactured from two cross-plied glass-epoxy prepreg skins bonded to a phenolic resole coated Nomex honeycomb with a specially formulated epoxy film adhesive (Fig. 96).

This adhesive had to be formulated to match the cure cycle of the composite skins as the panels were produced by a "one-shot" process, i.e. the curing of the prepreg skins and the bonding to the honeycomb core took place at the same time. It also needed to have some degree of flame retardancy to meet the requirements of the aircraft manufacturer's specifications, for example the BMS 4-17 [38] specification of Boeing.

This was achieved using a fairly conventional epoxy adhesive formulation (cf. Section 5.5.4.1.5) to make the matrix flame retardant, either a combination of brominated compounds and inorganic oxides or a synergistic blend of hydrated inorganic salts were used. The filleting performance of this adhesive was critical in enabling the loads experienced by the skins to be translated to the honeycomb structure without skin delamination.

So successful was the introduction of this lightweight flooring panel that the original design plus numerous variants are currently being used to meet the demands for aircraft interiors and cargo floors, ceilings and walls as well as floors for all passenger areas within the cabin: aisles, under seat flooring, galleys, toilets, etc. [39]. These variants include a range of core densities, to meet different compression strength requirements, carbon fibre reinforced composite skins, composite skins utilising epoxy-phenolic and phenolic matrices to meet current "Fire Smoke and Toxicity" (FST) regulations, carbon and glass hybrids and aluminium and glass hybrids.

These structures are a good example of the value of getting the adhesive formulation correct from the start as, for all these variants where a structural adhesive is used, the formulation has remained constant for 30 years. Some of the designs, however, where the application is less robust and high sandwich peel strengths are not required, can make use of the resin matrix in the prepreg as the adhesive.

Mention has often been made of the use of bonded-in fasteners. The manufacture of "drop-in" panels is a good example of the use of fasteners and can be used to illustrate the different types which are found [40].

"Drop-in" floor panels, as their name suggests, are panels which have been supplied ready cut to shape with the edges filled with a conventional syntactic paste adhesive and the fasteners (Fig. 97) fitted and bonded, usually with a conventional paste adhesive; the adhesive is often referred to as a potting compound.

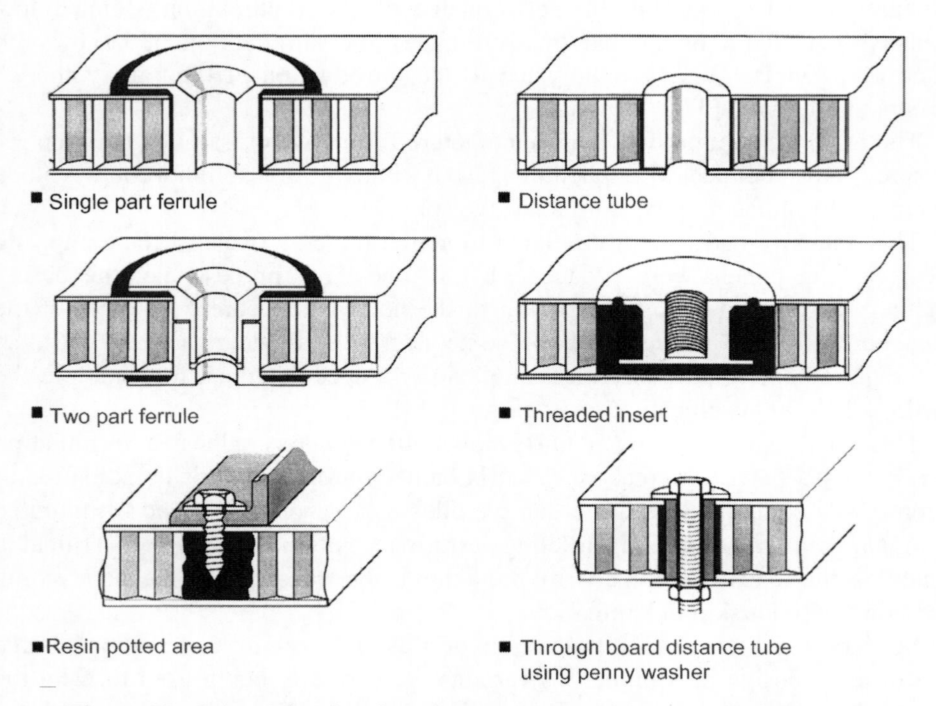

Figure 97: Schematic of various fasteners inserted into a sandwich panel.

In many instances, the fastener is supplied with injection holes so that the relatively low viscosity adhesive can be squirted into the body of the fastener to fill the gap between fastener and honeycomb and, when cured, secure the fastener in place.

It is also possible to apply the adhesive directly to the body of the fastener, by hand, and other fastener designs permit the use of film adhesive wrapped round the shaft or under the top flange.

5.6.4.5.3. Bonding of transmission rods and struts

Important components of any civil aircraft structure are the transmission rods, which drive the flaps and the slats and the floor-beam struts in the fuselage (Fig. 98). These are all Grade 1 structures, which are subjected to high levels of rotational, vibrational, bending and environmental stresses.

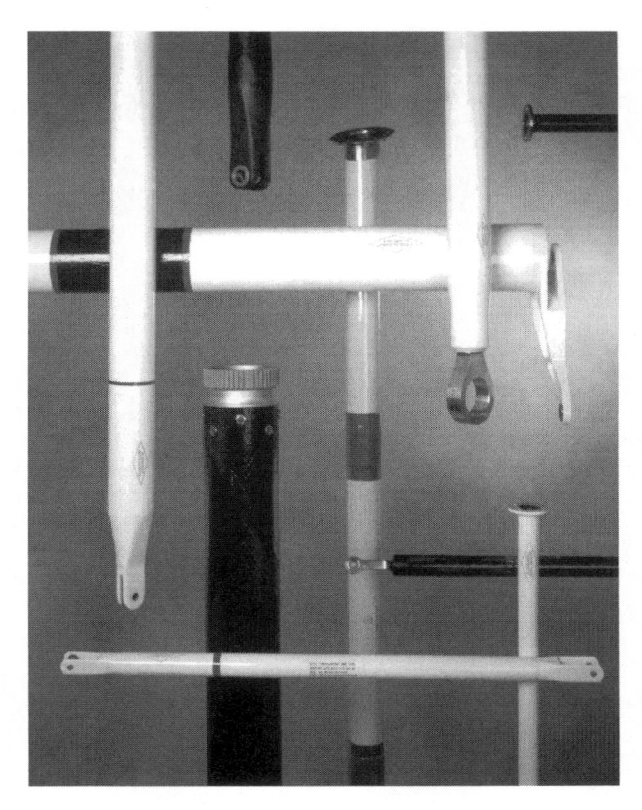

Figure 98: Various bonded rods and struts (courtesy of: FR-HiTEMP Limited).

For companies such as FR-HiTEMP, who manufacture a range of these components from various substrates such as aluminium, titanium, high chrome steel and fibre-reinforced composites, the use of adhesive bonding has made a considerable impact on reliability, weight saving and cost reduction. Their belief in adhesive bonding has enabled them to change from the production of assemblies, which relied on a combination of adhesive bonding and mechanical fixing, to processes utilising adhesives alone. This has resulted in an average component weight saving of over 20%, which has also enabled them to tighten up on their design criteria. The latter has had a significant knock-on effect on cost saving. Further, removing rivets from their components has meant that it has been possible to reduce the built-in safety margins, under "hot/wet" conditions, to a factor of two.

In these examples, the adhesive used is always a supported film adhesive and where primer is used – on anodised aluminium and titanium substrates – it tends to be BR 6743-1.

Composite components are pretreated using peel ply techniques and FR-HiTEMP has devised their own patented method for the pretreatment of high chrome steel.

These techniques have enabled Grade 1 steel-composite structures to be produced with a failure rate as low as 1.5×10^{-10} failures per flight hour.

Figure 99: Typical satellite structure. (A) Solar panels, (B) Antennae reflectors, (C) Main structure.

5.6.4.6. Bonding in Satellite Structures

Satellite structures are one of the best examples where weight saving is at an absolute premium. This means that the most effective method of joining many of the associated components must be by structural adhesive bonding.

Here, examples are drawn from Astrium Limited. This company is a wholly owned subsidiary of EADS and was formed by the merger of several long-established European space companies.

In essence, most satellites are boxes built round a central thrust component; composite struts stiffen this structure. Large (up to 2 m diameter) antennae reflectors and solar energy panels are then attached to this "structural box" (Fig. 99).

The equipment within these structures is mostly of a "bolt on" construction but there is considerable adhesive bonding seen in the composites struts, central thrust structures and in the reflectors.

The central thrust structure is, typically, a filament-wound, carbon fibre/epoxy construction. The side walls and floors comprise flat, bonded aluminium sandwich structures often incorporating other components. These are fastened to the central thrust structure by a combination of adhesively bonded and mechanically fastened brackets.

The sandwich panels comprise aluminium skins which have been chromic-sulphuric acid pickled and then primed with BR-127. Other structural components are made of 7000 series aluminium alloys so that the use of a corrosion-inhibiting adhesive primer is essential as these alloys are particularly prone to atmospheric corrosion. The pretreated aluminium skins are bonded to the honeycomb using ultra-lightweight film adhesives; again, weight saving is paramount. Redux 319L (175°C curing and 180 g/m^2 areal weight) and Redux 312UL (120°C curing and 100 g/m^2 areal weight) are the adhesives used.

Where further weight saving is required adhesive films at areal weights of 50–100 g/m^2 are also commercially available.

The attachments are bonded to the central core using epoxy-based paste adhesives. Some co-curing takes place but the usual approach is to bond the components to the already-cured carbon composite. In this case, the surfaces are prepared for bonding using conventional peel-ply techniques. The paste adhesive used depends on the usual factors, which affect any adhesive choice. These include thermal mismatch, service temperature range, temperatures encountered during production and also the innate conservatism of the industry. Typical adhesives used are Araldite AV 138M/HV 998 (a two-component paste curing in 2 h at 65°C or about 5 days at ambient) and toughened adhesives such as Hysol EA 9395 (cures in 1 h at 65°C or 7 days at ambient] and Scotch-Weld EC 9323 B/A (two-component adhesive curing in 2 days at ambient or 2 h at 65°C].

The struts are produced from carbon fibre/epoxy tubes of 30 mm diameter. At each end of each strut, there is a titanium end fitting. The titanium is pretreated using a simple grit-blasting technique and the end-fitting is bonded to the carbon tube using a toughened, two-part paste adhesive: Scotch-Weld EC 9323 B/A.

The antennae reflectors are large sandwich panel structures that combine Nomex honeycomb with Kevlar/epoxy prepregs. The film adhesive used in this application is AF 163-2. The use of this particular adhesive highlights one of the most important considerations when manufacturing bonded structural components for satellite construction, that of out-gassing. The total mass loss for satellite components, under high vacuum conditions, must be $<1\%$ and the total collected volatiles must be $<0.1\%$. To accommodate these requirements, the grade of AF 163-2 used is specially manufactured omitting the dyestuff usually included.

The problems of out-gassing are further illustrated when the essentially non-structural bonding of the satellite mirrors is considered. Here a silicone adhesive is used and, rather than using conventional acetic acid-based chemistry, which would fail the out-gassing criteria, oxime chemistry is employed.

The solar panels again comprise bonded sandwich panels where the skins are usually carbon/epoxy. These are bonded using film adhesives to the lightweight aluminium core.

5.6.4.7. Bonding in Missile Structures

Missile structures (Fig. 100) comprise three basic components: the body, the rocket motor and the radome. Radomes have been covered in detail in an earlier section, thus these examples will be confined to the missile bodies produced by MBDA and the motors manufactured by Roxel (UK Rocket Motors) Limited.

MBDA came into being after the merger of the following companies: Matra BAe Dynamics (50% BAE SYSTEMS and 50% EADS), EADS-Aerospatiale Matra Missiles and the missile activities of Alenia Marconi Systems (50% BAE SYSTEMS and 50% Finmeccanica).

Roxel (UK Rocket Motors) Limited is owned by MBDA and SNPE.

The body of the missile is, in essence, a tube which contains all the electronic equipment needed by the projectile and comprises two smaller tubes which are bolted together. Although structural adhesives have been evaluated for this application, mechanical fastening is still favoured as regular access for inspection is required. Another reason for bolting the assembly together is the difficulty in predicting the life of any structural bonds in view of the high stresses, over a considerable temperature range, to which any potential bondline would be exposed.

Some of the sub-systems within the body of the missile — these are usually made of aluminium, zinc-coated mild steel or stainless steel — can be structurally bonded but this is rare. When bonding does take place, the aluminium substrates

(a)

(b)

Figure 100: MBDA missile systems: (a) Seawolf, (b) Rapier. (Copyright: MBDA Limited.) (colour version of this figure appears on p. xxii).

are generally primed with either an epoxy silane (Silquest A-187) or a corrosion inhibiting epoxy-phenolic primer (Scotch Weld XB-3960 Primer). These primers are typically applied using a spray gun; thickness control is paramount. The choice of epoxy adhesive would depend on the end application.

For composite materials, both mechanical fastening and adhesive bonding would be considered. In the case of composites, adhesive bonding would be the norm. In these cases, the composite would be hand abraded prior to bonding and the choice of adhesive would depend on the bondline thickness tolerances and the temperature capability of the component itself. Thus, dependant on the component either ambient curing paste or elevated curing film adhesives can be used.

Two semi-structural bonding applications should also be mentioned. The magnets used in some of the electronic actuators are bonded into place often with an epoxy film adhesive. Amino silanes such as Scotch-Weld 3901 Primer are used in the bonding of plastic components.

The rocket motors, on the other hand, rely heavily on structural adhesive bonding; all the metallic substrates are pretreated using dry grit blasting techniques and solvent cleaning. Elevated temperature stoving is used for all elastomeric substrates prior to bonding. Mention has already been made of the use of phenolic film adhesives to bond the propellant to the motor lining and epoxy paste systems are also used in the construction of the motor casing and wing fittings.

The rocket motor is essentially a pressure vessel that has to withstand the manual and mechanical handling and environmental considerations associated with its manufacture and storage and then cope with the hot gases and pressure generated by the burning propellant.

The Thermopylae motor for the Rapier missile can be taken as a representative example. The motor casings are manufactured from strip steel (both stainless and carbon steels are used) which has been grit blasted prior to use. The strip steel is then coated with a simple, relatively brittle, dicyandiamide cured epoxy resin system (Araldite AZ 15/HZ 15) and the basic tube of the motor is formed [41,42] by winding a number of layers of this steel onto a mandrel, which is equal in cross-section to the internal diameter of the motor tube.

The tube is then cut to length and the “pressure vessel” is completed by fitting stainless steel or carbon steel end rings, which are coated with a Huntsman Advanced Materials one-part toughened epoxy paste adhesive [43]. The whole assembly is then oven cured to produce the final, integrally bonded, motor casing.

Of course, should the cure cycles of the strip steel adhesive and the end ring adhesive not be compatible, it is possible to cure the tube as a first step operation and then fit and cure the end cap.

The same toughened adhesive is also used to bond on the aluminium wing fittings on the Thermopylae motor (Rapier) and the centre-bands on ASRAAM; similar pretreatment and curing techniques are used.

5.6.4.8. Availability

Epoxy structural adhesives for the aerospace industry are available [33] from numerous companies. A representative list is as follows: 3Ms, Bryte Technologies, Cytec-Fiberite, Hexcel Composites, Henkel Corporation, Permabond, SIA Adhesives and Huntsman Advanced Materials (formerly Vantico Limited).

5.6.5. Bismaleimide Adhesives[7]

Formulations based on bismaleimide offer the highest temperature resistance and, hence, service temperatures of all the commonly encountered structural adhesives. However, at present, their use is rarely encountered. Those which are used are virtually all in aero engine structures.

5.6.5.1. Bonding in Military Aero Engine Structures

Military engines, for example those manufactured by Rolls Royce, call for little usage of adhesives, as they tend to be turbojets without bypasses. This means that the temperature build up is sudden and reaches levels not suitable for even BMI adhesives. What bonding there is generally uses structural paste adhesives to produce, essentially, non-structural (Grade III rating) or semi-structural bonds (see above). A few of these applications do, though, require high-temperature performance and hence, here, BMI paste adhesives are used.

However, the joint strike fighter (JSF) will require new engines with new design concepts and here, because of temperature demands, Rolls Royce are evaluating three novel BMI adhesive systems.

5.6.5.2. Bonding in Civil Aero Engine Structures

Of all the current civil engines, Rolls Royce's Trent 700 does have high temperature requirements. The IFS of the thrust reversers has to operate for considerable periods of time at elevated temperatures ($>140-150°C$). The component manufacturers, Hurel–Hispano, believe that this is asking too much of an epoxy adhesive and hence a Cytec-Fiberite BMI film adhesive (FM-2550) is used to bond this structure. Here the BMI adhesive is reticulated onto phosphoric acid anodised aluminium honeycomb. The core is then bonded directly onto pre-cured carbon-reinforced BMI laminates to form a sandwich structure, one of the skins being perforated to produce an "acoustic panel", which helps reduce engine noise.

[7] Detailed information on heat resistant adhesives (bismaleimide, cyanate esters, etc.) is available in the chapter "Heat stable adhesives" in Volume 2.

Hexcel Composites HP 655 is also qualified by Airbus Industries for use in structures either in or adjacent to the Trent 900 engines, which are to power the Airbus A.380.

5.6.5.3. Availability
Bismaleimide and imide structural adhesives for the aerospace industry are available [33] from Cytec-Fiberite, Hexcel Composites, Henkel Corporation and Monsanto.

5.6.6. Cyanate Ester Adhesives

5.6.6.1. Bonding in Satellite Structures
Within the last few years, cyanate ester adhesives have started to find applications in the space and, particularly, the satellite market. They are particularly suited to these applications due to their exceptionally low water uptake and out-gassing characteristics, good high-temperature performance and low dielectric constant and low loss tangent electrical properties.

As carbon/cyanate ester prepregs start to be used in the construction of antennae reflectors, cyanate ester adhesives will be used to bond the sandwich panel together.

Other current applications have been seen in bonding structural components in space telescope applications.

5.6.6.2. Bonding in Radome Structures
The fact that cyanate ester formulations offer outstanding electrical properties makes them very good candidate systems for both composite matrices and film adhesives for radome applications.

The approach taken by Chelton Radomes is that of a slow but steady replacement of the conventional epoxy adhesives by cyanate ester-based formulations (for example, EX 1516 from Bryte Technologies) in new designs, such as the EH101 Merlin helicopter mentioned above.

5.6.6.3. Availability
Cyanate ester structural adhesives for the aerospace industry are available [33] from Bryte Technologies, Cytec-Fiberite and Hexcel Composites.

5.6.7. Acrylic Adhesives

Virtually no structural bonding in the aerospace industry is achieved using acrylic adhesives. However, they have found use in the construction of the Airbus A.320 flap track fairings where acrylic adhesives are seen as easier to use and apply than the more conventional two-part epoxy paste adhesives. Bostik M 890, a two-component, modified acrylic adhesive is also used to bond the vortex generator to the tail of the Tornado.

Some secondary bonding is also encountered; refer to the paragraphs concerning helicopter and aero engine bonding in the section covering epoxy adhesives.

5.6.7.1. Availability
Acrylic structural adhesives for the aerospace industry are available [33] from 3Ms, Bostik, Henkel Corporation and Permabond.

5.6.8. Polyurethane Adhesives

Virtually no polyurethane systems are used for structural bonding applications within the aerospace industry with one notable exception — blast pipe manufacture in rocket motor structures.

The Roxel blast pipe for the Thermopylae motor on the Rapier missile can be taken as an example. The blast pipe comprises a chromic acid anodised aluminium outer component with a Durestos inner liner. It acts as a conduit for the gases, generated by the burning propellant, to atmosphere. Initially, the assembly was bonded with an acetal phenolic resin but in order to obtain a continuous bond which would meet the C-Scan criteria and, more importantly, allow the glueline to adsorb the stresses generated a compliant, relatively low modulus, two-part polyurethane structural adhesive was successfully used in place of the phenolic.

5.6.8.1. Availability
Polyurethane structural adhesives for the aerospace industry are available [33] from 3Ms, Huntsman Advanced Materials and Henkel Corporation.

5.7. Adhesive Costs and Market Share

The cost of structural adhesives will vary considerably depending on the format, the chemistry used, the end-user industry, the quantities being purchased and, to a certain extent, on the adhesive manufacturing company concerned. It is, therefore, not possible to give an indication as to cost per square metre or cost per kilogram, which is meaningful.

In Fig. 101, an attempt is made to address this difficulty. Overall market data from the United States of America (source Skeist [44]) has been plotted to try to show the position of the aerospace business within the structural adhesive market as a whole.

The *x*-axis shows the average selling price in $/lb; for the aerospace business this is in the region of $15–$25 per pound weight. It is important to remember that

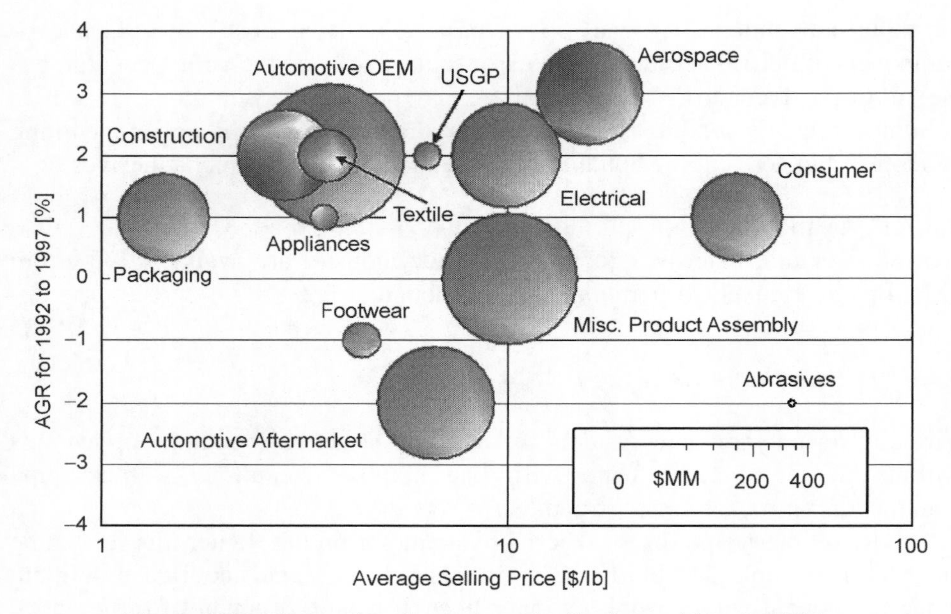

Figure 101: Market data for structural adhesives within the USA.

this is an average value; some adhesives in this area will be $\gg$\$100/lb. It does show, however, that apart from the consumer market, which is very small volume sales at relatively high prices, adhesives in this sector are significantly more expensive than in any other.

The *y*-axis shows the expected annual growth rate over a 5-year period to 1997. At 2.25–3.75% AGR, structural adhesives within the aerospace market show the most encouraging performance.

The width of each data point shows the size of the market in millions of dollars (the scale indicator is logarithmic); the aerospace market is shown to be about $150 MM. This is highly significant when compared against the market volumes for the two biggest users of adhesive, the packaging and the automotive industry: only 10 million pounds per annum as opposed to 80 and 60 million pounds, respectively.

Acknowledgements

Special thanks are due to Eddie van der Straeten for his help in understanding the history, the nomenclature and the technology of bonding in early wooden aircraft. Thanks are also due to George Newell, Barrie Hayes and Dennis Stevens, Debbie Wilkerson, Pat Brooks, Chris Price and Dean Bugg of Hexcel Composites, Andy

Preslent of the Shuttleworth Collection, Gary Critchlow of Loughborough University, Rob Davies of Westland Helicopters, Ian Simmons of FR-HiTEMP, David Bell of Rolls Royce, Ben Hawtin, Mike Turner, Tony Warren and Roger Digby of Airbus, Trevor Watmough of Hurel-Hispano, Ian Kitson of BAE SYSTEMS, Matt Bryanton of Chelton Radomes, Terry Ackerman of MBDA, Andy Bell of Roxel and Roger Ellis and Brian Webb of Astrium for their help with many aspects of the text, particularly the application case studies.

Finally, thanks are due to Hexcel Composites Limited for their sponsorship of the work as a whole.

Appendix A5.1. European (EN) Adhesive Test Methods and Specifications

Until fairly recently, each European country used to publish their own test methods and specifications. Thus, DTD specifications came from the UK, FN from France, LN from Germany, etc. This situation has now been unified with the production of EN or prEN (provisional) specifications; the actual designation often being slightly changed to indicate country of publication. Thus BS EN XXXX would emanate from the UK (BS being British Standard).

Test methods for structural adhesives and foaming adhesives are covered by the "mother" specifications EN 2243 and prEN 2667, respectively.

From EN 2642 onwards, a new format has been introduced. At the top level are the Technical Specifications (the mother documents). All other documents are intended to cascade down from these, i.e. the Material Specifications and finally the Test Method documentation

Designation	Tile
prEN 2243-1	Aerospace series: structural adhesives test methods part 1 – single lap shear
EN 2243-2	Aerospace series: structural adhesives test methods part 2 – peel test — metal
EN 2243-3	Aerospace series: structural adhesives test methods part 3 – peeling test — metal honeycomb core
EN 2243-4	Aerospace series: structural adhesives test methods part 4 – flatwise tensile test — metal honeycomb core
EN 2243-5	Aerospace series: structural adhesives test methods part 5 – ageing tests

(*Continued*)

Designation	Tile
prEN 2243-6	Aerospace series: structural adhesives test methods part 6 – determination of shear stress and shear strain
prEN 2243-7 (Yet to be issued)	Aerospace Series: structural adhesives test methods part 3 – determination of the flow of adhesive films
prEN 2642	Aerospace series: structural adhesives systems – adhesive films – technical specification
prEN 2667-1	Aerospace series: non-metallic materials – foaming structural adhesives test methods part 1 – tensile single lap shear
prEN 2667-2	Aerospace series: non-metallic materials – foaming structural adhesives test methods part 2 – compressive tube lap shear
prEN 2667-3	Aerospace series: non-metallic materials – foaming structural adhesives test methods part 3 – expansion ratio and volatile content
prEN 2667-4	Aerospace series: non-metallic materials – foaming structural adhesives test methods part 4 – vertical slump
prEN 2667-5	Aerospace series: non-metallic materials – foaming structural adhesives test methods part 5 – exothermicity
prEN 2667-6	Aerospace series: non-metallic materials – foaming structural adhesives test methods part 6 – determination of water absorption
prEN 2757	Aerospace series: test method for determining the drying and ignition residues of adhesive primers
prEN 2781	Aerospace series: non-metallic materials – structural adhesives test methods – determination of the primer thickness
prEN 4106	Aerospace series: structural adhesives systems – paste adhesives – technical specification
prEN 4107	Aerospace series: structural adhesives systems – foaming adhesives – technical specification

Appendix B5.1. American (ASTM) Adhesive Test Methods and Specifications

Although other organisations produce some methodologies for testing adhesive performance (for instance various MIL SPECS), the most comprehensive documentation is issued by ASTM International; many of these mirror their European counterparts

Designation	Title
B117-02	Standard practice for operating salt spray (fog) apparatus
C273-00e1	Standard test method for shear properties of sandwich core materials
C297-94	Standard test method for flatwise tensile strength of sandwich constructions
C393-00	Standard test method for flexural properties of sandwich constructions
D896-97	Standard test method for resistance of adhesive bonds to chemical reagents
D897-01	Standard test method for tensile properties of adhesive bonds
D903-98	Standard test method for peel or stripping strength of adhesive bonds
D905-98	Standard test method for strength properties of adhesive bonds in shear by compression loading
D1002-01	Standard test method for apparent shear strength of single-lap-joint adhesively bonded metal specimens by tension loading (metal-to-metal)
D1183-96e1	Standard test methods for resistance of adhesives to cyclic laboratory aging conditions
D1184-98	Standard test method for flexural strength of adhesive bonded laminated assemblies
D1780-99	Standard practice for conducting creep tests of metal-to-metal adhesives
D1781-98	Standard test method for climbing drum peel for adhesives
D1828-01	Standard practice for atmospheric exposure of adhesive-bonded joints and structures
D1876-01	Standard test method for peel resistance of adhesives (T-peel test)
D1879-99	Standard practice for exposure of adhesive specimens to high-energy radiation
D2293-96	Standard test method for creep properties of adhesives in shear by compression loading (metal-to-metal)
D2294-96	Standard test method for creep properties of adhesives in shear by tension loading (metal-to-metal)
D2295-96	Standard test method for strength properties of adhesives in shear by tension loading at elevated temperatures (metal-to-metal)

(*Continued*)

Designation	Title
D2557-98	Standard test method for tensile-shear strength of adhesives in the subzero temperature range from −267.8 to −55°C (−450 to −67°F)
D2918-99	Standard test method for durability assessment of adhesive joints stressed in peel
D2919-01	Standard test method for determining durability of adhesive joints stressed in shear by tension loading
D3165-00	Standard test method for strength properties of adhesives in shear by tension loading of single-lap-joint laminated assemblies
D3166-99	Standard test method for fatigue properties of adhesives in shear by tension loading (metal/metal)
D3167-97	Standard test method for floating roller peel resistance of adhesives
D3528-96	Standard test method for strength properties of double lap shear adhesive joints by tension loading
D3762-98	Standard test method for adhesive-bonded surface durability of aluminium (wedge test)
D5656-01	Standard test method for thick-adherend metal lap-shear joints for determination of the stress-strain behaviour of adhesives in shear by tension loading
D6465-99	Standard guide for selecting aerospace and general purpose adhesives and sealants
E229-97	Standard test method for shear strength and shear modulus of structural adhesives

Appendix C5.1. Adhesive Specifications

There is now a plethora of specifications relating to the qualification and use of adhesives. Thus all, or nearly all, aircraft and aircraft component manufacturers issue their own specifications to cover the adhesives that will be used in their applications. Very often these are written around a particular product although some are still generic.

In many cases, a mother specification is sub-divided into various related specifications designated as different types and grades: these cover cure temperature, service temperature, adhesive format (i.e. primer, paste, foam or film] and substrate (i.e. for metal–metal bonding, for composite bonding, etc.).

To produce a comprehensive list would be impossible but it is true to say that all these specifications relate to two original American documents which were at that time, effectively, the worldwide specifications for the industry.

Both are still in existence and are:

- Federal Specification MMM-A-132: "Adhesives, Heat Resistant, Airframe Structural, Metal to Metal"
- Military Specification MIL-A-25463: "Adhesive, Film Form, Metallic Structural Sandwich Construction"

  This last named specification has very recently been replaced by, effectively, the same document from a different authority:
- Society of Automotive Engineers, Aerospace Materials Specification SAE AMS-A-25463.

References

[1] Van der Straeten, E. (1993). Adhesive-bonded joints in wooden aircraft, 1920–1945: techniques and materials, Unpublished paper presented at a seminar on *The History of Composites*, at the Aeronautical Department of the Science Museum, London.
[2] British Standards Institute, BS EN 312-2: 2000, *Plywood — Classification and Terminology — Part 2: Terminology.*
[3] British Standards Institute, BS 6100: 1992/Confirmed 1998. *Glossary of Building and Civil Engineering Terms. Part 4 Forest Products, Section 4.1 Characteristics and Properties of Timber and Wood Based Panel Products.*
[4] Edwards, W. P. (2001). Why not period glue? *Journal of the Society of American Period Furniture Makers*, November.
[5] Houwink, R., & Salomon, G. (Ed.) (1967). *Adhesion and Adhesives.* Elsevier, Amsterdam.
[6] Knight, R. A. G. (1952). *Adhesives for Wood*, Monographs on metallic and other materials — published under the authority of the Royal Aeronautical Society. Chapman and Hall, London, pp. 20–30.
[7] Perry, T. D. (1948). *Modern Plywood,* 2nd ed. Pitman, pp. 42–48 and 301p.
[8] de Bruyne, N. A. (1987). *Pioneering Times.* Techne Inc., Princetown, NJ.
[9] de Bruyne, N. A. (1996). *My Life.* Midsummer Books, Cambridge.
[10] Aero Research Technical Notes (1946). *Aerolite in Aircraft Manufacture*, Bulletin No. 46, October.
[11] Aero Research Technical Notes (1960). *The Durability of Aerolite 300*, Bulletin No. 205, January.
[12] *A Short History of the DH98 Mosquito.* http://www.home.gil.com.au/~bfillery/mossie01.htm.
[13] Aero Research Technical Notes (1958). *Adhesives in Sailplane Building*, Bulletin No. 192, December.
[14] Taylor, E. D. (1987). *Tego Gluefilm in Walthamstow: The Story of a German Material Produced in Walthamstow in 1939 with German Collaboration for War-Time Use against Germany*, Private Publication.
[15] Bärmann, L. *The TA 154 Moskito: Everything You Wanted to Know about the TA 154 but Didn't Dare Ask.* http://warbirdresourcegroup.org/LRG/ta154bar.html.

[16] CIBA Technical Notes (1967). *Aerolite in Replica Veteran Aircraft*, February.
[17] Aero Research Technical Notes (1945). *War Work*, Bulletin No. 34, October.
[18] Bishopp, J. A. (1997). The history of Redux and the Redux bonding process. *International Journal of Adhesion and Adhesives*, **17** (4).
[19] Aero Research Technical Notes (1946). *Redux Bonding in the Hornet*, Bulletin No. 39, March.
[20] Aero Research Technical Notes (1953). *Structural Adhesives for Metal Aircraft*, Bulletin No. 130, October.
[21] B.P. 577790, N. A. de Bruyne of Aero Research Limited and the de Havilland Aircraft Company Limited (Inventors), Filed 1939, accepted 1946.
[22] Oosting, R. (1995). *Toward Compliant Aluminium: A New Durable and Environmentally Adhesive Bonding Process for Alloys*. Structures and Materials Laboratory, Delft University of Technology, Delft, 60 p.
[23] Karsa, D. R. (1990). *Additives for Water-Based Coatings*. The Royal Society of Chemistry, London.
[24] Fletcher, T. (1992). Ion-exchanged silicas as alternatives to strontium chromate in coil coating primer. *Polymers Paint Colour Journal*, March.
[25] Weast, R. C. (1988–1989). *Handbook of Chemistry and Physics*. CRC Press Inc., Florida, Vol. 69.
[26] Plueddemann, E. P. (1991). *Silane Coupling Agents*. Plenum Press, New York.
[27] van Ooij, W. J., Zhu, D. Q., Prasad, G., Jayaseelen, S., Fu, Y., & Teredesai, N. (2000). Silane-based chromate replacements for corrosion control, paint adhesion, and rubber bonding. *Surface Engineering*, **16**, 386 p.
[28] Adhesive selection chart: *Redux Film Adhesives, Foaming Films, Pastes and Primers*. Hexcel Composites.
[29] Box, E. P. G., Hunter, W. G., & Hunter, J. S. (1978). *Statistics for Experimenters: An Introduction to Design, Data Analysis and Model Building*. Wiley, New York.
[30] Juran, M., & Godfrey, M. (1999). *Juran's Quality Handbook*. MacGraw Hill, New York.
[31] Bishopp, J. A., Parker, M. J., & O'Reilly, T. A. (2001). The use of statistical experimental design procedures in the development and "robust" manufacture of a water-based corrosion-inhibiting adhesive primer. *International Journal of Adhesion and Adhesives*, **21** (6) 473–480.
[32] *Bonded Aircraft Structures*, Papers given to a conference in Cambridge, UK. Bonded Structures Limited (1957).
[33] Hussey, B., & Wilson, J. (1996). *Structural Adhesives Directory and Databook*. Chapman and Hall, London.
[34] Marks, N. G. (1989). *Polymer Composites for Helicopter Structures*. Metals and Materials, August, pp. 456–459.
[35] Hexcel Composites Limited (2000). *Redux Bonding Technology*.
[36] Pate, K. D. (2002). Applications of adhesives in aerospace. In: M. Chaudhury, & A. V. Pocius (Eds), *Adhesion Science and Engineering* (Chemistry and Applications, Vol. 2, pp. 1128–1192). Elsevier, Amsterdam.
[37] Armstrong, K. B., Stevens, D. W., & Alet, J. (1994). 25 years of use for Nomex honeycomb in floor panels and sandwich structures. In: J. Hognat, R. Pinzelli & E. Gillard (Eds.), *50 Years of Advanced Materials or Back to the Future, the Proceedings of the Fifteenth International Chapter Conference of SAMPE, SAMPE Switzerland*, (pp. 17–40).

[38] Boeing Specifications, BMS 4-17 (2001). *Unidirectional Fiberglass Faced Nomex Honeycomb Core Floor Panel Stock*, Issue T.

[39] Hexcel Composites (2000). *Fibrelam Aircraft Floor Panels: Qualified Products Selector Guide.*

[40] Hexcel Composites (2001). *Sandwich Panel Fabrication Technology.*

[41] Amos, R. J. (1999). On stiffness and buckling of adhesively bonded rocket motor tubes at high temperature. *Extended Abstracts to Adhesion '99, IOM Communications*, pp. 129–134.

[42] Amos, R. J. (2002). Mechanical properties of structural adhesives used in rocket motors. *Extended Abstracts to Euradh 2002/Adhesion '02, IOM Communications*, pp. 100–101.

[43] Kavanagh, G. M., & Tod, D. A. (2002). Elastic–plastic analysis of bonded joints in rocket motor structures. *Extended Abstracts to Euradh /Adhesion '02, IOM Communications*, pp. 208–211.

[44] Skeist Incorporated (1993). *Adhesives, VI: A Multiple-Client Study.*

Chapter 6

Adhesives in Electronics

Guy Rabilloud

In 1962, Guy Rabilloud got a doctorate in chemical and physical sciences from the Nobel Prize of physics Louis Néel. Two years later, he joined the department of fine chemicals at IFP, the French Petroleum Institute. For 30 years, he studied and developed new heat-resistant polymers for the advanced technologies, signing more than 150 patents and scientific publications. During the 1970–1980 decade, he was also associate professor in macromolecular chemistry at the universities of Grenoble and Lyon. In the early 1980s, he became assistant manager of Cemota, a subsidiary of IFP created to develop, manufacture and commercialize high-performance polymers for the electronic and aerospace industries. In 1990, he spent one year as a consultant attached to the Singapore Institute of Standards and Industrial Research. Dr. Rabilloud is the co-author of several books dealing with the applications of polymers, including adhesives, films, and composite materials. He recently published three books providing an overview on the use of polymers in electronics, in particular conductive adhesives and heat-resistant insulating films.

6.1. Introduction

6.1.1. Historical Background

In electronics, the definition of adhesives is quite versatile. It not only involves the formation of strong bonds between two rigid or flexible bodies, but also concerns the adhesion of thin films to metals, ceramics, and other inorganic materials. As polymers are universally employed to fabricate protective shields before packaging microcircuits, excellent adhesion is a prerequisite condition to achieve the long-term reliability of packaged devices. Although the electronic industry has been initially reluctant to incorporate organic compounds as permanent parts of functional systems, the situation has changed to face the challenge of low production costs for consumer-oriented products. As shown in Figure 1, photosensitive polymers (photoresists) have been used very early to transfer circuit

Handbook of Adhesives and Sealants
P. Cognard (Editor)

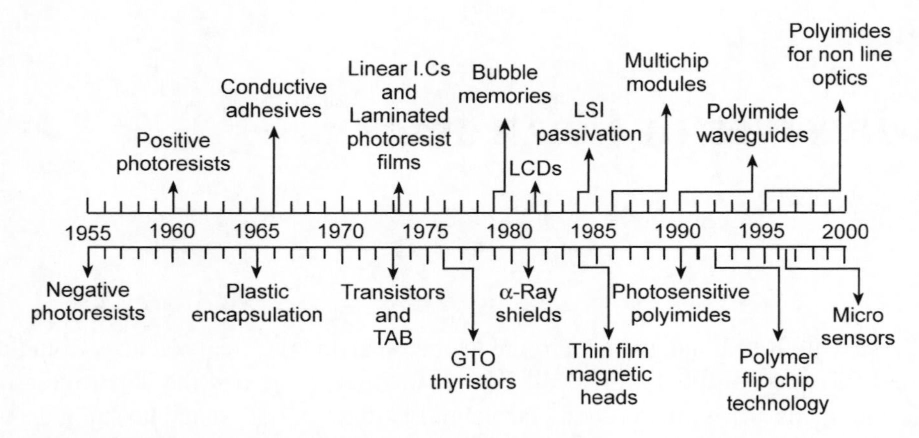

Figure 1: Dates of introduction of organic polymers in the electronic industry. Typical applications of heat resistant polymers, mostly polyimides, are indicated in the right part of the graph.

patterns from a mask to the underlying semiconductor substrate. Photoresists are, however, transient materials (processing aids) that are washed away at the end of each lithographic step.

During the 1960–1970 decade, plastics progressively replaced metals and ceramics to package the microdevices, and silver-filled epoxy resins were proposed by Epoxy Technology Inc. in 1966 for bonding integrated circuits to their supporting base (leadframe). Since Kapton® film became commercially available from du Pont de Nemours, new interconnection technologies have been developed to fabricate both simple flexible circuitry and more complex interconnects based on the tape automated bonding (TAB) process. After four decades of intensive research and development programs, Figure 1 shows that polymers are now undisputed materials to produce high-volume, low-cost electronic systems.

6.1.2. Semiconductor Technology

Integrated circuit chips (dice) are fabricated by a “collective technology” divided into front-end and back-end processes. Front-end processing refers to all steps involved in the formation of thousands of active devices such as transistors, diodes, capacitors, and resistors on a silicon wafer. Currently, design rules of transistor gates are only 0.13 μm wide and last-generation microprocessors or memory devices include 64 million transistors over a few square centimetres. The metal circuitry necessary to interconnect the active devices and to deposit the metal pads which are then bonded to packages is implemented by photolithography and etching in the back-end processes. Except photoresists, polymers are exclusively involved in back-end processing and packaging. After sawing, qualified chips are mounted on substrates such as chip carriers or the leadframes of dual-in-line packages (DIP). These metal or metallized substrates form the basic structures providing electrical interconnection with other integrated circuits (ICs) or passive devices. They are then encapsulated in plastic packages or sealed into hermetic metal or ceramic cases.

For high-end electronic products with large numbers of inputs/outputs (I/Os), alternative interconnection schemes are the use of TAB and flip chip technologies, where both chips and substrates are protected by an encapsulant. Hybrid circuits are another option constituted of a single ceramic interconnecting substrate for mounting unpackaged active devices and passive components. The highest packing density is, however, achieved with thin film multichip modules (MCMs). A brief overview of the back-end processes is requisite to summarize the most important applications of adhesives in the manufacture of electronic products. As stated earlier, adhesives have not restricted the meaning of organic materials supplied as pastes, tapes, or self-standing films that are used to bond stiff or flexible substrates. Virtually, all polymers involved in the back-end processing must exhibit strong adhesive properties.

With the advent of very large and ultra large integrated circuits (VLSI and ULSI), new interconnection schemes have been implemented to optimize the expensive real estate of doped silicon sites. The last step of the front-end process is the deposition of a thin layer of silicon dioxide etched over the regions where metal pads will be formed by evaporation or sputtering. In traditional processes, the dielectric layers of the multilevel interconnect system are composed of 1–2 μm thick films of silicon dioxide or silicon nitride. In the early 1970s, Hitachi and IBM extensively evaluated the replacement of inorganic dielectrics by polyimide films for the three-level metallization of complementary metal-oxide semiconductors (CMOS) and linear integrated circuits. Finally, a polyimide overcoat is deposited as the uppermost layer to reduce the interfacial stresses, thus preventing passivation cracks and electrode displacement. Thick polyimide films are also used to reduce the soft errors observed in random access memories (RAM) as a consequence of α-ray irradiation from trace amounts of uranium and thorium existing in the packaging materials. For example, a film of Kapton® bonded to the surface of integrated circuits by means of a poly(imide-siloxane) adhesive may cut α-ray transmission from 10^3 to $10^{-3}\ \alpha\ cm^{-2}\ h^{-1}$. After openings have been defined over the metal pads by photolithography, the individual dice are electrically tested and sawed.

Good dice are then mounted onto leadframes or chip carriers with a die attach material providing mechanical stability, heat transfer, and electrical grounding before first level interconnection is implemented. Although wire bonding is still the dominant technology, it is generally limited to chips with a maximum number of I/Os of about 200. This number depends on the circuit complexity and can be approximately calculated by applying Rent's rule $N = AG^r$, where N is the total number of inputs/outputs, G the number of logic gates, and A and r the empirical constants. Taking $A = 4.5$ and $r = 0.4$, the total number has been calculated as 500 pinouts for logic gate arrays with 10^5 gates. The alternate first level interconnection processes are TAB and flip chip techniques. Integrated circuits bonded to TAB substrates are either mounted onto chip carriers or directly surface mounted onto printed circuit boards (PCBs) to achieve higher packing density. In flip chip technology, a solder material is used to produce bumps on the bonding pads patterned on the top surface of the microcircuit. Once turned upside down, bumps are aligned to face directly the connecting leads of the substrate and the two parts are bonded by melting the solder.

After the first level interconnection has been completed by using one of these techniques, the assembly is protected from the environment by encapsulation into hermetic or non-hermetic packages. For applications requiring high reliability, chips or MCMs are fixed into metal cans or ceramic packages whose lids are sealed using either solders or glass sealants. However, more than 90% of the chips and their leadframes are encapsulated by an epoxy-based moulding material. Most dual-in-line packages made of plastic (P-DIP) or ceramic (CERDIPS), chip carriers, pin-grid arrays (PGA), or small-outline plastic packages (SOIC) include metal pins designed to plug directly into the plated through holes drilled into a PCB. Alternate metal layouts are used for chips that are surface mounted on wiring boards. This is the second level of interconnection where packages are the vehicles to connect the small bonding pads of the dice to the PCB circuitry.

By nature, polymers are insulating materials and some of them are employed without any added fillers. However, most applications require either electrical or thermal conductivity or both. According to the final use, thermal stability and resistance to aggressive environment can be important criteria to select a particular type of polymer. In fact, few adhesives have been specifically developed for the electronic industry. Instead, most materials were preexisting compositions that have been tested and incorporated in production processes. Specific requirements such as ionic contamination, water absorption, and better thermal stability appeared later. Major chemical companies and semiconductor manufacturers, in particular in the USA and Japan, launched variants of these polymers fulfilling the demand for organic materials with enhanced properties.

6.2. Chemical Compounds

Except a few thermoplastic materials that are processed by melting, most adhesives are formed in situ by polymerization or polycondensation of small molecules called "monomers". Free radical and ionic mechanisms are involved in the polymerization of unsaturated monomers containing carbon–carbon double bonds. Acrylates, cyanoacrylates, or UV-curable compounds are typical examples of such monomers. Rigid and flexible epoxies, epoxy-phenolic, and epoxy-silicone compositions are generally composed of the base resin, a co-reactant, and various additives. At moderate temperature, ring opening predominates, whereas condensation reactions occur when the temperature is raised. Silicone resins are formed by polycondensation of two reactive silicon-containing monomers at ambient temperature (room temperature vulcanization or RTV silicones) or by addition of silane groups (Si–H) to ethylenic double bonds. Cyanate resins and adhesive compositions containing them polymerize predominantly by cycloaddition. Finally, polyimides are synthesized by reacting, in solution, aromatic dianhydrides with aromatic diamines. The reaction easily proceeds at room temperature to give high molecular weight linear polymers called polyamic acids. These polymers are transformed into heterocycles (imidization) by chemical or thermal cyclodehydration.

The chemistry and reaction mechanisms involved in the formation of these macromolecules are not discussed in the present chapter. A number of books, reviews and review symposia were published, which provide deep insight of the different chemical processes that take place during the formation of the adhesive joint. As far as possible, relevant references are provided in the following sections for readers, scientists, and production engineers, searching extensive information about a particular class of polymer. When necessary, one figure displays the chemical formulae of the organic compounds engaged in adhesive compositions and other figures illustrate either the first "initiation" step when active species are formed or the final polymer structures.

6.2.1. Free-radical and Ionic Polymerization

Aliphatic polymers prepared by addition polymerization of unsaturated monomers containing –C=C– groups are broadly used in the fabrication of general purpose adhesives. They are, however, of marginal utility in electronics because of their low glass transition temperature (T_g) and poor thermal stability. Poly(methyl methacrylate) (PMMA), e.g. has a continuous service temperature limited to 90°C. Applications for this category of polymers are mainly moulding, casting, and coating formulations providing materials with excellent optical clarity. Light emitting diodes (LED) are often protected by a thick layer of PMMA moulded in the form of convergent lens. The principle of addition polymerization is sketched in Figure 2.

An initiator, decomposed under the influence of heat or UV-irradiation, generates two free radicals **1** containing highly reactive unpaired electrons. Radical **1** reacts with methyl methacrylate **2** to produce the intermediate free

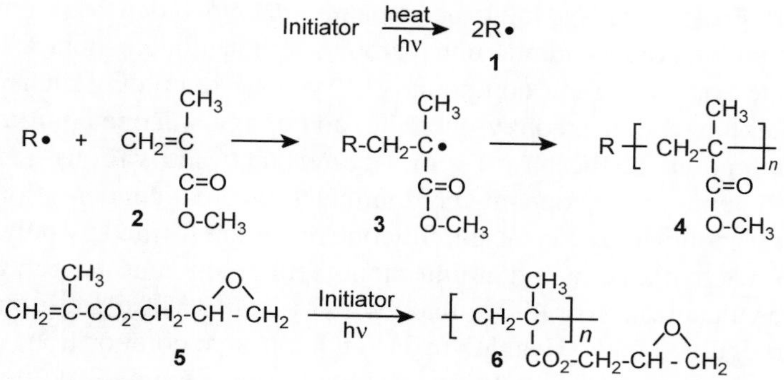

$$\text{Initiator} \xrightarrow[h\nu]{\text{heat}} 2\text{R}\bullet \quad (\mathbf{1})$$

$$\text{R}\bullet + \underset{\mathbf{2}}{CH_2{=}C(CH_3){-}C({=}O){-}O{-}CH_3} \longrightarrow \underset{\mathbf{3}}{R{-}CH_2{-}C\bullet(CH_3){-}C({=}O){-}O{-}CH_3} \longrightarrow \underset{\mathbf{4}}{R{-}[CH_2{-}C(CH_3)(C({=}O){-}O{-}CH_3)]_n}$$

$$\underset{\mathbf{5}}{CH_2{=}C(CH_3){-}CO_2{-}CH_2{-}\overset{O}{CH{-}CH_2}} \xrightarrow[h\nu]{\text{Initiator}} \underset{\mathbf{6}}{[CH_2{-}C(CH_3)(CO_2{-}CH_2{-}\overset{O}{CH{-}CH_2})]_n}$$

Figure 2: Simplified mechanism of free radical chain polymerization. On heating or by UV-irradiation, the initiator generates free radicals **1** whose reaction with methyl methacrylate **2** creates the active species **3**. Further reaction with monomer molecules leads to poly(methyl methacrylate) PMMA **4**. Applied to 2,3-epoxypropyl methacrylate **5** this process gives the linear aliphatic polymer **6** with pendent epoxy groups.

radical **3** which constitutes the initial element of chain propagation. Once this reactive species is formed, polymerization occurs extremely rapidly by opening the carbon–carbon double bonds of other monomer molecules to give finally PMMA **4** until termination takes place by radical recombination or disproportionation. Chain polymerization of 2,3-epoxypropyl methacrylate **5** leads to linear aliphatic polymer **6** with pendent epoxy groups. Further reaction with co-reactants can be used to increase the T_g. Addition polymerization does not have to occur only via free radicals. For example, the cationic attack of epoxy resins by photosensitive onium salts generates insoluble crosslinked polymers.

6.2.2. Epoxy Adhesives

Epoxy adhesives typically contain several chemical components, the most important being the epoxy resin constituting the base of the formulation. To the epoxy compound are added a variety of materials including catalysts, co-reactants, reactive diluents, solvents, flexibilizing or toughening agents, and fillers. All these additives have a considerable effect on the properties of the adhesive composition before, during, and after the curing process. For instance, fillers impart thixotropy, or provide electrical and thermal conductivities, whereas the Young's modulus can be increased by one order of magnitude with multifunctional co-reactants. Single or one-part systems are pre-mixed compositions of epoxy resins, hardeners, accelerators, adhesion promoter, and conductive filler. Latent hardeners and catalysts are chemical compounds that react with epoxies only on heating at temperatures higher than 100–120°C, and such compositions are stable at ambient temperature for at least 6 months. Conversely, frozen one-part adhesives include highly reactive catalysts and must be preserved at low temperature (−40°C) before use. In two-part systems, the epoxy resin loaded with conductive filler is delivered in one container and the hardener mixed with a solvent or reactive diluent combined with the same or a different filler forms the second component. The chemistry of epoxy resins is now well understood and pertinent information can be found elsewhere [1–3]. In the particular field of the conductive adhesives used in electronics, relevant references have been recently published [4].

6.2.2.1. Epoxy Resins*

All epoxy resins used in the fabrication of conductive adhesives are commercially available materials that were developed several decades ago to prepare structural adhesives. They can be aromatic, cycloaliphatic or aliphatic, monofunctional or polyfunctional, physically ranging from low-viscosity liquids to high-melting solids. These resins have a common feature: the three-membered oxygen-containing epoxy (oxirane) rings that are incorporated into organic molecules by using either condensation or oxidation reactions. Figure 3 shows the synthesis of epoxy resins by the two-step reaction of 1-chloro-2,3-epoxypropane (epichlorhydrin) **7** with 4,4′-(1-methylethylidene)bisphenol (bisphenol-A) **8**. The first stage is the formation of bischlorhydrin **9** which is subsequently dehydrochlorinated with sodium hydroxide to produce the diglycidyl ether of bisphenol-A **10** known as DGEBA. In practice, the reaction is not simple because the epoxy groups may react with another molecule of phenol to give, as by-products, variable amounts of low molecular weight oligomers **11**. According to the quantity of oligomers and their average molecular weight, the mixture of epoxidized material may change from viscous liquids to waxy or even hard solids.

In addition, the conversion of phenols to glycidyl ethers does not follow this ideal pathway, and side reactions or incomplete dehydrochlorination yield compounds containing chlorinated groups such as **12** and **13**. These reactions have a dual effect because they first reduce the number of epoxy groups, which is generally lower than two, and secondly they introduce covalently bonded chlorine atoms that can be hydrolyzed under exposure to high temperature and humidity. Thus, crude epoxy resins contain large amounts of alkali metal (Na^+) and free chlorine (Cl^-) atoms. Manufacturers now offer resins complying as closely as possible with the ionic purity, quality, and reliability requirements of the electronic industry. The high-purity epoxy resins produced for electronic applications are

* Editor's note: For more information about epoxy adhesives and their chemistry and formulation, the reader may refer to the chapter "Epoxy Adhesives", which will be published in Volume 5 of this handbook.

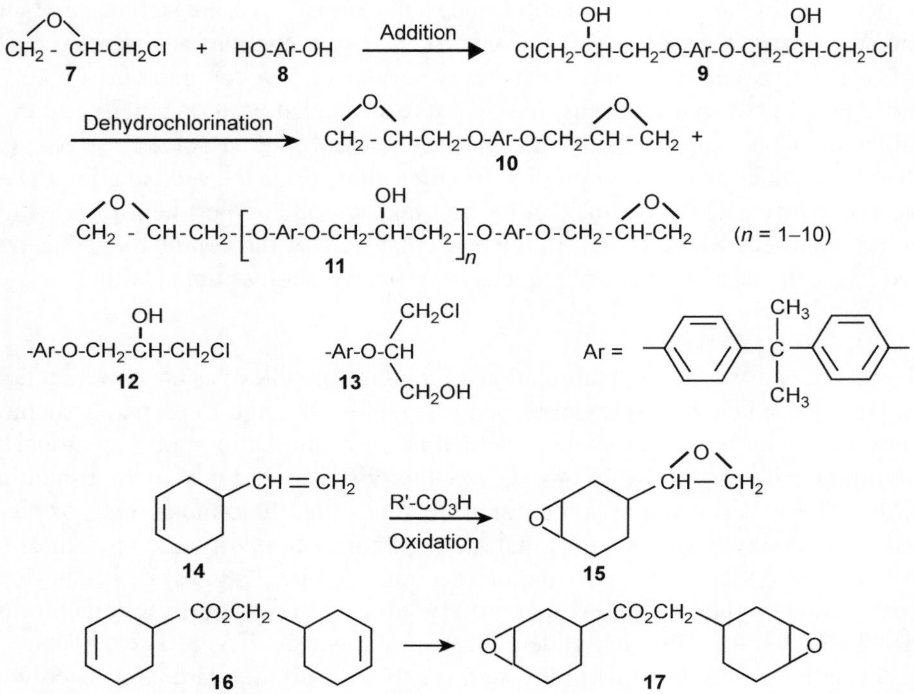

Figure 3: Synthesis of epoxy resins by addition of 1-chloro-2,3-epoxypropane **7** to bisphenol-A **8** providing intermediate compound **9**. Subsequent dehydrochlorination gives a mixture (e.g. Shell Epon® 828) of DGEBA **10**, low molecular weight oligomers **11**, and chlorine-containing species **12** and **13**. Cycloaliphatic epoxy resins **15** and **17**, such as Degussa F126®, are prepared by oxidation of unsaturated compounds **14** and **16** by means of organic peracids.

mostly used to make plastic packages and PCBs, the adhesive market accounting for only a small part, less than 2%, of the total consumption of these epoxies.

The second important category of epoxies, represented in Figure 3, is known as the cycloaliphatic series because they are prepared by a controlled oxidation of linear and cyclic carbon–carbon double bonds. Organic peracids ($R{-}CO_3H$), such as peracetic and perbenzoic acids, are used to introduce exocyclic oxygen bridges into the molecules. This is the only class of epoxies that does not contain residual chlorine. A second advantage lies in the lower viscosity of these compounds compared to that of glycidyl ether derivatives, making them very useful for decreasing the viscosity of high molecular weight epoxies in packaging formulations. Oxidation of 4-vinylcyclohexene **14** and 3-cyclohexenylmethyl 3-cyclohexenecarboxylate **16** gives 4-vinyl-1-cyclohexene dioxide

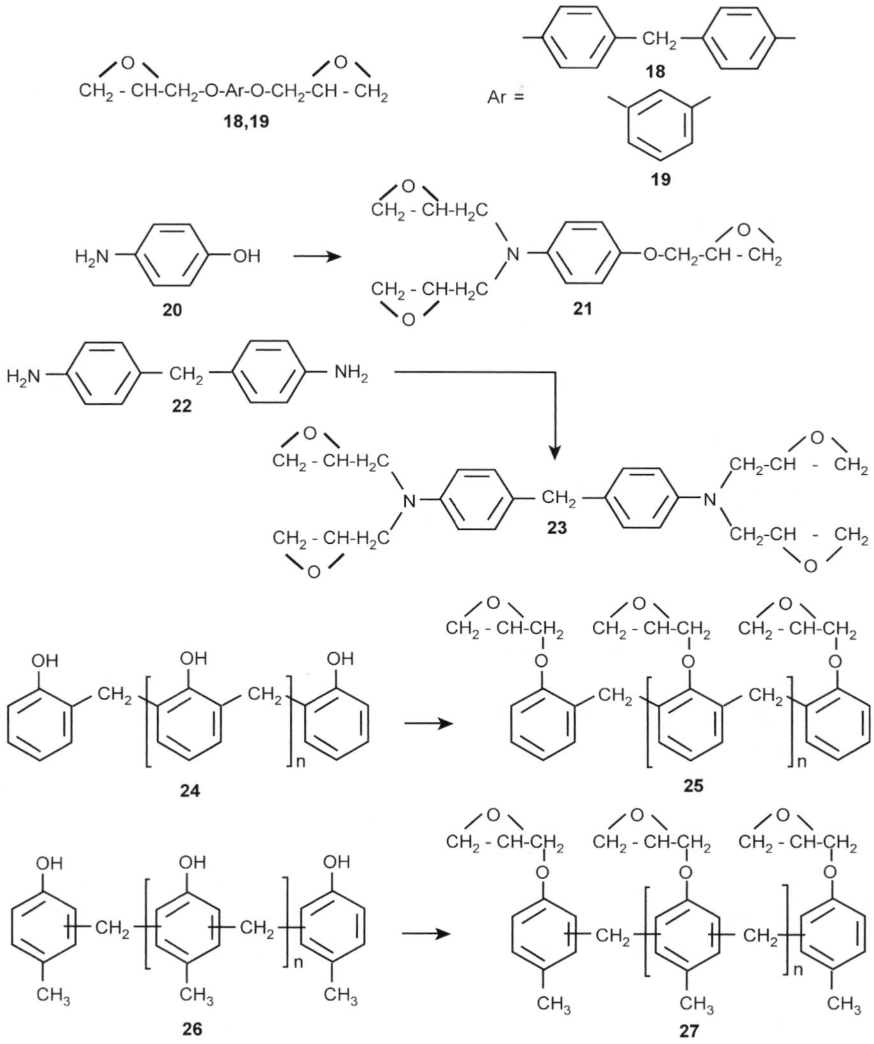

Figure 4: Chemical formulae of di- and polyfunctional epoxies commonly used to manufacture electrically conductive adhesives: Dainippon Epiclon 830 **18**, Ciba-Geigy ERE 1359 **19**, Ciba-Geigy MY 0510 **21**, Ciba-Geigy MY 720 **23**, Dow Chemical DEN 438 **25**, and Ciba-Geigy ECN 1280 **27**.

15 and 3,4-epoxy-cyclohexylmethyl 3,4-epoxycyclohexanecarboxylate **17**, respectively.

Figure 4 shows the chemical formulae of other epoxy resins commonly used to prepare conductive adhesive compositions. Representative of approximately difunctional compounds are the diglycidyl ether of bisphenol-F (DGEBF) **18** and the diglycidyl ether of 1,3-dihydroxybenzene **19**. Epoxy resins with a functionality higher than two are prepared by the reaction of 1-chloro-2,3-epoxypropane **7** with polyfunctional starting materials.

The reaction with amine derivatives such as 4-hydroxybenzeneamine **20** and 4,4′-methylenebisbenzeneamine **22** is used to produce the tri- and tetrafunctional epoxies *N,N,O*-tris(2,3-epoxypropyl)-4-hydroxybenzeneamine **21** and *N,N,N′,N′*-tetrakis(2,3-epoxypropyl)-4,4′-methylenebisbenzeneamine **23**, respectively. However, the polyfunctional epoxies that combine the most attractive properties for electronic applications are the resins produced by epoxidation of the phenol novolac **24** and cresol novolac **26**. Novolac resins are obtained by the condensation of a phenol with formaldehyde in the presence of acid catalysts in such conditions that the degree of polycondensation is in the range of 3–5. The epoxy novolacs **25** and **26** are produced by the reaction of epichlorhydrin with the corresponding phenol novolac and *ortho*-cresol novolac resins. Epoxy resins are generally characterized by their dynamic viscosity (η) at 25°C, expressed in millipascal second (mPa s), their melting temperature (mp) or dynamic viscosity measured at 50°C for solid materials, epoxy equivalent weight (*EEW*), which is the weight of resin corresponding to one epoxy group, number average of epoxy groups (*NAE*) per molecule, and number average molecular weight M_n.

Today, almost 50% of the semiconductors encapsulated in plastic packages are made for surface-mount assembly that subjects the devices to a considerable thermal shock during the soldering process. Within a few seconds, the internal package temperature rises to 215–260°C and the moisture absorbed by the plastic encapsulant and the organic adhesive evaporates explosively. This sometimes results in package cracks that start at the interface between the chip and the die pad or in delamination within the die attachment layer. To investigate the relationship between the chemical structure of epoxies and this so-called “popcorn effect”, a series of polyfunctional resins has been evaluated [5]. They include new experimental epoxy novolacs whose chemical formulae have been previously displayed [4].

Because of their excellent adhesive properties on most hard materials and relatively low production cost, poly(glycidyl ethers) have been used for more than 30 years to formulate high-volume consumer-orientated adhesives as well as high-performance structural adhesives. When not loaded with large amounts of inorganic fillers, these resins have dynamic viscosities that are convenient for most applications. However, the addition of metal or oxide fillers dramatically increases the viscosity to a level higher than 10^3 Pa s. To lower the viscosity of the adhesive

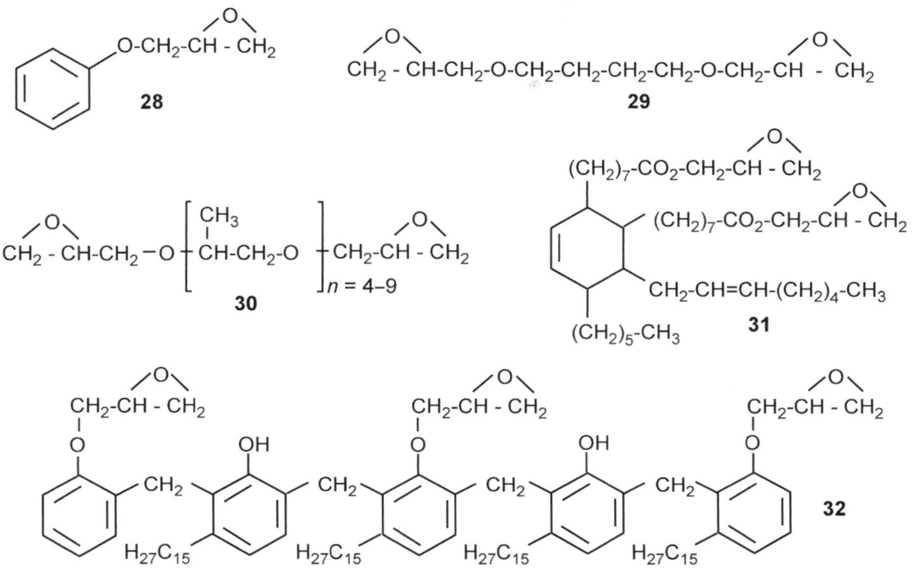

Figure 5: Chemical formulae of epoxy reactive diluents such as phenylglycidyl ether **28** and butanediol diglycidylether **29**; and flexible epoxy resins including polyoxypropylene-α, ω-diglycidyl ether **30**, diglycidyl ester of linoleic acid dimer **31**, and epoxy novolac resin of pentadecylphenol **32**.

compositions based on epoxy novolacs or other high-functionality epoxies, low-viscosity materials, also called reactive diluents, are prepared by the condensation of epichlorhydrin with either monophenols or diols such as butanediol or polyetherdiols. Figure 5 shows the formulae of phenylglycidyl ether **28** ($\eta = 6$ mPa s) and butanediol diglycidylether **29** ($\eta = 20$ mPa s).

Most of the epoxies used to prepare conductive adhesives are those described in the previous section because they allow the production of systems with high glass transition temperatures and good thermal resistance. However, the cured thermoset are rigid and brittle materials that develop significant thermal stresses when applied to large-area substrates. The versatility of the epoxy chemistry permits a substantial reduction of Young's modulus to the detriment of the thermal and mechanical properties. Some flexibility can be imparted to epoxy compositions by using flexible epoxy resins, flexible hardeners, plasticizers or flexibilizers, and elastomeric materials. A brief survey of the chemistry, properties, and applications of flexible epoxy adhesives was published by Edwards who emphasized that these materials have found limited use in the microelectronic industry [6]. When compared to rigid epoxies, flexibilized systems exhibit degradation in solvent resistance, moisture immunity, and thermal stability. The other drawbacks are low glass transition temperatures and high coefficients of thermal expansion.

Two examples of application are the encapsulation of bubble memory devices and LED where the flexible material provides the required stress relief. However, the use of flexible epoxy systems is expected to grow with the development of hybrid circuits bonded to large-area ceramic substrates. The three commercial flexible epoxies drawn in Figure 5 were reported by Hermansen and Lau [7]. The aliphatic backbone and a large distance between the epoxy groups explain the flexibility of both the epoxidized poly(oxypropylene)diol **30** (Dow Chemicals DER-732® and DER-736®) and the diglycidyl ester of the linoleic acid dimer **31** (Shell Epon® 871), whereas the epoxy novolac **32** (Cardolite NC 547®) carries long aliphatic chains acting as an internal plasticizer. The most promising chemistry for the production of high-quality flexible adhesives would be the combination of epoxies and silicones that are thermally stable up to 250°C. At the moment, these materials are extensively investigated as encapsulants for large-size integrated circuits which are more sensitive than small dice to the thermal stresses generated during the curing process.

6.2.2.2. Reactive and Non-reactive Solvents

The one-part and two-part commercially available conductive adhesives are often claimed as solventless formulations. In fact, many of them contain from 5 to 10% by weight of organic solvents mainly used to lower the viscosity. The hydroxyl-terminated glycol ethers and esters may possibly react with the epoxy groups during the curing process whereas non-functional glycol derivatives cannot. These organic compounds are commonly introduced into adhesive formulations, in particular 2-butoxyethanol (butylcellosolve), 2-butoxyethyl acetate (butylcellosolve acetate), diethyleneglycol monobutylether (butyl carbitol), diethyleneglycol monoethylether acetate (ethyl carbitol acetate), and γ-butyrolactone.

6.2.2.3. Curing Agents

The reactivity of epoxy groups toward both nucleophilic and electrophilic species is explained by the release of the ring strain inherent to the three-membered oxirane group. Both types of curing agents can take the form of catalysts, such as Lewis acids and tertiary amines, or co-reactants, such as primary amines, thiols, carboxylic acids, dicarboxylic acid anhydrides, and phenols. When the curing agents are catalysts, they initiate the homopolymerization of the oxirane rings so that the properties of the cured adhesive depend primarily on the chemical structure of the epoxy resin. The catalyst activity plays a role in the extent of polymerization that can be achieved under given cure conditions. In contrast, the co-reactants, containing mobile hydrogen atoms, offer greater latitude because they become an integral part of the final macromolecular network. The chemical and physical properties of the crosslinked epoxy matrix are equally influenced by the epoxy base and the co-reactant. Thus, epoxy adhesives can be produced with virtually any property, ranging from high glass transition temperature brittle thermosets to low-modulus flexible materials.

As a general guideline, one-part adhesives stable at room temperature utilize latent co-reactants or protected catalysts, while two-part epoxy compositions contain highly reactive curing agents or catalysts. The frozen one-part adhesives are mainly formulated with catalytic systems. Bauer summarized as follows the curing agents commonly used in electronic applications: dicyandiamide for PCBs and one-part adhesives, aliphatic polyamines for two-part die attach adhesives, anhydrides for glob top encapsulation and die attachment, epoxy novolac resins for transfer moulding compounds, and conductive adhesives [8]. Because of the overwhelming number of curing agents developed over the years, the following discussion is limited to the chemical compounds employed to formulate insulating and conductive adhesives. Figure 6 shows the chemical formulae of the catalysts and co-reactants commonly used to manufacture these adhesives.

Examples of catalysts are shown on the first line while all the other compounds are co-reactants including dicyandiamide, ureas, imidazoles, aliphatic polyamines, cycloaliphatic polyamides, and cycloaliphatic dicarboxylic acid anhydrides. As all the corresponding reaction mechanisms have been previously disclosed in detail [4], the following presentation is limited to the initial reaction steps leading to the active species involved in the polymerization or polycondensation processes. These primary attacks are enlightened in Figure 7, which displays only one epoxy group reacting with catalysts or co-reactants.

Polymerization of epoxy groups can be initiated by Lewis or Brönsted acids that lead to a cationic opening and polymerization of the oxirane rings. For example, the chemical complex **33** of boron trifluoride (Fig. 6) was used in the past as the catalyst of many one-part adhesive compositions. This acid–base complex is stable at room temperature but rapidly dissociates on heating, generating free boron trifluoride, which induces the polymerization of epoxy groups. In Figure 7, path A, BF_3 is represented as the electrophilic centre El^+ that transfers the positive charge to the oxygen atom of the oxirane ring. Subsequent polymerization is initiated by cationic attack of other epoxy groups by the active species **55**. One drawback of this process is the corrosivity of boron trifluoride. Ageing tests in humid conditions have demonstrated the destruction of aluminium conductors by this strong acid. The less aggressive trifluoromethanesulphonic acid (CF_3SO_3H), a strong Brönsted acid, is used in the form of an amine salt as a rapid curing agent at relatively low temperatures (5 min at 120°C) [9].

As represented in Figure 7, path B, catalytic polymerization of epoxies is also initiated by nucleophiles Nu^-, including inorganic and organic bases such as

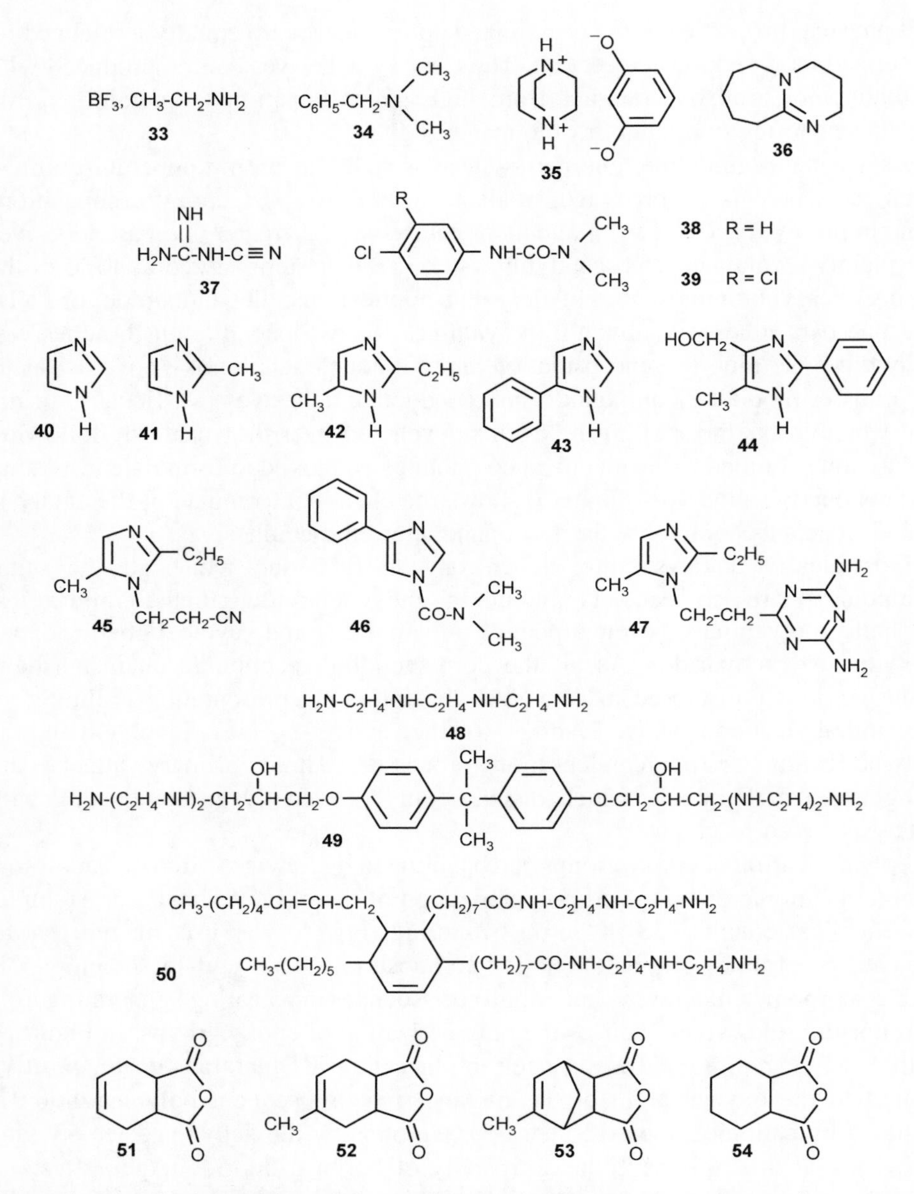

Figure 6: Chemical formulae of the catalysts and co-reactants used as curing agents of epoxy resins: electrophilic **33** and nucleophilic **34–36** catalysts, dicyandiamide **37**, ureas **38**, **39**, imidazoles **40–47**, aliphatic amines **48**, **49**, polyamide **50**, and cycloaliphatic anhydrides **51–54**.

tertiary amines, quaternary ammonium hydroxides, and some heterocyclic compounds. Reaction propagation develops from the initially formed oxygen anion **56**. These catalysts being highly reactive at ambient temperature, they are generally used as curing accelerators combined with less active hardeners. Thus, the reactions of epoxies with phenols, carboxylic acids, and anhydrides are often accelerated by small amounts of benzyldimethylamine **34**, 1,4-diaza-bicyclo[2,2,2]octane (DBO) **35**, represented as its resorcinol salt, and other compounds such as 1,8-diazabicyclo[5,4,0]undec-7-ene (DBU) **36**, tris-(dimethyl-aminomethyl)phenol, or tertiary amines prepared by adding secondary amines to epoxy resins. This latter curing agent would allow a pot life of 5 days at room temperature. To impart some latency to the one-part adhesives, the high reactivity of the tertiary amino groups can be masked by using their amine salts with diphenols such as resorcinol, pyrogallol, or hydroquinone.

The chemical formulae of dicyandiamide **37** and of two ureas: 3-(4-chloro-phenyl)-1,1-dimethylurea **38** and 3-(3,4-dichlorophenyl)-1,1-dimethylurea **39**, which are accelerators frequently used in combination with **37**, are drawn in

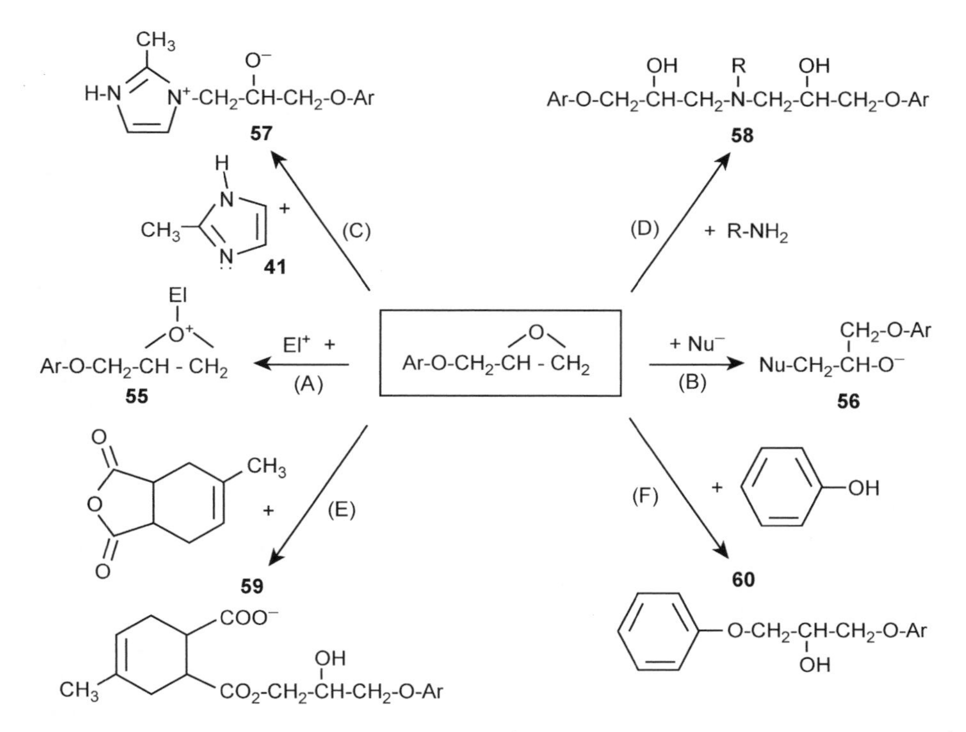

Figure 7: First step (initiation) of epoxy ring polymerization using: (A) electrophilic, El^+, catalysts; (B) nucleophilic, Nu^-, catalysts; or co-reactants including (C) imidazoles; (D) aliphatic amines; (E) cycloaliphatic anhydrides; (F) phenols.

Figure 6. The latency of these compounds is largely due to their insolubility in the epoxy resins at 20°C. On heating to temperatures higher than 120°C, dicyandiamide becomes soluble and reacts slowly with the epoxy groups so that the full cure requires temperatures around 170–180°C. The reaction is accelerated by adding ureas **38** and **39**, benzyldimethylamine **34,** or various imidazoles. The mechanism of addition of dicyandiamide to an epoxy group is complex and depends on the absence or presence of a tertiary base. In the former case, Zahir proposed a reaction sequence with the formation of an adduct leading to cyanamide and different oxazoline derivatives [10]. Adhesive compositions based on dicyandiamide have an extended pot life that can be as long as 1 year at ambient temperature. A disadvantage of dicyandiamide, which has been observed during outgassing tests in hermetic packages, is the continuous formation of ammonia (NH_3).

Despite the adverse effect of ammonia release, the patent literature offers many examples of one-part conductive adhesives formulated with dicyandiamide alone or mixed with 3-(3,4-dichlorophenyl)-1,1-dimethylurea **39** as crosslinking systems. Depending on the epoxy base, the amount of dicyandiamide, and co-catalyst, the pot life is at least 6 months at 25°C. Typical cure schedules range between 30 min at 170°C and 90 min at 120°C, yielding a volume resistivity of $5 \times 10^{-4}\ \Omega$ cm, and adhesive strengths in the range 5.9–19.6 MPa at 20°C. Some patents report that 2-phenyl-4-methyl-5-(hydroxymethyl)-imidazole **44** can also be used to increase the reactivity of dicyandiamide. The combination of dicyandiamide and imidazole **44** produces either a heat-resistant adhesive with a mixture of the tetra-epoxy **23** and 4-vinyl-1-cyclohexene dioxide **15** or a fast cure formulation based on bisphenol-F epoxy resins. These adhesives are solventless one-part compositions that can be cured for 1 min at 180°C, giving device shear strengths of 19.8 and 6.7 MPa at 25 and 350°C, respectively.

Imidazoles are strong organic bases and unusually good nucleophiles, these properties being explained by the resonance interactions, which increase the basicity of the 3-nitrogen atom. Imidazole chemistry is so versatile that most modern epoxy adhesives used in electronics contain an imidazole either as the unique curing agent or as an accelerator. Figure 6 shows the formulae of the imidazoles that have been cited in recent patents reporting the fabrication of adhesive compositions. Imidazole **40** was early utilized in one-part adhesive formulations because it presents some latency due to its insolubility at ambient temperature in epoxy resins.

The mechanism of homopolymerization of epoxy resins initiated by imidazoles is now well established. A first nucleophilic attack by the unsubstituted nitrogen atom of the imidazole ring forms zwitterion **57** (Fig. 7, path C), which rearranges to an adduct by internal proton transfer. This is followed by a nucleophilic reaction of the newly formed unsubstituted nitrogen, opening a second epoxy group to give the 2:1 adduct that promotes the anionic polymerization of the epoxy. FTIR spectroscopy and differential scanning calorimetry (DSC) have shown that two types of etherification reactions occur with the imidazoles unsubstituted at the 1-nitrogen atom [11]. When an epoxy resin (DGEBA) is cured with variable amounts of 2-ethyl-4-methylimidazole **42**, the N–H stretching vibration disappears within the first 10 min of cure. The decrease in epoxy concentration as a function of time indicates that this step corresponds to the formation of both the 1:1 and 2:1 adducts. The generation of the adducts is characterized by a slow initial rate of epoxy conversion followed by a sudden increase corresponding to the initiation of the polyetherification reaction by the oxygen anion.

Compounds such as 2-methylimidazole **41**, 2-undecylimidazole, 2-ethyl-4-methylimidazole **42** are highly reactive materials. It has been stated that soluble imidazoles are too reactive to permit their use in one-part adhesive systems stable at ambient temperature [12]. Among the other compounds, the less reactive 4-phenylimidazole **43** would be a convenient curing agent at elevated temperature, although the reaction still occurs over a few days of storage at 25°C. 4-Methyl-2-phenylimidazole is cited in different patents allowing a 3-day pot life at room temperature for silver-filled epoxy resins based on bisphenol-A, bisphenol-F, and phenol novolac hardener. Imidazole **44** mixed with bisphenol-F epoxy resin, an epoxidized phenol novolac, and silver flakes provides a conductive adhesive that can be cured in 20 s at 300°C. Die shear strengths of 15.7 N at 25°C and 7.1 N at 280°C have been claimed for compositions comprising imidazole curing agents, liquid cycloaliphatic epoxies, and diglycidyl ether resins.

A method known to slow down the degree of polymerization of epoxies at low temperatures while maintaining a satisfactory reaction rate at elevated temperature is the use of "protected" curing agents. In the case of imidazoles, this goal is achieved by masking the N–H proton. Because of the reversibility of the protection by the cyanoethyl group, 2-ethyl-4-methyl-1-(2-cyanoethyl)imidazole **45** is used in several two-part compositions. One package comprises bisphenol-F epoxy resin and silver flakes, while the second package includes imidazole **45**, γ-butyrolactone and silver flakes. After curing for 15–30 min at either 120 or 150°C, a shear strength of 9.8 MPa and a volume resistivity of 2.22×10^{-3} Ω cm are achieved. 1-(*N*,*N*-dimethylcarbamoyl)-4-phenylimidazole **46** is an imidazole-urea exhibiting the stability of the substituted ureas at room temperature. On heating to 175–185°C, it reacts with the residual hydroxyl functions of the epoxy resins to generate the freed imidazole available as a catalyst. Used in combination with resorcinol diglycidyl ether **19**, an epoxy novolac, and 1,4-butanediol diglycidyl ether **29**, it provides an adhesive curable at 175°C, exhibiting a volume resistivity of 10^{-4} Ω cm and a T_g in the range 83–93°C. One-part conductive adhesive with a pot life of 10 days at 25°C has been prepared by using 2,4-diamino-6-[2-(2-ethyl-4-methyl-1-imidazolyl)ethyl]-1,3,5-triazine **47**.

Amine curing agents include aliphatic amines, polyamide-amines of fatty acid dimers and trimers, and aromatic amines. In electronics, epoxy-amine compositions are used to make die attach and surface-mounting adhesives as well as encapsulants for discrete devices. According to Lyons and Dahringer, the first commercial conductive epoxy adhesive, marketed in the early 1960s, was a two-part system using triethylenetetramine **48** as the curing agent [13]. High reactivity, short pot life, moderate T_g, and very good adhesion at ambient temperature characterize the epoxy-amine systems. Aliphatic polyamines such as diethylenetriamine and triethylenetetramine **48** contain five and six active NH bonds that may react by addition to the oxirane ring (see Fig. 7, path D), This means that full cure is achieved with only 12–14 wt% of curing agent. An adhesive composition, prepared with DGEBA resin, phenylglycidyl ether, diethylenetriamine, and 75–90% silver powder, and cured at 80°C, provides a lap shear strength of 13.8 MPa and a volume resistivity of $5 \times 10^{-4}\ \Omega$ cm.

The ideal structure of polyamide **50** represents the product obtained by reacting diethylenetriamine with linoleic acid dimer. It can be seen that this compound contains two primary and four secondary amine functions, and two amide groups. The commercially available polyamides are in fact mixtures of oligomers that are characterized by their hydrogen equivalent weight. When used in combination with the epoxidized fatty acid dimers such as **31** (Fig. 5), the mixing ratio can be varied to achieve the desired balance between flexibility, adhesion, and chemical resistance. A typical formulation curable at room temperature is prepared as follows: a liquid DGEBA resin is mixed with a flexible epoxy resin based on dimer acid (11%), polyamide (89%), 2,4-bis[(dimethylamino)methyl]phenol catalyst, titanium dioxide, Cab-O-Sil thixotropic agent, and carbon black. The gel time is 45 min at 20°C but 24 h are required to achieve maximum strength.

Free carboxylic acids are not very popular for electronic applications but their anhydrides are widely used because of the long pot life that can be achieved. Epoxy resins cured with anhydrides have excellent dielectric properties, high temperature stability, and glass transition temperatures as high as 270–300°C. The formulae of commercial anhydrides are shown in Figure 6 including tetrahydrophthalic **51**, methyltetrahydrophthalic **52**, methyl-endomethylenetetrahydrophthalic **53**, and hexahydrophthalic **54** anhydrides. The uncatalyzed reaction of carboxylic acid anhydrides with epoxies is initiated by traces of hydroxyl groups (alcohols or water) leading to the formation of an *ortho*-acid-ester or a dicarboxylic acid. The free carboxylic acid group then adds to the oxirane ring to produce a hydroxy-ester **59** (Fig. 7, path E) that can react with another anhydride ring, and so on. The reaction is sluggish even at 150–200°C and the homopolymerization of the epoxy competes with the esterification reaction. The base-catalyzed addition of the anhydrides to the epoxies exhibits a greater selectivity toward polyester formation, even though conflicting mechanisms have been published so far [14]. A typical second-generation two-part epoxy adhesive comprises a liquid DGEBA epoxy loaded with silver flakes for the resin-based component and a co-reactant system formed of hexahydrophthalic anhydride **54**, benzyldimethylamine, and silver flakes [13]. The catalyzed epoxy-anhydride system has a pot life of 16–24 h and is cured at 150–180°C for 2 h.

6.2.3. Epoxy-phenolic Resins

In the electronic industry, phenols are widely used as co-reactants of epoxy novolac resins to encapsulate the integrated circuits. As shown in Figure 7, path F, the uncatalyzed reaction of phenolic compounds with an epoxy proceeds through the nucleophilic addition of the phenolic hydroxyl group to the oxirane ring leading to the hydroxy-ether **60**. The subsequent reaction is the etherification of a second epoxy group by the aliphatic hydroxyl function which generates another hydroxy-ether, and so on. Using model compounds, Schechter and Wynstra indicate that the reaction rate is very slow at temperatures up to 100°C whereas at 200°C the epoxy disappears at a faster rate than the phenolic hydroxyl functions [15]. In practical applications the reaction of phenols with epoxies is catalyzed by either organic bases or triphenylphosphine [$P(C_6H_5)_3$], or both.

Three main classes of polyphenols are used to prepare high-performance adhesives and moulding materials: phenol novolacs, resols, and poly(*para*-hydroxy-styrene). A few chemical formulae of these polyphenols are shown in Figure 8. The phenol novolac **61**, *ortho*-cresol novolac **62**, and *para*-xylene modified phenol novolac **63** are prepared by the reaction of phenol and substituted phenols with carbonyl-containing compounds (aldehydes and ketones) in the presence of acid catalysts. The *ortho–ortho*-resol of *para*-cresol **64** is an example of a resol stable at ambient temperature, produced by reacting formaldehyde with *p*-cresol using carboxylate salts as catalysts. In addition to phenolic hydroxyl groups, resols have highly reactive terminal hydroxymethylene ($-CH_2-OH$) functions. Poly(*para*-hydroxy-styrene) **65** was introduced by Toshiba Chemical Corp. to prepare flexible adhesives.

Novolac resins are broadly used in electronics because their functionality higher than two increases the crosslinking density and yields cured resins exhibiting enhanced chemical and physical properties. Mixtures of epoxy resins and phenol novolacs **61** are excellent structural adhesives in the aerospace industry. However, the phenolic hydroxyl groups are not very reactive at moderate temperatures and most systems include catalysts or accelerators. Classical adhesive compositions are prepared by mixing a solid epoxy resin, typically an epoxidized phenol novolac resin (60 parts), a phenol novolac resin (40 parts), a solvent such as 2-butoxyethanol or butylcellosolve acetate, an imidazole catalyst, and silver flakes.

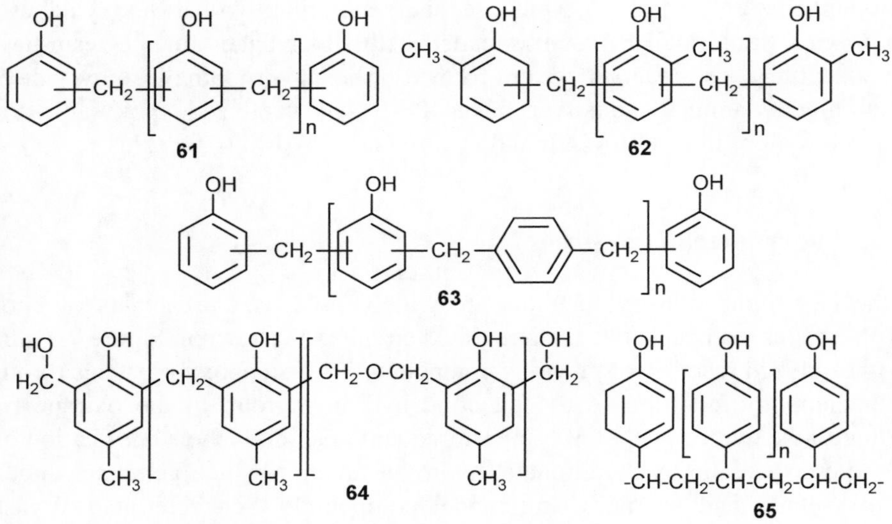

Figure 8: Chemical formulae of phenolic resins used to prepare epoxy-phenolic adhesives based on: Gunei Kagaku phenol novolac **61**, Dow Chemical OCN *ortho*-cresol novolac **62**, Mitsui Toatsu *para*-xylene-modified phenol novolac **63**, phenolic *ortho–ortho*-resol of *para*-cresol **64**, and poly(*para*-hydroxystyrene) **65**.

Various curing schedules are reported in the patent literature, the best volume resistivity (5.7×10^{-5} Ω cm) being achieved after curing for 24 h at 120°C and 5 h at 180°C. It has been claimed that a die attach adhesive composed of an epoxy-phenol novolac system and a catalyst of the Ph_4P^+ BPh_4^- complex type provides die shear strengths of 43 and 9 N at 20 and 350°C, respectively, after curing for 30 s at 250°C. Thermally conductive adhesives are also prepared by loading epoxy-phenolic compositions with crystalline silica powder.

Because of their inherently high reactivity, resols generally provide epoxy adhesive compositions with a limited shelf life at ambient temperature. In the coating industry, epoxy-resols combine good heat and moisture resistance along with improved flexibility. A typical adhesive paste contains a phenolic resol (molecular weight 750 g mol^{-1}), an epoxy resin, phenyl glycidyl ether, and silver flakes [16]. After curing at 170°C for 20 min, the adhesive affords a die shear strength of 4 MPa and a volume resistivity of 5×10^{-5} Ω cm. Poly(*para*-hydroxystyrene) **65**, commercialized by Toshiba Chemical Corp. under the trade mark Maruzen® resins, has been employed to prepare either conventional or fast cure conductive adhesives. To enhance the reactivity of the phenolic hydroxyl groups, boron trifluoride–ethylamine complex **33** is used as a catalyst in most formulations and norbornene resins are added to improve the resistance to hydrolysis. Semiconductor devices are bonded to lead frames by using a two-step cure schedule. The assemblies are first heated at 130–170°C for 20–30 s and then at 300°C for 30 s. According to the chemical composition, the die shear strength is in the range 49–62 N at 25°C and 4.9–7.9 N at 350°C while the volume resistivity is of the order of 2.1×10^{-4} Ω cm.

6.2.4. Epoxy-silicone Compositions

Multifunctional epoxy resins provide brittle materials while silicones have a high degree of flexibility. The glass transition temperature, Young's modulus, and coefficient of linear thermal expansion (CTE) of epoxy resins are high enough to generate significant mechanical stresses. In contrast, some stress relief is expected for silicone resins, in particular the polydimethylsiloxane series, because of their low modulus and low glass transition temperature. However, silicones exhibit poor adhesive properties and large coefficients of thermal expansion. The incorporation of silicone rubber particles into brittle thermoset networks is commonly used to increase the impact resistance and reduce the crack propagation within the matrix of composite materials. Another approach consists in producing elastomeric particles by a phase segregation during the polymerization process to produce a block copolymer with alternate epoxy hard segments and polysiloxane soft segments. This implies a judicious control over the polymerization conditions to achieve phase separation. Figure 9 displays some of the chemical reactions that have been studied to produce polydimethylsiloxanes terminated with either reactive groups or oxirane rings, most of these siloxane telechelic oligomers being based on the commercially available α,ω-dihydropolydimethylsiloxane **66**.

All the difunctional compounds are prepared by the hydrosilylation reaction, which is a selective addition of the Si–H terminal groups to carbon–carbon double bonds in the presence of noble metal catalysts such as chloroplatinic acid. This general reaction is shown at the first line of Figure 9. The addition of hydrogen-terminated silane **66** to protected 2-propenoic acid **67**, 2-propenylamine **69**, and allyl glycidyl ether **71** leads to α,ω-bis(3-carboxypropyl)polydimethylsiloxane **68**, α,ω-bis(3-aminopropyl)polydimethylsiloxane **70**, and α,ω-bis[(2,3-epoxypropyloxy)propyl]polydimethylsiloxane **72**, respectively. It has been shown that tris(triphenylphosphine)rhodium chloride and polymer-bound Wilkinson's catalysts are able to promote selectively the hydrosilylation reaction of allyl glycidyl ether **71** without any side reaction with the oxirane ring [17]. The other approach illustrated in Figure 9 is the "sea–island" concept developed to reduce the stress of epoxidized *ortho*-cresol novolacs for the encapsulation of electronic devices [18]. Islands of silicone rubber particles are dispersed in a sea of uncured

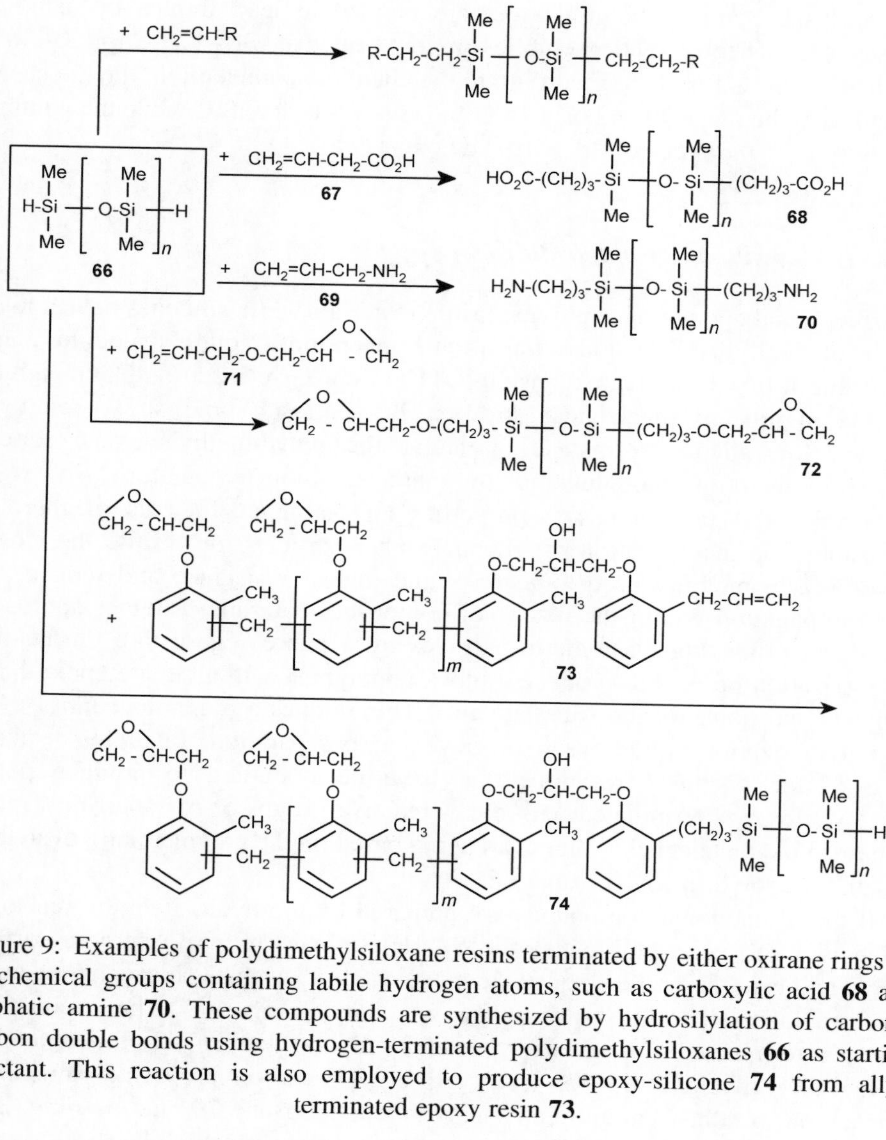

Figure 9: Examples of polydimethylsiloxane resins terminated by either oxirane rings **72** or chemical groups containing labile hydrogen atoms, such as carboxylic acid **68** and aliphatic amine **70**. These compounds are synthesized by hydrosilylation of carbon–carbon double bonds using hydrogen-terminated polydimethylsiloxanes **66** as starting reactant. This reaction is also employed to produce epoxy-silicone **74** from allyl-terminated epoxy resin **73**.

epoxy matrix. In the first reaction step, 2-(2-propenyl)phenol is allowed to react with an epoxy *ortho*-cresol novolac to prepare the partially functionalized compound **73** containing both oxirane rings and unsaturated allyl substituents. Hydrosilylation of this "macromonomer" is then performed with a series of hydrogen-terminated polydimethylsiloxanes **66** with weight average molecular weights of 4×10^2–6.2×10^4 g mol^{-1}. Polydimethylsiloxane carrying either lateral hydrogen atoms on the main chain or both terminal and lateral hydrogen atoms can be used as well. Because of the incompatibility between the epoxy novolac and the silicone phases, the epoxy-silicone compound **74** is an epoxy resin containing about 10 wt% of homogeneously dispersed silicone rubber particles.

*6.2.5. Silicone Resins**

Silicon-containing organic polymers are used to formulate adhesive compositions with low elastic modulus and high elongation at break. The main chain of polydimethylsiloxane is made of silicon–oxygen bonds with methyl groups stemming from the silicon atoms. According to their molecular weight, these polymers are either liquids, greases, rubbers, or hard solids with a glass transition temperature of -123°C. They have reasonably good thermal resistance but poor adhesive properties. Commercial silicones are available either as one-part or two-part systems that can be cured at room temperature (RTV-resins) or on heating to 160–180°C depending on the monomer structures. Figure 10 illustrates one of the different RTV-silicone cure process using methoxysilyl-terminated polydimethylsiloxane **75** and methyltrimethoxysilane **76**.

These siloxanes are easily hydrolyzed by ambient humidity and water bound to the substrates to generate transient silanols **77** and **78**. Subsequent dehydration initially provides a precursor **79** of crosslinked polydimethylsiloxane network, which is formed by further condensation of the silanol groups. With many variants, this process constitutes the base of the RTV silicones that are mainly used as protection shields in the glob top encapsulation process. Low-stress conductive adhesives have been prepared by the condensation reaction of ethylphenylsilanediol, 1,4-phenylenebis(dimethylsilanol), and trimethylsilyl-terminated polydiethylsiloxane. When loaded with carbon particles and cured at 180°C, this composition exhibits a volume resistivity of $1.2 \times 10^{-3}\ \Omega$ cm and a thermal conductivity of 2.1 W m^{-1} K^{-1}.

Hydrosilylation in the presence of platinum catalysts is used to prepare heat-curable silicone adhesives and encapsulants. Polydimethylsiloxanes carrying terminal and eventually side vinyl groups **80** are blended with polymers containing silane (Si–H) linkages **81** and chloroplatinic acid. On heating, addition of Si–H to vinyl unsaturations provides interchain connections by formation of silicon–carbon bonds leading to polymer **82**. Infrared spectroscopy and microdielectrometry show that complete curing is achieved after 30 min at 175°C [19]. This technique has been used to prepare thermally conductive adhesives by mixing vinyldimethylsiloxy-terminated poly(methylphenylsiloxane), α,ω-(dihydrosilyl)-

---

* Editor's note: For more information about silicones and their chemistry, refer to the chapter "Silicone Adhesives and Sealants" in Volume 3 of this handbook, which provides a very comprehensive study.

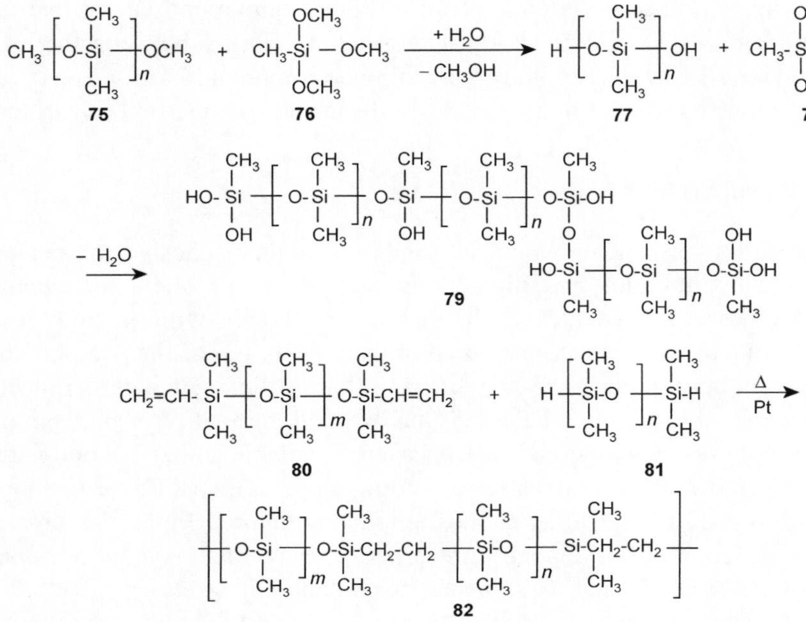

Figure 10: Formation of linear and crosslinked polydimethylsiloxane networks. Room temperature vulcanization (RTV) is performed by reacting methoxysilyl-terminated polymer **75** and methyltrimethoxysilane **76**. Hydrolysis produces transient silanols **77** and **78** whose dehydration yields silicone precursor **79**. Two-part systems are based on the platinum-catalyzed addition reaction between vinyl-terminated polysiloxane **80** and polydimethylsiloxane carrying either only terminal **81** or multiple silane (Si–H) pendent groups. Silicone main chains of the resulting polymer **82** are linked through silicon–carbon bonds.

polydimethylsiloxane, chloroplatinic acid, 2-(3,4-epoxycyclohexyl)ethyltrimethoxysilane, tetrabutoxytitane, and alumina powder. Electrically conductive adhesives utilizable from −60 to 180°C in bonding sensitive electronic components to substrates were also prepared by addition of hydrogen siloxanes to vinyl-terminated siloxanes in the presence of hydrosilylation catalysts.

6.2.6. Cyanate Ester Resins

Cyanate esters are organic compounds containing –OCN groups. Figure 11 shows the thermal cyclotrimerization of cyanic acid phenyl ester **83** yielding 2,4,6-triphenoxy-1,3,5-triazine **84**. The commercial cyanate esters **85–89** have glass transition temperatures in the range 190–350°C [20]. Biscyanates **85** and **86**

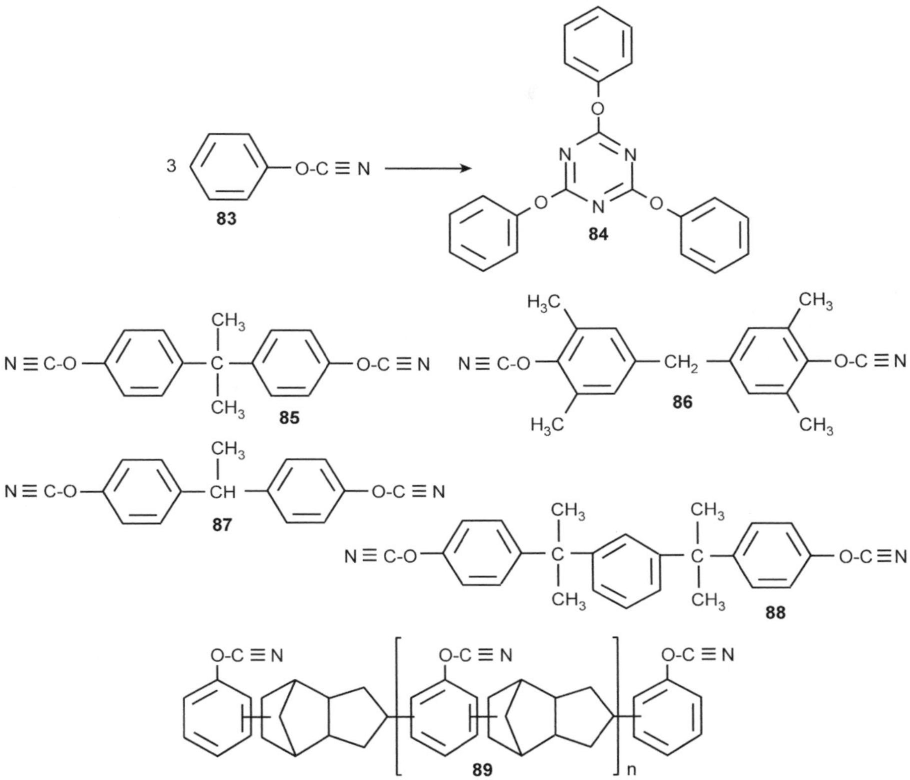

Figure 11: Thermal cyclotrimerization of cyanic acid phenyl ester **83** yielding 2,4,6-triphenoxy-1,3,5-triazine **84** and chemical formulae of commercial cyanate esters **85–89** used as base resins to produce heat-resistant adhesives and moulding materials.

are crystalline materials melting at 79 and 106°C, respectively, whereas compound **87** is a liquid with a dynamic viscosity of 100 mPa s, and products **88** and **89** are semi-solid waxes.

The trimerization reaction of the polyfunctional cyanates is catalyzed by zinc or manganese octoate, copper carboxylates, and acetylacetonate metal chelates with a synergistic effect when the catalyst is mixed with 4-nonylphenol [21]. The final cure or postcure temperature depends on the glass transition temperature of the tridimensional network. This means that the semi-solid biscyanate **88** is cured at 180–195°C (T_g 192°C) while the biscyanate **85** is cured at 250–300°C (T_g 290°C). However, the maximum cure temperature can be reduced to 120°C in the former example by adding 6–10% of 4-nonylphenol. When compared to epoxies, cyanate esters have higher glass transition temperatures, better thermal stability, and improved dielectric properties. Cyanate esters have been primarily developed as a substitute for epoxies in the fabrication of multilayer PCBs. They have been more recently incorporated in the formulation of adhesives to attach silicon dice in hermetic and non-hermetic packages [22]. High-modulus materials are intended to be used for die attachment in hermetic packages with temperature exposures up to 370°C.

Two silver-filled adhesives, Johnson Matthey JM® 7000 and JM® 7800, have respective elastic modulus of 10 and 5.8 GPa, T_g 250 and 210°C, thermal conductivity 1.1 and 1.6 W m^{-1} K^{-1}, volume resistivity 2×10^{-3} and 5×10^{-5} Ω cm. According to the manufacturer, the thermal stresses, determined by measuring the radius of curvature of silicon chips, are far from the level of stress where the risk of cracking and delamination increases considerably. In 1994 Johnson Matthey proposed a low-stress cyanate adhesive (JM® 2500) exhibiting a modulus of elasticity of only 0.4 GPa. The low stress properties of this material were demonstrated by the large radius of curvature (1 m) of a 15×15 mm^2 die bonded to a 0.15 mm thick leadframe. This value remained constant after 1000 thermal cycles from −65 to 150°C.

6.2.7. Polyimides

In electronics, polyimides are now extensively used in the form of self-standing films for flexible circuitry, deposited films for interlayer dielectrics, passivation and buffer coatings, moulding thermoplastic powders for PCBs, and adhesive pastes or tapes. The basic polyimide chemistry has been adapted to fulfill the specific requirements of these applications. A series of books provides complete information not only on the chemistry of polyimides but also on their utilization in electronics [4,23,24]. The following figures summarize the chemical formulae of the most important categories of polyimide precursors or precyclized polymers that are commonly used in electronics.

6.2.7.1. Polyimide Precursors

In electronics, thin polyimide films are formed on the surface of silicon wafers by depositing a solution of polyamic acid, which is then subjected to a thermal treatment to perform the cyclodehydration reaction yielding ultimately the heterocyclic polymer. Figure 12 illustrates this general technique patented by Du Pont de Nemours in the early 1960s. Here, 1,2,4,5-benzenetetracarboxylic acid dianhydride (pyromellitic acid dianhydride, PMDA) **90** is opposed to 4,4′-oxybisbenzeneamine (4,4′-oxydianiline, ODA) **91** to produce the intermediate high molecular weight linear PMDA–ODA polyamic acid **92**. Solutions of this polymer in *N*-methylpyrrolidone (NMP) are marketed by Du Pont as Pyralin® PI 2540 and PI 2545. Once deposited on the substrate, polymer **92** is imidized by

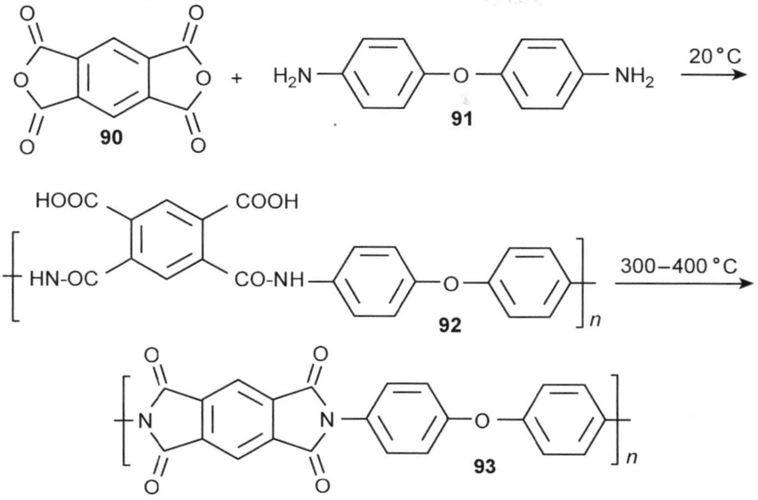

Figure 12: Reaction of 1,2,4,5-benzenetetracarboxylic acid dianhydride (PMDA) **90** with 4,4′-oxybisbenzeneamine (ODA) **91** yielding high molecular weight polyamic acid **92**, which is then subjected to thermal cyclodehydration to give the PMDA–ODA polyimide **93**.

a multistage heating ending at 300–400°C to produce films of PMDA–ODA polyimide **93**.

6.2.7.2. Self-standing Polyimide Films

Self-standing polyimide films with thicknesses ranging between 25 and 150 μm are commercially available either as non-oriented amorphous materials primarily used as adhesives or oriented films with high mechanical properties. They are manufactured by using two main processes which differ by the reaction temperature. The first technique consists in coating a viscous solution of polyamic acid on the surface of a conveyor belt and then converting it into polyimide. Imidization is accomplished either by heating the film to temperatures in excess of 300°C, or chemically with a mixture of acetic anhydride and triethylamine. This latter process is followed by a heating stage to remove all volatile materials and to complete the formation of imide rings. In the second method, developed by Ube Industries, the reaction is performed in 4-chlorophenol at 160°C to produce a viscous solution of 95% imidized polymer, which is subsequently deposited on a substrate heated to 100°C, and then cured at 300°C.

Figure 13 shows the repeating units of commercial polyimide films whose trade marks, commonly accepted acronyms, and mechanical properties including tensile

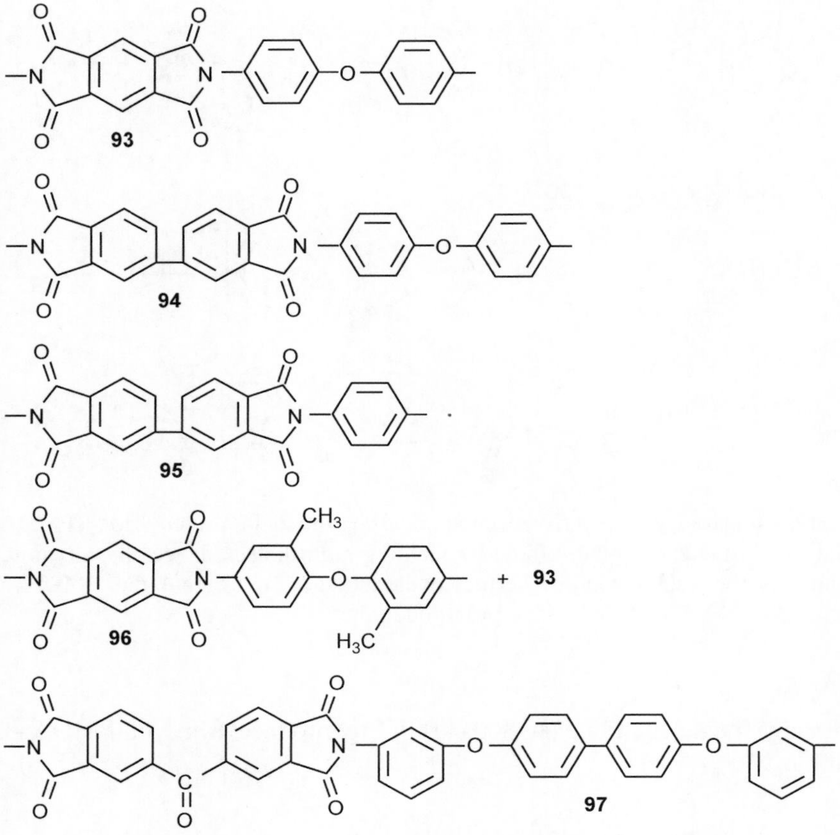

Figure 13: Repeating units of commercial polyimide films: Kapton® and Apical® **93**, Upilex R® **94**, Upilex S® **95**, Novax® **96 + 93**, and Regulus® **97**.

modulus E, tensile strength σ, and elongation at break ε are the following: Kapton® and Apical® **93** (PMDA–ODA), $E = 2.97$ GPa, $\sigma = 173$ MPa, $\varepsilon = 70\%$; Upilex R® **94** (BPDA–ODA), $E = 2.62$ GPa, $\sigma = 172$ MPa, $\varepsilon = 130\%$; Upilex S® **95** (BPDA–PPDA), $E = 6.20$ GPa, $\sigma = 275$ MPa, $\varepsilon = 30\%$; Novax® **96** + **93** (PMDA–2,2′-dimethyl ODA + PMDA–ODA), $E = 6.86$ GPa, $\sigma = 324$ MPa, $\varepsilon = 40\%$; Regulus® **97** (BTDA–3,3′-BP–ODA), $E = 3.04$ GPa, $\sigma = 118$ MPa, $\varepsilon = 110\%$. Glass transition temperatures increase from 285°C (Upilex R) to 350°C (Novax), 385°C (Kapton), and 500°C (Upilex S). For drawn films, the rigid structure of Upilex S explains the high T_g, a low CTE (8×10^{-6} K^{-1}), a water uptake of 1.2%, and a hygroscopic linear expansion coefficient of 1×10^{-5}% RH^{-1}, where RH is the relative humidity. The other films exhibit a twofold increase for the thermal expansion coefficient and water uptake. All of them have good dielectric properties with permittivity 3.0–3.5, dissipation factor 0.001–0.003, dielectric strength 2.0–2.8 MV cm^{-1}, and volume resistivity 10^{15}–10^{16} Ω cm.

6.2.7.3. Polyimide Adhesives

Compared to epoxy adhesives, polyimides share a very small part of the global market, limited to the military, aerospace, and geothermal applications requiring long-term stability at elevated temperatures. Owing to their aromatic heterocyclic structure, virtually all polyimides are stable at 300°C. When loaded with inorganic particles such as alumina, silica, silicon nitride, or aluminium, the thermal stability is still better. By contrast, some metals used in the composition of conductive adhesives, in particular silver and nickel, dramatically decrease the thermal resistance. This means that the adhesive strength of most polyimides is excellent at 200°C, good at 250°C, but limited to short-term uses at 300°C. Information retrieval indicates that, for the past three decades, only three series of conductive adhesives have been developed and manufactured. The open market has been covered by the polyimide precursor **99** shown in Figure 14.

A marginal production is based on the acetylene-terminated oligoimides presented in another chapter (see Volume 2, Chapter 7: Heat-Resistant Adhesives), and a proprietary composition has been used at National

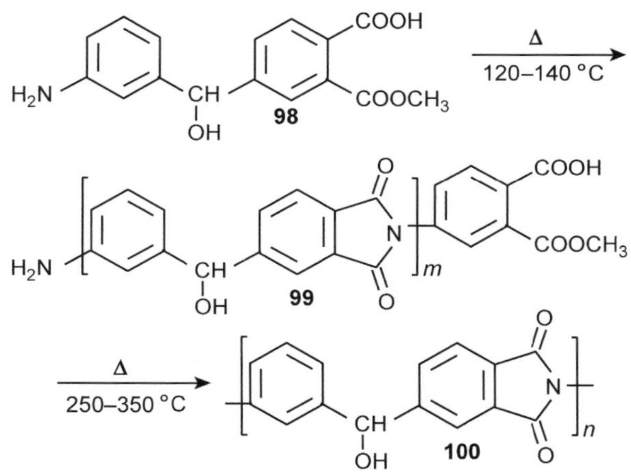

H_2N CH OH COOH $COOCH_3$ **98** Δ 120–140 °C

H_2N CH OH N O O COOH $COOCH_3$ *m* **99**

Δ 250–350 °C CH OH N O O *n* **100**

Figure 14: Polycondensation of self-condensable monomer **98** providing low molecular weight oligomer **99**, which, on heating at high temperature, gives high molecular weight polyimide **100**.

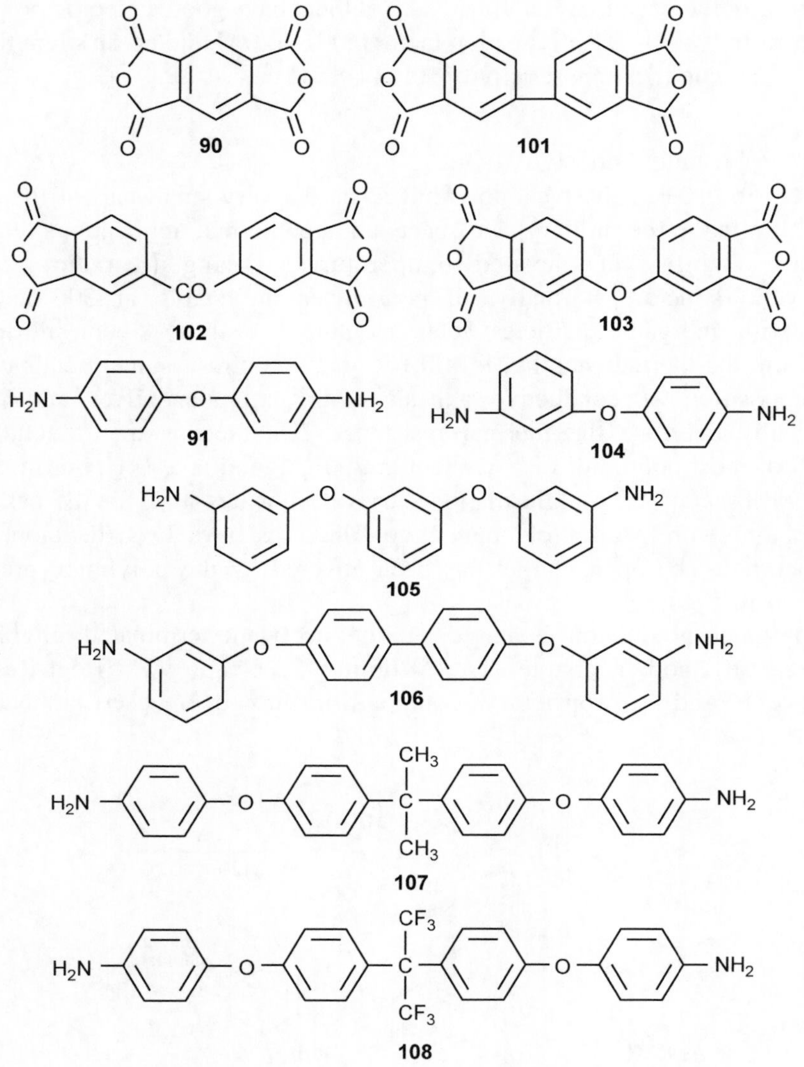

Figure 15: Chemical formulae of the dianhydrides and diamines used to synthesize thermoplastic adhesive films of the poly(ether-imide) series.

Semiconductors [25]. In this later example, the die attach material is prepared from a PMDA–ODA polyamic acid solution in NMP (Du Pont PI 2561) mixed with γ-aminopropyltriethoxysilane and loaded with silver flakes. The adhesive composition is used to bond semiconductor dice in ceramic packages with a first cure at 125 and 180°C for 2 h each with a final imidization occurring under vacuum when the ceramic lid is sealed with a glass frit at 400–450°C.

For die attachment, polyimide adhesives have reached a peak and now the market is progressively slowing down. But, for other applications, there is a trend in the direction of thermoplastic polyimide films that can be used as interlayer adhesives. Flexible circuits and MCMs seem to be the main target for these materials. In the polyimide chemistry, high plasticity is obtained by introducing either molecular symmetry disruption or flexible links, in particular ether bridges. In 1974, General Electric (GE) patented both the synthesis of a new dianhydride with two ether-linking units and its use to produce poly(ether-imides). The production process, chemical, mechanical, and electrical properties of Ultem® 1000 have been recently reviewed [26]. Research at Occidental Chemical has focussed on the synthesis of 4,4′-oxybis(1,2-benzenedicarboxylic acid) dianhydride (OPDA, **103**) and on the polyimides prepared from this compound and various aromatic diamines [27]. For the last 15 years new poly(ether-imides) have been proposed as thermoplastic materials to produce adhesives, self-standing films, and interlayer dielectrics. Chemical formulae of typical ether-linked aromatic diamines are illustrated in Figure 15.

The initial goal has been the production of melt-processable polyimides without taking into account the enhanced solubility provided by two oxygen atoms and other flexible bonds. In electronics, poly(ether-imides) are extensively studied as interlayer dielectrics for MCMs in which electrical conductors are grown by electroless plating. Another important application is the fabrication of adhesive layers for flexible circuits and TAB laminates. The patent literature has been previously reviewed [23,24], so only a brief information is given hereafter. A thermoplastic polypyromellitimide has been synthesized by reacting PMDA **90** with 3,3′-[1,3-phenylenebis(oxy)]bisbenzeneamine **105** alone or combined with ODA **91** or 1,3-benzenediamine. The homopolymer, with inherent viscosity $0.66\ \mathrm{dl\ g^{-1}}$, has crystalline melting temperature 307°C and glass transition temperature 218°C, similar to that of Ultem® 1000. 3,3′-[[1,1′-biphenyl]-4,4′-diylbis(oxy)]bisbenzeneamine **106** was used to develop a series of polyimides commercialized by Mitsui Toatsu under the trade marks Aurum® for moulding resins and Regulus® for self-standing films. The polyamic acid prepared from dianhydride OPDA **103** and diamine **105**, mixed with aluminium powder, has been proposed as an adhesive for sealing ceramic packages. Polymerization of diamine **105** and 4,4′-[(1-methylethylidene)bis(4,1-phenyleneoxy)]bisbenzeneamine **107** with BTDA **102** and OPDA **103** gives polyimide films for bonding Upilex® S tapes to copper foils, whereas the combination of BPDA **101** and diamine **105** provides an adhesive for bonding Upilex to Upilex with peel strength $2.94\ \mathrm{kN\ m^{-1}}$.

6.3. Electrical and Thermal Conductivities

6.3.1. Inorganic Fillers

Conductive adhesives are prepared by loading organic resins with small particles of inorganic fillers that are generally metals, metal oxides, and metal nitrides. Silver is the metal of choice for the manufacture of electrically conductive adhesives. The fabrication and characterization of silver flakes have been previously reported [4]. It is worth noting that the manufacturers of conductive adhesives try to exceed the percolation threshold with minimum amount of silver flakes. This can be achieved by incorporating flakes with broad particle size distributions (3–10 μm). Nevertheless, stable electrical conductivity requires approximately 30% by volume (70–80 wt%) of silver particles. Gold is reserved to military and aerospace applications requiring long-term reliability in severe thermal and aggressive environments. Spherical nickel particles, which are commercially available with tightly controlled diameters up to 50 μm, are mainly used in the fabrication of low-cost anisotropic adhesive films. Copper-filled adhesives do not retain stable electrical conductivity after exposure to elevated temperatures because surface copper oxide is easily formed. Tin and solders have also been proposed to prepare low-cost conductive adhesives. Removable adhesive layers have been experimented by using powdered solders with a solidification temperature and a eutectic melting point higher than the polymer curing temperature, but lower than the onset of polymer thermal degradation.

Metal oxides and ceramic precursors are commonly used to enhance the thermal conductivity and decrease the thermal expansion coefficient (CTE) of the organic binder. Fused silicon dioxide and crystalline quartz are mainly employed in the production of plastic encapsulant packages exhibiting CTE values of $2.5 \times 10^{-5}\ K^{-1}$. Most commercial adhesive pastes include a small amount (0.5–2.5 wt%) of fused or colloidal silica as thixotropic agent. Aluminium oxide powders are the predominant materials in the manufacture of low-cost thermally conductive adhesives with a thermal conductivity of approximately $1.5\ W\ m^{-1}\ K^{-1}$ at the maximum loading level of 75 wt%. Crystalline boron nitride is an expensive material with a thermal conductivity of $1300\ W\ m^{-1}\ K^{-1}$, but the particles commonly used to produce the adhesive compositions are of pyrolytic quality ($30\ W\ m^{-1}\ K^{-1}$). Sintered ceramic sheets of aluminium nitride have a thermal conductivity in the range 170–$220\ W\ m^{-1}\ K^{-1}$ and various grades of powdered material are commercialized. The highest thermal conductivity ($2000\ W\ m^{-1}\ K^{-1}$) is that of diamond which was expected to provide the material of choice for high-end applications. It will be seen later that diamond-filled adhesives are not better than pastes loaded with boron or aluminium nitrides.

6.3.2. Electrical Conductivity

ASTM designation D257 draws the test methods and procedures to determine the direct current resistance or conductance of insulating materials. The volume resistance R_v between two electrodes in contact with a specimen is directly proportional to the thickness h (cm) and inversely proportional to the area A (cm^2) of the sample according to the equation $R_v = \rho_v(h/A)$ where ρ_v is the volume resistivity expressed in ohm cm (Ω cm). Figure 16 shows the methods used to determine the volume resistivity of conductive adhesive films or pastes deposited on the surface of substrates carrying metallic electrodes.

A current of intensity I is applied between two electrodes and the electrical resistance R of the adhesive strip, measured with an ohm-meter, provides the volume resistivity by the relation $\rho_v = Rhw/l$.

Table 1 lists the volume resistivity values measured for some metals, eutectic solders, electrically conductive adhesives, oxide-filled resins, and unfilled polymers.

Although all metals have a good electrical conductivity and might be used to make conductive adhesives, most of them readily form an insulating oxide layer by air oxidation in normal ambient conditions. Moreover, in the case of copper, nickel, and silver, the metal oxides are chemically reactive and may catalyze the decomposition of the organic matrix when the adhesive is subjected to isothermal ageing or temperature cycling. Silver-filled adhesives are prepared by dispersing silver particles in an insulating polymer matrix to form a metallic network within the organic binder. Electrons can circulate between the two surface boundaries only across the points of contact between adjacent particles, and the current flow through the conductive adhesive layer is generally described by the percolation theory. As regards to electrical conduction, the percolation model predicts a critical volume fraction φ_c of the conducting material randomly distributed within the insulating medium, below which the overall conductivity is zero. In other words, for volume loadings φ_f smaller than the percolation threshold, that is for

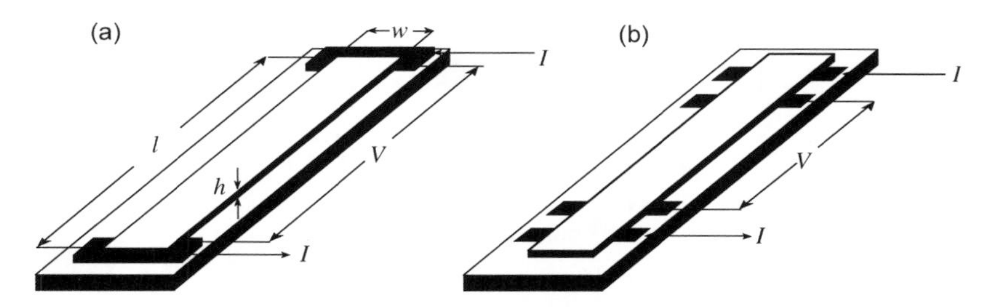

Figure 16: Methods used to determine the volume resistivity of conductive adhesive strips with thickness h, width w, and length l, printed on (a) two-point and (b) four-point probes.

Table 1: Volume resistivity, ρ_v, of metals, electrically conductive adhesives, and insulating materials.

Materials	ρ_v (Ω cm)	Materials	ρ_v (Ω cm)
Silver	1.6×10^{-6}	Silver-filled inks	1.0×10^{-4}
Copper	1.7×10^{-6}	Silver-filled polyimides	5.0×10^{-4}
Gold	2.4×10^{-6}	Silver-filled epoxies	1.0×10^{-3}
Aluminium	2.8×10^{-6}	Graphite	1.3×10^{-3}
Molybdenum	5.4×10^{-6}	Low-end silver epoxies	1.0×10^{-2}
Tungsten	5.5×10^{-6}	Graphite-filled coatings	$10-10^{2}$
Nickel	7.8×10^{-6}	Polyamide	10^{13}
Palladium	1.1×10^{-5}	Oxide-filled epoxies	$10^{14}-10^{15}$
Platinum	1.1×10^{-5}	Unfilled epoxies	$10^{14}-10^{15}$
Tin–lead solder	1.5×10^{-5}	Dielectric polymers	$10^{16}-10^{17}$
Au–Si eutectic	2.5×10^{-5}	Poly(tetrafluoroethylene)	10^{18}

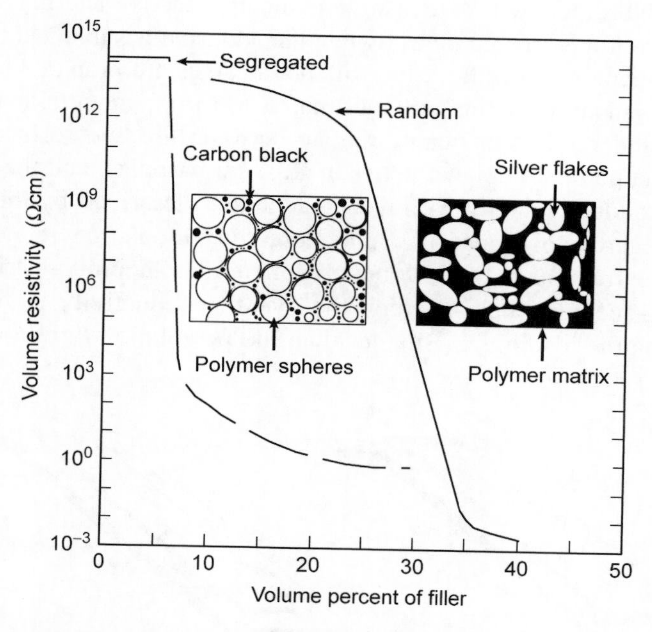

Figure 17: Variation of volume resistivity as a function of volume percent of filler for random and segregated conductive adhesive compositions. The exploded designs reflect the arrangements of the conductive particles in the two networks.

$\varphi_f < \varphi_c$, the probability that a conductive path will cross the thickness of the sample is zero, because the formation of such conductive channels requires uninterrupted contacts between the particles of metallic filler. The critical concentration corresponds to a transition from isolated aggregates of particles to an "infinite" cluster spanning the whole lattice. A continuous electrical path can be produced through either a random network of conductive particles or a segregated system, the former being the most common morphology encountered with metal-filled epoxy or polyimide adhesives. Figure 17 illustrates the geometry of the random and segregated structures and the change of the volume resistivity as a function of the percentage by volume of filler.

The conditions necessary for the segregation to appear in a given filler-matrix combination were reviewed by Kusy who underlined the importance of the respective radii of the two materials [28]. When spherical particles of polymer (rubber or thermoplastic resin) of radius R_p are mixed with smaller spheres of a conductive filler of diameter R_f, the higher the ratio R_p/R_f the lower the critical volume fraction of filler. This can be explained by the restricted volume that the conductive particles can occupy and by the fact that a partial coverage of the surface of the polymer spheres by a thin layer of conducting material is sufficient to build a conductive path. The particle size ratio R_p/R_f plays a significant role in the case of the segregated model because of the theoretical hyperbolic function $R_p/R_f = f(\varphi_f)$ indicating that the conductivity can be achieved at a normalized filler content of about 0.05 when the ratio R_p/R_f is in the range of 40–80. In contrast, the electrical continuity of the random packing is governed by the probability of forming a continuous network, which is related to the statistical average number of contacts between contiguous particles. This means that the same conductivity level requires 25–40 vol% of randomly distributed conductive particles in most die attach adhesives. The conductivity threshold depends on at least seven factors: the particle size distribution, particle shape, metal penetration, thickness of the oxide layer, electrostatic attraction, viscosity of the organic resin, and polymer bead shear [29].

Except for some anisotropic adhesive films, the resins used to prepare most silver-epoxy adhesives are low or medium-viscosity fluids that act as interstitial binding materials preserving the integrity of the closely-packed random conductive lattice. An adequate resin viscosity and a clean filler surface are prerequisite to produce homogeneous and highly conductive adhesive pastes. The former condition can be managed to avoid metal sedimentation by adjusting the initial viscosity and eventually adding thixotropic agents. On the other hand, the native oxide layer that forms on the surface of most metals contributes to the reduction of electrical conductivity, and this requires that the electrons be able to jump this insulating gap between the particles. Also, the good wetting properties of epoxy resins and polyimide solutions should provide a 5–10 nm thick insulating coating of organic material that impedes actual contact between the conductive particles. It has been suggested that metal particles and fibres mainly conduct by electron hopping although physical contact between contiguous particles is also possible in highly loaded adhesives [30].

Whatever the mode of conduction, the electron mobility augments when the ratio of the surface area to the volume of filler particles is increased. In this regard, the critical volume fraction of filler for electrical continuity decreases when elongated flakes, needlelike, or metallic fibres are used, instead of spherical or cubic particles. However, as the current flow only occurs over the very small area of each point of contact, the volume resistivity of the adhesive composition is higher than that of the metal filler. The following data are typical: silver, $\rho_f = 1.6 \times 10^{-6}\ \Omega$ cm; epoxy resin, $\rho_p = 5 \times 10^{14}\ \Omega$ cm; conductive epoxy loaded with 75 wt% silver flakes, $\rho_a = 1 \times 10^{-3}$–$5 \times 10^{-4}\ \Omega$ cm. The curves plotted in Figure 17 show that the composite materials change from insulators to conductors over a narrower range of filler concentration for the segregated systems than for the randomly dispersed materials. This sharp decrease in resistivity reflects the formation of a conductive network that has been treated as a percolation process.

The relation between the statistical probability P_c of forming a conductive lattice, the critical number of contacts N_c between contiguous particles, and the coordination number Z, which is the maximum number of possible contacts allowed by an ideal crystallographic arrangement, is $N_c = P_c Z$ [31]. It has been pointed out that N_c remains virtually constant ($N_c = 1.5$) whatever the geometry of the conductive particles, even for a random dispersion of spheres in the matrix. Obviously, the average number of contacts N per particle is related to the volume fraction of filler by a general function $N = f(\varphi_f)$ which depends on Z, P_c, and the maximum packing fraction φ_m. At the critical loading for the formation of a conductive lattice, Bigg shows that Jantzen's model

$$\phi_c = \frac{1}{1 + \frac{Z}{N}\left[\frac{1-\phi_m}{\phi_m}\right]} \tag{1}$$

leads to a critical volume fraction φ_c equal to 0.305 representing the percolation threshold at which the drop in resistivity starts, when the values of the different parameters are $N_c = 1.5, Z = 6$, and $\varphi_m = 0.637$ [30]. As a general guideline, it can be expected that the value of φ_c is in the range of 0.35–0.38 when the formation of the conductive network is completed and the lowest resistivity is achieved. The different approaches considering the electrical conductivity under qualitative and quantitative aspects have been previously discussed [4]. A percolation threshold at about 25 vol.% silver is a typical value for the commercial silver-filled epoxy adhesives, which exhibit a volume resistivity lower than 10^{-3} Ω cm at a silver content of 25–30 vol.%.

6.3.3. Thermal Conductivity

For a flat slab specimen of thickness h (m), the thermal conductivity λ, expressed in W m^{-1} K^{-1}, is the heat flux per unit of temperature gradient in the direction perpendicular to an isothermal surface and is a material constant defined by the one-dimensional Fourier equation $\lambda = Qh/A(T_1 - T_2)$ where Q is the time rate of the heat flow (W), A the area (m^2) on a selected isothermal surface, and T_1 and T_2 the temperatures of the hot and cold surfaces, respectively, and h the sample thickness (m). Several steady-state or transient methods are available to measure the thermal conductivity of organic and inorganic materials [4]. The guarded-hot-plate (ASTM F433), heat-flow (ASTM C518), and Colora thermoconductometer methods are accurate for thick plastic samples with thermal conductivity of 0.1–10 W m^{-1} K^{-1}. In electronics, steady-state techniques suffer from inaccuracies, primarily due to the small thickness of the adhesive layer and to the thermal resistance in the two interfacial regions. Contact and non-contact transient systems have been developed to measure the time–temperature dependence of the heat flow through the adhesive film. Flash radiometry, transient hot wire, and heat-generating test chips are well adapted to measure the bulk thermal conductivity and thermal resistance of thin adhesive layers. The thermal conductivities λ of various metals, oxides, ceramics, and filled polymers are listed in Table 2.

The data of this table show that unfilled polymers and plastics have rather low λ values and are very good thermal barriers or insulators. Furthermore, the thermal conductivity of most polymers falls in the tight range indicated for unfilled epoxy resins, about 0.2 W m^{-1} K^{-1}, which is more than three orders of magnitude less than that of silver. Currently, the heat dissipation of advanced CMOS silicon chips approaches 10–30 W cm^{-2} and exceeds 100 W cm^{-2} for the power devices used to control heavy electrical machines. This means that dissipation of heat is a major concern because semiconductors have a maximum operating temperature. In the case of plastic-encapsulated dice bonded with organic adhesives, the polymers used in intimate contact with the electronic devices require a high thermal conductivity to transfer the heat generated at the junction to the surrounding atmosphere. Polymers filled with metal powders or flakes give adhesives that conduct both electricity and heat with a 10-fold improvement in heat dissipation capabilities. However, the thermal conductivity of silver-filled epoxies and polyimides is still at least one order of magnitude below the values of solder alloys. This means that the thermal resistance $R\Theta_{JA}$ from the die surface to the ambient of a silver-filled adhesive is 1–2°C W^{-1} higher than that of a solder [32].

Table 2: Thermal conductivity λ of metals, oxides, ceramics, conductive adhesives, and unfilled polymers.

Materials	λ (W m^{-1} K^{-1})	Materials	λ (W m^{-1} K^{-1})
Diamond	2000	Steel	43–70
Boron nitride	1200	Eutectic solders	34–51
Silver	420–430	Sapphire	35
Copper	380–400	Boron carbide–Al_2O_3	30
Gold	318	Boron nitride pyrolytic	29
Au–Si eutectic	294	Aluminium oxide 94%	28
Beryllium oxide	294	Si_3N_4–Al_2O_3	25
Aluminium	200–237	80% Ag/epoxies	2.5–5.0
Aluminium nitride	170–220	50% Al/epoxies	3
Silicon	149	Silicon dioxide	1.67
Si–SiC	130	75% Al_2O_3/epoxies	1.4–1.7
Silicon carbide	90–120	50% Al_2O_3/epoxies	0.5–0.7
Nickel	91	25% Al_2O_3/epoxies	0.3–0.5
Boron carbide	80	Unfilled epoxies	0.1–0.3
Silver-glass	79	Low-density foams	0.02–0.05
Palladium	72	Air	0.0255

This is not a major concern in the case of plastic packages because they already exhibit values of $R\Theta_{JA}$ between 30 and 100°C W^{-1}.

Some applications, however, require adhesives that conduct heat but not electricity in order to bond power devices to heat sinks, metal substrates, or ceramic plates. The thermal transfer efficiency obviously depends not only on the thermal conductivity of the adhesive but also on the thickness of the bond line, and on the homogeneity of the adhesive layer, which must be free of voids and other defects. The addition of fillers to polymeric materials increases the thermal conductivity by at least one order of magnitude and, as shown in Table 2, the conductivity increases when the volume fraction of filler is increased. No organic material yet exists that combines the high thermal conductivity and the low electrical conduction of diamond, beryllia, and aluminium nitride. The inherent thermal conductivity of the filler is an important parameter, but the particle shape is often more significant in the sense that high aspect ratio particles increase the λ value more effectively than small cubic or spherical particles. For most applications requiring only thermal conductivity, the filler selected should be a good electrical insulator in order to avoid any ohmic contact. The commonly used fillers that meet these requirements are aluminium oxide or alumina (Al_2O_3), silicon oxide or silica (SiO_2), mica, and beryllium oxide or beryllia (BeO). As an example, the thermal conductivity of an Epon 828 epoxy resin-diethylenetriamine formulation is improved three to fourfold by adding silica, aluminium, or beryllium oxide fillers, but it is still far below the thermal conductivity of the filler alone.

Many equations have been proposed for the transport properties of two-phase systems and in-depth details of the existing models are discussed elsewhere [4]. Noticing that virtually all the early theories neglected the effects of the particle shape, their packing density, and the possible formation of anisotropic clusters, Lewis and Nielsen modified the Halpin-Tsai equation for the elastic modulus of composite materials by incorporating the maximum volume fraction of filler φ_m while still maintaining a continuous matrix phase [33,34]. Transposed to thermal conductivity Lewis and Nielsen's equation becomes

$$\frac{\lambda_a}{\lambda_p} = \frac{1+Ab\phi_f}{1+\Psi b\phi_f}; \quad A=k_E-1; \quad B=\frac{\frac{\lambda_f}{\lambda_p}-1}{\frac{\lambda_f}{\lambda_p}+A}; \quad \Psi = 1+\frac{1-\phi_m}{\phi_m^2}\phi_f \tag{2}$$

where λ_a, λ_p, λ_f are, respectively, the thermal conductivities of the adhesive, polymer matrix, and inorganic filler, A a function of the generalised Einstein coefficient k_E and depends primarily upon the geometry of the dispersed particles and B a factor which takes into account the relative conductivity of the two components. The factor ψ is related to the maximum filler content possible φ_m, and φ_f is the volume fraction of the filler particles in the adhesive. The authors calculated the typical values for the shape factor A and showed that for spherical or cubic particles the volume fraction of filler determines the thermal conductivity of the composite whereas for flakes, rods, or fibres, increasing the aspect ratio of the filler increases the thermal conductivity values of the filled polymer. Nielsen outlines that Eq. (2) includes virtually all laws of mixture from the ordinary "rule

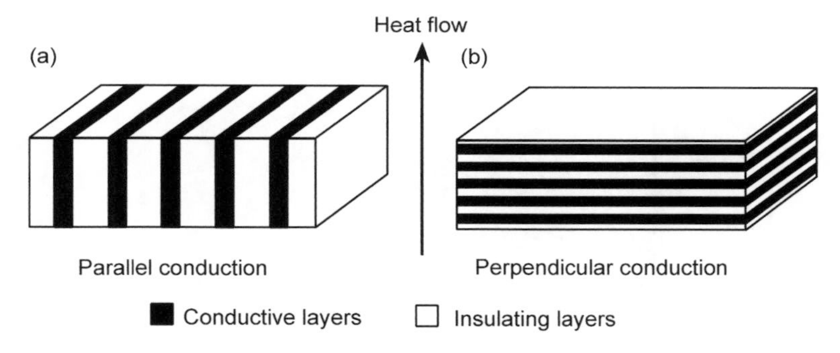

Figure 18: Schematic representation of the concept of parallel and perpendicular arrangements as ideal planes of alternate conductive and insulating layers either parallel or perpendicular to the heat flow.

of mixture" often referred to as the parallel conduction model described by the equation $\lambda_a = \lambda_f\varphi_f + \lambda_p\varphi_p$ to the inverse rule of mixture also known as the perpendicular plate model given by the relation $1/\lambda_a = \varphi_f/\lambda_f + \varphi_f/\lambda_f$. Figure 18 sketches the concept of the parallel and perpendicular arrangements as ideal planes of alternating conductive and insulating layers either parallel or perpendicular to the heat flow. These equations represent the upper and lower bounds of the thermal conductivity, respectively.

Being inexpensive and providing an excellent shear strength, alumina is broadly used to formulate thermally conductive adhesives. The best solventless epoxy adhesives contain about 70% of aluminium oxide and give thermal conductivities in the range of 1.4–1.7 W m^{-1} K^{-1}. These values are 8–10 times greater than for the unfilled epoxy resins, but are still much lower than for pure metals or solders. Nevertheless, Bolger and Morano point out that the heat flow is adequate for bonding most electronic components [32]. For example, an adhesive with a thermal conductivity of 1.6 W m^{-1} K^{-1} and a bond thickness of 75 μm would be able to transfer about 20 W cm^{-2} with a ΔT of 10°C above the heat sink temperature. In the same conditions, an unfilled epoxy resin, with a heat conductivity of 0.17 W m^{-1} K^{-1}, would cause a ΔT of almost 100°C. For at least two decades alumina and crystalline silica have been used to boost the thermal conductivity

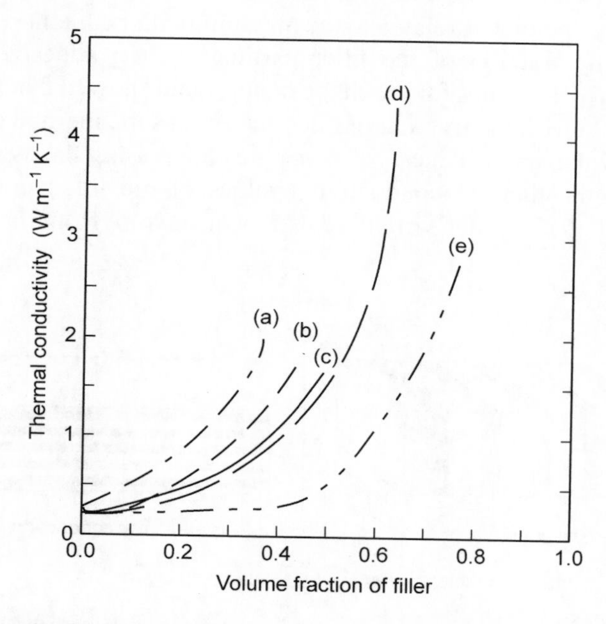

Figure 19: Comparison of the thermal conductivity data published in the literature: (a) adhesives filled with boron nitride [37,40]; (b) aluminium; (c) diamond [37–39]; (d) aluminium nitride [37–42]; (e) crystalline silica.

of epoxy resins. When highly conductive fillers such as boron nitride, aluminium nitride, and diamond powders become commercially available, these materials have been incorporated in adhesive compositions. The expected target was the attainment of λ_a values of at least 10 W m^{-1} K^{-1}, if not better. Such high values have been claimed for diamond-filled adhesives [36] but they remain currently questionable with regard to the experimental results summarized in the graph of Figure 19.

Curves of Figure 19 compare the data published for (a) boron nitride [37,40]; (b) aluminium; (c) diamond-[37–39]; (d) aluminium nitride [37–42]; (e) crystalline silica. It can be seen that, at 45 vol.%, the maximum thermal conductivity achieved with diamond powder is 1.5 W m^{-1} K^{-1}, while crystalline boron nitride at 35 vol.% affords 2.0 W m^{-1} K^{-1}. The thermal conductivity of silver-filled adhesives was studied by using silicon test chips attached to copper and molybdenum substrates [43]. The authors outline the importance of the shape factor A, related to the aspect ratio of the particles, to achieve the highest level of thermal conductivity. Another study reports the variation of the effective thermal resistance, between a test chip and the chip carrier, in relation to the volume fraction of silver and the thickness of the bond layer [44]. The ultimate value of bulk thermal conductivity is 2 W m^{-1} K^{-1} at 25 vol.% silver. However, the effective thermal conductivity, calculated from the thermal resistance measurements, is only one-fifth of the bulk value when the silicon chip is bonded to a copper substrate.

Another work provides numerical examples using epoxy and silicone resins with thermal conductivity 0.15–0.2 W m^{-1} K^{-1} [45]. The addition of 0.25 volume fraction of silver to these resins results in an increase of the bulk thermal conductivity up to 2.46 and 1.86 W m^{-1} K^{-1} for the epoxy and silicone adhesives, respectively. In contrast, the normalized thermal resistance declines from 1.00 to 0.52 cm^2 K W^{-1} for the epoxy but increases from 1.00 to 1.01 cm^2 K W^{-1} for the silicone. These calculated data are confirmed by a measurement of the thermal resistance of silver-filled epoxy adhesives showing that the initial thermal resistance of the unfilled resin is lower than that of many of the silver filled materials. There seems to be an optimum volume fraction of approximately 0.15 corresponding to a minimum thermal resistance lower than 1 cm^2 K W^{-1}. Somewhat better values were obtained with two commercial adhesives with thermal conductivities of 3.4 and 1.6 W m^{-1} K^{-1}, resulting in thermal resistances of 0.75 and 0.95 cm^2 K W^{-1}.

The conclusion that can be drawn from these experiments is that the use of highly priced fillers such as diamond powder does not improve the thermal conductivity better than less expensive materials such as aluminium nitride, boron nitride, boron carbide, or silicon carbide. Within certain limits, the higher the λ value of the filler particles, the higher the thermal conductivity of the adhesives with respect to the λ_f/λ_p ratio that exhibits a favourable optimized value at about 100. This means that fillers with a thermal conductivity in the range of 20–50 W m^{-1} K^{-1} are particularly suitable for the preparation of good thermally conductive adhesives. The results of experiments performed by loading an epoxy resin with silver flakes, aluminium spheres, diamond powder, and alumina particles help to clarify the actual thermal conductivity achievable with adhesive films [46]. The thermal conductivity measurements in decreasing order are 3.8, 2.6, 2.4, and 0.8 W m^{-1} K^{-1} for silver, aluminium, diamond, and alumina, respectively. From these data, the author suggests that the extrapolated λ value of the neat diamond powder is not 2000 but only 300 W m^{-1} K^{-1} because of the impurities included in the synthetic diamond. Maximum electrical and thermal conductivities are achieved by using silver flakes with a relatively high aspect ratio. With silver-filled epoxies and polyimides, an electrical conductivity of 1 to 5×10^{-4} Ω cm can be reliably obtained. For electrically insulating adhesives, boron nitride and aluminium nitride particles provide a thermal conductivity of 2.5 W m^{-1} K^{-1} at a loading level of 55 vol.%.

6.4. Material Properties

The first specification for the selection and use of organic adhesives in hybrid microcircuits was reported in 1983 [47]. Application of this qualification method to commercially available gold-filled epoxies was described 3 years later [48]. MIL-A-87172 of MIL-STD-883C, method 5011-2 released in 1989, establishes the evaluation and qualification requirements for the polymeric adhesives used in military hybrid circuits. The last version of MIL-STD-883D published in 1991 is currently used as a guide for the die attach materials utilized for non-military applications of hermetic packages as well [49]. This specification does not include any test methods for the devices encapsulated in plastic packages which, according to the final use, rely upon JESD-22 and JESD-26 test standards of the Joint Electronic Devices Engineering Council. The former specifies the procedures for plastic packaged devices intended for car and transportation markets, whereas the latter reports the sequences of tests applicable to high-reliability plastic encapsulated devices for commercial applications in severe environments. In the following sections, some important qualification standards are discussed, even if some criteria, such as the 15,000*g* acceleration test, are not used to qualify general-purpose adhesives.

6.4.1. Properties of Uncured Adhesives

6.4.1.1. Materials

Being composite materials, prepared by loading an organic matrix with very large amounts of inorganic fillers, typically from 70 to 80% by weight, the appearance and

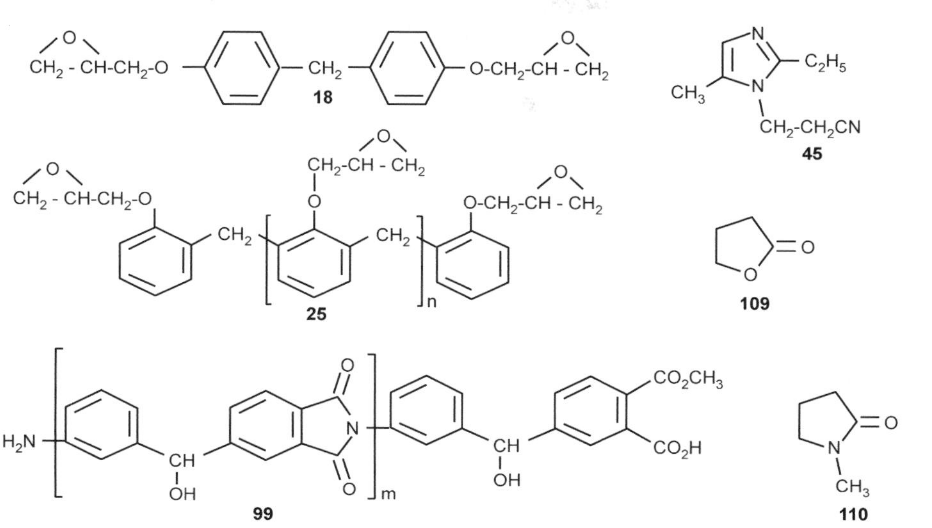

Figure 20: Chemical formulae of the organic compounds used to prepare a two-part silver-filled epoxy adhesive and a series of polyimide adhesives loaded with aluminium, silver, gold, diamond, or aluminium nitride.

uniformity of the adhesive pastes may change during storage. In particular, the components of two-part adhesives pastes have to be of uniform consistency, free of foreign materials, and any filler must remain homogeneously dispersed and suspended during the required pot life when inspected at 30 times magnification. The electrically conductive fillers are gold, silver, or alloys of silver or gold, or other precious metals. The specification excludes metals such as copper, nickel, tin, and solder alloys. The implementation of these tests to actual adhesives has been illustrated by using the two proprietary compositions sketched in Figure 20 [4].

The epoxy adhesive is a two-part system whose first container comprises equal weights of bisphenol-F diglycidyl ether **18** and epoxydized novolac resin **25**, and 144 parts per hundred resin (phr) of silver flakes with average particle diameter 5 μm. The composition is then homogenized for a few hours in a three-roll mill. The second container is prepared by mixing 100 parts of *N*-(2-cyanoethyl)-2-ethyl-4-methylimidazole **45**, 108 parts of γ-butyrolactone **109**, and 739 parts of silver flakes (average particle diameter ≤ 2 μm). The two containers are degassed in partial vacuum if the adhesive is dispensed with an automatic mixing-metering device. The two compositions are designed to provide a convenient 40:60 mixing ratio leading to the adhesive composition referenced IP 670 in the following discussion. A series of electrically and thermally conductive polyimide adhesives are prepared by using a solution containing 60% by weight of polyimide oligomer **99** (IP 605 with $m = 9$, corresponding to a number average

Table 3: Composition and properties of electrically and thermally conductive adhesive pastes based on polyimide oligomer 99 (IP 605).

Trade marks[a]	IP 675	IP 680	IP 685	IP 690	IP 695
Filler	Al	Ag	Au	AlN	Diamond
Polyimide IP 605	100	100	100	100	100
Solvent (phr)	106	100	100	133	133
Fillers[b] (phr)	80	400	547	400	400
Solid content[c]	63.5	77.5	86.5	79.1	79.1
Density (g cm^{-3})	1.34	2.65	3.6	2.0	2.5
Viscosity (Pa s)	6	25	60	80	100
Filler after cure[d]	46	81	83	80	80

[a] Trade marks of IFP-Cemota polyimides.
[b] This item comprises the main filler and, if any, silicon dioxide thixotropic agent.
[c] Percent by weight of polyimide and filler in the adhesive composition.
[d] Percent by weight of fillers after curing the adhesives for 30 min each at 150 and 275°C.

molecular weight of 2.56 kg mol^{-1}) in *N*-methylpyrrolidone **110**. When cured at 300°C, the degree of polycondensation increases to about 100 and the molecular weight rises to 25 kg mol^{-1}. According to the application, the base resins are loaded with aluminium, silver, gold, aluminium nitride, or diamond and the pastes are processed in a three-roll mill for 10 h. The composition and physical properties of these polyimide adhesives are listed in Table 3.

6.4.1.2. Viscosity and Rheology

The viscosity of adhesive pastes, which is specified in the supplier's document, is determined using any acceptable method. For example, it can be measured with a cone-and-plate viscosimeter at a constant temperature and a given shear rate. The thixotropy of the paste is not taken into account in the qualification standard. Viscosimetric measurements are used to control the raw materials, the changes due to incorporation of metal or inorganic fillers, and the effect of thixotropic agents. The final goal for manufacturers of conductive adhesives is to supply materials with constant batch-to-batch quality. The rheological behaviour of silver-filled adhesives has been analyzed in terms of frictional forces between the flake surfaces [50]. This means that the larger the flakes, the higher the frictional forces, and the higher the viscosity. Experimental data show that the viscosity increases from 1 to 7 Pa s when the tap density of silver flakes decreases from 4.3 to 3.1 g cm^{-3}. Also, the electrical conductivity increases by a factor of 4–5 for an increase of the apparent density of 25–30%. Another study suggests that both the flake size and the specific surface area have a combined effect on the viscosity [51]. Thus, a viscosity of 26 Pa s can be achieved with a median flake size of 7 μm and a surface area of 1.3 $m^2 g^{-1}$ or with a flake size of 12 μm and surface area of 0.9 $m^2 g^{-1}$.

Adhesive pastes are non-Newtonian fluids whose viscosity depends upon temperature, time, and shear rate. They are applied to the substrates by means of either stamping (pin transfer), screen printing, or dispensing. The performance of the last technique depends on a key factor: the paste rheology. There exists at least six semi-empirical models to describe the rheological response of non-Newtonian fluids [4]. For shear-thinning fluids (thixotropic materials), it has been reported [52] that the relation between viscosity and shear rate is best described by Carreau's model, which can be written as $\eta = \eta_0[1 + (\lambda\gamma)^{0.5}]^{(n-1)/2}$ where η_0 is the zero shear viscosity, γ a characteristic time constant, and n a dimensionless power law index.

The incorporation of fillers into organic resins results in increased viscosity with the maximum loading of filler being restricted by the permissible working viscosity, which in turn depends on the dispensing equipment and method [53]. The decrease in viscosity of conductive adhesives with increased shear rate is a time-dependent property, known as shear-thinning or thixotropy. Some fillers, in particular silica particles, exert an effect that enhances the thixotropy already provided by many silver flake varieties. The metal and inorganic particles dispersed in the organic resin affect the rheological properties by their nature, size and shape, thus allowing the desired thixotropy to be engineered.

Thixotropy is generally determined using the thixotropic loop test, where the shear rate is continuously ramped from zero to a maximum value, depending on the measuring device and on the rheological properties of the adhesive as well, and then ramped back to zero. The competition between the kinetics of structural breakdown, due to the increasing shear rate, and the kinetics of re-aggregation is such that the shear stresses are higher in the increasing than in the decreasing ramp. The region in between the two ramps is the thixotropic loop. It provides valuable information on the rheological behaviour of the materials that need some time at rest to recover their original viscosity. This is illustrated by comparing in Figure 21 the Newtonian behaviour of the polyimide oligomer **99** in solution in NMP (straight line (a)) to the thixotropic loop exhibited by its silver-loaded version IP 680 (curve (b)).

The rheological curves of thixotropic fluids, such as IP 680 in Figure 21, show that there is a yield stress τ_0 to overcome before shear can take place. This phenomenon is taken into account in Casson's model represented by the equation $\tau^{0.5} = (\tau_0)^{0.5} + [\eta_0(dy/dt)]^{0.5}$, where the yield stress τ_0 is usually determined by extrapolating to zero shear the curve representing the variation of the shear stress at different shear rates. For the sake of convenience, conductive adhesive pastes are often characterized by a thixotropy index, which is the ratio of the viscosities

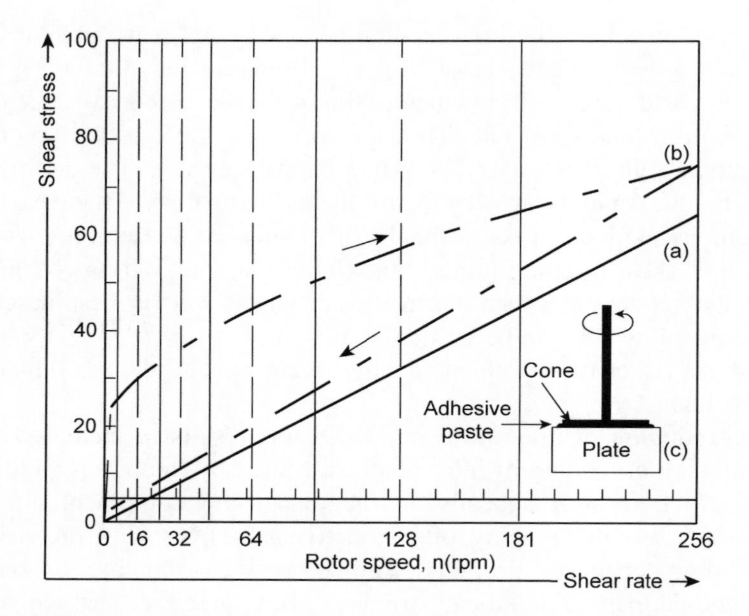

Figure 21: Rheological behaviour of: (a) Newtonian solution of oligoimide 99 (IP 605) in NMP and (b) thixotropic adhesive paste IP 680 prepared by loading this solution with silver flakes. Insert (c) represents the cone and plate viscosimeter principle.

measured at low shear rate, typically at a rotation speed of 0.5 rpm, and at high shear rate, e.g. 5, 10, or 30 rpm.

A feature related to viscosity is the resin bleed effect, which has been observed with several die attach materials. The event occurs when low molecular weight epoxy resins, reactive diluents, or low-viscosity curing agents migrate away from the adhesive layer to the surface around the chip. A definite degradation in the bond strength has been observed when the gold balls are wedged through the bleed area, compared to bonds made on pure metal surfaces [54]. In the resin bleed test, the silver-filled adhesive paste is applied on the surface of the lead frames or ceramic substrates as a drop, 2 mm in diameter, and the length of the resin bleed is measured after leaving the paste at rest for a few hours. Particularly susceptible to resin bleed-out are ceramic chip carriers, and the severity of the phenomenon was investigated with epoxy and polyimide adhesives coated onto different ceramic substrates [55]. The most serious phase separation was observed with epoxy adhesives deposited on a ceramic with a porous surface that appeared to contribute substantially to the resin spreading. Resin bleed-out can be eliminated by combining high temperature (200–250°C) and high vacuum to clean the substrates.

Other distinctive elements connected to thixotropy are tailing and cobwebbing. The processability of silver pastes depends to a large extent on their thixotropy because high viscosity at low shear rates improves the sagging property, while low viscosity at high shear rates helps to limit the defects due to tailing or cobwebbing. The formation of a cobweb not only gives irregular drops of silver paste but very often includes air bubbles under the tail after sagging. In the case of multineedle dispensers (16–25 hole nozzles), cobwebbing may cause adjacent drops of silver paste to be connected by the tail formed during the operation. Tailing and cobwebbing tests can be performed with a simple set-up. A pin, 2.8 mm in diameter, is plunged into the paste to a depth of 5 mm. Then, the pin is pulled up at a constant speed and the height where the cobweb is cut is measured and drawn versus the pull-up speed. In a recent study, an oscillating rheometer was used to measure the tangent of the phase angle (tan δ) between the sinusoidal stress applied to uncured adhesives and the resulting sinusoidal strain [56]. In these experiments stringiness was estimated by separating the plates of the rheometer and measuring the distance between the plates where the string broke.

These two aspects, bleeding and tailing, have also to be considered for using adhesives in surface-mount technology (SMT). Various non-conductive adhesives are currently used to hold the components during the assembly and wave soldering operations required for bonding surface-mounted devices to PCBs. For this application, the bleeding and stringing effects have to be carefully controlled because they are prominent factors for producing dots of adhesive with consistent size and shape [57]. Stringing is the result of a combination of factors including the viscosity of the adhesive, its wetting properties, and the relevant machine parameters, i.e. the nozzle diameter, the distance between the nozzle and the board, the dispensing pressure, and the dispensing time. Stringing can be circumvented when the diameter of the dispensed dot of adhesive is twice the internal diameter of the nozzle with a further parameter being the ratio of the internal diameter of the nozzle to the distance between the nozzle and the board. The wet adhesion of the glue with the PCB terminations must be better than its adhesion to the nozzle to avoid the formation of the string as the nozzle lifts. Additionally, the larger the glue dot diameter, the higher the surface tension of the dot and the sooner the adhesive dot will pull off the nozzle. This can be achieved by using nozzles with a smaller internal diameter and by increasing the dispensing pressure, time, and temperature. By changing the nozzle distance, it is possible to dispense, all other parameters being constant, dots of the same volume but with different diameters and shapes.

6.4.1.3. Pot Life

One-part epoxy adhesives generally need to be kept frozen and allowed to come to room temperature before use. In this case, the pot life is the time where the adhesive is workable at the ambient. With two-part systems, the epoxy resin and hardeners are mixed just before use and the pot life is taken from that time. Any change in parameters such as viscosity, skin-over, or loss of bond strength may be used to determine the pot life. The qualification requirement is 1 h minimum and the change in viscosity as a function of storage time at a given temperature is often used to predict the period of time during which the adhesive may be safely used.

6.4.1.4. Shelf Life

According to their chemical composition, conductive adhesive formulations have to be preserved either at low temperature, typically − 40°C for one-part epoxies, or at room temperature. For example, IP 680 polyimide or two-part epoxies can be kept at 20°C for at least 6 months, whereas one-part epoxies have a shelf life of 2 months at − 10°C and 6 months at − 40°C. The minimum shelf life requirement is 6 months at the temperature indicated by the supplier. Material inspection includes the measurement of the pot life, corrosivity, volume resistivity at 25°C, and lap shear strength at 25°C. Typically pot life, bond strength, and volume resistivity have been used by adhesive manufacturers to determine the shelf life of the formulated materials. However, other methods such as DSC can predict the remaining shelf life by a quantitative measurement of the heat of reaction [58].

6.4.1.5. Infrared Spectroscopy

Infrared spectroscopy (IR) provides rapid access to the changes induced by chemical reactions within the organic matrix. The absorption bands characteristic of the epoxy groups and curing agents are altered in position or intensity if they react during storage. Any shift, disappearance, or introduction of absorption bands throughout the specified wavelength range (2.5–15 μm) mean that the whole batch will be rejected, but minor changes in peak intensity are allowed.

6.4.1.6. Corrosivity

It has been shown that some latent hardeners of the epoxy resins may corrode the aluminium wires or bonding pads. This corrosivity has been demonstrated for the complex of boron trifluoride and monoethylamine. In the specification NSA 77-25A, small dots of adhesive are applied to the aluminized side of a Mylar® polyester film and allowed to stand in the room ambient without cure. After 48 h the dots are removed from the film by washing with acetone and the requirement is that there are no changes in the light transmission of the film.

6.4.2. Properties of Cured Adhesives

All the other test requirements deal with the performance and reliability of the adhesive joint, starting with the cure schedule that very often conditions the quality of the attachment. The bond strength requirements have been established to ensure that the assembly can be subjected to the subsequent manufacturing steps such as wire bonding, moulding of plastic packages, or sealing of hermetic packages. Volume resistivity has to be determined for type I adhesives (electrically conductive), whereas thermal conductivity is the most adequate measure of the particle-to-particle contacts. Finally, a series of environment tests have been issued to check the resistance of the bonded devices used in the aerospace industry. However, these tests are also employed to evaluate the behaviour of all die bonding materials currently in development.

6.4.2.1. Adhesive Cure

The adhesive cure temperature of MIL-STD-883 is limited to a maximum of 165°C for a maximum of 4 h. This specification constitutes the requirements for the adhesives used in space applications, and the upper limit is set at 165°C to minimize the degradation of the wire bonds. The polyimide adhesives requiring temperatures as high as 275–350°C are therefore excluded from this specification. The cure schedule, reported by the supplier, has to be identical for all tests. DSC can be used to observe the exothermic cure reaction of thermosetting resins. Figure 22, curve (a), shows a typical DSC profile in the scanning mode. The area under the curve represents the total heat of polymerization, ΔH_T. A partially cured adhesive will exhibit a less intense signal (curve (b)) with a residual heat of reaction, ΔH_R. The degree of cure D_c is determined by the relation $D_c = (\Delta H_T - \Delta H_R)/\Delta H_T$.

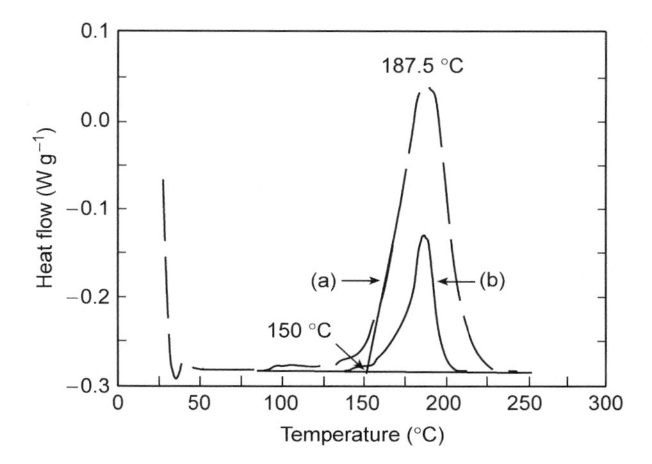

Figure 22: Differential scanning calorimetry profile of a silver-filled epoxy adhesive using a temperature ramp of 10°C min^{-1}. The area under the first scan (a) provides the total heat of polymerization ΔH_T, while the area under the second scan (b) gives the value of the residual heat of reaction ΔH_R of a partially cured material.

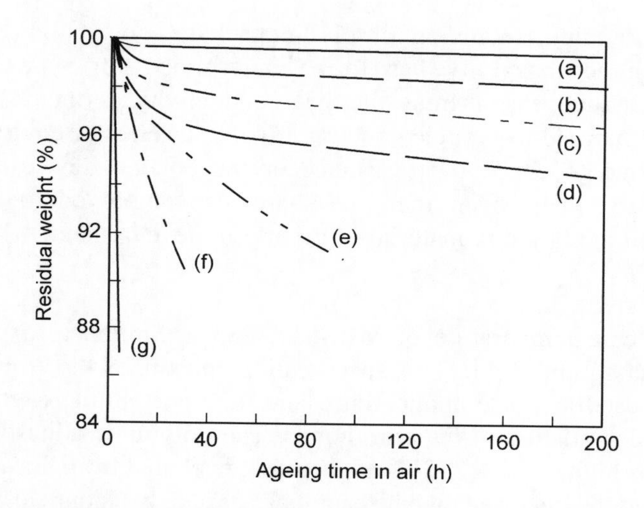

Figure 23: Isothermal ageing of conductive adhesives in air: gold-filled polyimide IP 685 at (a) 250, (b) 300, and (c) 350°C; silver-filled polyimide IP 680 at (d) 250, (e) 300, and (g) 350°C; silver-filled epoxy IP 670 at (e) 250 and (f) 300°C.

6.4.2.2. Thermal Stability

A convenient way of assessing and comparing the thermal stability of the conductive adhesives is to measure the total weight loss as a function of time and temperature. The recommended temperature is 150°C because this is the maximum that hybrid circuits generally encounter in either screen testing or actual operation. A plot of the isothermal weight loss as a function of time up to 1000 h gives a good picture of the thermal stability and the amount of volatile materials that can be expected to outgas. It has been demonstrated that solventless epoxies pass this test with a weight loss of approximately 0.5% [59]. Polyimide adhesives are generally subjected to isothermal gravimetric analyses at much higher temperatures. Figure 23 compares the behaviour of IP 670 silver-epoxy, IP 680 silver-polyimide, and IP 685 gold-polyimide when they are heated in air at 250, 300, and 350°C.

The first observation is that gold-filled polyimide IP 685 is more stable than the silver-filled version IP 680 at any temperature. After 200 h at 250°C, the weight loss of IP 685 is negligible, while the residual weights are 98 and 96% at 300 and 350°C, respectively. During the same period, the weight loss of IP 680 is 6% at 250°C, 9% after 100 h at 300°C, and 14% in less than 1 h at 350°C. For the silver-filled epoxy adhesive IP 670, the residual weights are 91% after 90 h at 250°C and 89% after 38 h at 300°C.

Because of the long duration of this test, the thermal stability is determined by thermogravimetric analysis (TGA) in accordance with ASTM D3850 using 10 mg

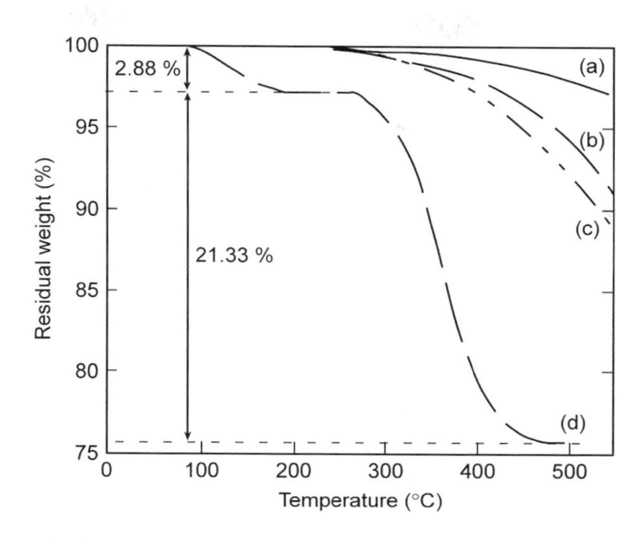

Figure 24: Dynamic thermogravimetric analysis of conductive adhesives: (a) Al-filled IP 675 in air and Au-filled IP 685 in argon; (b) Ag-filled IP 680 in argon; (c) IP 680 and IP 685 in air; (d) Ag-filled IP 670 epoxy adhesive in air.

samples tested from 25 to 350°C at a heating rate of 10°C min^{-1}. The specification requires that the weight loss must not exceed 0.3% at 250°C and 1% at 300°C. The TGA curves provide a quick picture of the onset of thermal decomposition of organic materials. It indicates the temperature at which volatile compounds are released, but it does not provide the same information as the isothermal weight loss of the previous test. A complete thermal assessment should therefore employ both tests and also DSC. Figure 24 shows the results obtained with aluminium- (IP 675), gold- (IP 685), and silver-filled (IP 680) polyimide adhesives, compared to silver-filled epoxy IP 670. The same trends have been observed by other research groups at the Sandia National Laboratories [60], Ablestik Laboratories [61], Epoxy Technology [62], and Amicon Corporation [63,64].

6.4.2.3. Filler Content

The filler content is determined using thermogravimetric analysis by heating the sample from the ambient temperature to 600°C in air. The temperature is maintained at 600°C until a constant weight is achieved. Figure 24 shows the TGA profile of uncured IP 670 silver-filled epoxy adhesive heated from 25 to 500°C. The initial weight loss (2.9%) between 100 and 250°C gives information on the amount of volatile materials evolved during the curing stage. The onset of rapid thermal decomposition is approximately 270°C and the residual weight left at

500°C (75%) corresponds to the amount of silver loaded into the epoxy adhesive composition.

6.4.2.4. Outgassing

A major concern in using organic adhesives to assemble semiconductor dice and hybrid microcircuits is the effect that the outgassed products have on the electrical performance of the devices and circuits. The main concern is water, although other constituents, released in smaller amounts, may be even more damaging than water. TGA curves of Figure 24 do not provide clear information on the outgassed products. The damages initiated by moisture on the chip devices and microcircuits are due to a combination of factors. They include the sensitivity of devices and circuits, the nature and integrity of the device passivation, the amount of moisture, the time and temperature of exposure, gases other than water, and ionic contaminants on the surface. In view of the number of parameters and the difficulty in controlling them, the industry and government agencies have settled on specifying the amount of moisture that is allowable in a hermetically sealed circuit package. The moisture requirement for Class B circuits has been set at 5000 ppmv (part per million in volume equivalent to 0.5% v/v) maximum and for Class S circuits at 3000 ppmv. Although hybrid circuits can meet these requirements, it is still not a guarantee that the circuit will be reliable. Other constituents that may evolve, such as ammonia, amines, ketones, alcohols, chlorinated hydrocarbons, hydrogen chloride, and boron trifluoride, must be reported, if detected, but quantitative requirements for these gases have not been established.

Moisture in itself, if pure, is probably not harmful but, in a packaged microcircuit, it is unlikely that water would remain pure because there are other constituents from the epoxy resin that could contaminate the circuit, including chlorides, metal ions, and amines. Moisture then acts as a medium dissolving and transferring these ions and contaminants to other portions of the circuit. The specific effects may be electrochemical corrosion of the aluminium metallization causing electrical shorts, deterioration of the wire bonds with increases in resistance or bond lifts, device leakage current, and metal migration. To avoid these failure mechanisms, other tests have been specified in MSFC-SPEC-592 including weight loss at elevated temperature, corrosivity, total ionic impurities, and concentrations of the Cl^-, Na^+, and K^+ ions. In recent years the manufacturers of adhesives have made significant improvements in reducing the amounts and types of outgassing products and ionic species in their formulations. Both epoxy resins and hardeners have been purified through distillation or extraction with solvents to remove chloride and other ionic contaminants. The manufacturers of hybrid circuits have also reduced outgassing by optimizing the vacuum bake schedule used prior to hermetically sealing the circuits. Vacuum baking at 150°C for 16–96 h has been found effective in removing most of the moisture and other volatile materials from the adhesive and the package as well. The analysis of the materials outgassed per Method 1018 of MIL-STD-883, following a 168 h bake at 125°C, shall be made by the users.

6.4.2.5. Ionic Content

Epoxies of the first generation used in electronic assemblies as adhesives and moulding compounds were commercial grades containing large amounts of sodium and chloride ions that are byproducts of the synthesis of diglycidyl ethers. It has been outlined that ionic contaminants not only adversely affect device performance, but are also the most frequent cause of component failure, through either direct shorting or corrosion effects [65]. The epoxies that are produced today for semiconductor and hybrid applications are purified to reduce the ionic content or are synthesised by procedures that avoid the generation of sodium chloride. The quantitative analysis of the individual ions in the adhesive may be performed by atomic absorption spectrophotometry or ion chromatography. Analysis is conducted using a powdered sample of cured adhesive that is extracted in deionized water for 24 h either at 100°C with full reflux or in a Teflon-lined Parr bomb heated at 121°C. The aqueous extracts are then analyzed with a Dionex ion chromatograph using a column for anions and a column for cations. In another method, the sample is burnt in the flame of a Wickbold oxyhydrogen blowpipe at a temperature of 2000°C, the combustion gas flow being flushed through the chromatographic medium. The mineralization is made according to ASTM D-2785 standard. Anions such as Cl^- and NO_3^- are analyzed with a Vidac-Anions column, whereas cations (Na^+, K^+, NH_4^+) are separated with a Mitsubishi SCK01 column.

Typical values for Cl^-, F^-, Na^+, and K^+ are less than 10 ppm for purified silver-filled epoxies and 1–5 ppm for polyimides. In addition, the following data have to be reported: the total ionic content as the specific electrical conductance, which must be $\leq$ 4.5 millisiemens per metre; the hydrogen content, which must be in the range of $4.0 \leq pH \leq 9.0$, and the presence of NH_4^+ cations if their level is more than 5 ppm. The determination of the total ionic content may be obtained by measuring the electrical resistivity of a water extract of the adhesive since ions are readily soluble in water. The test involves 3 g of ground sample that is heated for 20 h in 100 ml of boiling deionized water of known resistivity, and measuring the decrease in resistivity. The total ion content is then calculated according to Method 7071 of FED-STD-406 and reported as NaCl in ppm.

6.4.2.6. Bond Strength

A major requirement for devices attached with an organic material is an adhesive strength sufficient to last the life of the circuits and to withstand the environmental, thermal, and mechanical exposures during processing and testing. Generally shear strength is not a problem with epoxies because values over 20 MPa are easily attainable. However, to obtain reliable results, the properties of the materials and the process parameters must be carefully controlled. In particular, the surfaces to be bonded are thoroughly cleaned with both polar and non-polar solvents to remove all organic and inorganic residues. Because of its general use in the semiconductor industry to remove traces of organic materials, plasma cleaning is now widely accepted to produce clean surfaces before die attachment. In the case of gold-plated surfaces, the adhesive strength is improved either by abrading the gold layer or by using a primer or a coupling agent. The qualification of an adhesive for a given application requires a series of tests to determine the process parameters, the resistance to organic solvents, and the behaviour of the assembly during thermal ageing and environmental stressing. The procedures and conditions for the bond shear strength are given in MSFC-SPEC-592 for dice, substrates, and

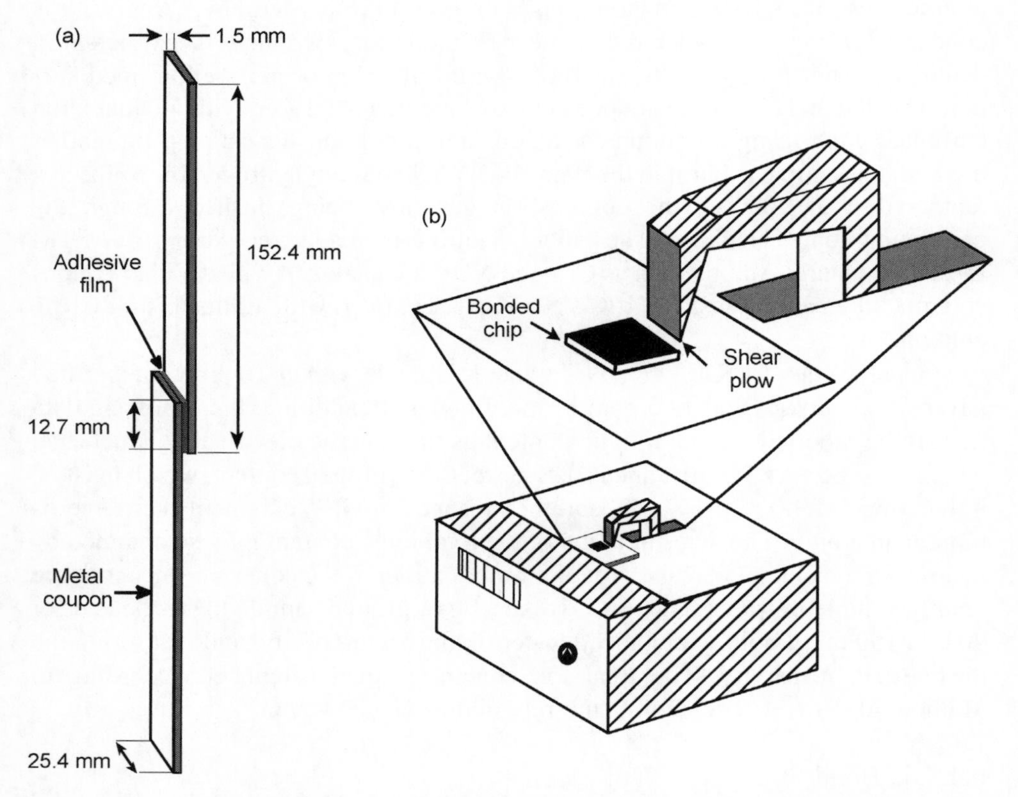

Figure 25: (a) Single lap shear configuration recommended in ASTM D1002 specification. (b) Principle of the die shear tester.

capacitors bonded with organic adhesives. It specifies measurements at room temperature, at 150°C, after solvent immersion, after temperature cycling, and after ageing for 1000 h at 150°C. For the last three conditions a 70–80% retention of the initial bond strength is required.

In electronics, the adhesive strength of bonded parts is determined primarily by applying an in-plane shear stress using both lap shear and die shear tests. The lap shear trial is performed by the vendor using standard aluminium tensile specimens in accordance with ASTM D1002. Figure 25(a) shows the single lap shear configuration recommended by this specification. Requirements for the lap shear strength are values $\geq$6.9 MPa at 25°C, $\geq$3.5 MPa at 150°C, and $\geq$80% of the initial strength after 1000 h at 150°C. The device shear tests are conducted by the user in accordance with MIL-STD-883, Method 2019 and performed with the die shear tester illustrated in Figure 25(b). Requirements for the device shear strength are at least 6.9 MPa at 25°C for dice and substrates, and 10.3 MPa for capacitors. These values may decrease to 3.5 and 5.2 Mpa, respectively, at 150°C. A 70% retention of the initial device shear strength is required after solvent immersion and after temperature cycling, and 80% after 1000 h ageing at 150°C.

Specimens for lap shear strength are prepared by applying the adhesive paste of tape between two aluminium coupons (2024-T3 alloy) whose dimensions are given in Figure 25(a). The assembly is cured in an air-circulating oven according to the appropriate thermal schedule. The lap shear strength is measured with a tensile machine at a pull rate of 1–2 mm min^{-1} and expressed in MPa on average of five specimens at least. Most commercially available silver-filled adhesives exhibit a lap shear strength in the range 6–20 MPa at ambient temperature, dropping to 5–10 MPa at 150°C for materials with a glass transition temperature higher than 180°C. Typical lap shear strength data for IP 670 silver-filled epoxy and IP 680 silver-filled polyimide adhesives are plotted in Figure 26, which shows that the adhesive strength considerably drops when the temperature approaches the T_g of the epoxy resin at 110–120°C.

Representative values obtained for the lap-shear strength of polyimide adhesives are: 8.5 MPa (IP 680-Ag), 7.9 MPa (IP 685-Au), 8.5 MPa (IP 690-AlN), 8.9 MPa (IP 695-diamond), and 16.1 MPa (IP 675-Al), all adhesives being cured at 275°C. It has been reported that the lap shear strength is a function of the adhesive thickness [66]. A maximum value is attained when the thickness is in the range of 35–75 μm and then it significantly decreases beyond 200 μm. In addition, the lap shear strength is reduced by 5–10% when the die size increases from 25 to 100 mm^2. The adhesive strengths, measured with different substrates commonly encountered in the semiconductor industry, indicate that bare silicon chips with rough surface give better values than metallized devices.

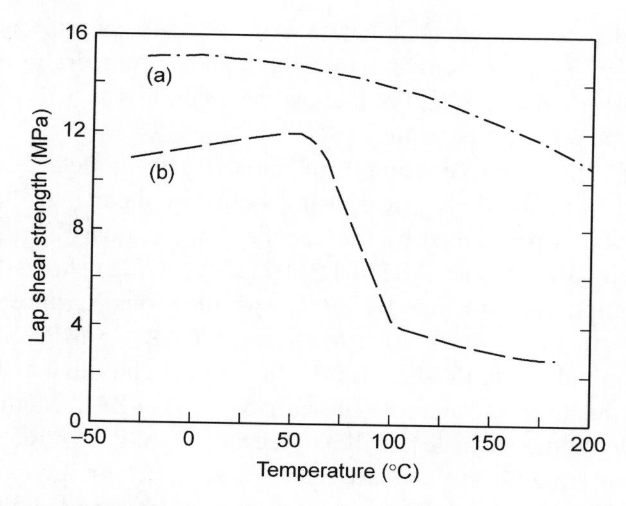

Figure 26: Variation of the lap shear strength as a function of temperature for conductive adhesives: (a) IP 680 silver-filled polyimide cured at 300°C; (b) IP 670 silver-filled epoxy cured at 120°C.

Lap shear test conditions are not really convenient in electronics involving small dice and two substrates with different thermal expansion coefficients (CTE). The stresses generated during the cure cycle by this CTE mismatch account for possible crack formation and delamination, which in turn may degrade the adhesive strength. The die shear strength method consists in the measurement of the force required to shear the die from the chip carrier. Although any die size can be used, a standard technique is to bond $1.27 \times 1.27\ mm^2$ non-functional silicon dice to silver-plated lead frames.

The dice and lead frames are first cleaned in stirred freon and dried at 100°C for 15 min. Drops of adhesive paste are dispensed onto the lead frames (20 positions) with an automated pneumatic syringe. The dice are then placed on top of the adhesive drops with a pick-and-place machine and the assemblies are cured either in a forced-air oven or on a heating block. The die shear strength of each individual die is measured with a die-shear tester working as shown in Figure 25. The die shear strength is calculated by averaging the results obtained from the 20 individual dice tested and expressed either as a force (N) or as a stress (MPa) if the bonded area is introduced into the calculation. The average values determined with polyimide adhesives are: 22 MPa (IP 680-Ag), 16.3 MPa (IP 685-Au), 22 MPa (IP 690-AlN), and 15 MPa (IP 695-diamond). For IP 680, the initial value of 22 MPa regularly decreases to 20.7 MPa at 250°C and then dramatically drops to 8 MPa in the region of the glass transition temperature (230°C).

The effect of isothermal ageing at 200 and 300°C on the die shear strength of commercial polyimide adhesives has been studied with small dice ($1.12 \times 1.12\ mm^2$) bonded to silver-plated copper lead frames, using the usual cure cycle of 30 min each at 150 and 275°C [67]. The results indicate that the die shear strength variation in the temperature range 20–250°C for the two adhesives exceeds the requirements of MIL-STD 883B. Using the same cure cycle with IP 680-Ag to bond $1.27 \times 1.27\ mm^2$ dice to silver-plated copper lead frames, it has been shown that the die shear strength, measured at ambient temperature, remains virtually constant after 100 h of ageing in air at 200°C whereas the adhesive loses 50% of the initial shear strength after 20 h at 300°C [4]. The adhesive strength of one-part and two-part silver-filled epoxies has been measured by using semiconductor devices and chip capacitors, ranging in size from 0.508×0.508 to $2.286 \times 2.286\ mm^2$, and bonded to gold-plated alumina substrates [68]. The general trend is that the die shear strength drops to about 92% of the initial value after 150–200 h at 150°C, and then decreases at a lower rate for 2000 h of ageing in air. The two

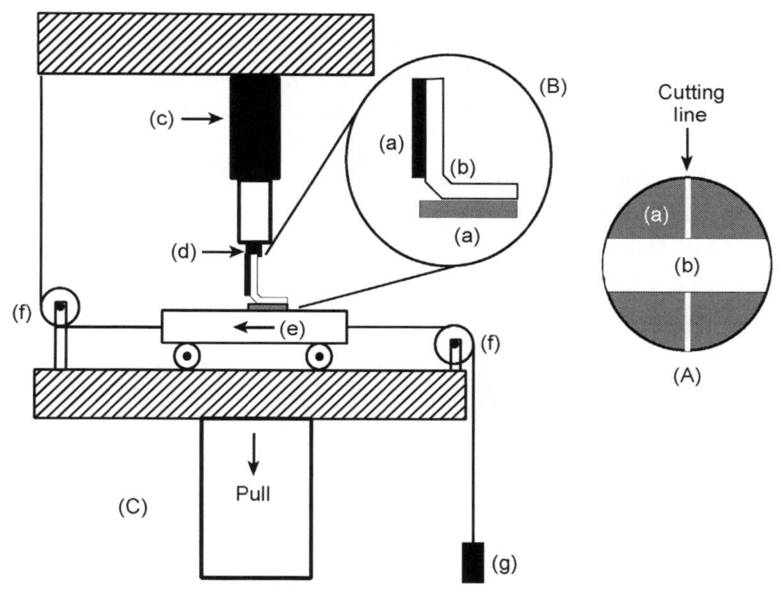

Figure 27: Measurement of adhesion by the peel test method using thin film specimens coated on silicon wafers and cured at 350–400°C. (A) A 1 cm wide strip is cut in the polyimide film. (B) The wafer (a) is cut and broken perpendicularly to the film (b). (C) One end of the specimen is fixed with a clip (d) into the upper jig (c) of an Instron testing machine. The other part lies on a mobile carrier (e) designed to maintain a 90° peeling by means of pulleys (f) and counterweight (g).

adhesives pass the strength requirements for long-term uses at 150°C with a safety margin of 2–3. A study performed with four epoxies and one polyimide shows that the die shear strength decreases during long-term ageing in air at 200°C with, however, the lowest strength degradation for adhesives with low glass transition temperature [69]. The authors suggest that this behaviour is related to the smaller residual stresses generated in low-T_g adhesive layers which keep 63–80% of their initial mean shear stress after 504 h at 200°C. For the epoxy with a glass transition temperature of 250°C, only 24% of the initial value remains after the same ageing test, while the polyimide with T_g of 200°C works better with a retention of 62%.

As stated earlier, adhesion is a major concern in electronic applications involving thin polyimide films either coated on hard substrates or laminated with metal ribbons. In these cases, neither lap shear nor die shear techniques allow the determination of the adhesion strength. This can be done by using either the 90° peel test or the island blister test whose principles are sketched in Figures 27 and 28. The 90° peel test provides reliable data for the measurement of "practical adhesion", especially useful for comparing the effect of surface treatments on the interfacial adhesion. The standard peel test procedure has been modified to determine the adhesive strength of thin polyimide films coated onto 10 cm silicon wafers. The equipment illustrated in Figure 27 maintains a 90° peel effort during the test conducted at room temperature with a constant rate of crosshead displacement of 2 mm min^{-1}.

A more sophisticated peel tester has been implemented to evaluate the adhesion of polyimide films at elevated temperature and in high-humidity environment [70]. The peel strength values measured in "dry" atmosphere rapidly degrades when the devices are exposed to humidity, temperature–humidity, or thermal cycle tests.

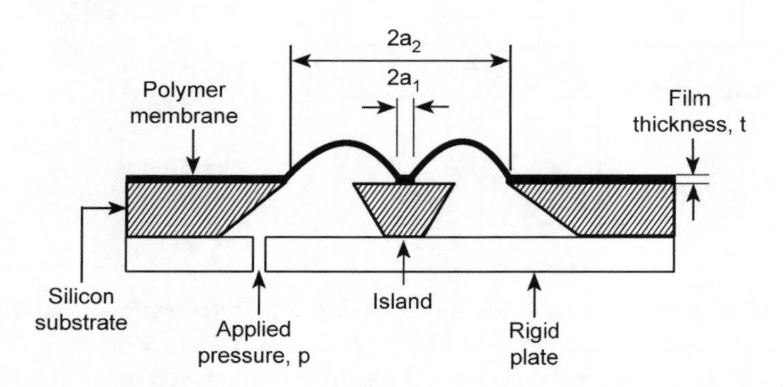

Figure 28: In the island blister test, the values of the diameters of the coating adhering part $2a_1$ and suspended membrane $2a_2$ are measured as a function of the pressure p necessary to peel the polymer.

As stated above, peel tests provide practical adhesion values that include polymer and substrate mechanical properties, stored stresses, plastic deformation, and other parameters. It has been demonstrated that an analysis of the peel test mechanics allows to extract the "fundamental adhesion" from the experimental data [71]. A method to calculate the interfacial fracture energy of a polymer bonded to a rigid substrate by using peel tests has also been presented [72].

Microfabricated structures based on suspended membranes have been developed to determine in situ the values of adhesive strength, film modulus, and residual stress [73,74]. Figure 28 shows that the island blister test consists of a suspended membrane of polymer with an island of substrate attached to the centre. Pressure is applied through holes patterned into the supporting plate and the pressure p at which the film begins to peel is observed with an optical microscope as a function of the radius of film still adhering to the island. Film thickness, residual stress, radius of suspended film a_2, and radius of adhered polymer a_1 are introduced in constitutive equations to calculate the debonding energy.

The blister test applied to metal/polyimide and polyimide/silicon interfaces has provided self-consistent values for the adhesion energy G_a calculated from five different equations [75]. G_a can be reliably determined from the pressure data alone, without using the blister geometry. The effect of cure temperature on the adhesion energy of PMDA–ODA polyimide **93** coated over silicon wafers illustrates the need for high temperature curing. G_a increases from 1.37 to 25 J m^{-2} for final cure temperature of 300 and 450°C, respectively. By contrast rigid BPDA–PPD polyimide **95** exhibits poor adhesion ($G_a = 0.1$–0.3 J m^{-2}) weakly dependent on the final cure temperature except on silicon-modified surfaces [76]. Experimental data also show that adhesion energies of PMDA–ODA are only 0.03, 0.52, and 13.8 J m^{-2} for Au, Cu, and Al, respectively. Further studies have shown that most of the energy expended is responsible for interfacial fracture, allowing the determination of a fracture energy that represents the true adhesion strength [77].

6.4.2.7. Coefficient of Linear Thermal Expansion

Silicon dice, lead frames, and ceramic or metal substrates have different coefficients of linear thermal expansion α that can cause significant normal and shear stresses after the adhesive layer has been cured, as well as throughout the life of the device during power cycling. The coefficient of thermal expansion must be determined from -65 to 150°C or to the glass transition temperature and above the T_g. The specification requires the determination of the CTE in accordance with ASTM D3386. This is done to incorporate the use of thermomechanical analyzers. The requirements are a CTE $\leq 6.5 \times 10^{-5}$ K^{-1} from -65°C to the glass transition temperature and a value $\leq 3.0 \times 10^{-4}$ K^{-1} above the T_g. The coefficients of linear

Table 4: Coefficient of thermal expansion α at 25°C of the metals, oxides, ceramics, and polymers commonly used in the fabrication of microcircuits and electronic components.

Materials	α (K^{-1})	Materials	α (K^{-1})
Silicon dioxide	4.2×10^{-7}	Beryllium oxide	7.5×10^{-6}
Silicon nitride	1.8×10^{-6}	Iron	1.2×10^{-5}
Silicon	3.0×10^{-6}	Gold	1.4×10^{-5}
Silicon carbide	3.8×10^{-6}	Silver glass adhesives	1.5×10^{-5}
Aluminium nitride	4.1×10^{-6}	Copper alloy	1.7×10^{-5}
Alloy-42	4.2×10^{-6}	Silver	2.0×10^{-5}
Tungsten	4.5×10^{-6}	Aluminium	2.1×10^{-5}
Si/SiC	4.9×10^{-6}	Low CTE polyimides	10^{-6}–10^{-5}
Molybdenum	5.1×10^{-6}	Silver-filled epoxies	3–5×10^{-5}
Boron carbide/Al_2O_3	5.3×10^{-6}	Al-filled epoxies	4–6×10^{-5}
Silicon nitride/Al_2O_3	5.5×10^{-6}	Polyimides	4–6×10^{-5}
Kovar alloy	5.5×10^{-6}	Unfilled epoxies	6–8×10^{-5}
Aluminium oxide	6.0×10^{-6}	Poly(imide-siloxanes)	6–12×10^{-5}
Boron nitride	7.3×10^{-6}	Silicone rubbers	1–2×10^{-4}

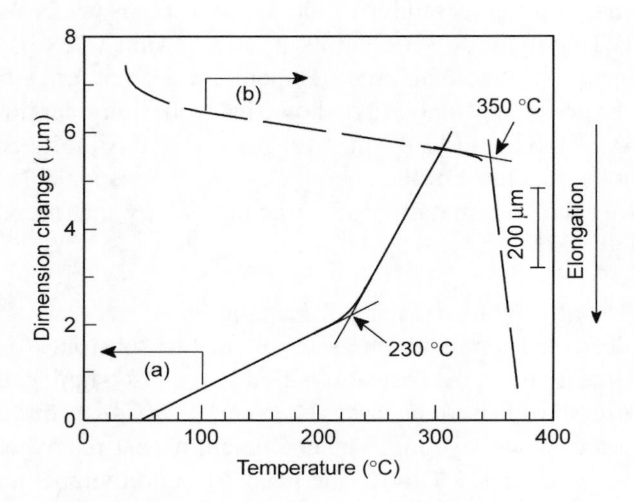

Figure 29: Determination of thermal expansion coefficients and glass transition temperatures by thermomechanical analysis of conductive adhesives: (a) IP 680 silver-filled polyimide analyzed with a dilatation probe (T_g 230°C); (b) self-standing polyimide film studied with an extension probe (T_g 350°C).

thermal expansion of the materials commonly used in the manufacture of electronic components are listed in Table 4.

Thermal expansion coefficients are determined by thermomechanical analysis using a dilatation–penetration probe for adhesive pastes and an extension probe for self-standing films. The output of the thermal analyzer equipped with a dilatation probe is a curve plotting the variation of adhesive thickness as a function of the temperature.

For thin films, thermomechanical analysis is performed by measuring the length variation of specimens ($15 \times 5\ mm^2$) under load (0.02 N) applied to the free end of the film. Extension is recorded when the temperature is increased at a rate of $10°C\ min^{-1}$. These techniques are illustrated in Figure 29 which shows (curve b) the dilatation of IP 680 silver-filled polyimide adhesive cured at 275°C. The first linear region with a slope of $4.5 \times 10^{-5}\ K^{-1}$ (left *Y* axis) is associated with the glassy state and is followed at 230°C by a second linear region of higher slope of $1.2 \times 10^{-4}\ K^{-1}$. This change in slope is a second-order thermodynamic transition of the material from the glassy to the rubbery state, which is one definition of the glass transition temperature. An example of film elongation (right *Y* axis) is also shown in the figure (curve a) for a high T_g self-standing polyimide film exhibiting a thermal expansion coefficient of $6.9 \times 10^{-5}\ K^{-1}$.

6.4.2.8. Thermal Conductivity

The methods used to measure the thermal conductivity and the most significant data are discussed in Section 6.3.3. When a high level of thermal transfer must be assured, soft solders and eutectic alloys present the best choice. Adhesives filled with metals and certain oxides generally have sufficient thermal conductivity to transfer the heat generated by metal oxide semiconductors and other low power devices. Thermal conductivity is measured by ASTM C117 or C518 at 121°C. The requirements are $\geq 1.5\ W\ m^{-1}\ K^{-1}$ for electrically conductive adhesives and $\geq 0.17\ W\ m^{-1}\ K^{-1}$ for insulating adhesives.

6.4.2.9. Volume Resistivity

The electrical conductivity of metal-filled adhesives is discussed in detail in Section 6.3.2, which provides the theoretical models and the necessary conditions to build a continuous electrical path. In summary, it is a function of the number of physical contacts between contiguous particles, so that the higher the filler content the higher the conductivity. However, the maximum volume fraction of metal particles that can be dispersed in an organic resin depends on the flow properties required by the dispensing equipment. Thus, to maximize electrical conductivity, the size and shape of the filler particles must be selected to provide the highest filler content allowed by the processing characteristics. The volume resistivity of the best silver-filled epoxies is of the order of 10^{-3}–$10^{-5}\ \Omega\ cm$, about two orders of magnitude lower than pure silver metal but convenient for most applications. Licari and Enlow reported the results of tests conducted with five commercial adhesives [59]. They showed that the electrical conductivity of the silver-filled epoxies is relatively stable, decreasing slightly at elevated temperatures and increasing after ageing for 1000 h at 150°C. Although these characteristics are excellent, many examples have been reported that indicate the formation of a resistive path during accelerated environmental testing. This increase of volume resistivity can be explained by the formation of an oxide layer at the interfaces as well as voiding, cracking, and delamination within the adhesive joint. The specification uses a four-point probe measurement of an adhesive strip (0.254 cm wide and at least 5.08 cm long) on a glass slide. The tests are performed at 25, 60, and 150°C and at 25°C after 1000 h at 150°C in nitrogen. The volume resistivity measurements for electrically insulating adhesives must be performed at 25 and 125°C. The requirements are 5.0×10^{-4} Ω cm for silver-filled adhesives and 1.5×10^{-3} Ω cm for gold-filled materials, while it would be $\geq 10^{12}$ and $\geq 10^{8}$ Ω cm at 25 and 125°C, respectively, for the electrically insulating adhesives.

6.4.2.10. Dielectric Properties

The measurement of the dielectric constant and dissipation factor for electrically insulating adhesives is performed in accordance with ASTM D150 at 1 kHz and 1 MHz at 25°C. The requirement for the permittivity is $\varepsilon' \leq 6$ and the dissipation factor must be ≤ 0.03 at 1 kHz and ≤ 0.05 at 1 MHz.

6.4.2.11. Electrical Stability

The stability of the electrical parameters after ageing at high temperature under power is of prime importance in the selection of die attach adhesives. The test requires a current density of 139.5 A cm^{-2} applied to five gold-plated Kovar tabs, each tab being adhesively bonded to thin film gold conductors and series bonded with gold wires. The biased test specimens are heated at 150°C in nitrogen for 1000 h, and the electrical resistance of the bonds is measured at 25°C at 200 h intervals. The specification requirement states that the resistance of the five tabs does not increase more than 5%.

6.4.2.12. Sequential Environmental Testing

To complete the selection trials applied to the conductive adhesives, MIL-STD-883 has issued a series of sequential environmental tests mainly aimed at devices used in the aerospace industry. The three tests are humidity-induced stress tests at elevated temperatures with and without bias, temperature-induced stress tests with and without bias, and a mechanical shock comprising acceleration 3000*g*, 0.3 ms and constant acceleration 15,000 *g* (MIL-STD-883, Method 2001). The most severe tests are those involving both humidity and temperature, because they simulate the harshest conditions that a plastic package would be subjected to. An excellent overview of the two former test procedures was published by Khan who outlined that the tests listed under the environmental category are not all related to the environment [78]. The moisture–temperature testing procedures include the following:

Moisture resistance: 10 cycles from 25 to 65°C under 90–100% relative humidity (RH) with an applied voltage, 3 h at 65°C, cool at 25°C, and repeat the cycle.
Cycled temperature–humidity–bias life: the test devices are subjected to a temperature cycle between 30 and 65°C, with a relative humidity of 90–98% and the operating voltage turned on and off at 5 min intervals. The specification requires a minimum of 63 cycles (total of 1008 h).
Steady temperature–humidity–bias life (85°C/85% RH): the test specimens are stored for 1000 h or more at 85 ± 2°C and a relative humidity of 85 ± 5% with, depending on the device, a voltage applied either constantly or intermittently.
Unbiased autoclave: in the steam cooker test, the packaged devices are stored at 121°C under a saturated steam pressure of 0.103 MPa in a sealed autoclave. Depending to the intended application, the test conditions vary from 24 to 48 or 96 h of storage. The electrical measurements are performed at ambient temperature after the devices have been stored at room temperature for 48 h.
Biased autoclave: the test procedures are similar to the unbiased, except that a bias is applied through hermetic electrical connections.
Highly accelerated stress testing: unlike the pressure cooker tests, this test uses unsaturated steam, varying from 50 to 100%, at temperatures of 105–150°C, the experimental data being extrapolated to any temperature–humidity conditions by using theoretical models based on the usual Arrhenius equation.

The temperature stressing refers to burn-in, steady-state life, endurance life, high temperature storage, and temperature cycling. In the burn-in test, the devices are biased at the maximum rated operating voltage and a minimum temperature of 125°C for 160 h (class B) or 240 h (class S). The duration of the test decreases with increasing temperature and is, for instance, 12 h at 250°C. In the steady-state life test, the temperature and duration as well as the conditions of applying the voltage comprises six categories from A to F. The failure data are recorded in any of the 10 time–temperature couples varying from 1000 h at 125°C to 30 h at 190°C or more. In the high-temperature storage test, the plastic packaged devices are stored at 150°C for 1000 h without bias. Finally, in the temperature cycling test for military applications, the lower limit temperature is either −55 or −65°C whereas the upper limit can be 85, 125, 150, 175, 200, or 300°C. For commercial applications in the automotive industry, the temperature

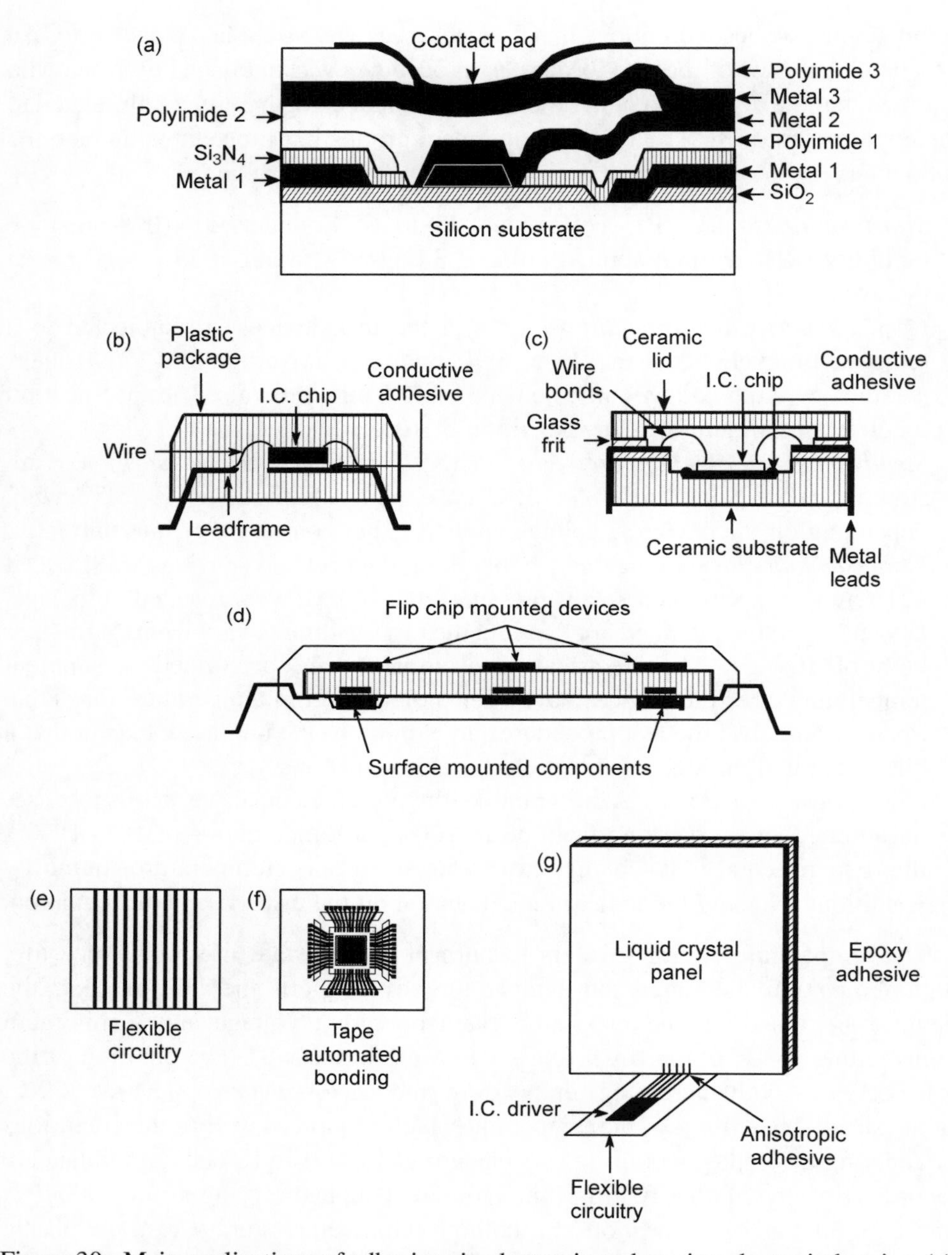

Figure 30: Main applications of adhesives in electronic and semiconductor industries: (a) polyimide interlayer dielectric films in the fabrication of integrated circuits; (b) die attachment in plastic packages; (c) die attachment in ceramic packages; (d) multichip module with flip chip and surface-mounted devices; (e) flexible circuitry; (f) tape automated bonding; (g) liquid crystal display panels.

limits are −40 and 125°C. The number of cycles is 10 for screening and 1000 for qualification.

6.5. Applications of Adhesives

The different applications of adhesives in electronics can be extracted from Figure 1 in their order of introduction on the market place. Drawings of Figure 30 summarize these applications that are described using a more logical presentation which starts from the integrated circuit fabrication, the formation of bumped connectors on the chip pads, the manufacture of substrates for MCMs, the production processes of flexible circuitry and copper-polymer films for TAB, the conventional die attachment of microcircuits into plastic and ceramic packages, the flip chip bonding technology, and finally the development of anisotropically conductive adhesives for surface-mounted components and liquid crystal display devices.

6.5.1. Integrated Circuit Fabrication

Polyimides are now commonly used in the fabrication of integrated circuits. They cannot be considered as adhesives in the usual sense but they must be adhesively bonded to the metals and inorganic insulators. IBM's 20-year experience in multilevel interconnect technology can be used to illustrate the advantages of polymer insulation compared to inorganic dielectrics [79]. ULSI circuits incorporating polyimide insulation and passivation are being used across the entire spectrum of IBM's product lines. A typical three-level metal interconnection using a dual-insulator scheme is shown in Figure 30(a). The silicon substrate is protected with a thin layer of thermally grown silicon dioxide, which is etched over the doped regions. The film of metal-1 is then deposited, patterned by photolithography, and insulated with thin films of silicon nitride and polyimide-1. This dual inorganic–organic insulation is used between metal-1 and metal-2 levels, whereas only polyimide film 2 is employed between metal-2 and metal-3.

Once all integrated circuits are implemented on the silicon wafer, a thicker uppermost layer of polyimide is coated over the silicon dioxide or silicon nitride passivation. After openings have been defined over the metal pads by photolithography and plasma etching, the individual dice are separated, die attached to their leadframes, electrically connected by wire bonding, and then coated with a second layer of polyimide. This final buffer coating absorbs the interfacial stresses, thus preventing passivation crack and electrode displacement when the integrated circuit is subjected to the pressure cooker test. In addition,

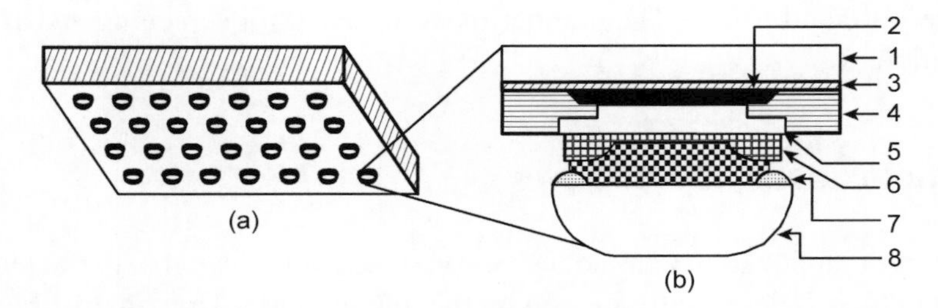

Figure 31: Structure of IBM's solder bumps fabricated at the wafer scale before integrated circuits are connected to wiring boards by the flip chip technology: (a) integrated circuit with an area array of solder bumps; (b) exploded view showing the bump structure with silicon substrate 1, aluminium bonding pad 2, silicon dioxide 3, silicon nitride passivation 4, titanium–tungsten alloy 5, sputtered copper 6, plated copper 7, and reflowed plated solder 8.

thick polyimide films are excellent α-ray barriers protecting the devices from soft errors caused by the radioactive elements contained in ceramic packages or carried out by silica and other inorganic fillers in epoxy moulding resins.

6.5.2. Fabrication of Bumped Connectors

The solder bump flip chip process was developed and patented by IBM more than 25 years ago [80]. It is characterized by an even distribution of connecting bumps over an area array instead of the more conventional placement along the periphery of the IC chips. As shown in Figure 31(a), bumps are formed at the wafer scale using "collective" processes. This means that the connecting metallurgy is constructed over all integrated circuits at the same time. The process begins by primarily making a thick layer of silicon nitride passivation to provide better protection to the circuit. Then the bumps are built by placing three layers of metal over the bonding pads (Fig. 31(b)). A first titanium–tungsten alloy film is deposited on the bonding pads, followed by sputtered and plated copper layers overcoated with a 94Pb–6Sn binary soft solder alloy. Finally, the solder layer is reflowed at a temperature exceeding 200°C to form hemispherical bumps. Despite the complicated manufacturing sequence and its capital-intensive aspect, IBM's flip chip technology has been licensed to many multinational companies for both direct bonding of integrated circuits or delayed bonding via bumped plastic packages (ball-grid array).

To reduce the cost of flip chip interconnection, an "organic" process was developed at Epoxy Technology Inc. [81]. The polymer flip chip technology (PFC®) uses a polymer dielectric paste and an electrically conductive adhesive, which can be deposited by screen printing. It deserves to be discussed because of its potential for the interconnection of semiconductor devices on MCM carriers. The PFC bumping process is an additive method in which the wafers are coated with a dielectric passivation layer and conductive bumps using screen printing techniques. The formation of the conductive polymer bumps can be performed with or without dielectric coating. This insulating layer, a low α-particle-emitting polyimide, is especially useful for memory devices. As shown in Figure 32(a), the initial step is to define on the wafer the bumping sites on the metallized pads by screen printing the silica-filled polyimide paste to a resultant thickness of 25 μm [82]. High-resolution screen printers have a registration accuracy of $\pm$ 10 μm and offer automatic alignment and printing. In one pass thousands of bumping sites are patterned with the required geometrical arrangement.

After the polyimide layer is cured, the conductive polymer paste is printed to the substrate in order to fill the holes over the metal pads and then cured (Fig. 32(b)). Using the same stencil and conducting paste, the polymer bumps are then screen printed on top of the conductive polymer pads and cured (Fig. 32(c)). Current printing techniques allow production of 50–100 μm diameter bumps, 30–50 μm high, with a centre-to-centre pitch of 125–150 μm, and flat, conic, or hemispherical shapes. Once the bumps are cured and the dice sawed,

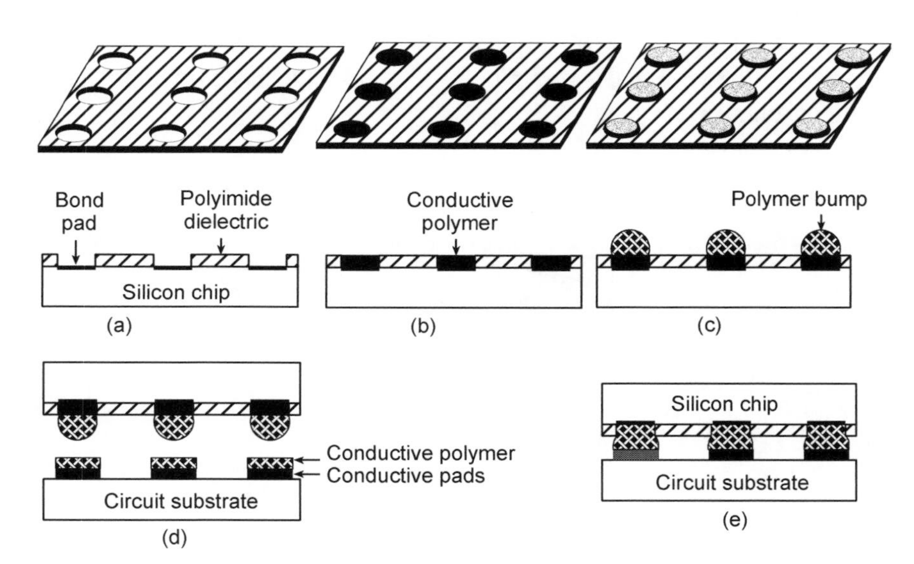

Figure 32: Process for the fabrication of polymer bumps: (a) bumping sites delineated by screen printing silica-filled polyimide paste; (b) silver-filled epoxy resin composition is printed over the metal pads; (c) polymer bumps are formed by screen printing a second layer of conductive epoxy. Steps (d) and (e) sketch the subsequent flip chip process using a layer of conductive adhesive coated on the substrate bonding pads.

the integrated circuits are ready to be bonded to the printed wiring board, hybrid circuit, or MCM carrier (Fig. 32(d–e)).

6.5.3. Substrates for Multichip Modules

Although integrated circuits supplied in standard packages remain the workhorse of the electronic industry, new packaging technologies have emerged to match the increasing performance of VLSI and ULSI semiconductor devices. By eliminating individual packages, MCM technologies enable the construction of smaller systems that exhibit higher speed performance. Depending on the carrying substrate and fabrication processes, MCMs are divided into four main groups, but only one of them is presented hereafter. MCM-D is made by deposition of thin films of metals and dielectrics over rigid materials such as silicon, glass, ceramic, diamond, or metal substrates. The manufacturing tools and processes employed to produce MCM-D substrates are comparable to those implemented in the semiconductor industry. All technologies involved in the fabrication of MCMs are extensively discussed in previously published books [24,83–87]. The packaging density achieved with MCM structure increases from 30 to 75% of the package area as the conducting line width is reduced from 50 to 10 μm. Thus, the wiring efficiency attained with MCM-D structures is so high that all connections can be obtained with only two signal layers. Figure 33 depicts the basic principle of MCM of the silicon-on-silicon type.

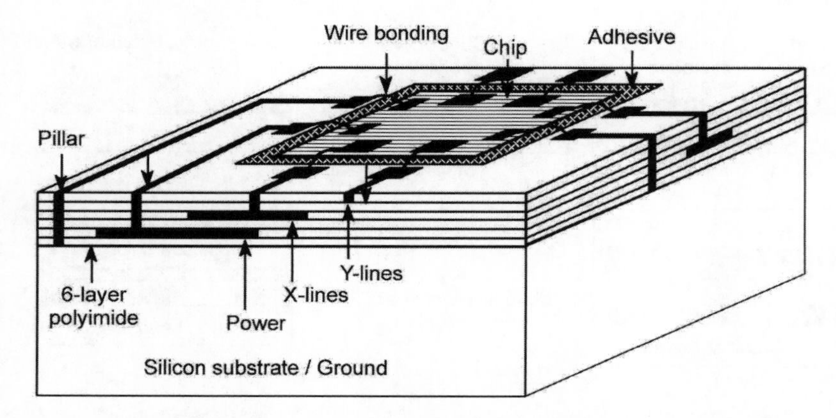

Figure 33: Hybrid multichip module structure with five copper conductor levels and six polyimide insulating films deposited by spin coating onto a doped silicon wafer. In this process, the integrated circuits are inserted in cavities that are created by laser drilling or anisotropic chemical etching of the silicon substrate. IC chips are bonded to the substrate by means of a heat-resistant insulating adhesive.

Commonly used carrier substrates are ceramics, silicon wafers, metals, and more recently glass for high-density large-area processing. As MCM-D construction is based on lithographic techniques, these supporting materials must present controlled surface roughness, flatness, and reduced camber. Other constraints are high mechanical and thermal properties, in particular thermal conductivity and coefficient of thermal expansion, along with excellent resistance to process chemicals. The fabrication of copper/polyimide modules using Du Pont Pyralin® PI 2555 and alumina substrate was reported 20 years ago [88]. Ceramics are currently used either as simple supporting polished plates with size culminating at $25 \times 25\ cm^2$ or as co-fired ceramic–metal multilayer structures that contain power and ground distribution planes. Advantages of ceramic materials are high flexural modulus (240–260 GPa), low CTE ($4-7 \times 10^{-6}\ K^{-1}$), and good thermal conductivity increasing from $20-30\ W\ m^{-1}\ K^{-1}$ for alumina to $200-300\ W\ m^{-1}\ K^{-1}$ for beryllium oxide and aluminium nitride.

Silicon substrate is attractive because of its coefficient of thermal expansion matching that of integrated circuit chips ($3 \times 10^{-6}\ K^{-1}$), thermal conductivity ($150\ W\ m^{-1}\ K^{-1}$), highly polished surface, and the possibility of implanting planar active and passive components. Conversely, the flexural modulus is one order of magnitude lower than that of ceramics and significant warpage may result from the stresses imposed by the successive deposition of metal and polyimide films. Metallic carriers — mostly aluminium, copper, or Cu/Mo/Cu sandwich — have excellent thermal properties with thermal conductivity in the range of $135-400\ W\ m^{-1}\ K^{-1}$, high mechanical characteristics, but relatively high CTE values of 2.1 and $1.7 \times 10^{-5}\ K^{-1}$ for Al and Cu, respectively, decreasing to $5.1 \times 10^{-6}\ K^{-1}$ for Cu/Mo/Cu composite.

Compared to inorganic insulators, polymeric materials are more convenient to create MCM-D interconnects. Lower dielectric constant, better planarization, thicker layers, and high deposition speed are the most important factors explaining the shift from inorganic to organic dielectrics. Polyimides, photosensitive polyimides, parylene, fluorinated parylene, cyanate esters, benzocyclobutenes, and fluoropolymers have been evaluated and some of them are currently used to produce thin film MCMs. Most MCM-D interconnects have been elaborated by using existing commercial polyamic acids such as PMDA–ODA **93** (Du Pont Pyralin® PI 2545, Hitachi PIQ® 13), BTDA–ODA–MPD (PI 2555), BPDA–PPD **95** (PI 2611 and PIQ® L-100), low CTE polyimide based on *para*-terphenyl derivative, polyamic esters from Du Pont, Hitachi, and Ciba-Geigy Probimide 600, precyclized polyimide Amoco Ultradel® 4212 (6FDA/3,3′-[[1,1′-biphenyl]-4,4′-diylbis(oxy)]bisbenzeneamine) or photosensitive polymers. The AT&T's "Polyhic" technology is based on photoimageable cyanate/acrylate esters. As previously indicated, the use of photosensitive polyimides has been extensively detailed by IBM, NTT, NEC, Boeing, and AT&T [24]. Module design and performance strongly depend on polymer properties that in turn rely on their chemical structure.

In summary, critical parameters are: (1) a low, frequency-independent, dielectric constant allowing for faster signal propagation and decreased film thickness; (2) low level of water absorption that degrades both adhesion strength and permittivity, which may increase by 25–35% between 0 and 100% RH; (3) good planarizing properties because lack of planarity leads to poor step coverage resulting in thinned zones that are prone to cracking; (4) tolerance to processing chemicals including organic solvents and aqueous acids; (5) excellent adhesion at the four interfaces existing in multilevel interconnection structures. Although the six polyimide layers of the MCM shown in Figure 33 cannot be considered as adhesive films in the usual sense, strong adhesion is one of the most important parameter for long-term reliability. A complete description of the adhesive forces developed between polyimides and inorganic materials can be found elsewhere [24].

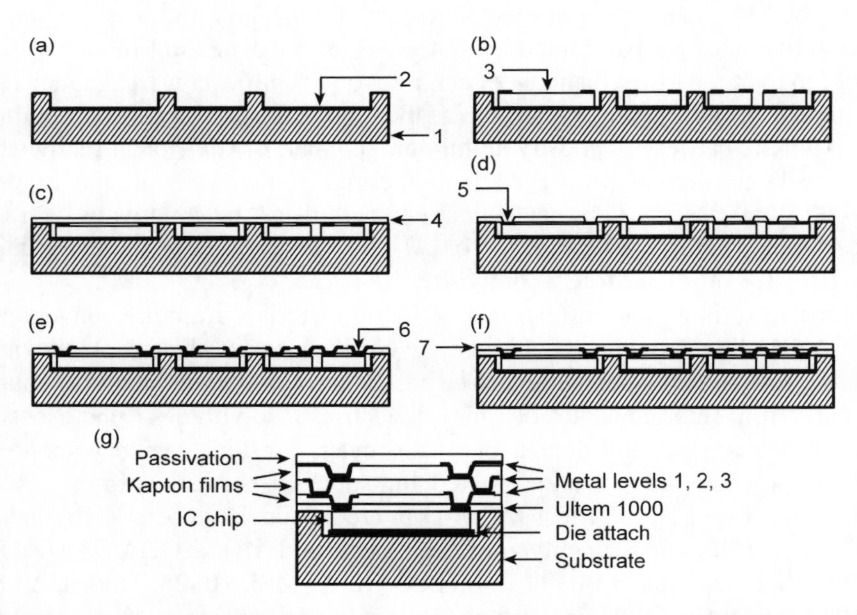

Figure 34: General Electric interconnection process sequence: (a) milling ceramic substrate 1 to make recessed areas that are covered with a thin film of aluminium 2; (b) placing and bonding chips 3; (c) laminating Kapton film 4 by means of an adhesive film; (d) laser drilling vias 5; (e) sputtering Ti/Cu, electroplating Cu, sputtering Ti, and etching to form conductor pattern 6; (f) applying dielectric level-2 7 and repeating the cycle to obtain the typical cross-section (g).

Table 5: Dielectric, thermal, and mechanical properties of commercial polyimides including dielectric constant ε', glass transition temperature T_g, thermal expansion coefficient α, tensile modulus E, tensile strength S, and elongation at break ε.

Polyimides	ε'	T_g (°C)	$\alpha \times 10^6$ (K^{-1})	E (GPa)	S (MPa)	ε (%)
Amoco Ultradel 4212	2.9	295	50	2.8	101	30
Amoco Ultradel 5106[a]	2.8	400	24	1.3	122	70
IFP-Cemota IP 200	2.9	350	55	2.2	120	10
Du Pont PI 2540	3.5	400	26	1.3	160	60
Du Pont PI 2611	2.9	350	5	6.6	600	60
Du Pont PI 2730[a]	2.9	350	15	–	170	–
Du Pont WE 1111	–	385	19	4.4	300	55
Hitachi PIQ 13	3.4	290	50	3.0	116	10
Hitachi PIQ L 100	3.2	360	3	10	320	22
Hitachi PL 2135[a]	3.3	270	40	3.6	124	10
OCG Probimide 400[a]	3.0	350	39	2.2	140	56
OCG Probimide 500[a]	3.2	400	7	11.6	144	28
OCG Probimide 600[a]	3.2	366	23	2.7	194	104
Toray UR 3800	3.3	280	45	3.4	145	30
Toray UR 5100	–	–	25	–	200	20

[a] Photosensitive polyimides of the ester, ionic, or intrinsic type [24].

There exists at least one example of solderless chip mounting using a thermoplastic polyimide as an adhesive for bonding Kapton films. In this process, General Electric (GE) has employed a "chip first" approach using either silicon or ceramic substrates [89,90]. As shown in Figure 34, the bare chips are inserted into cavities that have been created in the ceramic substrate by laser direct writing. Gaps between chips and substrate are filled with an epoxy resin, which is cured prior to thin film processing. A thin polyimide film is laminated over the substrate to form the first dielectric layer. Via holes are made by laser ablation down to the I/O pads [91]. Connection to the chip pads and the interconnect traces are formed by a 4 μm Ti/Cu seed sputtering and plate-up process. The metal interconnect pitch is typically 100 μm. Multiple layers are formed by repeating the PI film lamination, via formation, and metallization processes. Two orthogonal signal layers and power and ground layers are processed.

Thermoplastic poly(ether-imide) Ultem® 1000 (T_g 217°C) is used as overlay adhesive for a Kapton film on which the interconnect is created [92]. This adhesive layer is disclosed to produce void-free lamination at approximately 300°C. Actual dielectric layers therefore consists of 25 μm thick Kapton over 12.5 μm thick adhesive film. The self-relaxation properties of Ultem 1000 tend to release the thermomechanical stresses generated in this multilayer laminate. The chip first approach using ceramic substrates and laser cuttings has also been reported by researchers of the TUB (Technical University of Berlin) [93]. Interconnections of the entrenched chips are generated by laser direct writing of thin copper lines produced from copper formate, followed by chemical copper deposition. The data listed in Table 5 summarize some properties of commercial polyimides employed in the fabrication of MCMs.

6.5.4. Flexible Circuits and Tape Automated Bonding

Basically, tapes intended to the flexible circuit market are fabricated by using two- or three-layer construction schemes. Two-layer tapes are made by coating solutions of polyimide precursors over copper foil or by plating copper onto polyimide films. Typical substrates include 35 μm thick copper bonded to 15–50 μm thick polyimide. Three-layer tapes are formed of polyimide films (75–150 μm thick) coated with organic adhesive and laminated to 35 μm copper foil. Information retrieval shows that more than 1000 patents have been published over the past 10 years in the field of copper–polyimide systems. Patents dealing with two-layer flexible circuits can be subdivided into a few main classes. Historically, the first process has been developed from commercial solutions of polyamic acids that are coated on copper foils and subsequently thermally imidized to produce copper–polyimide substrates. Circuit patterns are then transferred to the copper layer by photolithography followed by chemical etching. This technique is particularly suited with highly rigid, low thermal expansion polyimides. It has been applied to polymer prepared from pyromellitic dianhydride **90** and 2-methyl-1,4-benzene-diamine, as well as copolymers of [1,1′-biphenyl]-3,3′,4,4′-tetracarboxylic acid dianhydride **101**, PMDA, 1,4-benzenediamine, and 4,4′-oxybisbenzeneamine **91**. Incorporation of 3,3′-dihydroxy[1,1′-biphenyl]-4,4′-diamine as a co-reactant enhances the peel strength while maintaining a low thermal expansion coefficient of $2.7 \times 10^{-5}\ K^{-1}$. In one variant of the process, the polyamic acid film is only dried for a few minutes at 130 and 150°C, copper is etched with a $CuCl_2$–HCl mixture, and imidization is performed at 400°C.

The second method is based on the use of polyimides with lower glass transition temperatures that can be melt-processed. For example, intrinsically photosensitive preimidized polymers formed by reacting 3,3′,4,4′-benzopheno-netetracarboxylic acid dianhydride (BTDA) with methyl-substituted 4,4′-methylenebisbenzeneamine (MDA), blended to epoxy resins, give high strength laminate with copper. Poly(isoimides), which exhibit good melt-flow properties before thermal isomerization to imides, are also used to make flexible circuits. In the last process copper is electroless plated on polyimide film such as Kapton 200H.

As indicated above, three-layer systems use organic adhesives for bonding polyimide films to copper. Acrylic, polyester, polyamide, epoxy, and polyimide adhesives have been used in the process. However, the main trend, perceptible in the patent literature, is the replacement of acrylic and polyester resins by more performing materials. Epoxy adhesives based on DGEBA resins are used either in combination with epoxy novolacs, dicyandiamide, and imidazole derivatives or blended with aromatic diamines and CTBN rubber. Epoxy-nylons give excellent adhesive layers for bonding copper foils to Kapton films. Various aliphatic polyamides (nylons) or carboxyl-terminated polyamides have been incorporated to di- and trifunctional epoxies before being applied on polyimide film, followed by copper lamination. Multilayer flexible circuits have been made by using self-standing films composed of B-staged ether-linked poly(amide-imide) blended with an epoxy resin and dicyandiamide. The films are laminated on both sides of 25 μm thick Kapton and the resulting material is inserted between two 35 μm thick copper foils to give a laminate with good solder resistance and high peel strength. However, the optimum thermal stability is achieved by using thermoplastic polyimide adhesives. Thus, polyamic acid precursors of thermoplastic poly(ether-imide-siloxanes) or poly(ether-imides) have been used to coat Upilex® S film, followed by thermal imidization and lamination with copper foil.

Wire bonding remains the predominant technique for connecting integrated circuits to the outside world. In recent years, it has been supplemented with two other techniques, TAB and flip chip technology. Developed in the early 1970s, TAB initially found little industrial acceptance facing the progress of high-speed wire bonders meeting the chip-to-package interconnection requirements. Thirty years later, TAB is successfully applied to the packaging of VLSI chips for a variety of electronic products. An excellent review of the TAB technology has been previously published [94]. Figure 35 illustrates one process used to produce the copper lead pattern onto polyimide film.

Starting from a 35–70 mm wide blank tape (a), sprocket holes and window patterns are mechanically fabricated in the film by die punch or laser cut (b). The film is then laminated to 35 μm thick copper foil (c). Photoresist film is deposited on copper and patterned through a mask (d), before beam leads are formed by chemical etching (e). In the final stage, the connecting pads along the periphery of the integrated circuit are bonded to copper leads (f).

Bonding integrated circuit chips to tape lead frames, called inner lead bonding (ILB), is usually accomplished by using bumped bonding pads and thermo-compression tools (thermodes). Figure 36(A) shows an ILB sequence for bumped chips **2** automatically aligned by the indexing table **1**, whereas the copper leads **3** supported on polyimide tape **4** are matched with the bonding bumps under the thermode head **5**. In the "gang" or mass bonding process, all TAB beam leads are bonded simultaneously. At the completion of the bond cycle, the thermode

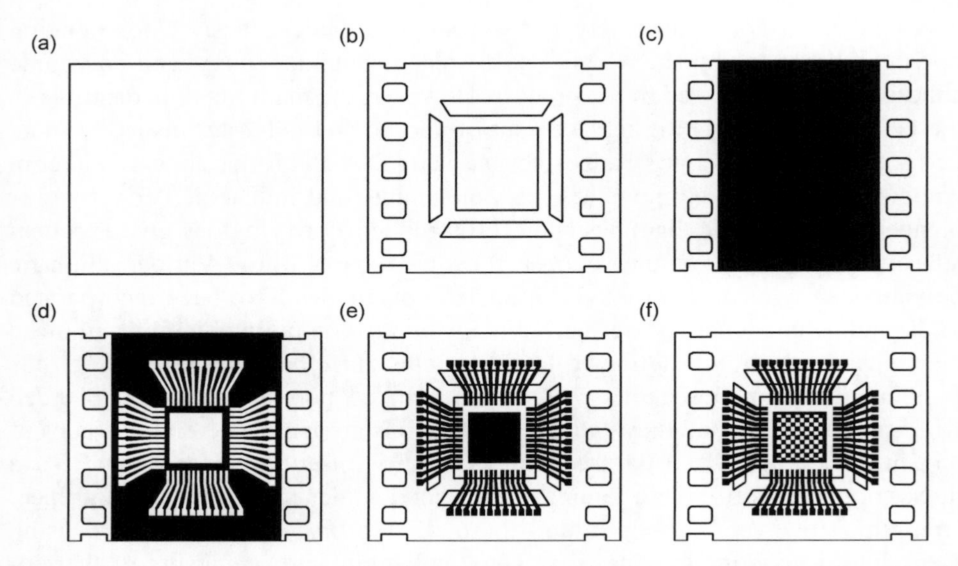

Figure 35: Fabrication stages of flexible circuitry for the tape automated bonding (TAB) process: (a) blank polyimide film; (b) punching sprocket holes and windows; (c) bonding copper foil to the tape; (d) coating and patterning photoresist film deposited over copper; (e) copper etching to delineate conducting lines; (f) gang bonding integrated circuit chip to copper leads.

retracts and the tape moves to carry the finished assembly **6** away. The process illustrated in this figure is employed when bumped wafers are available from semiconductor manufacturers or are produced in-house for captive markets.

The exploded view drawn in Figure 36(B) illustrates the fabrication of the bumps as the last step of wafer processing on the top of a semiconductor (a) whose passivation layer (b) is opened over the aluminium bonding pads (c) before gold or solder bumps (d) are formed. Another process, called bumped tape TAB, is used when bumped wafers are not available. Here, bumps are built on the inner end of the beam leads, rather than on the chips. Figure 36(C) shows a gold-plated bump (e) positioned over the aluminium bonding pad (c).

The primary differences between Kapton H® and Upilex S® as free-standing films for TAB applications have been compared using test specimens of two-layer (adhesiveless) circuitry prepared with a chrome/copper seed layer followed by copper plating [95]. Parts fabricated with Upilex show smaller dimension changes than those made with Kapton. When subjected to temperature–humidity testing (80°C/80% RH) for 2500 h, the peel strength of metal/Upilex assembly decreases from 1.77 to 1.03 kN m^{-1}, whereas that of metal/Kapton laminate decreases from 1.57 to 0.20 kN m^{-1}.This difference is explained by the lower moisture regain and smaller thermal expansion of Upilex. In addition, solvent absorption of assemblies

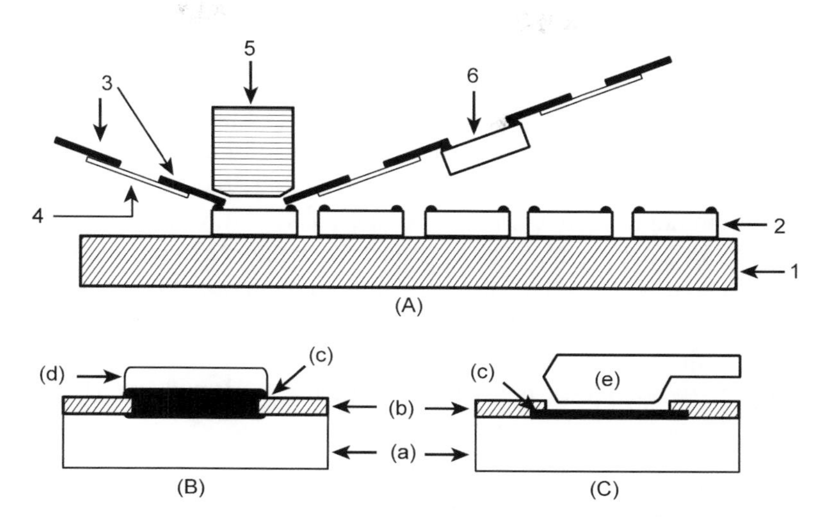

Figure 36: Schematic representation of the TAB process (A) in which bumped chips 2 are aligned by the indexing table 1, whereas the copper leads 3 supported on polyimide tape 4 are matched with the bonding bumps under the thermode head 5. Exploded views (B) and (C) represent bumped chip and bumped tape, respectively, with: (a) silicon substrate; (b) passivation layer; (c) aluminium bonding pad; (d) multimetal chip bumps or (e) gold-plated tape bumps.

immersed in methylene chloride is negligible for Upilex but reaches about 22% for Kapton. Therefore, rigid polyimides such as Upilex or similar materials seem to be more desirable for TAB and other thin film applications.

6.5.5. Die Attachment

The term die attachment designates the operation of bonding integrated circuits (dice) to copper leadframes or to hybrid circuits. In the hermetic packages encapsulating the microprocessors and high-performance devices used in telecommunication, aerospace, and military applications, the dominant method of die attachment is by the formation of a metallic bond between the dice and the gold-plated substrates. For the multilayer ceramic and CERDIP packages, the bonding material is almost always the gold–silicon eutectic melting at 363°C. Organic adhesives cannot be used in CERDIP packages because of the high temperature attained during the frit seal process which requires heating to 400–450°C for 30 min. In 1971, the hybrid industry was becoming very active in replacing Au–Si eutectic bonding of chips with silver-filled epoxies. It was 4 years later that the semiconductor industry decided to switch over to epoxy bonding. The reasons for doing this was the price of gold and especially the fact that the performance and reliability of epoxy adhesives has been well established over the 9 preceding years.

Conductive adhesives are less expensive than the gold–silicon eutectics, silver replaces gold for plating the lead frames, the backside of the die does not need gold plating, and finally a very high throughput is attainable with automated die attachment equipment. Combined with plastic packages, organic die attachment has permitted low-cost production of billions of integrated circuits. Conductive adhesives are also used in the hybrid technology to bond integrated circuits as well as other components such as LED, resistors, capacitors, and transistors to various substrates. Figure 37 sketches the most important packaging techniques, i.e. moulded plastic package (A), hermetic ceramic CERDIP package (B), and glob

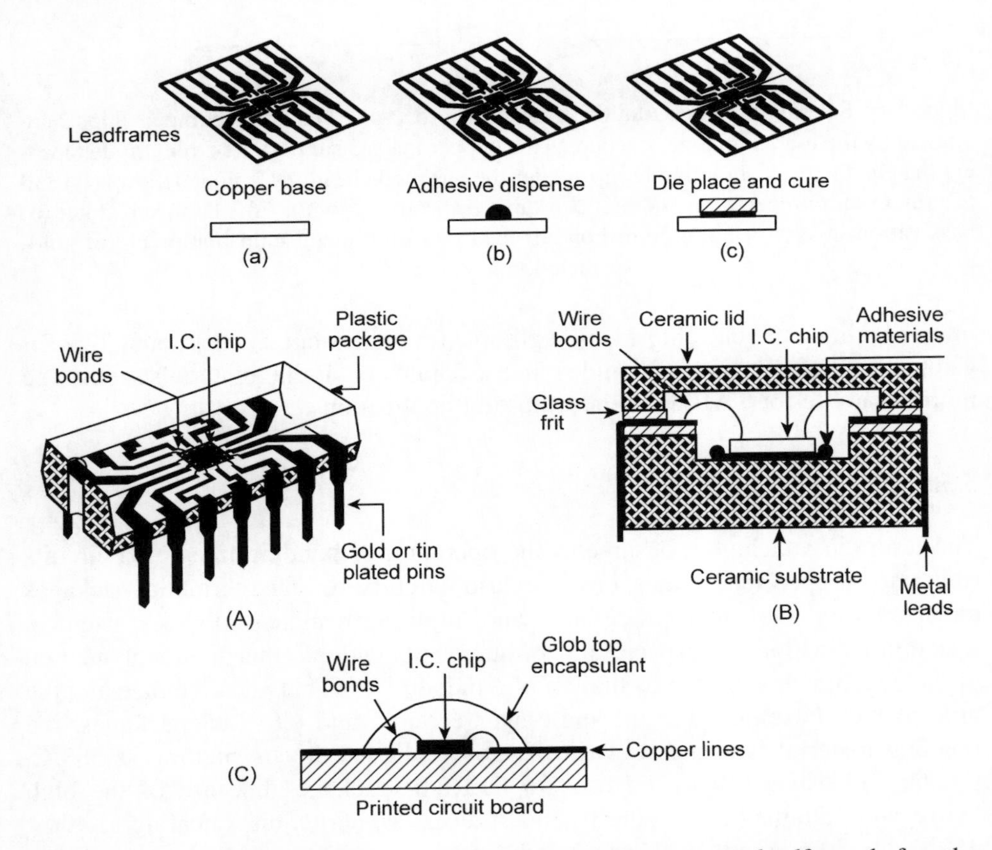

Figure 37: General scheme of the die attachment process on copper leadframe before the assembly is encapsulated into an epoxy moulded package (A). Other packaging techniques are the hermetic Cerdip package (B) and glob top encapsulation of surface-mounted devices (C).

top encapsulation (C). The top design shows a bare copper leadframe (a) with 14 gold or tin-plated connecting pins and a silver-plated central base. A drop of conductive adhesive paste is dispensed on the base surface (b). A die is picked up by means of a pick-and-place machine and the assembly is cured into an air circulating oven (c).

6.5.5.1. Adhesive Pastes

Electrically and thermally conductive adhesives are available in the form of pastes, films or tapes. Adhesive pastes are more sensitive to the processing conditions because errors may occur in dispensing too much or too little material. With small amounts a complete continuous coverage may not be attained, therefore reducing the bond strength. Conversely, an excess of adhesive tends to flow up and around the periphery of the die or the substrate and can contaminate the top circuit metallization. It has been shown that an excessive flow of silver-filled adhesives can short the circuit because of a bias-induced silver migration. Adhesive tapes are more adapted for bonding large devices whereas, with the advent of highly automated pick-and-place machines and dispensers, the adhesive pastes are more convenient for the attachment of small dice. Paste adhesives are formulated to have optimum flow properties so that they can be dispensed either by screen printing, stamping, or with automated dispensing machines. The thick film screen printing process is mainly used in the manufacture of surface-mounted devices and hybrids but it has not been successfully applied to die bonding technology, which requires a better volume control and process flexibility. A typical stamping equipment consists of a rotating cup containing the adhesive formulation, which is maintained in a low-viscosity state by exerting a high shear rate with a lateral squeegee. The adhesive is transferred to the pins of the stamping tool by both surface tension and thixotropy. The arm carrying the stamping tool is then moved in the programmed direction and the dots of adhesive are transferred to the substrate by contact printing.

The technology of dispensing, in which the adhesive paste is applied through a single- or multiple-tube nozzle, is divided in two categories: the "write" and the "print" methods. The write approach, represented in Figure 38(a), has the significant advantage of using a single-tube nozzle with a software programmable dispenser equipped with a high pressure pump. This type of material has been previously described and it has been shown how the volume distribution can be exactly controlled through the pattern design and additional parameters [96]. In the case of square chips, the cross pattern of Figure 38(b), with a simple central volume, is successfully employed for chips with edge lengths between 2 and 25 mm while the pattern of Figure 38(c) is designed for rectangular chips.

However, the write method is a relatively slow process in comparison to the print approach in which the complete pattern is printed in one shot through

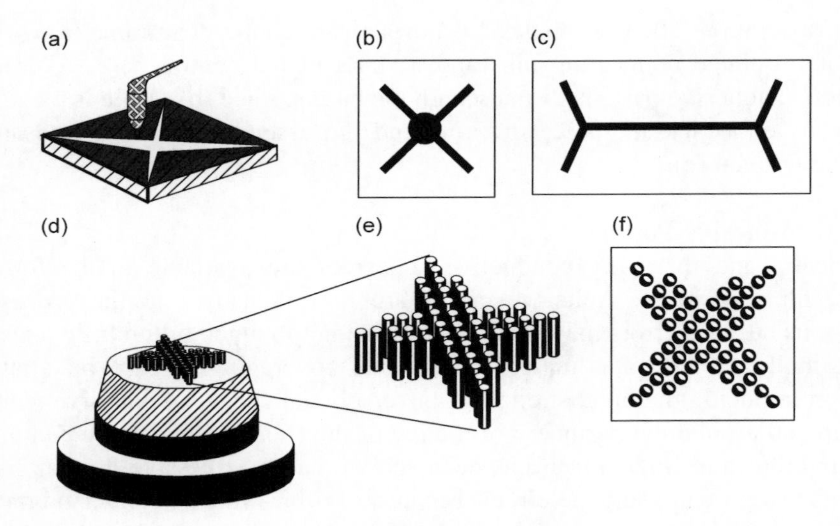

Figure 38: Dispensing technologies used to apply adhesive pastes to the surface of chip carriers. The write method (a) leads to cross pattern (b) for square chips and (c) for rectangular dice. For large dice, the print method uses a dispenser (d) equipped with a multitube star-shaped nozzle (e) providing pattern (f) in one shot (adapted from Refs. [34,35]).

a dedicated multitube dispensing tool. The design rules that have to be considered to optimize a multitube star-shaped nozzle, as illustrated in Figure 38(d) and (e), for the attachment of large dice were discussed by Sela and Steinegger [97]. The star pattern shown in Figure 38(f) was designed with an advanced CAD system using a mathematical model considering that, to produce a void-free bond, the adhesive has to flow from the centre outward. This means that the first contact between the chip and the adhesive must occur at the geometrical centre where the dots must be thicker than those at the end of the radiating beams. Figure 38(d–f) shows the design of an optimised $9 \times 9\ mm^2$ star-shaped multitube nozzle used for the die attachment of large dice. The star design alleviates the tailing problem, discussed in Section 6.4.1.2, because the printed dots are kept at a safe distance from the boundaries, thus avoiding the risk of a tail creating a short to a lead.

6.5.5.2. Adhesive Films

The polymers used to manufacture the film or tape adhesives are either thermoplastic materials including acrylic, polyester, and fully imidized polyimides or thermosetting resins based on epoxy and polyimide chemistries. The tapes are often made with partially cured (B-staged) epoxies similar to those used in the fabrication of the PCBs. Some commercially available films are also produced by blending thermoplastic and thermosetting resins. The first adhesive films for the attachments of dice encapsulated in plastic packages were applied to the backside of the wafer prior to sawing the individual dice, which were then bonded to the heated lead frames under pressure. Now, for application with the existing automated pick-and-place equipment, the adhesive tapes can be purchased either precut to the size of the devices or in large sheets and rolls cut by end-users. The device-sized preform is placed between the two surfaces, held in place with an applied pressure, and cured by heating. In principle, adhesive films have two production advantages over pastes. On one hand, they allow control over the amount of adhesive that is used and on the other hand, they make it possible to obtain a constant thickness over the whole bonding area. In addition, adhesive films may be theoretically free of voids, even if they are prepared by casting solutions of polymers, because the conditions to evaporate the solvent can be tightly controlled. These factors are especially important in attaching large dice to chip carriers or in bonding large-area substrates to metallic packages.

6.5.6. Flip Chip Bonding Technology

Most of the 60-billion integrated circuits produced each year are bonded to their leadframes with conductive adhesives and then electrically connected to the individual copper pins by soldered gold or aluminium wires. However, the trend to higher pin counts and improved microcircuit speed, combined with the demand for smaller package sizes, leads to newer methods of interconnection. One of the most convenient is the flip chip technology, in which the bonding pads are distributed over the totality of the die surface, in contrast to the conventional peripheral geometry of wire bonded circuits. In terms of manufacturing and reliability, the advantages of the flip chip method are: (1) a smaller size because flip chip bonded circuits do not need lead frames and plastic encapsulation, and therefore take up less space than conventional plastic packages; (2) better thermal performance due to the number of thermally conductive paths directly connected to the substrate by the conductive bumps; (3) improved reliability resulting from the elimination of wire bonding with its two bonding interfaces between the chip and the leadframe; (4) less inductance and faster circuit speed due to the shorter distance the signal has to travel. Figure 39 shows the cross-section of an integrated circuit connected to the bonding pads of a silicon-on-silicon MCM by means of solder bumps. Using photosensitive polyimides, AT&T has implemented this process to produce large volumes of consumer-oriented telephones [98].

Next to the reflow of solder bumps, two other processes have been developed for bonding bumped integrated circuits to MCM substrates. A first approach uses

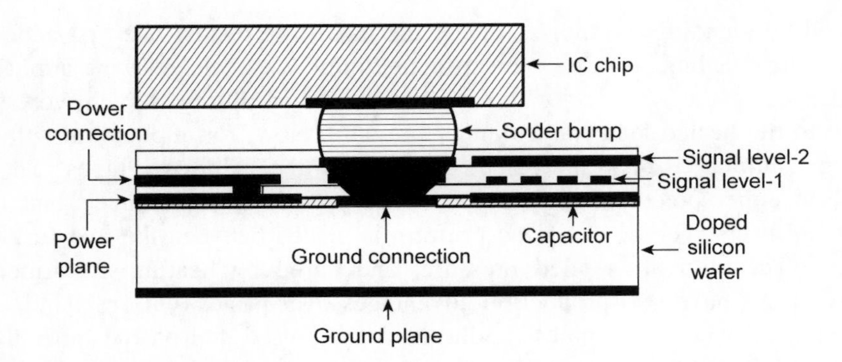

Figure 39: Cross-section of the silicon-on-silicon AT&T's advanced VLSI packaging process comprising four copper layers for ground, power, and routing signal planes with polyimide insulating films. (reprinted from Ref. [98], Copyright© 1987 IEEE.)

a non-conductive epoxy adhesive coated on the bump electrodes. Bumps are aligned over the conducting pads of the board and a pressure is applied as the adhesive is cured. Electrical continuity is provided by the direct metal-to-metal contact and maintained by the shrinkage of the epoxy adhesive. This technology was evaluated for bonding gold bumped dice in chip-on-flex and chip-on-board interconnections [99]. For flip chip interconnects with a pad size of 100 μm^2 on the chip site and a contact area of about 60 μm in diameter, the resistance is less than 8 mΩ. The second technique is illustrated in Figure 40.

A conductive epoxy adhesive is screen printed over the connecting pads of the wiring board on top of which the integrated circuit carrying gold stud bumps is flip

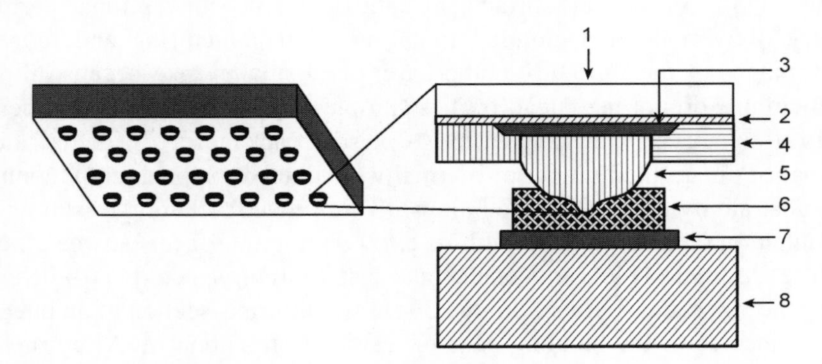

Figure 40: Flip chip process using gold stud bumps and epoxy conductive adhesive: (1) silicon substrate; (2) silicon dioxide; (3) metal bonding pad; (4) silicon nitride passivation; (5) stud gold bump; (6) silver-filled epoxy adhesive; (7) metal bonding pad; (8) printed wiring board.

chip bonded [100]. Reliability tests show that after 500 thermal cycles between −40 and 100°C, the maximum contact resistance between the bumps and the substrate does not exceed 10 mΩ. In addition, the rate of function errors with a gate array measuring 10 mm^2 is 0% after the following tests: 500 h at 100°C with a 5.5 V bias, 500 h at 85°C/85% RH with a 5 V bias, and 30 cycles (−40 to 100°C) followed by 144 h at 105°C and 100% RH.

The low-cost polymer flip chip technology, depicted in Figure 32, uses only a conductive adhesive to build the bumps on the integrated circuits and a bonding layer on the circuit board pads. Then, the devices are flip chip attached by applying a pressure during the cure schedule. Whatever the technology, the flip chip process is completed by introducing an underfill material to fill the gap between the chip and the board. As part of the assembly, the interfilling non-conductive resin primarily enhances the adhesion strength and provides environmental protection against corrosion, silver migration, and potential hydrolytic reactions during the accelerated ageing tests. Underfill materials are solventless low-viscosity epoxy resins to enhance the capillary flow under the chip. Low coefficient of thermal expansion and high T_g are also desirable to minimize bumps fatigue, hold the chip in compression, and provide dimensional stability.

6.5.7. Surface-mounted Components

For the past few decades microdevices encapsulated into ceramic and plastic packages have been implemented on printed wiring boards by inserting the connecting pins into the metallized through holes of the boards. Electrical continuity is then warranted by a soldering operation using tin–lead alloys. The number of components that can be inserted on a given board area is limited by the distance between holes and their diameter. The advent of SMT initially implied major changes in the design and fabrication of the components and PCBs as well, but only minor changes in solder bonding techniques. In SMT, the connecting pads are patterned by lithography on the board surface. However, the trend to a denser packing of components carrying a large number of small closely positioned terminal pads was an incentive to develop new conductive adhesives offering reliability levels as high as those achieved with the soldering process. Drawings of Figure 41 illustrates a high-density printed wiring board carrying both flip chip and surface-mounted components bonded by means of isotropic or anisotropic conductive adhesives.

The conventional method using screen printed solder pastes or isotropic conductive adhesives is difficult to apply when the distance between adjacent connections is less than 0.5 mm. Thus, the main impetus for developing anisotropic conductive adhesives has been the fine pitch interconnection of electronic devices

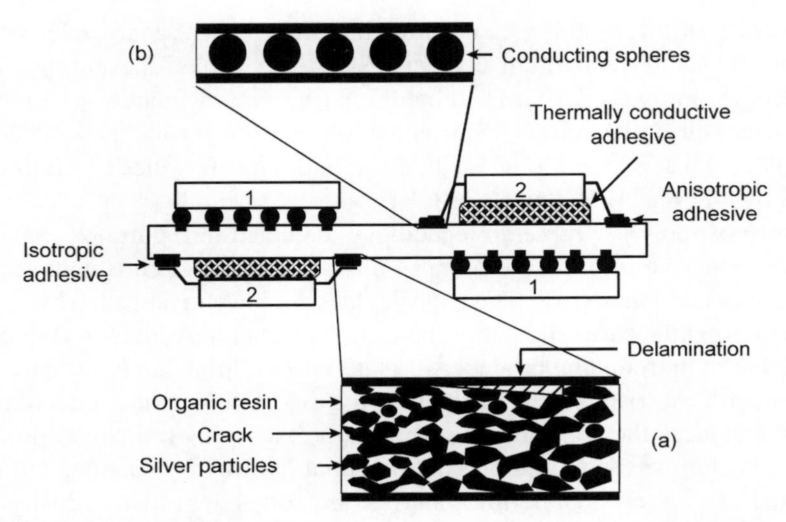

Figure 41: Schematic representation of microdevices attached to thick film metallization pads printed on a ceramic carrier in a multichip module carrying both flip chip 1 and surface-mounted 2 components. Enlarged views show the bonding area: (a) silver-filled isotropic adhesive with possible irreversible changes due to thermal cycling; (b) anisotropic adhesive made of conducting spheres embedded in thermoplastic or thermosetting resins.

used in SMT [101]. Most papers issuing from the scientific community deal with the theoretical models associated to the electrical and thermal conductivities and to the stresses in relation with thermal expansion. By contrast, technologists and manufacturing companies are more engaged in the evaluation of conductive adhesives in the assembly processes and during the expected working life of the devices. Most of the available data concern the behaviour of isotropic adhesives but it appears that a significant part of the current research is focussing on anisotropic adhesives in order to achieve finer pitch interconnections.

The development of anisotropic adhesives was initiated by the LCD industry but an extensive survey of the patent literature indicates other general trends [102]. The resins and fillers intended for use as anisotropic adhesives must satisfy new requirements compared to those previously defined for isotropic materials. In particular, the geometrical factors influence the formulation and the processing technique in relation to the final application: LCD, surface mounting or flip chip bonding.

In terms of technology, anisotropically conductive adhesives are either materials that are anisotropic before processing or those that develop anisotropic conductivity during the curing process. Figure 42(a) shows the former material supplied in the form of a tape manufactured by laser-drilling

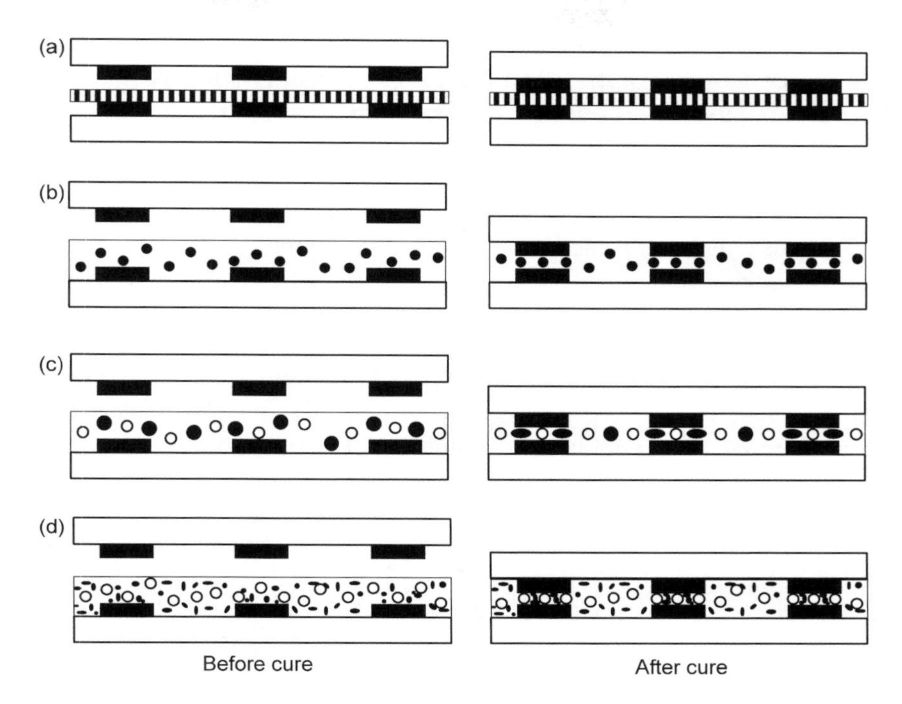

Figure 42: Main categories of anisotropically conductive adhesives: (a) materials anisotropic before processing; (b) adhesives containing metallic spheres with controlled diameter; (c) mixture of collapsible conductive filler and lower size non-conductive spacers; (d) the reverse situation with spacers larger than the electrically conductive particles.

or etching before filling the through holes with electrically conducting paste. This is a complex process with the advantage that the number of electrical contacts can be easily predicted in any geometrical arrangement of the pads. However, the majority of commercial anisotropic adhesives belong to the second class of materials, in which the vertical organization appears during the processing step by applying local pressure to the device leads positioned over the electrically conductive tracks.

Three technical approaches are used to do that. Figure 42(b) illustrates the behaviour of adhesives made by dispersing into an organic matrix spherical conductive particles with a tightly controlled diameter, e.g. nickel spheres with an average diameter of 50 μm. The process sketched in Figure 42(c) employs a mixture of large diameter collapsible conductive filler mixed with lower size insulating spacers. Examples are compositions prepared by loading either thermoplastic or thermosetting resins with polymer microspheres electroless plated with nickel, gold, or silver, used in combination with inorganic spacers.

The third process drawn in Figure 42(d) shows the reverse situation in which the spacers are larger than the electrically conductive particles. In this case, the loading level of silver is adjusted just below the percolation threshold so that any significant reduction of the film thickness results in the formation of elongated clusters ensuring the electrical continuity.

Many alternatives exist and the choice between thermoplastic and thermosetting resins depends on the expected lifetime and maximum operating temperature. Styrene–butadiene block copolymers and acrylic resins can be used to produce low-end adhesives with an acceptable stability up to 100°C, whereas epoxy and silicone thermosets are preferred for their robustness at 150–200°C. In the most severe environments, poly(imide-siloxanes) and polyimides can sustain medium-term exposures to 250 and 300°C, respectively. Various conductive fillers are cited in the literature, including noble metals such as gold or silver, and low-cost metals such as copper, nickel, chromium, and soft solders.

A survey of commercial conductive adhesives supplied by 15 leading vendors identifies about 95 relevant products, roughly 30% of them being anisotropic materials including 12% pastes and 18% films [102]. These adhesives are primarily proposed for low-temperature applications (100–150°C), which are representative of the LCD market. Thermoplastic anisotropic adhesive films are presented as either custom preforms or roll tapes, the polymer carrier being made of styrene-based block copolymers, acrylate or methacrylate resins, or other low-melting plastics. Because of the relatively low temperature required for the bonding process, these films present the advantage of avoiding any thermal damage to the integrated circuits and an easy rework when necessary.

Conversely, the main drawbacks are that the upper operating temperature must be kept below the polymer melting point and that the onset of thermal degradation is as low as 150–170°C. The upcoming class of adhesive films is based on thermoplastic polyimides, which can sustain temperature bursts at 350°C. Conversely, anisotropic thermosetting adhesive cements are mainly composed of epoxy resin systems similar to those used to prepare isotropically conductive pastes. They can be applied by screen printing, pin transfer, or syringe dispensing with, however, the necessity of applying a pressure to establish the electrical contacts along the vertical axis. The theoretical aspects of electrical conductivity in relation with the morphology of the conductive particles have been previously discussed [4].

The surface-mount assembly of fine pitch integrated circuits is a potential market for anisotropic adhesive films [103,104]. The best conductive adhesives exhibit a volume resistivity of 2 to 5×10^{-4} Ω cm with a contact resistance of 10 mΩ, although AI Technology claims volume resistivities of 5×10^{-5}–5×10^{-6} Ω cm, similar to that of metallic solder joints. This allows a current carrying capability of 50–60 A mm^{-2} compared to 20–30 A mm^{-2} for typical silver-filled adhesives. This series of materials is based on epoxy compositions with low glass transition temperature (-25°C) and rubberlike properties. The die shear strength at room temperature is about 10 MPa, decreasing to less than 2.8 MPa at 150°C, a value which is far from achieving the requirements of MIL-STD-883D. In contrast, the elastomeric properties are well adapted to improve the fatigue resistance of the adhesive layers used to bond alumina substrates to aluminium cans up to 12.7×12.7 cm^2. Obviously, the flexible molecular backbone of the low T_g adhesives allows a rework at temperatures as low as 80–100°C.

The degree of flexibility as well as the T_g and adhesion strength of epoxy adhesives can be engineered by varying the ratio of flexible and rigid epoxy resins. Tacky versions of rubber-like epoxy films are primarily designed to attach large size modules to various substrates, while tack-free films are best suited for the automated assembly of integrated circuits. Low-stress thermally conductive epoxies have also been applied to MCMs [105]. A typical MCM comprises 30 silicon dice dissipating a power level of 300 W with some dice generating 30 W cm^{-2} of electrical energy. In this approach, soft solders are replaced by *Z*-axis adhesives for the interconnection of fine pitch integrated circuits, jumpers, flexible circuitry, and surface-mounted components.

Both thermoplastic and thermosetting resins can be used to produce films or screen-printable pastes, the former materials making easier the repair operations, whereas the latter provide better adhesion strengths [106]. Computer programs have been developed to predict the limiting pitch, the minimum pad size for different filler concentrations and the minimum pad spacing as a function of the particle sizes [107]. The important process parameters are the pressure applied during cure and the cure time and temperature. A pressure lower than 1.38 MPa leads to a high resistance with some open circuits, the optimum pressure being around 2.07 MPa. For cyanate ester films, the resistance decreases to a minimum value when the adhesive is cured between 180 and 200°C and then increases over two orders of magnitude between 200 and 240°C. Computer simulation shows that the transition from failure to reliability is in a narrow region. As an example, for an adhesive film with a loading density of 500 particles mm^{-2}, the width of the conduction pads must be at least 0.18×0.18 mm^2. Conversely, the particle loading and size can be determined by the program for a given pad size and pitch.

The electrical properties and the reliability of connections made of an anisotropic adhesive loaded with both deformable conductive particles and non-deformable dielectric particles used as spacers have been reported [108]. The amount of deformation of the conductive particles is calculated from their diameter, 10 μm, relative to the diameter of the spacers, 4, 6, 8, and 10 μm. Plotting the electrical resistance of the connection as a function of the number of particles, the authors show that the resistance decreases when the particle count expands from 5 to 60 and when the amount of deformation increases from 20 to 60%. The second observation is that the current–voltage characteristics are linear up to a certain current, whose value augments with the number of conductive particles for a given amount of deformation. Thus, the greater the current-carrying capacity, the greater the number of conductive particles and the larger their deformation. This can be explained by the enlarged contact area due to the deformation of the conductive particles. When the connections were subjected to the 85°C/85% RH ageing test for 2000 h, it was found that the lowest electrical resistance and the highest reliability were obtained with 20 particles deformed by 40%.

These experimental results have been obtained primarily with anisotropic adhesive tapes. However, tests have also been performed with anisotropic pastes based on epoxy resins and thermoplastic polyimides loaded with 50 μm gold-plated nickel spheres [109]. The thermal resistance, determined with power MOSFETs attached to ceramic tiles coated with thick film silver–palladium metallization, is 33°C W^{-1} compared to 10–15°C W^{-1} for isotropic adhesives. A possible reason for this difference is the relatively low volume of metallic filler. Chip capacitors bonded with the anisotropic paste without any pressure applied to the components do not exhibit any conductivity. In contrast, when a compressive force of 0.2 N is applied during the cure cycle, the contact resistance of the capacitor bonds decreases to 0.24–0.73 Ω. The initial bond strength of the anisotropic adhesive is 5.3 MPa for the "2200" capacitors and 30.6 MPa for the "1206" resistors, compared to 57.9 MPa for a typical isotropic adhesive. One of the problems connected with the random dispersion of conductive spheres is the non-uniformity of the particle distribution. More uniform particle organization can be obtained by applying a vertical magnetic field of 7.96×10^3–3.2×10^4 A m^{-1} to adhesives loaded with ferromagnetic nickel spheres electroless plated with a 120 nm thick layer of gold [110].

Anisotropic adhesives prepared from epoxy resins loaded with 5–20 vol.% of lead-free powdered solder alloys perform well when they are subjected to hot and humid environment tests for more than 2000 h. Scanning electron microscopy reveals that a metallurgical bond is formed between the Sn–Bi solder particles and the copper conductors because the bonding temperature is above the melting point of the filler [111]. The contact resistance also remains constant during 300 cycles between −40 and 100°C, but increases from 0.1 to 2 Ω after 300 h at 85°C/85% RH. As expected, the shear strength is slightly better (12.5 MPa) for the lowest amount of filler than for adhesives containing more than 10% filler (7.5 MPa).

6.5.8. Liquid Crystal Display Devices

The simplified drawing of Figure 30(c) shows that two categories of adhesives are used to fabricate LCD panels. The liquid crystal cell is formed of top and bottom multilayered glass plates separated by a gap of 5–10 μm and sealed together [24]. The perimeter seal attaching the two glass plates at the edge of the active area is made of UV or heat curable epoxy resins. To switch the liquid crystals between on and off states, the glass plates are coated with a thin film of transparent conducting material formed of tin oxide (SnO_2) or indium–tin-oxide (In_2O_3–SnO_2), known as TO and ITO, respectively. A series of conducting columns and rows are patterned by photolithography and then connected to the integrated circuit driver by means of a flexible circuitry. Both isotropic and anisotropic adhesives are used for bonding the flexible circuit to the ITO electrodes and driver. When compared to the die attachment technology, the assembly of LCDs imposes new requirements such as low bonding temperatures, an accurate placement over fine pitch leads, and lesser production costs. The manufacture of LCDs involves different interconnection techniques which have been rapidly changing in terms of technological advances [112]. The survey of the packaging technologies developed for active matrix liquid crystal displays (AMLCD) is also valid for less demanding panels using passive addressing and either twisted or super-twisted nematic liquid crystals [113]. The main packaging methods are illustrated in Figure 43, which includes the chip-on-board (COB), the chip-on-film (COF) and the chip-on-glass (COG) techniques to bond the integrated circuit driver and the control circuit to the liquid crystal panel.

In COB packaging (Fig. 43(a)) the electronic module is made of bare chips wire bonded to the PCB, which is connected to the liquid crystal panel with a flexible heat seal connector (HSC). Flexible connectors are fabricated by screen printing on a polyester film two layers of conductive pastes, containing either silver, or graphite, or silver-graphite particles, the conductors being then insulated by printing a hot-melt resin between the traces and over the entire top surface. The advantage of COB packaging is a reduction of the peripheral area of the display panel because the electronic module can be placed on the side or on the backside of the light module. The HSC is accurately positioned on the substrate and melted (thermoplastics) or cured (thermosets), the process being repeated to bond the second substrate. Thermosetting adhesives offer better reliability whereas thermoplastic materials present processing advantages because they can be reworked [114–118]. Mixtures of thermoplastic poly(vinyl butyral) and thermosetting epoxies are also employed to make anisotropic conductive films loaded with indium alloy particles with an average diameter of 10 μm. Reliability tests have been performed with an HSC made of a supporting polyester carrying carbon tracks and a 30 μm thick thermoplastic adhesive film loaded with conductive carbon particles [119]. The measured values of the peel strength, contact resistance, and seal length indicate highly reliable joints for low-end applications, although all the ageing tests do influence the quality of the interconnections.

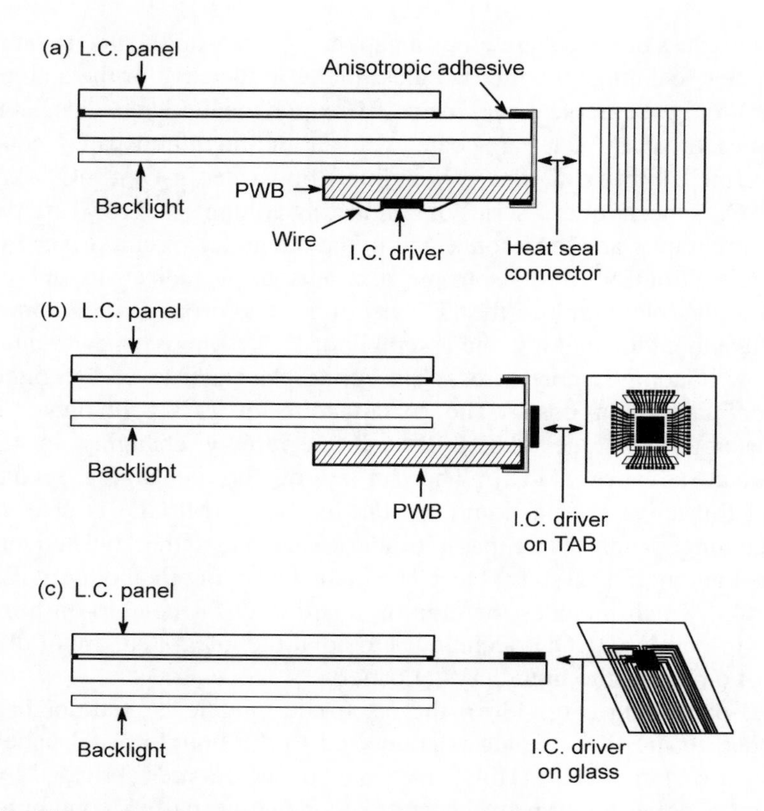

Figure 43: Illustration of the main packaging technologies of liquid crystal display devices: (a) chip-on-board; (b) chip-on-film; (c) chip-on-glass. (adapted from Ref. [113], Copyright© 1993, Penwell Publishing Company.)

The chip-on-film packaging (Fig. 43(b)) is the most popular because it uses, as a bridge between the display panel and the printed wiring board, a TAB interconnect (see Section 6.5.4) carrying the bare integrated circuit driver. In chip-on-glass packaging (Fig. 43(c)), the integrated circuits are directly connected to the metal pads patterned on the peripheral area of the liquid crystal panel. The electrical interconnection is performed by using different methods including wire bonding, soldering, isotropic and anisotropic conductive adhesives, and flip chip bonding. An obvious trend is the development of anisotropic adhesives for chip-on-glass or flip chip applications in which the bond pads are much smaller than the bonding areas encountered with flexible connectors. It has been observed that the number of conductive particles in standard anisotropic adhesives is such that only a few particles are present on a $100 \times 100\ \mu m^2$ bond pad [120]. However, the layout of the pads has a significant effect on the contact resistance, which decreases from 23 to 14 and 6 Ω by changing the layout. Conversely, both the bonding time and bonding force are not critical although the contact resistance decreases by 10% when the bonding force is increased from 1 to 2 N per bump.

A recent publication describes a pragmatic approach for developing a thermosetting anisotropic conductive adhesive film to electrically connect a flexible circuit to ITO terminal pads [121]. The basic components are a solid high molecular weight bisphenol-A diglycidyl ether, a liquid bisphenol-F diglycidyl ether, a phenoxy resin ($M_w = 4.6 \times 10^4$; $T_g = 94°C$), (2,3-epoxypropyl) trimethoxysilane, a microencapsulated 2-methylimidazole hardener providing a shelf life of 1 month at 40°C, and 3 vol.% of conductive particles made of crosslinked polystyrene beads plated successively with nickel and gold. The adhesive film is prepared by mixing all these ingredients in a 3:1 toluene/2-butanone mixture, coating the solution on a polyethylene terephthalate support and evaporating the solvent. The adhesive properties of the anisotropically conductive film are determined by first positioning the film on the ITO glass at 80°C for 5 s under a pressure of 4.9×10^5 Pa, then placing the polyimide flexible circuit in the same conditions, and finally curing the assembly at 170°C for 30 s with an applied pressure of 2.94 MPa. Before curing, the peel strength of the adhesive film is 0.94 N cm^{-1} increasing to 11.8 N cm^{-1} after the fast cure cycle. The initial contact resistance of 0.3 Ω increases to about 0.9 Ω after 2000 h at 85°C/85% RH and remains stable after leaving the assembly for 2000 h at 100°C.

A process for chip-on-glass bonding of bumped integrated circuits to ITO conductors has been developed by using a two-part isotropic epoxy adhesive loaded with silver flakes [122]. The first observation is that the volume resistivity and contact resistance depend on the type of flakes and the percentage by weight of particles dispersed in the adhesive composition. According to the kind of flakes, the volume resistivity of adhesives loaded with 75–80 wt% silver varies from 1×10^{-4} to 1.8×10^{-3} Ω cm. Other vital parameters include viscosity of the adhesive, screen printing characteristics, and cure cycle. Theoretically, silver-filled isotropic epoxy adhesives may exhibit a volume resistivity as low as 1.5×10^{-4} Ω cm, which ensures a contact resistance of less than 2 Ω per bump between the bump and the ITO pad. In practice, this value cannot be achieved because of the defects arising from the thermomechanical stresses and the characteristics of the ITO films. However, a contact resistance of about 6–8 Ω is easily obtained and remains virtually unchanged after a series of ageing tests at 85°C, at 85°C/85% RH, at 60°C/90% RH, and after thermal shocks between −40°C and 85°C.

Currently, the limit of the connection pitch of anisotropic adhesives is held to about 100 μm to circumvent possible shorting between neighbouring conductors. An ideal process would be to avoid the use of a conductive adhesive by performing a direct metal-to-electrode contact between the conductive tracks of the flexible circuit and the electrodes of both the electronic driver and the display panel.

This option is currently being evaluated to determine the reliability of the direct electrode contact concept when the assembly is subjected to accelerated ageing tests. This can be illustrated by the interconnection technique using copper tracks etched on the surface of a polyimide carrier as the flexible circuitry [123]. A lattice of vertical microholes is etched in the polyimide film and the holes are injected with a hot fluid adhesive to ensure the requisite insulation when the conductor lines and the electrodes are bonded under pressure to create the electrical connection. In this process, the shrinkage occurring during crosslinking of the adhesive material is used to bring the copper tracks and the terminal electrodes into intimate contact. When subjected to the environmental tests generally accepted to check the reliability of LCD connectors, this interconnection system exhibits a stable contact resistance.

6.6. Reliability of Adhesive Bonds

6.6.1. *Integrated Circuits*

As stated in Section 6.5.1, polyimide films are now permanent parts of microcircuits as interlayer dielectrics, protecting overcoats, α-ray barriers, and buffer coatings. The design criteria can be summarized as follows. Thermal stability up to 400–450°C and high mechanical properties are afforded by a number of linear aromatic polyimides but their planarizing properties are limited to 13–15%. Better planarization (30–80%) is obtained by using low molecular weight prepolymers with low melting temperature. Moisture uptake is reduced by suppressing water absorbing groups. Low dielectric constant and dissipation factor are obtained by increasing molecular symmetry and incorporating fluorinated monomers. Lithographic sensitivity is carried out by using photosensitive polyimides. Once the best polymer profile has been delineated, it remains that low thermal expansion coefficient and good adhesive properties are also prerequisite conditions. The simplified drawing of Figure 30(a) shows the different interfaces existing in high-density interconnect schemes.

Starting from a semiconductor or ceramic substrate, patterned with metal steps, polyimide level-1 exhibits two interfacial regions defined as polyimide-on-substrate and polyimide-on-metal. Subsequent deposition of metal layers generates metal-on-polyimide and polyimide-on-polyimide interfaces. The specific characteristics of these four main interfaces have been previously summarized with a special attention given to the leading role of aminosilanes [24]. These coupling agents are particularly effective to maintain a good adhesive strength when the integrated circuits are subjected to temperature–humidity tests. The excellent review by Matenzio and Unertl is highly recommended for those with keen interest in the subject of metal–polyimide interfaces [124]. The same adhesion problems are encountered in the fabrication of MCMs.

6.6.2. Multichip Modules

Adhesion of polyimide to silicon, metals, and ceramics has been studied by means of the island blister test sketched in Figure 28 [75]. Different equations provide self-consistent values for the adhesion energy G_a, which can be reliably determined from the pressure data. The effect of cure temperature on the adhesion energy of PMDA–ODA polyimide **93** coated over silicon illustrates the need for high temperature curing. G_a increases from 1.37 to 25 kJ m^{-2} for final cure temperatures of 300 and 450°C, respectively. The same polymer exhibits adhesion energies of 0.03, 0.52, and 13.8 kJ m^{-2} over gold, copper, and aluminium, respectively. By contrast rigid BPDA–PPDA polyimide **95** exhibits poor adhesion to silicon ($G_a = 0.1–0.3$ kJ m^{-2}). Unreactive noble metals do not inhibit the thermal imidization of polyamic acids, provide relatively clean interfaces, and generate the weakest adhesion strength, typically 200 J m^{-2} for gold [125]. Interaction of polyamic acids with copper is characterized by the formation of copper carboxylates, which inhibit imidization, and copper oxidation. These reactions are not observed with preimidized polymers. Over nickel surfaces, the native nickel oxide is primarily in hydrated form whose dehydration at 300°C would generate Ni^0 species that are powerful oxidation catalysts. On such surface, the polyimide backbone is rapidly oxidized, while it remains unaffected on thermally grown nickel oxide. Adhesion energies measured at the chromium–polyimide interface are of the order 1.0–1.2 kJ m^{-2} without observable metal migration [126]. Accordingly, chromium constitutes an excellent adhesion and passivation layer between polyimide films and high conductivity metals such as copper and silver.

Another source of potential failure is the thermal expansion coefficient mismatch between polyimides and silicon, ceramics, or metals (Table 4 in Section 6.4.2.7). First generation polyimides, such as Hitachi PIQ® 13 and Du Pont Pyralin® PI 2545 and PI 2555 have high CTE, typically lying in the range $4–6 \times 10^{-5}$ K^{-1}. The effect of chemical structure on the coefficient of thermal expansion has been previously discussed [24]. Polyimides synthesized by reacting rigid (PMDA) or semi-rigid (BPDA) dianhydrides with rigid aromatic diamines exhibit lower CTE values dropping to $1.4–2.9 \times 10^{-6}$ K^{-1}. Stresses at the silicon–polyimide interface have been calculated for flexible BTDA–ODA–MPDA (PI 2525), semi-flexible PMDA–ODA (PI 2540), and rigid BPDA–PPDA (PI 2611) polyimides. The residual stresses generated in these films coated over silicon wafers are 54, 43, and 4 MPa, respectively, whereas the stress in silicon is less than 0.6 MPa. To achieve long-term reliability, the most desirable properties are low dielectric constants allowing faster signal propagation. Also low level of water absorption is a prerequisite because humidity degrades both adhesion strength and permittivity. It has been shown that the dielectric constant increases linearly by increasing the water content in polyimide films. The adhesion strength in humid environment is enhanced by coating the substrates with organosilane coupling agents such as (3-aminopropyl)triethoxysilane.

6.6.3. Flexible Circuits

As indicated in Section 6.5.4, the industry of flexible circuitry proposes various substrate–metal combinations including polyester and epoxy resin coated on glass fabrics, high T_g polyimides (Kapton®, Upilex®, and Novax®), and thermoplastic polyimides (Regulus®, Larc TPI®, and Ciba-Geigy XU 218®). For the two-layer copper–polyimide construction scheme, low thermal expansion polymers are preferred to avoid curling or other geometrical deformation. In the three-layer system, the adhesives inserted between copper and polyimide film have to be thermally resistant while exhibiting elastomeric properties. They are generally composed of aliphatic polyamide copolymers mixed with epoxy resins. For simple flexible circuits, blistering and deformation at soldering temperature are the main concerns, while TAB-mounted devices are subjected to additional tests to determine the reliability of the die attachment.

6.6.4. Die Attachment

The reliability of die attach adhesive has been extensively investigated for the past three decades. A complete survey of these studies has been previously published [4]. The following sections underline the critical factors that have an impact on the electrical, mechanical, and thermal properties of the adhesive bonds. For process engineers, the main concern is the evaluation of die attach integrity in relation with the assembly technologies and after a series of accelerated ageing tests.

6.6.4.1. Porosity

The formation of voids or non-wetted areas is observed with virtually all adhesive pastes as a result of incomplete removal of solvent and unreacted species. A visual appreciation of porosity has been done by bonding integrated circuits to glass slides and examining the bond line under a microscope with a 100 times magnification [127]. By reducing the effective bonded area, porosity has a significant effect on the adhesive strength which decreases by a factor of 7–8 when large voids are formed on curing. Experiments performed with $1.016 \times 1.016\ \mathrm{mm}^2$ dice bonded to silver-plated copper leadframes show that the die shear strength achieved with porous adhesives (16 MPa) is about half the value obtained with less porous material (34 MPa). This study also demonstrates that the formation of voids is due to volatile materials present in the silver-filled epoxies. All adhesives cured in "open-to-air" conditions over Kapton® films exhibit a volume resistivity of $1.3-1.6 \times 10^{-4}\ \Omega$ cm. However, when they are cured between two polyimide strips, the volume resistivity remains constant for non-porous adhesives but is an order of magnitude higher ($2.1 \times 10^{-3}\ \Omega$ cm) for the porous materials. The surface resistivity is even more sensitive to the porosity with respective average values of 6.25×10^{-3} and $4.3 \times 10^{-3}\ \Omega\,\square^{-1}$ for the porous and non-porous adhesives cured open-to-air, increasing, respectively, to 6.75×10^{-1} and $1.1 \times 10^{-2}\ \Omega\,\square^{-1}$ in the covered surface cure.

There is a general agreement to consider that the lifetime of a semiconductor device is an exponential function of the junction temperature, decreasing by approximately a factor of two for every 10°C increase in temperature. Depending on the package design, heat generated at the transistor junctions is transferred to the surrounding atmosphere mainly by the header, the lead frame, or the packaging material. In all cases the quality of the bond line between the integrated circuit and the chip carrier is of prime importance. The effect of voids on the heat transfer capability of different packages has been studied by using thermal transient tests and steady-state infrared scanning technique [128]. The different package constructions employed in this study are illustrated in Figure 44, which shows a general view of the packaging structures frequently encountered. In this drawing, T_J and T_C are, respectively, the temperatures of the junction and the case while the thermal conductivity λ of the materials is given in $\mathrm{W\,m^{-1}\,K^{-1}}$. Either small randomly dispersed voids or contiguous large voids have been produced into both the soft solder and silver-filled epoxy adhesive.

The TO-3 steel package is composed of a copper header **2a**, electroless plated with nickel **3** and brazed to a 1020 steel case **1a**. Semiconductor chip **5** is bonded to this substrate by means of a soft solder **4a**. The TO-3 aluminium package comprises an aluminium header **1b** with a nickel plating **3** on the top surface. Silver-filled epoxy adhesive **4b** is used for bonding the integrated circuit **5**. The 24-lead CERDIP package is formed of a nickel-plated **3** alumina substrate **1c** on the top of which the active device is bonded with soft solder **4a**. The 40-lead P-DIP package includes a silicon chip **5** bonded to a copper leadframe **2b** encapsulated in an epoxy moulding material **1d** and **6b**.

In all packages, the heat generated at the junction primarily flows through the silicon die, the solder or adhesive bond, the nickel plating when existing, and the base substrate. In the 40-lead P-DIP, thermal dissipation occurs in any direction through the moulded epoxy in contact with the chip. The experimental data

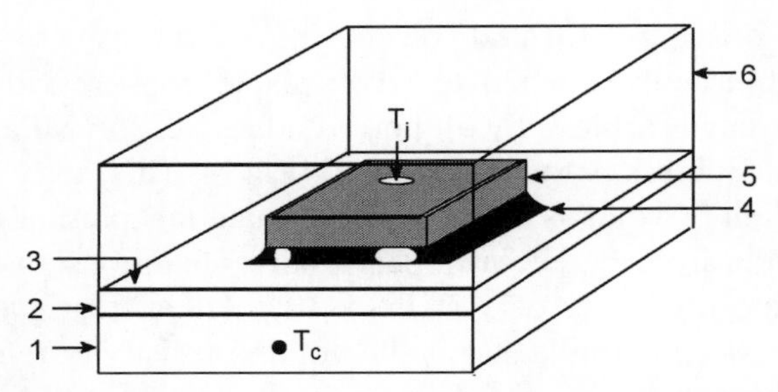

1a: 1020 steel: $\lambda = 60\ \mathrm{W\,m^{-1}\,K^{-1}}$
1b: 5062 aluminium: $\lambda = 140\ \mathrm{W\,m^{-1}\,K^{-1}}$
1c: Alumina: $\lambda = 25\ \mathrm{W\,m^{-1}\,K^{-1}}$
1d: Epoxy moulding: $\lambda = 0.8\ \mathrm{W\,m^{-1}\,K^{-1}}$
4a: Soft solder: $\lambda = 20\ \mathrm{W\,m^{-1}\,K^{-1}}$
4b: Ag-epoxy: $\lambda = 1.7\ \mathrm{W\,m^{-1}\,K^{-1}}$

2a: OFHC copper: $\lambda = 390\ \mathrm{W\,m^{-1}\,K^{-1}}$
2b: Copper alloy: $\lambda = 260\ \mathrm{W\,m^{-1}\,K^{-1}}$
3: Ni-plating: $\lambda = 8\ \mathrm{W\,m^{-1}\,K^{-1}}$
5: Silicon: $\lambda = 110\ \mathrm{W\,m^{-1}\,K^{-1}}$
6a: Steel, aluminium or ceramic lid
6b: Epoxy moulding: $\lambda = 0.8\ \mathrm{W\,m^{-1}\,K^{-1}}$

Figure 44: Package constructions used to evaluate the thermal transfer properties of the die attachment materials and substrates. TO-3 steel package is composed of steel case 1a, OFHC copper 2a, nickel film 3, soft solder 4a, and integrated circuit 5. TO-3 aluminium package is formed of aluminium header 1b plated with nickel 3 on the top of which a silver-filled epoxy adhesive 4b is employed to bond the semiconductor device 5. The 24-lead CERDIP package is based on nickel-plated alumina substrate 1c with soft solder bonded silicon chips. In 40-lead plastic package, integrated circuits 5 are bonded to leadframes 2b by using silver-filled epoxy adhesive 4b and then encapsulated in moulding epoxy resins 1d.

are presented by plotting the thermal resistance from the junction to the case, Θ_{JC}, versus the percentage of voids in the die bond. For the steel TO-3 package, the values separate into two bands corresponding to the random and contiguous voids, respectively. As the void fraction rises from 10 to 50%, Θ_{JC} augments by nearly 20% (from 1 to 1.2°C $\mathrm{W^{-1}}$) and 400% (from 1 to 4°C $\mathrm{W^{-1}}$) for the random and contiguous voids, respectively. Computer calculation clearly shows that the effect of small random voids on Θ_{JC} is much smaller than for large single voids and that, for the same level of voiding, the dissipation of heat is more effective in the former case. The lower thermal conductivity of both the aluminium case and epoxy adhesive increases the thermal resistance of the aluminium TO-3 package. At near zero void percent, initial Θ_{JC} ranges from 3 to 5°C $\mathrm{W^{-1}}$, being quite stable for less than 30% voids and increasing slightly ($\Theta_{JC} = 6$–7°C $\mathrm{W^{-1}}$) for void concentrations above 50%. In the case of 24-lead CERDIP combined with soft solder, Θ_{JC} increases by about 40%, from 5 to 7°C $\mathrm{W^{-1}}$ when the percentage of voids increases from 10 to 50%. As expected, the thermal performance of the assembly is poorer than that of the TO-3 metal can but the thermal resistance rises less than the random small voids of the TO-3 package. For the 40-lead P-DIP, the λ values are 0.8 and 260 W m^{-1} K^{-1} for the moulding epoxy and the copper leadframe, respectively. The experimental results, obtained with this package in which the die is bonded with a silver-filled epoxy resin containing small, randomly dispersed voids, show that void concentration has a very little effect on thermal resistance. The value of Θ_{JC} remains virtually constant at about 6°C W^{-1} when the void concentration varies from 10 to 90%.

To conclude this study, the authors outline the strong interplay among the package construction, the die attach material, the overall void concentration, and the specific characteristics of the voids. The packages in which the die bond is the dominant route for the overall heat flow are more adversely affected by the level of voiding than are poorer performing thermal packages. The metal cans used to house high-power devices are the most sensitive to the amount of voids in the die bond layer. Conversely, in CERDIP and P-DIP packages, small concentrations of random voids have little effect on the peak junction temperature. However, high concentrations of large contiguous voids degrade the thermal performance of all the packages investigated. This study shows that the voids have a significant effect on the thermal resistance of the different adhesive layers. However, Feinstein points out that a level of voids as high as 50% has little influence on the total thermal transfer of most plastic packaged devices [32]. To illustrate this point, he provides the example of a P-DIP package mounted on a printed wiring board, in an ambient temperature of 30°C, with a typical value of Θ_{JC} of 30°C W^{-1}. If the temperature of the junction must be restricted to 150°C, the maximum admissible power is about 4 W. A level of voids of 50% would increase the thermal resistance of the package by 2°C W^{-1}, thus decreasing the power handling capability by only 6% to 3.75 W.

6.6.4.2. Water Uptake

The kinetics of water vapour diffusion through polymer films is a fast process that leads to equilibrium between the bulk of the organic matrix and the surrounding atmosphere within minutes. At equilibrium, the water content of a polymer depends on both the chemical structure and the degree of porosity. The water uptake of non-polar polymer, such as polyethylene and poly(tetrafluoroethylene) is as low as 0.5 wt%, whereas it can be of the order of 14% for hygroscopic polymers such as polybenzimidazoles. According to their chemistry, epoxy resins and polyimides can absorb up to 6% of water, which plasticizes the matrix and reduces its glass transition temperature. On the other hand, the absorbed water has a swelling effect, resulting in an expansion of the matrix similar to the thermal expansion in the sense that the conductive particles could be separated from each other with a loss of electrical conduction.

The precise mechanisms by which moisture degrades a conductive adhesive bond are not well established because it is likely that many effects interact simultaneously. On a macroscopic scale, the influence of water must be separated into three main categories: the effects of water retained within the matrix, those of water diffusing to the interfacial boundaries of the adhesive layer, and the transport properties of water. The water content of a polymer is linearly related to the ambient relative humidity, and crack propagation is faster in humid atmosphere than in dry ambient [129]. The diffusion of water to the interfaces between the adhesive layer and both the die and the chip carrier can lead to complete debonding if a continuous film of water is formed on the surfaces of the adherends. The most efficient method to counteract this adverse influence of water is the use of silane coupling agents, which not only reduce the hydrophilic character of the metal and oxide surfaces but also considerably enhance the adhesive strength in humid conditions. The reliability data published so far indicate that chip lift-off is rarely observed in the relative humidity of the ambient atmosphere, whereas the phenomenon often occurs in the pressure cooker test at 121°C in saturated water vapour.

6.6.4.3. Ionic Contamination

The development of silver-filled polyimide adhesives for the die attachment of integrated circuits in plastic packages has been explained by their good performance — longer time to failure — in pressure cooker and other accelerated temperature–humidity tests. Unlike epoxy resins, which are manufactured by a process producing Na^+ and Cl^- as undesirable byproducts, polyimides are prepared by using highly purified reactants which do not introduce significant amounts of ions. Silver-filled polyimide adhesives may, therefore, be made to a very high degree of ionic purity (less than 10 ppm of Cl^- and Na^+). Accordingly, the use of conductive polyimides for die attachment, together with cleaner epoxy moulding compounds, has decreased the possibility of failure by corrosion of the integrated circuits encapsulated in plastic packages. First-generation epoxy die attach materials were characterized by high levels of ions, typically 600 ppm in extractable Cl^- and 100 ppm or more in Na^+ and K^+. In the presence of water vapour, the chloride ions can migrate to the die surface and initiate reactions with the aluminium metallizations (conducting lines) or attack the aluminium bond pad areas by a catalytic mechanism. Low-chloride epoxy resins of the second-generation were then developed by companies involved in the market of die attach adhesives such as Ablestik, Amicon, Epoxy Technology, Hitachi, Sumitomo, and others. These new materials are proposed with ionic content of 5–6 ppm for Cl^-, 2–7 ppm for Na^+, and 2–4 ppm for K^+. When ion extraction is done in the pressure cooker (121°C, 0.1 MPa), acetate, formate, and other anions of organic acids are formed by hydrolysis of the epoxy resins. There is, however, no evidence of corrosion of the aluminium metallization caused by these organic acid anions.

6.6.4.4. Silver Migration

Another concern with electrically conductive adhesives arises from the presence of silver. Plastic packages are permeable to moisture and, in a moist environment, silver has been seen to migrate under bias to the surface of the die, producing electrical shorts. Silver migration has been observed from a silver-filled adhesive acting as the anode towards any nearby conductor if a constant d.c. voltage is applied between the two conducting lines. The voltage gradient required for silver to migrate and to finally form a conductive path of metallic silver between the conductors is of the order of 0.04 V μm^{-1}. An effective solution to obviate the problem of silver migration in military hybrids is the use of gold-filled adhesives or to substitute a silver–palladium alloy for pure silver. However, even with a palladium content in the alloy exceeding 30%, silver migration is only retarded but not totally suppressed [130]. It is worth noting that migration of silver occurs only if droplets of water completely bridge the interval between the two conductors. The phenomenon does not appear when the specimens are exposed for 2 weeks to 80–100% relative humidity at 80°C under an applied voltage up to 400 V. Comparative experiments performed with silver-filled epoxy and polyimide adhesives have shown that neither the level of extractable chlorine nor the glass transition temperature has a significant effect on the kinetics of silver migration [131]. Water-repellent silicone resins deposited over the metallic conductors as a 75–125 μm thick coating impede the formation of a continuous liquid water path at the interface. Therefore, as long as the silicone film adheres to the substrate and the metallic conductors, the migration of silver is prevented.

6.6.4.5. Electrical Properties

The saturation voltage, V_{CE}(Sat), of a transistor is a measure of the voltage between the collector and the emitter when the base current is saturated. In a bipolar transistor, the voltage between the collector and the emitter is the product of the collector current and the collector–emitter resistance. As an example, saturation occurs for the Motorola NPN 2N-3055 bipolar transistor when a current of 400 mA d.c. is applied to the base and 4 A d.c. to the collector at a 2% duty cycle. The die attach adhesive acts as a resistor in series with the intrinsic resistance of the semiconductor, and accordingly contributes to the total collector–emitter resistance. Tests have been done to compare the electrical stability of six silver-filled epoxies to that of a soft solder by measuring the changes in V_{CE}(Sat) during thermal ageing at 150°C for 1000 h [132]. The transistors attached with the soft solder exhibit the lowest saturation voltage (43 mV) which does not change after thermal ageing. Most of the tested epoxies have an initial V_{CE}(Sat) approaching the value of the soldered devices but all these silver-filled adhesives show electrical instability over the ageing test. However, some of them are fairly stable with an increase of only 6.5% over 1000 h at 150°C.

The electrical conductivity of gold- and silver-filled adhesives is addressed in Section 6.3.2 and the qualification standards in Section 6.4.2.9. Although this latter directive limits the thermal ageing conditions to 150°C, different studies have been published on the behaviour of organic adhesives at higher temperatures [133,134]. The volume resistivity of commercial adhesives is of the order of 2.25×10^{-4} Ω cm dropping to a low 2.74×10^{-5} Ω cm for the best solventless two-part material. After the conductive adhesives have been heated at temperatures between 100 and 300°C the resistivity of all epoxies decreases slowly at or below 150°C and rapidly at 300°C to reach 2×10^{-6} Ω cm after 1000 h at that temperature. The authors' hypothesis is that this increase in conductivity is associated to the thermal degradation of the epoxy resin and a possible sintering of the silver flake. These results consequently imply that the volume resistivity of the conductive epoxies subjected to high-temperature annealing does not increase and that the changes in V_{CE}(SAT) of the devices attached with these epoxies relate to something other than variations in the bulk resistivity of the adhesive layer. It has been suggested [135] that the fluctuating contact resistance observed when the electronic devices are bonded with silver-filled epoxies would be explained by the difference between the coefficients of thermal expansion of the conductive filler and of the resin.

The ageing behaviour at 300°C contrasts with the changes observed at lower temperatures. At 300°C, the normalized bulk resistance of the specimens increases by a factor of 30 within a few hours, and then drops rapidly in less than 100 h. At lower temperatures there is a long period where the resistivity remains approximately constant, followed generally by a decrease in resistance. The onset of variation of the initial resistivity depends on the ageing temperature and ranges from about 100 h at 250°C to 7000 h at 150°C. The electrical resistance changes during long-term thermal ageing, from 150 to 300°C, have also been investigated by using simulated transistor specimens adhesively bonded with silver-filled compositions. At 250 and 300°C the response of the adhesive layer is similar to that of the bulk resistivity specimens with an increase in resistance from 5 to 50 mΩ in 5 h at 300°C and from 5 to 10 mΩ after 1000 h at 250°C. At lower temperatures, the resistances of the simulated transistors are relatively stable with a deviation less than 50% from their initial values after 3000 h at 200°C and 7000 h at 150°C. It has been noted that in the temperature–humidity test (85°C/85% RH) significant differences can be observed according to the metallization investigated, i.e. gold, copper, or silver–platinum [136]. On copper terminal pads, the test circuits exhibit less variation in electrical resistance when they are coated with a silicone resin.

6.6.4.6. Adhesive Strength

Conductive adhesives should provide adequate electrical and mechanical connections over the expected lifetime of the assembly. Consequently, the effects of long-term ageing on the reliability of die attach materials have been studied since the early 1970s. For many applications, the bond integrity should be maintained at operating temperatures of 125–150°C. Although the adhesive strength of polymers degrades with high-temperature storage, silver-filled epoxies exhibit no decrease in die shear strength and bulk resistivity with high-temperature storage, even after 10,000 h at 150°C. This means that the adhesion between the silver-filled polymer and both the die and the substrate is maintained. The key experiments were conducted during the 1970–1980 decade to establish the behaviour of conductive adhesives subjected to accelerated ageing tests. This corresponds to the period where the consumption of polymeric materials significantly increased. A few other papers were published in the early 1980s when conductive adhesives had been generally accepted by the industry as reliable die attach materials. The most recent studies mainly deal with the other applications of silver-filled polymers.

Figure 45 shows the change in the lap-shear strength of a one-part silver-filled epoxy during thermal ageing at 100, 150, 200, 250, and 300°C [133]. The important effect of the annealing temperature is evident, with a drop in shear

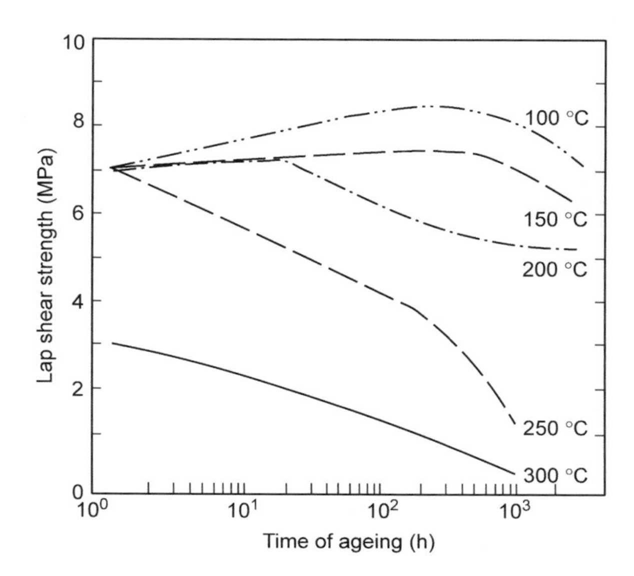

Figure 45: Variation of the lap shear strength of one-part silver-filled epoxy adhesive as a function of ageing time at different temperatures. (adapted from Ref. [133], Copyright© 1976 IEEE.)

strength from 7 to 2.8 MPa within 1 h at 300°C. In the 100–200°C region, the increase in lap-shear strength, which is commonly observed with epoxy resins, indicates the occurrence of some additional curing. The authors outline that the mechanical stability during long-term exposure at 150 and even 200°C of several silver-filled epoxies appears to be more than that required for reliably bonding integrated circuits to lead frames.

The behaviour of commercial silver-filled epoxy and polyimide resins is illustrated by the curves of Figure 46 which show the variation of the die shear strength, measured at 25°C, as a function of the ageing time at different temperatures. These results have been obtained with $0.41 \times 0.41\ mm^2$ silicon devices bonded to gold-plated nickel lead frames [134]. Plotting the shear data in the usual Arrhenius form, the authors predicted a 50% strength retention for 40 years at 105°C, whereas on the basis of their lap-shear experiments, Mitchell and Berg estimated a lifetime of 40 years at 120°C [133]. According to the authors, this discrepancy can be explained by the difference of the bonded areas in the die shear and lap shear tests.

For solvated polyimide adhesives the lap shear strength test is not appropriate because the large overlap bond area prevents the full release of the solvent [61]. The authors underline that the die shear strength data are also highly scattered and

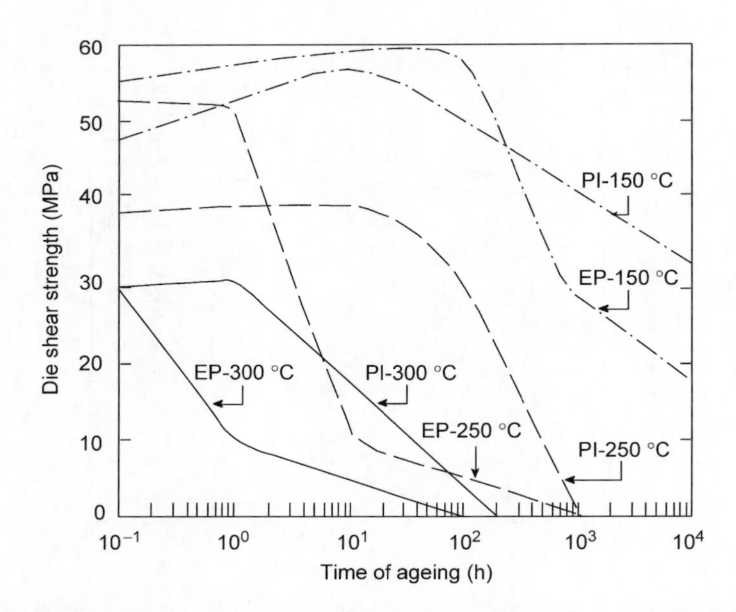

Figure 46: Variation of the die shear strength of silver-filled epoxy (EP) and polyimide (PI) adhesives as a function of ageing time at 150, 250, and 300°C. (adapted from Ref. [134], Copyright© 1977 IEEE.)

unreliable because the fillet formed along the periphery of the die has a significant influence on the shear strength, which can double for small dice. Nevertheless, they employed the die shear technique to compare the behaviour of three silver-filled polyimide adhesives during isothermal ageing. The die shear specimens were prepared by bonding gold-plated $1.52 \times 1.52\ mm^2$ non-functional silicon chips to alumina substrates coated with a thick film of gold. The adhesives show the expected shear strength lowering when tested at elevated temperatures (150–350°C) but the strength retained at all test temperatures is adequate to withstand any load applied during hybrid assembly processing. This can be illustrated with Ablebond® 71–1 which exhibits an initial die shear strength of 15.1 MPa at 25°C decreasing to 14.8 MPa after 1000 h at 150°C. At higher temperatures, the results are similar to those shown in Figure 46 with die shear values of 10.7 and 4.68 MPa after 138 h ageing in air at 250 and 350°C, respectively.

Some experiments show that the device shear strength of commercial silver-filled adhesives increases during thermal ageing (1000 h at 150°C) or 5000 cycles at limit temperatures of −65 and 150°C. For instance, the shear strength rises by 7% for capacitors and 44% for semiconductor dice [137]. Other trials performed with semiconductor devices and chip capacitors with size ranging from 0.508×0.508 to $2.286 \times 2.286\ mm^2$ indicate a continuous reduction of the die shear strength throughout the ageing test [68]. The authors note that the storage time before using one-part (Epo-Tek® H31) and two-part (Epo-Tek® H20E) epoxy adhesives is an important parameter. A twofold increase in device shear strength is observed when the adhesives have been stored for a period of time corresponding to 20% instead of 80% of the shelf life indicated in the manufacturer's data sheet.

6.6.4.7. Thermal Resistance

The effect of porosity on thermal resistance is addressed in Section 6.6.4.1 but other factors influence the value of Θ_{JC}. The results of tests conducted by bonding Motorola 2N-3055 power devices to TO-3 packages by means of commercial silver-filled epoxies show that Θ_{JC} varies from 0.99 to 4.90°C W^{-1} [133]. The lowest value is close to that of the soft solder used as a comparison standard ($\Theta_{JC} = 0.9$°C W^{-1}). The authors advance that the broad range of thermal resistance is related to the size of the silver particles, the smallest flakes yielding the thinnest bond line and the lowest thermal resistance. As shown in Figure 47, no changes are observed in thermal resistance after annealing for 1000 h at 200°C or below, although a significant jump is noted after short exposures at 250 and 300°C. This increase of Θ_{JC} primarily relates to the degraded adhesion of epoxy adhesives to the nickel header or to the nickel layer of the transistor backside.

With the same 2N-3055 bipolar transistors, other authors have found an average thermal resistance as low as 0.3°C W^{-1} for the devices attached with a solder [132].

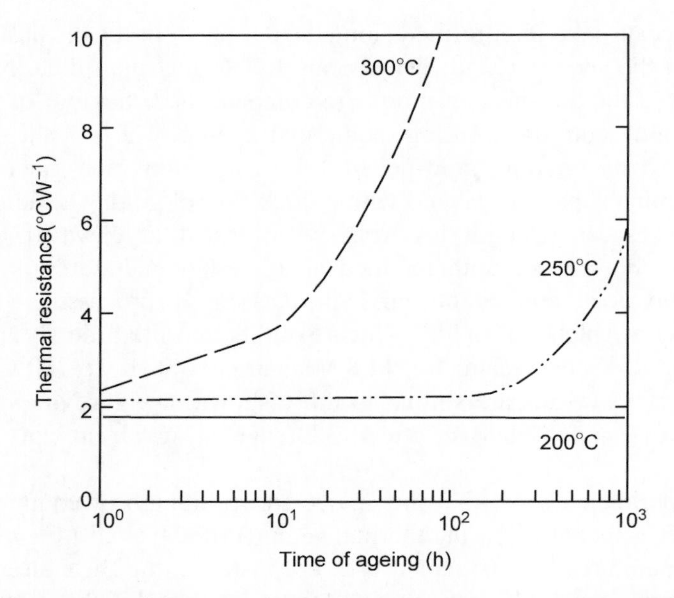

Figure 47: Variation of the thermal resistance Θ_{JC} as a function of ageing time at 200, 250, and 300°C for power devices bonded to TO-3 packages by means of two-part silver filled epoxy adhesive. (reprinted from Ref. [133], Copyright© 1976 IEEE.)

This value remains virtually constant after ageing at 150°C for 1000 h. Conversely, only two of the six epoxies evaluated in this study have an initial thermal resistance approximating that achieved with the solder. Two other adhesives exhibit high Θ_{JC} (1.40–1.45°C W^{-1}) increasing to 1.7°C W^{-1} after thermal ageing. Silver-filled epoxy adhesives have also been tested to attach power chip components to alumina substrates, some of them being bonded to 3 mm thick copper cooling plates [136]. In these attempts, two insulated gate bipolar transistors (IGBT) and two power diodes bonded to the substrates allow to study the effects of the die bonding materials on the thermal resistance of the diodes. After the initial screening tests, three silver-filled epoxies with glass transition temperatures of −20, 79 and 160°C, volume resistivity $1-4 \times 10^{-4}$ Ω cm, and thermal conductivity 2–8 W m^{-1} K^{-1} have been subjected to ageing tests [138]. A 62Sn–36Pb–2Ag solder alloy is used as comparative standard material.

The ageing procedure was conducted as follows: the test modules were subjected to 100 thermal cycles (−40°C, +125°C) and then to either humidity exposure (85°C/85% RH, reverse bias, 1000 h) or 10,000 operational cycles between 25 and 100°C performed by providing a heating current of 8.5 A for 15 s sequentially through the diodes and the IGBT transistors. At the end of these tests, some modules received 700 additional thermal cycles. As expected, the metallic solder exhibited the best properties with a thermal resistance of 4.03°C W^{-1} measured without cooling plate and 1.72°C W^{-1} with a cooling copper plate bonded to the hybrid module. The thermal resistance of the three epoxy adhesives ranged from 4.51 to 4.57°C W^{-1} and from 2.46 to 3.07°C W^{-1} without and with the cooling plate, respectively. This means that a power of 50 W applied to a die bonded with solder caused the same rise in temperature in the chip as did a power of 30 W for dice attached with conductive adhesives.

The other results are worth noting because, whatever the adhesive, the voltage–current relationship of the transistors were not affected by thermal or operational cycling. In contrast, humidity–temperature ageing had a significant effect on the bonded transistors because many of them ceased to function. Humidity stressing was disastrous for the adhesive with the highest glass transition temperature (160°C) but the low T_g (−20°C) adhesive did not perform better. Only the adhesive with a T_g of about 80°C exhibited good properties in most trials. The thermal resistance measured from the diodes, 4.5°C W^{-1} for the three adhesives compared to 4°C W^{-1} for the solder, did not increase during thermal cycling and exposure to humidity but most of the diodes bonded with the high and low T_g adhesives ceased to function.

The chip-on-board (COB) technology is one of the assembly and packaging techniques used to satisfy the need for a higher density of large dice directly bonded to interconnecting boards such as a glass–epoxy FR-4 laminate or a thick film ceramic substrate. The thermal transfer property of the conductive adhesive, as well as its coefficient of linear thermal expansion, play a crucial role to make reliable mechanical joints between the chip and the substrate. A paper by Rörgren and Tillströn affords experimental data obtained with large-area bare dice (12 × 12, 16 × 16, and 20 × 20 mm^2) and dedicated test chips bonded to interconnecting substrates [139]. Two silver-filled epoxies, one thermally conductive silicone, and one thermosetting tape adhesive are compared to an indium–lead solder for bonding the chips to FR-4 printed wiring board (PWB) and alumina substrates. Once bonded to the substrates, the assemblies were subjected to temperature cycling tests of either 1000 cycles from −40 to 100°C or 1300 cycles from −55 to 100°C. X-ray analysis and scanning electron microscopy of cross-sections revealed the formation of micro-cracks penetrating the surface of many dice. These cracks occurred more frequently on the FR-4 PWBs than on alumina. The total thermal resistance of the assembly, including the die, the die attachment layer, and the PWB, was measured before and after the thermal cycles. For all assemblies, the contribution of the PCB to the total thermal resistance was 8°C W^{-1} compared to 1°C W^{-1} and less than 3°C W^{-1} before and after thermal cycling, respectively, for the best performing material, the In–Pb solder. The epoxy and silicone adhesives showed comparable thermal resistances slightly higher than that of the solder but considerably better than that of the tape adhesive, which exhibited a total resistance of 16°C W^{-1} before and 17.5°C W^{-1} after ageing.

6.6.4.8. Comparison of Leadframe/Die Attach Systems

The reliability of copper leadframe combined with silver-filled polyimide adhesive has been compared to the initial P-DIP packaging based on alloy-42 lead frames and a gold eutectic die attachment [140]. The same tests were run on a 16-lead 1K static RAM and on three 40-lead P-DIP packages. Alloy-42 has a relatively low thermal conductivity of about 10 $m^{-1} K^{-1}$ but a coefficient of thermal expansion ($4.2 \times 10^{-6} K^{-1}$) close to that of silicon ($3 \times 10^{-6} K^{-1}$). Conversely, the selected copper alloy has a thermal conductivity of 200 W $m^{-1} K^{-1}$ and a CTE of $1.7 \times 10^{-5} K^{-1}$. Accelerated operating life testing was run by stressing the devices at 125°C at a nominal voltage of 5 V for 1000 h, extended to 3000 h for several groups. The evaluation of the results and failure analysis of the defective devices showed that the operating life of the copper–polyimide system was equivalent or slightly better than that of the alloy-42 system.

The biased temperature–humidity test (85°C/85% RH) and the pressure cooker test (121°C/100% RH and 0.1 MPa pressure) were used to assess the resistance of the devices to a humid environment. The results obtained in the first biased humidity test for 1000 h and in the pressure cooker test for 336 h showed the Olin-195/polyimide die attach system had a slightly better stability than the alloy-42 system. The ability of lead frames to withstand bending fatigue was evaluated according to MIL-STD-883, Method 2004.2, and the results were as expected. The alloy-42 lead frames had a higher lead bend resistance (7.7 bends to break) than the Olin-195 copper alloy (4.4 bends to break). The other environmental tests applied to check package integrity were thermal series, salt atmosphere (48 h), solderability, extended temperature cycle (−65 to 150°C, 1000 cycles), and extended thermal shock (−55 to 125°C, 1000 cycles), all defined in MIL-STD-883B. No significant differences were seen between the two systems at the end of the environmental testing.

In addition to these tests of reliability, the die shear test conducted on the samples prior to plastic moulding showed that the die shear strength was greater than 49 N for both systems. The thermal resistance from the junction to the ambient (Θ_{JA}) is dependent upon the thermal conductivity of the adhesive layer and the quality of the silicon–polyimide and polyimide–lead frame interfaces. As expected, the thermal conductivity was improved by approximately 50% on the 16-lead devices (Θ_{JA} = 140–150°C W^{-1} with alloy − 42 and 68–78°C W^{-1} with Olin-195) and by approximately 30% on the 40-lead devices (Θ_{JA} alloy-42 = 50–60°C W^{-1}; Θ_{JA} Olin-195 = 40–45°C W^{-1}). The 16-lead devices for each packaging system were subjected to 25,000 cycles of power cycling (5 min "on", 5 min "off") and neither alloy-42 nor Olin-195 showed a significant change in Θ_{JA}. In addition, Θ_{JA} was used to assess the quality of the die attachment over rapid thermal shocks from −55 to 125°C, for 1500 cycles and no significant changes were seen during the stress period.

6.6.5. Flip Chip Bonding

It has been shown that the reliability of polymer flip chip technology in consumer electronic products depends on the chip size and the substrate materials [141]. The test chips were daisy-chains with pads arranged in a matrix with a pitch of 0.3 mm. The basic size of 2.4×2.4 mm^2 could be extended to larger chips, for instance 4.8×4.8, 7.2×7.2 mm^2, and so forth. The test chips were bonded to various substrates by means of the Epo-Tek PFC® bump technology described in Section 6.5.2 and the transition resistance of the bumps with the under-bump metallization was measured after exposure to environmental stress. On silicon substrates (Al and Ni–Au metallizations) there was no change in resistance after 250 h at 150°C and after 100 cycles from −25 to 125°C even for 6×6 units. On glass substrates, there was no deterioration up to 130 cycles up to 6×6 units and for 30 h exposure to humidity (85°C/85% RH) up to the 4×4 units. On alumina substrates, the resistance was stable after 250 h at 150°C or 85°C/85% RH, but a high failure rate was observed for the 6×6 units after 10 cycles between −25 and 125°C, whereas the 4×4 units withstood this stress for 60 cycles. Finally, on FR4 substrates carrying copper pads electroless plated with Ni–Au metallization, all the bumps were broken after five thermal cycles for the units equal to or larger than 4×4. An increase in transition resistance was also observed after 100 h of storage in humid environment for chips smaller than 3×3 units. The conclusion of these experiments is that the mortality rate of polymer bumps is very low for samples bonded to silicon substrates, but that polymer flip chip technology can be used to bond small chips on FR4 PCBs only for a moderate environmental demand.

In addition to the previous technical approaches assessed to lower the cost/performance relationship of flip chip interconnection, the potential of anisotropically conductive adhesives has been investigated to design new interconnecting systems [142]. In the case of isotropic adhesives, all available data show that the volume resistivity is not impaired when the assembly is subjected to thermal excursions at temperatures below the T_g of the adhesive. This is possibly due to the number of conducting paths allowed by the high level of silver loading. Most anisotropic systems are conversely based on the concept that electrical continuity is provided by a few conducting particles working in compression along the *Z*-axis and may vanish at high temperatures. The authors derived a set of simultaneous linear equations allowing for a quantitative estimate of the behaviour of the assembly passing from the compressive state to stress-free and even tension conditions. It is worth noting that they observed considerable changes in the apparent Z-axis modulus of the adhesive as a function of the thickness for typical epoxies, rising from 6.9 to 41.4 GPa when the thickness increased from 25 to 100 μm. In agreement with this, the apparent coefficient of thermal expansion was found to increase from $3.4 \times 10^{-5}\,K^{-1}$ for a 0.15 mm thick film to $1.27 \times 10^{-4}\,K^{-1}$ for a 25 μm thick film. Thus, the thermal expansion partly counteracts the effect of the change in compressive modulus but a compressive contact force is still attainable at temperatures in excess to 100°C.

6.6.6. Surface-mounted Components

Most electronic components are bonded to printed wiring boards by using tin–lead alloys because tin readily forms intermetallic layers with copper, whereas lead decreases the soldering temperature. The driving forces for using conductive adhesives in SMT are to overcome lead toxicity and the need for fluxing and cleaning steps. An additional advantage of conductive adhesives is a low cure temperature that does not damage the heat-sensitive electronic components during the die attachment process and the step of electrical interconnection to circuit boards. Moreover, the trend to thinner plastic packages with an increasing number of output leads make the assembly prone to cracking failure during the solder reflow at 250°C. The effects of the physical properties of the die attach layer on the mode of cracking have been studied with 52- and 80-quad-flat-pack (QFP) assemblies comprising the plastic package, the die, the conductive adhesive, and the lead frame [143]. When subjected to temperature–humidity test (85°C/85% RH) for 24–96 h, the number of cracked packages increases with the adhesive strength and the flexural modulus of the die attachment material. Moreover, there is a relationship between the adhesive strength of the silver-filled epoxies, the package size, and the mode of cracking. For the 52-lead QFP, all the cracks run according to a mode where delamination occurs at the interface between the moulding material and the die pad, followed by crack extension from the corners of the die towards the bottom of the package. Conversely, for larger packages delamination occurs at the interfaces between the die attach layer and either the die or the die pad. In certain cases, the cracks run towards the inner leads or extend to the bottom of the package for high-strength adhesives.

The long-term reliability of isotropic and anisotropic conductive adhesives for surface-mount applications has been extensively studied [144–148]. The data accumulated with die attached chips cannot be transposed to the hybrids or surface-mounted wiring boards because of the dissimilar requirements in mechanical and electrical properties [149]. As shown in Figure 41(a) and (b) of Section 6.5.7, the active and passive devices can be attached to thick film metallization pads printed on a ceramic carrier by means of isotropic or anisotropic conductive adhesives. Both types of materials have been subjected to extensive reliability evaluation [4]. Thermal cycling from − 50 to 150°C show that the adhesives can be arranged into three general classes. In the first category are the adhesives that do not survive mechanical strength testing at or below 250 thermal cycles. The second group comprises materials that fail between 1000 and 2000 cycles; and finally, the adhesives of the third class survive 3000 thermal cycles without significant loss in adhesive strength [150]. This last group is formed of highly compliant materials that can accommodate the shear strain as a result of their viscoelastic properties. Conversely, the electrical resistance of the adhesives investigated in this study increases 10 times in less than 1000 thermal cycles. For the first group, the electrical conductivity decrease is due to the formation of microcracks and voids with a progressive disintegration of the bonding layer. In the second group, the phenomenon is explained by oxidation of the metal filler, mainly copper and nickel, during thermal cycling. The increase in resistance of the last group is attributed to the high viscoelasticity of the compliant adhesives. As an example of this latter family, the initial resistance of about 15–20 Ω raises to 1–2 kΩ after 800 thermal cycles. Although the low-modulus adhesives exhibit superior mechanical strengths under thermal cycling, their electrical conductivity greatly decreases during the test.

The materials originally formulated for use as die attach adhesives are relatively brittle and are not necessarily suitable for SMT. For this application, the requirements are a fast cure at low temperature, typically 4 min in an IR oven at 130°C, a high thixotropy which permits a finer deposition pattern, the possibility of using the existing automatic pick-and-place and dispensing equipment, and a better toughness compared to tin–lead solders. Experiments performed with a one-part silver-filled adhesive demonstrates that the adhesive strength remains stable during all ageing tests if the initial strength is sufficient [151]. This is true for bonding on printed wiring boards, SOT-23 and SO-8 components, as well as "1206" resistors and capacitors, and large transistors, However, with components offering a small contact area, a very low adhesion is observed after cure that can be overcome only by changing the component design.

It has been shown that, with respect to the chemical structure, the very flexible epoxy-urethane and epoxy-silicone adhesives exhibit the largest resistance values and the lowest die shear strengths. However, material compliance is only one of the parameters governing the reliability of organic adhesives in the surface mounting of electronic components. The die shear strength and electrical conductivity also depend on the metals to be bonded. One study outlines that conductive adhesives give good reliable electrical connections if used in combination with Ag–Pd terminations on the devices, and copper or gold metallizations on the PCB [152]. Conversely, with 60Pb–40Sn metallization, the contact resistance seriously increases during high temperature and climate testing. In fact, reliable adhesive joints are achievable by using a material properly matched to the final application. For example, small chips and large quad flat packages require completely different adhesives, whereas copper, gold-plated boards and tin–lead surfaces need another adhesive chemistry.

The reliability of Ablebond® 8175 has been evaluated by using Kovar tabs whose lead terminations are made of copper, silver, nickel, or 60Sn–40Ni alloy [153]. In this study, the thick film metallizations printed on the surface of alumina panels are 3:1 Ag–Pd, 6:1 Ag–Pd, 100% Ag, 100% Cu, and 100% Au. The die shear strength, determined during thermal cycling from −65 to 150°C for 1000 cycles, shows that Ablebond 8175 performs as well as the 62Sn–36Pb–2Ag solder paste on all metallizations. In addition, the epoxy adhesive outperforms the solder on gold metallization with die shear strengths in the range of 24–41 MPa after 1000 thermal cycles. The specimens subjected to the 85°C/85% RH test for 1000 h performs similarly although the adhesion level is not as high as that of the solder. On each metallization, regardless of the lead termination, the epoxy adhesive maintains an asymptotic value of 13.8 MPa. Finally, after 1000 h of thermal ageing at 150°C, the performance of the epoxy is superior to the solder paste on the 100% Ag, Cu, and Au thick film substrates.

In general, silver-filled adhesives have filler volume fractions that exceed the percolation threshold and, therefore, any change in electrical resistance requires an understanding of the conduction characteristics at the interfaces. At the contact points the current flow is constricted because of the small contact area and the bending of the current flow path requires an added voltage called "constriction voltage". Another additional voltage must be applied if an insulating film, oxide or polymer, separates the conductive particles or their junction with the substrates. Scanning electronic microscopy (SEM) and Auger microscopy have been used to characterize the structure and composition of the interfaces between the organic matrix and silver particles, and at the junctions between the adhesive layer and various metallizations such as Cu, Sn, Ag, Au, Ni, Pd, and Sn–Pb alloy [154]. SEM photographs show that the silver flakes with aspect ratios greater than three are oriented with their long axis parallel to the substrate surface. All adhesives have voids randomly dispersed within the matrix and at the adhesive/substrate interface, and in these regions the flakes are oriented around the periphery of the voids as a circular boundary. It is worth noting that very few silver flakes are in direct contact with each other and with the electroplated substrate. In addition, none of the flakes penetrate contiguous particles and, when contacts are observed, they are very small (tens of nanometres). Therefore, the constriction resistance in isotropic conductive adhesives should be significant.

Volume conductivity measurements indicate that the composition of the electroplated substrates is of prime importance with an increase in resistivity covering 5–6 orders of magnitude. The conductive adhesives have volume resistivities in the range of 6.5×10^{-5}–1.6×10^{-4} Ω cm but the values measured for the joined assemblies extend from 9×10^{-4} to 1×10^{3} Ω cm. The general trend observed with all adhesives is that the increase in resistivity is related to the ease of oxidation of the plated substrates, in particular when nickel and aluminium are compared to noble metal plating. Other studies support these results indicating that the best silver-filled adhesives have a volume resistivity of 5×10^{-5} Ω cm approaching the value of solders, typically 2×10^{-5} Ω cm. The contact resistance at the interfaces with the metallic conductors is, however, more important than the bulk conductivity [155–157].

The electrical contact resistance of several isotropic adhesives has been evaluated using a copper comb pattern test vehicle electroplated with various metal combinations including palladium alloy over nickel, gold over nickel, nickel, and tin [158]. The experimental data are presented in the form of a contact resistance distribution with the contact resistance along the x-axis and a normalized probability scaling of the y-axis. After accelerated ageing tests including 2000 thermal cycles at 0–100°C, storage at 120°C for 2000 h, and 1000 h at 85°C/80% RH only one adhesive compares favourably to the solder with a mean value of 4.4×10^{-3} versus 4×10^{-3} Ω for the solder on the palladium alloy surface. This adhesive had not only the best initial contact resistance but also the best performance during environmental stressing, with a change of 2×10^{-4} Ω for the joints subjected to the three environmental tests. The other electroplating surfaces do not perform as well as the palladium alloy which is better than gold and far better than nickel and tin. In another study, optical and scanning electron microscopies reveal the formation of cracks after the cure schedule and crack propagation during humidity exposure [159]. Here, "0805" chip components with tin-plated terminations are attached to PCB test boards with passivated copper and tin/lead metallizations. After exposure to 85°C/85% RH, microstructure investigation by transmission electron microscopy shows different degrees of oxidation of the PCB pad metallizations due to moisture penetration.

Anisotropic adhesives are more sensitive than isotropic conductive pastes to the coplanarity of the connecting pads and leads of the substrate and the device because many of them are based on the single-particle bridging concept. The typical properties of a screen-printable isotropic conductive adhesive paste are a silver content of 70–75% by weight, a viscosity of about 80 Pa s, and a curing time of 15 s at 175°C. This fast cure cycle produces an adhesive layer with T_g 120°C, ρ 3.3×10^{-4} Ω cm, λ 7.1 W m^{-1} K^{-1}, and a die shear strength of 45 MPa at 25°C and 0.6 MPa at 250°C. In comparison, the silver content of anisotropic conductive pastes is in the range of 5–15 vol.% with a resulting viscosity of 20 Pa s. When cured for 30 s at 175°C on a hot plate, the adhesive has a T_g of 123°C, a coefficient of thermal expansion of 6.7×10^{-6} K^{-1}, an out-of-plane (Z-axis) volume resistivity of 1×10^{-4} Ω cm, and an in-plane volume resistivity of 2×10^{14} Ω cm [160]. The reliability of isotropic and anisotropic adhesive pastes and tapes has been investigated by means of different quad-flat-packs with a lead pitch of 0.635 mm [161]. Before temperature cycling and humidity testing, the electrical resistance and the adhesive strength of all adhesives are similar to those of a soft solder. After thermal cycling, most samples show an increase in electrical resistance with 7% of the connected leads losing contact for the isotropic adhesive compared to 44% for the anisotropic materials. As previously discussed, the effect of temperature cycling on the mechanical properties was to increase the strength of adhesion because of further curing. After the humidity tests the values of the electrical resistance increased considerably with many open connections. The percentage of leads that lose electrical continuity after humidity testing falls in the range of 10–75%.

A methodology has been developed at the AT&T Bell Laboratories for evaluating the mechanical and electrical properties of commercial anisotropic adhesives at any stage of the assembly process [104]. FTIR spectroscopy discriminates the chemical structure, i.e. epoxies, silicones, polyurethanes, or other polymeric materials. The overall surface morphology, observed with optical microscopy, provides information on the shape, size, and distribution of the particles. Scanning electron microscopy, coupled with energy dispersive X-ray analysis, gives more details on the conducting particles and their surface composition. Thus, the four examples selected for this study were: (A) 5–7 μm gold-plated nickel spheres, (B) 8–12 μm silver-plated glass beads, (C) 12 μm nickel-plated polymer spheres, and (D) 20 μm gold-plated nickel particles. The cure cycle was optimized for each adhesive by using DSC and microdielectrometry. The results are discussed in terms of thermal stability, adhesive strength, and assembly parameters including placement accuracy, temperature, load, tacking time, and bonding time. The alternating current carrying capability, measured with a bipolar integrated circuit bonded to a flexible circuit, was evaluated at 20 mA root mean square (RMS) at 2 MHz frequency. The direct current carrying capability is at the 200 mA level with the anisotropic films containing silver-plated nickel particles but is limited to less than 5 mA when the adhesive film is loaded with silver-plated polymer spheres.

To complete these results, the researchers applied accelerated temperature, humidity, and bias tests (THB) to study the time-to-failure and the associated conduction and failure mechanisms [146]. The experiments were performed using the same test vehicle and adhesive materials as in the previous work. The accelerated life test conditions were 85°C and 85% RH with three different voltages: 10, 50, and 100 V. The results and observations of the authors can be recapitulated as follows: for an applied voltage of 100 V across a 0.1 mm conductor spacing, the average uniform electric field stress is 1 kV mm^{-1}, but the field enhancement factor around some conductive particles and protrusions can exceed 10 kV mm^{-1}, approaching the breakdown voltage of the polymer matrix. This field enhancement results in strong anodic corrosion and silver migration for the adhesives containing silver-plated particles.

6.6.7. Liquid Crystal Displays

The use of heat seal connectors (HSC) in the fabrication of liquid crystal displays is described in Section 6.5.8. This flexible circuitry is bonded to the ITO electrodes of the glass panels and to the IC driver by means of anisotropically conductive adhesives. Heat and pressure are applied to the components to bring the conductive particles into continuous contact in the vertical direction. This is very often performed by using a thermode providing a pulse heat bonding method [162]. For high-density interconnection of high-resolution liquid crystal display panels, the flexible printed circuit is made of copper lines terminated by nickel and gold-plated pads. ITO electrodes can be used as bare conductors or coated with thin films of different metals with an aluminium or gold finish. The electrical characterization includes the measurement of the lateral insulation resistance between adjacent contacts and the vertical resistance of the three-layer assembly comprising flexible circuit, conductive adhesive, and TAB-packaged display driver chip. As no international test procedures exist to qualify heat seal connectors, various ageing conditions have been employed to determine the reliability of the anisotropic adhesive joints.

A paper by Hampel proposes a specification including the peel strength, the electrical resistance of the conductor tracks, and a series of environmental tests [163]. The requirement for adhesive strength, determined by a 90° peel test using 10 mm wide strips, should be set at 5 kN m^{-1}. The electrical resistivity obviously depends on the conductive filler with expected values lower than 35 $\Omega\,\square^{-1}$ for carbon powder, 0.1 $\Omega\,\square^{-1}$ for silver particles and 0.01 $\Omega\,\square^{-1}$ for 17 μm thick copper lines. For accelerated ageing tests, the heat seal connectors are made of 25 μm thick poly(ethylene terephthalate) strips carrying the conductor tracks, which are produced by printing a silver paste with carbon on top. The anisotropic adhesives are loaded with either 15 μm nickel spheres or 15–25 μm graphite particles or 15 μm gold-plated polystyrene beads. The experimental results show that the lowest value ($< 15\ \Omega$) is obtained with the gold-filled adhesive, remaining unchanged after 25 h at 85°C. In the same conditions, the resistance of the nickel-filled adhesive slightly increases whereas that of the graphite-filled material slightly decreases. The heat seal connectors exhibit excellent stability when the samples are subjected to ageing tests including 88 h at 65°C/90% RH, 22 cycles between the ambient temperature and 55°C/90% RH, 16 h at 110°C, and 65 h at 95°C/95% RH.

Another example includes a heat shock cycle from −40 to 100°C for 500 h, storage at 125°C for 1000 h, and the 85°C/85% RH temperature–humidity test for 1000 h [164]. The insulation properties of the three thermosetting anisotropic adhesives investigated are excellent with an insulation resistance larger than 10^{13} Ω before and after the stressing tests. However, one adhesive performs better than the other two on ITO and aluminium terminations with initial values of about 600 and 0.7 Ω mm^2, respectively, remaining virtually constant after 1000 h at 85°C/85% RH. The same adhesive exhibits a stable contact resistance of less than 0.15 Ω mm^2 on the gold-plated pads slightly higher than that obtained with another adhesive exhibiting a resistance of 0.08 Ω mm^2 over the 1000 h testing. This means that commercially available anisotropic adhesive films can fulfil the requirements for high-density interconnection of the LCD drivers to the matrix of conductors coated on the glass substrate.

The reliability of thermosetting anisotropically conductive adhesives has been determined for joining LCD modules to rigid wiring boards through flexible circuitry and for bonding these modules directly to a flexible circuit in a camera [165]. The purpose of this work was the replacement of thermoplastic heat seal connectors with thermosetting adhesives. The evaluation of the samples was performed using the following environmental tests: 1000 h at 60°C/95% RH, temperature cycling from −40 to 85°C for 1100 h (1000 cycles), and humidity cycling from 5 to 75°C at a nominal relative humidity of 95% for 1008 h (126 cycles). The contact resistance of the control thermoplastic heat seal connectors increases from 33 to 300 and 226 Ω after the constant humidity and the humidity cycling tests, respectively, with some catastrophic failures. In contrast, temperature cycling has less influence, with a final value of 52 Ω, corresponding to an increase in contact resistance of only 53%. The tape thermosetting adhesives perform very well with moderate changes in resistance after all environmental tests, the humidity cycling being more severe than the others. To explain the change in resistance induced by temperature–humidity ageing, the authors suggest that the formation of an insulating layer of metal oxide is a time-dependent process related to the diffusion parameter D of oxygen in the oxide layer. On the basis of this concept, the relationship between the time for diffusion and the resistance change can be expressed as $\Delta R/R = 1 + K(t)^{0.5}$ where $\Delta R/R$ is the relative change in resistance, t the time of exposure to humidity, and K a parameter related to the total resistance R_0 of the interconnection before exposure, the volume resistivity of the oxide layer ρ_{ox}, the diffusion rate of oxygen, and the surface area of the metal filler A. All these parameters are included in the relation $K = \rho_{ox}(2D)^{0.5}/AR_0$. Comparing the experimental data to this theoretical model,

Liu and Rörgren show that, except for one of the adhesives tested, a reasonably good fit is obtained for K values in the range of 0.47–1.59.

6.7. Thermal Stresses

The determination of stress in multilayer structures including polymer films can be conveniently achieved using different techniques such as bilayer bending beams, wafer curvature measurements, X-ray diffraction, laser interferometry, strain gauges embedded in bulk parts, or semiconductor surface transducers. These methods are compiled in a book which provides extensive information on the measurement of the stresses [166]. The total stress σ measured using these techniques is the sum of the intrinsic σ_i and thermal σ_t stresses, i.e. $\sigma = \sigma_i + \sigma_t$. The intrinsic stress is the addition of contributions including solvent evaporation for polyimides and cure shrinkage for epoxy resins. The origin of the thermal stress is now well understood and its value can be calculated by using the complete or simplified equations previously reported [4].

Since polymers generally exhibit high thermal expansion coefficients in comparison to the inorganic materials commonly employed in microelectronics, significant stresses are caused by the temperature transients. This occurs after polymer processing and also when in use because of the thermal cycling between the on and off states. The viscoelastic properties of the polymers make them compliant materials that are not prone to lethal failures such as chip cracking and rapid debonding. However, the repeated straining of the very small conductor lines implemented on the top surface of the integrated circuits may alter their electrical behaviour through geometrical distortion. Whatever the application of polymers in electronics, dimensional changes arise as a result of polymerization shrinkage and difference in thermal expansion coefficients, which generally do not match the substrate CTE. The resulting strain generates significant tensile and compressive stresses in both the polymer and the substrate. During the assembly and packaging stages, the integrated circuits are bonded to the lead frames with organic conductive adhesives and new stresses appear from the mismatch of the coefficients of thermal expansion of the three-layer system, namely, the die, the adhesive layer, and the lead frame. Finally, the integrated circuits are embedded in a polymer matrix during the packaging operation, which develops stresses generally in the opposite direction because of shrinkage of the epoxy moulding compositions.

Expanding the lap-joint theories previously discussed by other researchers, Chen and Nelson proposed different analytical models of stress analysis [167]. Figure 48 depicts the two schemes that can be used to compare the existing models for tri-material assemblies.

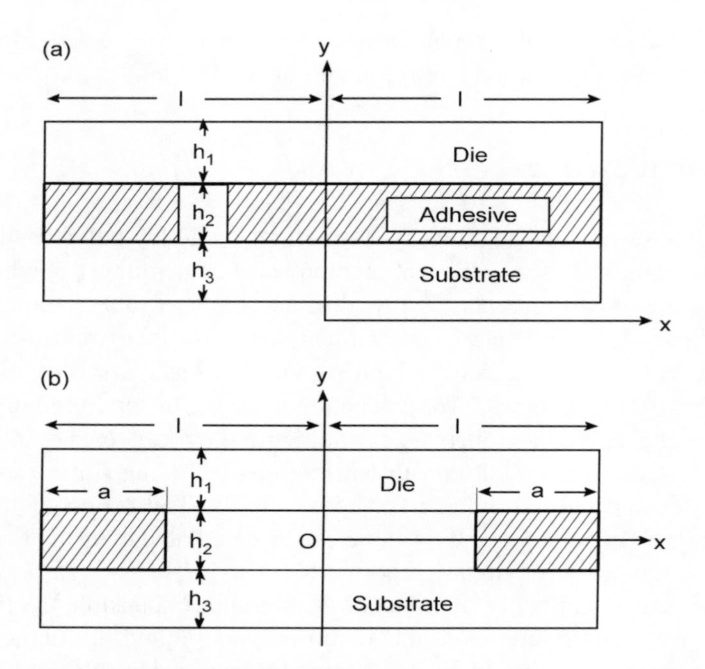

Figure 48: Stress analysis model for tri-material assemblies: (a) completely filled joint representative of backside die attachment; (b) bonding technique using small dots of adhesive as in flip chip technology.

Basically, two elastic layers denoted by the subscripts 1 and 3 are bonded by either a completely filled joint (Fig. 48(a)) or small dots of adhesive as in flip chip technology (Fig. 48(b)), both being denoted by the subscript 2. The three layers have uniform thicknesses h_1, h_2, h_3, elastic moduli E_1, E_2, E_3, shear moduli G_1, G_2, G_3, Poisson's ratios ν_1, ν_2, ν_3, and coefficients of linear thermal expansion α_1, α_2, α_3. The other parameters are the die half-length l, the distance from the centre of the die x, and the temperature differential ΔT between the zero-stress temperature T_0 and the operating temperature T. The model proposed by Chen and Nelson demonstrates that the shear stress generated by the temperature differential is given by

$$\tau = \frac{G_2}{\beta h_2} (\alpha_1 - \alpha_3)\, \Delta T \, \frac{\sinh \beta x}{\cosh \beta l} \tag{3}$$

with:

$$\beta^2 = \frac{G_2}{h_2} \left[\frac{1}{E_1 h_1} + \frac{1}{E_3 h_3} \right] \tag{4}$$

Eq. (3) indicates that the shear stress is zero at the centre of the die ($x = 0$), and increases gradually to a maximum at the edge ($x = l$) where:

$$\tau_{max} = \frac{G_2}{\beta h_2} \Delta\alpha \, \Delta\tau \tanh \beta l \quad (5)$$

In Eq. (5), the die and substrate properties are introduced through $\Delta\alpha = \alpha_1 - \alpha_3$ and the β factor. The authors point out that it is often sufficient to take tanh $\times$ $\beta l = 1$ and use Eq. (6) to obtain an accurate estimate of the actual shear stress:

$$\tau_{max} = \frac{G_2}{\beta h_2} \Delta\alpha \, \Delta\tau \quad (6)$$

Figure 49 represents in semi-logarithmic coordinates one of the numerical examples provided by Chen and Nelson for a joint width of 51 mm.

The materials used to calculate the shear stress and the tensile force have the following properties: $E_1 = 117$ GPa, $h_1 = 1.57$ mm, $\alpha = 1.6 \times 10^{-5}$ K^{-1}; $G_2 = 1.23$ GPa, $h_2 = 0.051$ mm, $l = 25.5$ mm; $E_3 = 275$ GPa, $h_3 = 1.52$ mm, $\alpha_3 = 6.5 \times 10^{-6}$ K^{-1}, and ΔT is 100°C. As expected from Eq. (5), the maximum

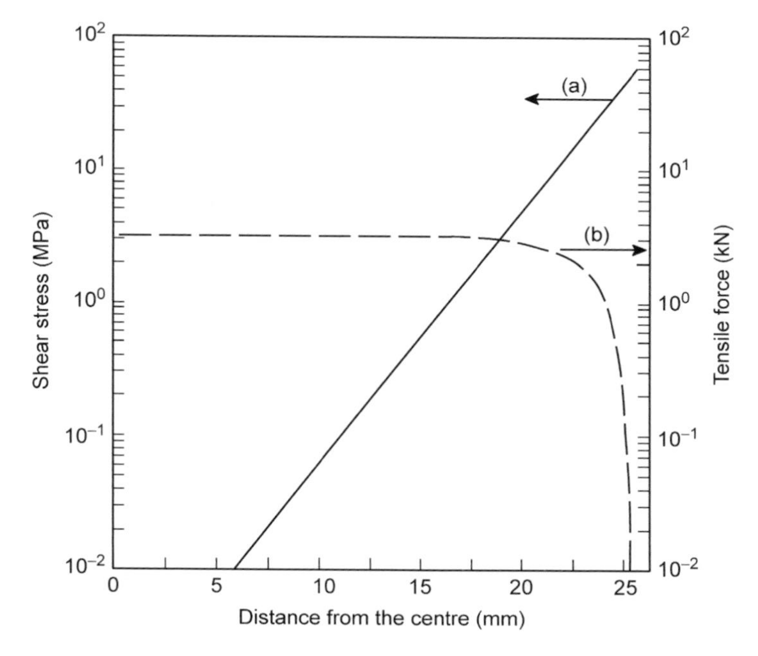

Figure 49: Variation of the calculated shear stress (a) and tensile force (b) as a function of the distance from the centre of the assembly. (adapted from Ref. [167], Copyright© 1979, International Business Machine Corp.)

shear stress occurs at the periphery and is found to be 52.71 MPa with an associated shear strain of 0.043. At 10 and 20 mm from the centre of the assembly, the values of the shear stress are 0.068 and 5.25 MPa, respectively, the latter being less than 10% of the maximum stress. Figure 49 also shows that the tensile force in the layer is constant in the centre area and drops to zero near the edge of the joint. Because of the large size of the specimen, tanh βl is close to unity and Eq. (6) gives a very good estimate (52.22 MPa) of the shear stress.

The stress analysis model developed by Suhir includes the normal stresses acting in the die as well as the interfacial stresses responsible for the cohesive and the adhesive strengths of the bonding layer [168]. The effects of the die size, attachment compliance, and non-linear behaviour of the attachment material are discussed and analyzed to assess the relationship between the compliancy of the attachment, the die fracture, and the fatigue life of the assembly. Starting from a complete tri-material assembly, the author elaborates a set of 90 equations which can be substantially simplified in the case where the adhesive layer has a very small thickness or Young's modulus in comparison to the die and substrate values. The axial λ and interfacial κ compliances of a tri-material assembly, reduced to a die–substrate assembly in the case of the axial compliance, are, respectively,

$$\lambda = \frac{G_2}{h_2}\left[\frac{1-\nu_1}{E_1 h_1} + \frac{1-\nu_3}{E_3 h_3} + \frac{h_2}{4D}\right] \tag{7}$$

$$\kappa = \frac{h_1}{3G_1} + \frac{2h_2}{3G_2} + \frac{h_3}{3G_3} \tag{8}$$

where h is the total thickness ($h = h_1 + h_2 + h_3$) and D the total flexural rigidity of the assembly expressed as the sum of the rigidities of the layers 1 and 3 in the form:

$$D = \frac{E_1 h_1^3}{12(1-\nu_1^2)} + \frac{E_3 h_3^3}{12(1-\nu_3^2)} \tag{9}$$

In the bonded area (the a region of Fig. 48(b)), the shearing stress function $\tau(x)$ is related to the axial and interfacial compliances through the parameter of assembly stiffness $k = \lambda/\kappa$ by the formula

$$\tau = \frac{\lambda}{\kappa}\,\Delta\alpha\,\Delta t\,\chi'(x) \quad \text{with} \quad \chi'(x) = \frac{\sinh kx}{\cosh kl} \tag{10}$$

where $\chi'(x)$ characterizes the distribution x of the interfacial stresses and reflects the effect of the die size l on the shear stresses. The function $\chi'(x)$ is maximum at the edges of the die ($x = l$) where:

$$\chi'_{\max} = \tanh kl \quad \text{and} \quad \tau = \frac{\lambda}{\kappa}\,\Delta\alpha\,\Delta t\,\chi'_{\max} \tag{11}$$

The hyperbolic functions of Eqs. (3) and (10) are similar, therefore the interfacial shearing stresses in the inner part of the assembly are very small and exponentially increase to become maximum at the periphery. In the contact areas, the force $F(x)$ is evaluated by the formula

$$F(x) = \frac{\Delta\alpha\,\Delta t}{\lambda}\chi(x) \quad \text{with} \quad \chi(x) = 1 - \frac{\cosh kx}{\cosh kl} \tag{12}$$

where the function $\chi(x)$ represents the longitudinal distribution of the forces and of the resulting normal stresses. In the case of great kl values, the function $\chi(x)$ is close to unity and therefore at any cross-section in the inner part of the assembly the stresses in the die are independent from the location and can be calculated under the assumption that the die is infinitely large. Near the edges, where the coordinate x is of the same order as the die half-length l, the normal stresses in the die rapidly drop to zero.

These equations have been used to estimate the maximum shearing stresses in the attachment and the maximum normal stresses in the die when the silver-filled epoxy IP 670 and polyimide IP 680 adhesives are employed to bond silicon dice to ceramic substrates and to copper lead frames [4]. The temperature differential is assumed to be 150°C for the epoxy resin and 280°C for the polyimide (from -50°C to their respective glass transition temperatures of 100 and 230°C). For the silicon–epoxy–ceramic assembly the maximum shearing stress in the attachment τ_{max} is 23.2 MPa while the maximum normal stress in the die σ_{max} reaches 73.2 MPa. On copper leadframe, whose CTE is three times larger, the values of τ_{max} and σ_{max} are 66.1 and 127.3 MPa, respectively. Although the elastic moduli are similar, the stresses calculated for the IP 680 polyimide adhesive exhibit the same general trends but the asymptotic values are larger than those of the epoxy because of the temperature differential which is 280 against 150°C. The maximum normal stresses in the die and the maximum shearing stresses in the attachment are, respectively, 137 and 45 MPa on a ceramic substrate and 238 and 66 MPa with a copper lead frame. These values are calculated for 50 μm thick adhesive layers. The stress distribution over the half-length of 1 cm dice bonded to ceramic substrates is depicted in Figure 50 for 25 and 100 μm thick silver-filled epoxy adhesive IP 670.

The effect of the die size can be discussed with the assumption that small dice have a length culminating at 5–6 mm while large dice are 10 mm long, or larger. For small dice, the factors χ_{max} and χ'_{max}, which reflect the effect of die size on the maximum normal stresses and the maximum shear stresses, increase when the die size is increased. In contrast, the maximum stresses in the die are virtually independent from its size for large backside attached dice with stiff attachments ($kl > 4$). On the other hand, the maximum interfacial stresses become independent from the die size when $kl > 2.5$. This means that, with respect to these kl values, if

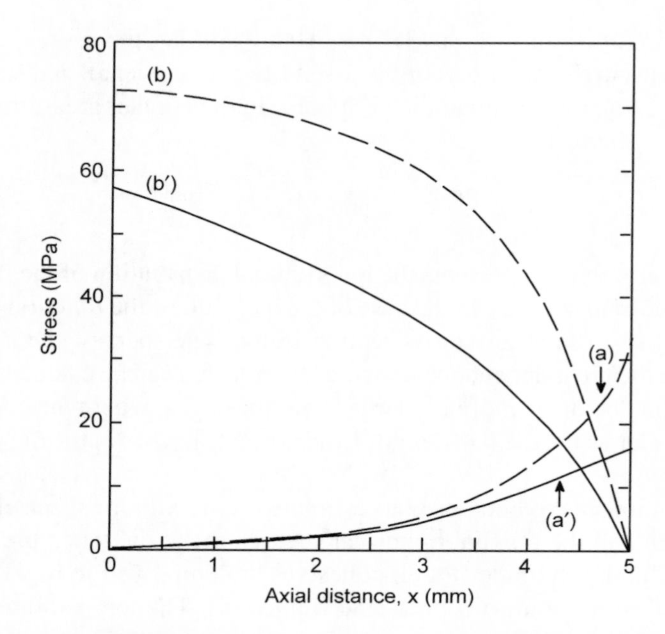

Figure 50: Stress distribution over the half-length of 1 cm dice bonded to ceramic substrate. Curves (a) and (a′) represent the maximum shearing stress in the attachment for 25 and 100 μm IP 670 adhesive layer, while curves (b) and (b′) show the maximum normal stress in the die for the same thicknesses.

a small die does not fracture under the test conditions, any assembly of larger size will also be safely utilized. Conversely, a compliant attachment can be used to reduce the stiffness of the attachment and the associated stresses if the above assembly fails under testing. Compared to the classical backside attachment, the maximum stresses are smaller in dice attached with multiple dots such as in flip chip bonding, but these stresses also increase with the size of the integrated circuit and they are maximum in the bumps located at the periphery of the die.

Thus, in theory the thermal stresses become rapidly independent of the component size. However, it has been pointed out that the shearing stresses are concentrated in the periphery of the bonded area along a distance of the order of magnitude of the device thickness [169]. In this paper, Riemer presents the quantitative data calculated for a gallium arsenide chip ($h_1 = 0.1$ mm, $l = 2.54$ mm) bonded to an aluminium nitride substrate ($h_3 = 0.508$ mm) through a temperature excursion of 100°C. For the theoretical case of laminated components (no adhesive), the tensile stress is 22 MPa in the centre of the die and remains constant along a distance of 1.9 mm outward. This means that the edge region where the stress drops to zero is 0.6 mm wide. With an Au–Sn preform, the stress is constant over a distance from the chip centre of 1.63 mm and the length of the edge region increases to 0.9 mm. Finally, the calculation shows that the initial stress generated by an epoxy adhesive remains constant only over 0.7 mm, therefore the edge region covers 72% of the half-length of the die. These results are similar to those presented in Figure 50, which shows that the greater the adhesive compliance, the smaller the initial stress and the shorter the distance of constant stress in the central part of the die.

Stress reduction parameters are implicitly contained in the above equations used to calculate the thermally induced stresses. Low-stress adhesives can be obtained by matching the thermal expansion coefficients, increasing the bond line thickness, using low-modulus compliant adhesives, and lowering the glass transition temperature. Table 4 shows that only some ceramics and a few metals, metal laminates, and alloys such as tungsten, molybdenum, copper/invar/copper, Kovar, and alloy-42 have coefficients of thermal expansion that can reasonably match the CTE of silicon (3×10^{-6} K^{-1}). However, most solders and adhesives have far higher thermal expansivity: 1.3×10^{-5} K^{-1} for silver-glasses, 1.7×10^{-5} K^{-1} for the common tin-based solders, more than 2×10^{-5} K^{-1} for gold eutectics, and over 4.5×10^{-5} K^{-1} for silver-filled epoxies. Only rigid polyimides exhibit low CTE values but they do not flow and wet the substrates even at 400°C.

The three-dimensional models predict that the stress level can be reduced if the bond line thickness is increased from 25 to 75 or even 175 μm. However, the curves of Figure 50 show, e.g. that the maximum shearing stress decreases by a factor of two, from 33 to 17 MPa, when the thickness of the adhesive layer increases from 25 to 100 μm. A bond line thickness of 50–75 μm is generally recommended for the die attachment because of the negligible thermal impedance penalty. The experimental results indicate that, between 20 and 80 μm, the thickness of the adhesive joint does not greatly affect the thermal transfer capability. This behaviour has been explained by the fact that the interfacial thermal resistances between the adhesive and both the die and the substrate are much higher than that contributed by the bulk thermal conductivity of the adhesive materials.

An alternative and effective way to reduce the level of the shearing stress of bonded assemblies is the use of extremely flexible adhesives. The elastic modulus of highly compliant epoxy adhesives is of the order of 0.1–0.2 GPa compared to about 3 GPa for IP 670 and IP 680 adhesives. However, low-modulus polymers have coefficient of thermal expansion values increasing up to 1.2×10^{-4} K^{-1}. This means that the maximum normal stress in the die is not significantly reduced by lowering the elastic modulus. Conversely, a low modulus significantly reduces the maximum shearing stresses from the 65 MPa level to about 14 MPa because of the reduction of the parameter of assembly stiffness k in Eq. (11). The curve in Figure 51 represents the relationship between the elastic modulus of the adhesive and the maximum shearing stress occurring at the interface when the substrate

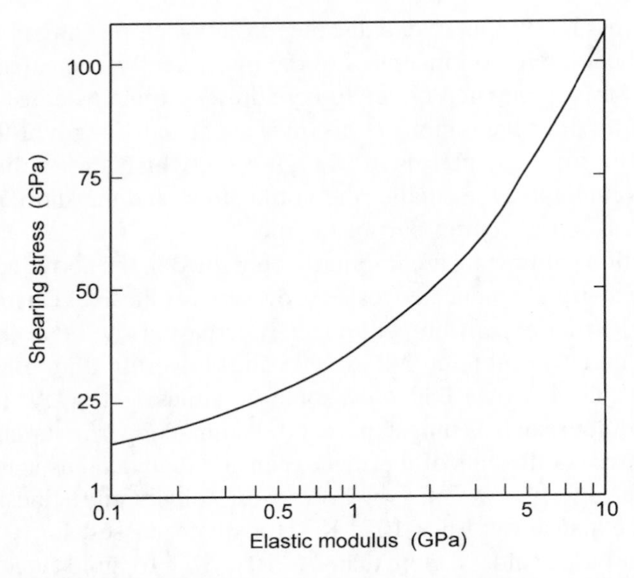

Figure 51: Variation of the maximum shearing stress occurring in the die attachment layer as a function of the elastic modulus of the epoxy adhesive.

is a copper leadframe. The calculation shows that reducing the elastic modulus by two orders of magnitude, from 10 to 0.10 GPa, decreases the shearing stress from 123 to 12.3 MPa. In fact, the shearing stress developed at the edges of the die is less than that because a compliant adhesive allows the stress to be averaged on a larger distance inward from the periphery.

The factors implied in stress reduction have been evaluated by measuring the warping of $8 \times 4 \times 0.38\ mm^3$ test chips bonded to 0.25 mm thick copper leadframes [170]. A die attach composition with a low elastic modulus can be obtained by lowering either the crosslinking density of the organic binder, or the silver content, or both. With regard to the effect of silver, the die warping decreases from about 15 μm to less than 10 μm when the silver content is decreased from 85 to 58%. However, lowering the silver content deteriorates the electrical conductivity by at least one order of magnitude from 5×10^{-5} to $1 \times 10^{-3}\ \Omega$ cm. Secondly, some phenol novolac resins enable the elastic modulus to be lowered from 9.8 to 6.9 GPa without loss in heat resistance. This means that the chip warping is reduced from 22 to 14 μm. Further reduction of the surface stresses can be achieved by decreasing the cure temperature. In the conventional silver-filled one-part epoxies, the amine, imidazoles and other curing agents need either a short term excursion at 200°C or longer cure in the range of 150–175°C. By using an undisclosed phenol curing agent, the authors claim the possibility of achieving a 120°C cure, reducing chip warping to half of that cured at 200°C.

Both the maximum normal and shearing stresses are dependent on the factor ΔT, which is the difference between the stress-free temperature and the operating temperature. In many publications, the upper limit is considered to be the curing temperature of the adhesive, typically 150–200°C for epoxy resins and 275°C for polyimides. However, all stresses virtually vanish when the polymer is heated above its T_g and it is better to introduce the T_g as the stress-free temperature in the calculation of ΔT. Theoretically, low T_g polymers must perform better than highly crosslinked materials from the shear stress viewpoint. All experiments demonstrate that a low glass transition temperature, associated with a low elastic modulus, significantly decrease the delamination of the adhesive layer and the formation of cracks during accelerated ageing tests. The general trend is that a very flexible polymer can sustain many thermal cycles from −50 to 125°C without any detrimental effect on its adhesive properties which are, however, initially lower than those of rigid epoxies. Conversely, the electrical conductivity of flexible adhesives is severely degraded during these tests, probably because of the natural tendency for a viscoelastic material to flow and form an insulating layer between the conductive particles.

Heat resistance associated with low elastic modulus and low glass transition temperature are easily achieved by using poly(imide-siloxanes). These copolymers combine high T_g, high modulus polyimide and polysiloxane repeating units with low mechanical properties, and T_g between 20 and −125°C. For instance, patents granted to General Electric describe a series of copolymers based on diamine **113** in Figure 52 with $m = 1$ and 4,4′-methylenebisbenzeneamine (MDA) **112** opposed to BTDA **102**. The resulting copoly(amic acid) formed of repeating units **114a** and **114b** is thermally imidized to provide copolyimide **115** whose physical properties depend on the molecular ratio **112/113** which controls the number of recurring units **115a** and **115b**.

The methods used to synthesize poly(imide-siloxanes) have been reviewed by Lee who points out that electronic and microelectronic applications are key factors for the development of low-modulus, low-stress new materials [171]. The patent literature indicates that copoly(imide-siloxanes) have been synthesized with all available dianhydrides such as PMDA, BTDA, BPDA, OPDA, and the diether dianhydride used to produce the thermo-plastic polyimide Ultem® 1000. These compounds were opposed to aromatic diamines in combination with siloxanediamine **113** with a degree of polycondensation m varying from 0 to 8 or more. Commercial polymers generally include only 5–10 mol% of siloxane units in order to maintain a glass transition temperature higher than 200°C. For example, preimidized thermoplastic copolymers have T_gs in the range of 200–220°C and can be extruded at temperatures below 300°C

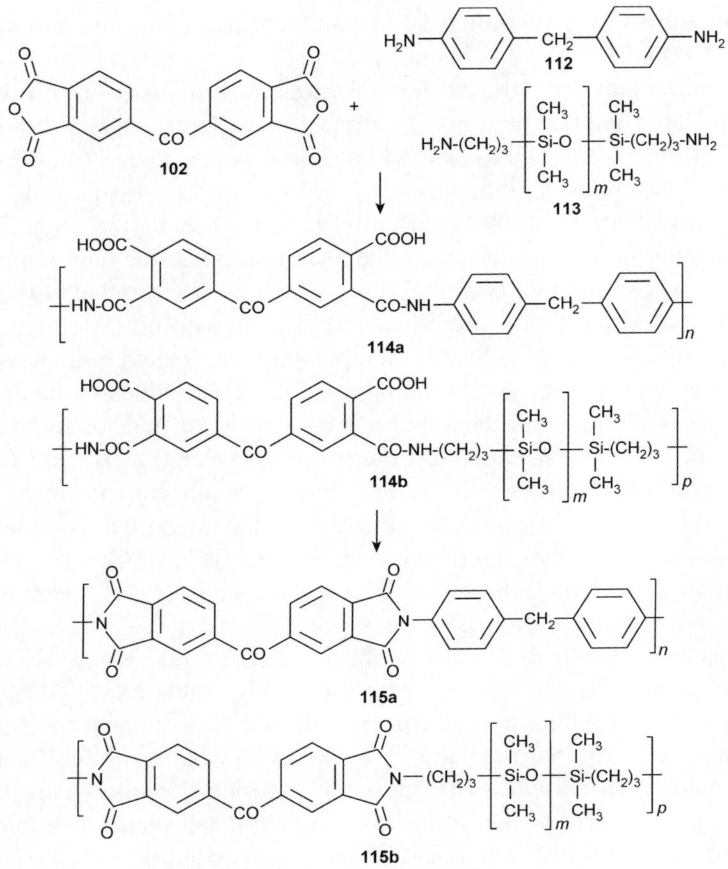

H_2N — CH_2 — NH_2
112

CO
102
+
$H_2N\text{-}(CH_2)_3$ — [$Si(CH_3)_2\text{-}O$]$_m$ — $Si(CH_3)_2\text{-}(CH_2)_3\text{-}NH_2$
113

HOOC, COOH, HN-OC, CO, CO-NH, CH_2, n
114a

HOOC, COOH, HN-OC, CO, $CO\text{-}NH\text{-}(CH_2)_3$ — [$Si(CH_3)_2\text{-}O$]$_m$ — $Si(CH_3)_2\text{-}(CH_2)_3$, p
114b

N, CO, N, CH_2, n
115a

N, CO, $N\text{—}(CH_2)_3$ — [$Si(CH_3)_2\text{-}O$]$_m$ — $Si(CH_3)_2\text{-}(CH_2)_3$, p
115b

Figure 52: Polycondensation of 4,4′-benzophenonetetracarboxylic acid dianhydride **102** with 4,4′-methylenebisbenzeneamine **112** and α,ω-diaminopolydimethylsiloxane **113** leading to intermediate polyamic acid recurring units **114a** and **114b** whose thermal cyclodehydration leads to a poly(imide-siloxane) formed of repeating units **115a** and **115b**.

when thermoplastic poly(ether-imides) must be processed at 380°C. However, the data listed in Table 6 show that many variants are proposed with lower T_g values and elastic modulus close to 1 GPa.

Many patents claim that poly(imide-siloxanes) exhibit very good adhesive properties and the component mostly used to prepare these copolymers is the siloxane diamine **113** with a degree of polycondensation m generally lying between 0 and 9. The copolymer prepared from BTDA, 4,4′-[(1-methylethylidene)-bis(4,1-phenyleneoxy)]bisbenzeneamine, and diamine **113** ($m = 7.4$) has a T_g

Table 6: Elastic moduli, E, and glass transition temperatures, T_g, of poly(imide-siloxanes) commercialized as conductive adhesives for electronic applications.

Trade mark[a]	***E* (GPa)**	T_g **(°C)**
Rely-imide® 660	1	85
Rely-imide® P560	1.5	120
Altisil® SPI 135	1	135
OxySim® 2020M	0.8	150
OxySim® 2030M	1.5	225
JM® J20M	1.3	150
JM® J30M	1.7	225
Imide-Sil® 2160	0.4	140
Imide-Sil® 2180	0.5	160
Imide-Sil® 217	0.8	175

[a] The registered trade marks are: Rely-imide for National Starch; Altisil for Microsil Inc., a subsidiary of Shin Etsu; OxySim for Occidental Chemical Co.; JM for Johnson Mattey; and Imide-Sil for ITK Inc.

of 226°C and provides assemblies with a lap shear strength of 30.6 MPa at room temperature. Adhesive tapes were made by coating Upilex-50S films with a polyamic acid prepared from BTDA, 1,4-benzenediamine, and diamine **113** ($m = 0$) as the major component. Other compositions prepared from different dianhydrides, diamine **113**, and ether-linked aromatic diamines provide adhesive films laminated to copper alloys with peel strength of 1.77–1.93 kN m^{-1}.

6.8. Conclusion

In the domain of advanced technologies, each application has a set of particular requirements with a definite order of importance. For example, the structural adhesives used in the aeronautics and space industry (see Volume 2, Chapter 7) have to withstand thermal shocks from −50 to 230°C and operate for long periods of time at high temperatures. In this case, the thermal stability is a selective parameter even if other properties are also of great concern. The adhesives used to bond electronic devices do not have to endure such stringent thermal conditions but the chemical structures must be chosen by considering other prominent factors. They are high adhesive strength, low humidity uptake, stress relief properties, low coefficient of thermal expansion, high ionic purity, and excellent resistance to high temperature bursts at the assembly and packaging stages. Both industries have in common that relatively high-cost materials can be tolerated if the adhesive film performs well for the expected lifetime.

The literature in the field of organic conductive adhesives is widely scattered because of the dual character of these materials. As sketched in Figure 53, the chemistry dominates the behaviour of the organic part of the composites before cure and during the curing cycle, which depends on the chemical reactants. The physicochemical properties of cured adhesives are also strongly connected to the basic chemistry — epoxies, polyimides, silicones, cyanates, etc. — which determines the thermal and mechanical characteristics of the organic phase. On the other hand, the nature (metal, oxides, ceramics) and the shape (spheres, flakes, or fibres) of the inorganic filler are factors of large changes in viscosity, thixotropy, conductivity, and thermal-oxidative resistance.

An extensive investigation of the scientific and technical literature shows that the chemical compositions are mainly described in patents summarized in the *Chemical Abstracts* or the *Derwent Patent Index*. However, most of the recent patents have been granted to Japanese chemical companies whereas the conductive adhesives were first introduced in the United States and are currently produced by a number of US manufacturers. Thus, non-patented proprietary adhesive compositions certainly share the greatest part of the market. Accordingly, the chemistry discussed in Section 6.2 must be considered only as a general guideline to understand the effects of the organic materials on the final properties of conductive adhesives. In contrast, and because of the fundamental aspect of the problem, the introduction of inorganic fillers into an organic binder is well documented in the scientific journals. Some coalescence between chemistry and

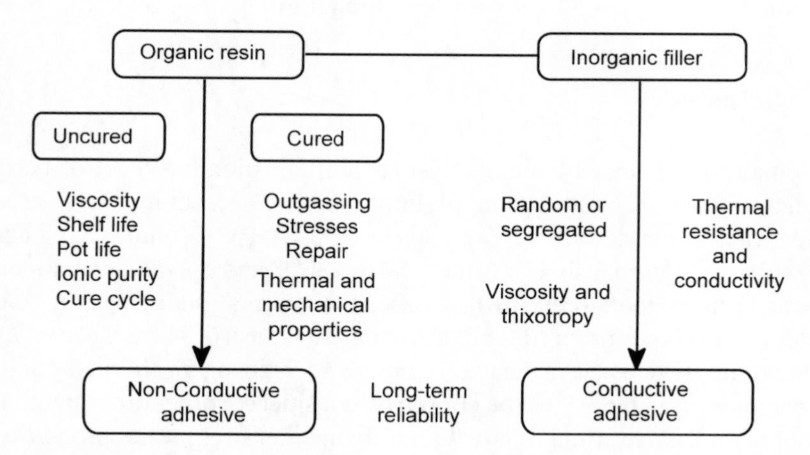

Figure 53: Main effects of the organic resins and inorganic fillers on the overall properties of conductive adhesives.

physics, i.e. the behaviour of conductive adhesives, can be found as sections or chapters of several books [13,59,172–177].

Figure 53 shows that conductive adhesives are composite materials prepared by the incorporation into an organic matrix of metal powders, flakes or fibres, or other fillers, such as metal oxides and ceramic powders. Such materials are now commonly used in electronics (hybrid technology) and in microelectronics for the attachment of integrated circuits to various carriers. Like conventional adhesives, the primary function of these materials is to provide a structural bond between two high-energy surfaces with the added property of assuring an electrical interconnection or heat transfer, or both, between the adherends. The only significant use of electrically conductive adhesives is actually the attachment of semiconductor chips to leadframes made of alloy-42 or copper. The market surveys indicate that about 90% of the 50 billions integrated circuits manufactured each year are encapsulated in plastic packages. Because of the similar chemistry used to make the packages and the conductive adhesives, most of these dice are now assembled with silver-filled, and marginally gold-filled, epoxy adhesives. All the epoxy resins used in the fabrication of adhesives for electronics are commercially available materials that were developed, often several decades ago, to prepare structural adhesives. The major difference is that a great effort has been made by the manufacturers to offer resins complying as closely as possible with the ionic purity, quality, and reliability requirements of the electronics industry. The high-purity epoxy resins produced for electronics applications are mostly used to make plastic packages and PCBs, the adhesive market accounting for only a small part, less than 1%, of the total consumption of these epoxies. The world production of epoxy resins for the electronic packaging is of the order of 40,000 metric tons per year of which approximately 50% is manufactured by Sumitomo Chemical Corp. at its Niihama factory, followed by Nippon Kayaku, Dainippon Ink and Chang Chun Plastic (Taiwan). However, because of the high added value offered by conductive adhesives, various specialty epoxy resins are produced in small volumes by different manufacturers.

For organic compounds, the market size estimates depend on what materials are included in the electronic category, which is described as a truly dynamic area by industry analysts. The sales of adhesives and sealants to the electronic industry have shown consistently strong growth during the past decade and will continue to outpace the other segments of the adhesive market during the next several years. The turning point was observed in the early 1990s when Business Communications indicated that the die attach segment set aside, the world market for electronic adhesives and sealants reached about US$300 million. In the particular domain of adhesives for electronic applications, Strategic Analysis indicated that the growth of this sector was 7.5% in 1991, to $215 million and 9.3% in 1992, to $235 million.

Frost and Sullivan forecast an increase in sales of 73% between 1992 and 1995 for the coatings used to protect integrated circuits to $138 million and for the same period a 67% sales progression for conductive die attach adhesives to about $100 million. This was approximately twice the rate growth expected up to 1995 for the totality of adhesives sales to the electrical/electronics sector. In the branch of die attach adhesives, 1992 sales reached $40 million, and Strategic Analysis predicted at least a 6% annual growth during the next 10 years. That market, the consultant said, represented about 40% of the world market for electronic adhesives. One year later, the same analyst estimated that the sales growth of epoxy-based electrically conductive adhesives was 7% in 1993, increasing to 10% in 1994. The consultant pointed out that, in value, the US producers would share about 45% of the world market of die attach adhesives. At the same time, the marketing manager of Olin Hunt Conductive Materials estimated that the market for epoxy die attach products alone was $50 million per year in 1992 with growth rates projected at 13–15% per year. These data confirm a market survey we did in South-East Asia in 1991 by visiting local and multinational companies in Singapore, Malaysia, Indonesia, Philippines, and Taiwan. At that time the die attach market was relatively well balanced with a consumption of about 30% by the three major countries: the United States, Japan and South-East Asia, and 10% in Europe. For this latter region, a similar picture emerged from the estimates of Johnson Matthey (JM), a major player in the domain of silver–glass adhesives. JM's marketing manager said the company shared about 60% of the European market for the adhesives used in hermetic ceramic packages and 20%, with sales of the order of 2 tons per year, of the same market for silver filled die attach adhesives dedicated to plastic packages. In term of production volume, we estimated in 1993 the world market at about 100 tons of conductive adhesives for the assembly of semiconductors and hybrids. This corresponds to about 25 tons of polymers loaded with 75% by weight of silver flakes. At an average price of $1000 per kg for large supplies, the market value should be worth $100 million with a 95% share for silver-epoxy adhesives. This probably underestimated the actual income because of the added value offered by the marginal small amount deliveries priced at $2000 per kilogram. The growth we forecast at that time has to be reconsidered now because of the unexpected increase in the demand for the leading devices of the semiconductor industry: high-density dynamic and static memory devices (DRAM and SRAM), flash memories, microprocessors, microcontrollers, and others. Taking into consideration the average growth of semiconductor production between 1992 and 1996, in particular for the integrated circuits manufactured in high volumes for the computer, telecommunication, and car industries, we estimated that the actual market for organic die attach materials was of the order of $200 million in 1998 increasing at the same rate for the forthcoming years. Industry analysts forecast a constant growth of 6% per year from $1.1 billion in 1992 to $2.1 billion in 2003 for the US sales of adhesives, sealants, and coatings to the electrical and electronic industries. During this period, the polyimide coatings have been increasing fastest, followed by silicones and epoxies and these three segments have extended their share of the total market revenues from 50% in 1992 to 55% by 1999.

At this point, it is worth noting that if the conductive adhesive market increases in parallel with the growth of semiconductor production, the share of polyimide adhesives slowly decreases in relative value. At the end of the 1970s, polyimides were introduced by the US suppliers mainly because of their high ionic purity compared to the epoxy resins available at that time. Thermal stability was the other incentive for using polyimide binders in the preparation of electrically conductive adhesives. If condensation polyimides can withstand more than 3000 h at 300°C without significant degradation, the addition of silver flakes decreases the thermal-oxidative resistance to 250°C at best, and 200–220°C for long-term use in air. This is due to the catalytic effect of silver on the degradation of the organic materials, which is traditionally obviated by choosing gold-filled polyimides for applications requiring a good thermal stability at 300°C and higher. However, the heat resistance of silver-filled polyimides remains very good in an inert atmosphere such as the nitrogen flow used to encapsulate the integrated circuits in hermetic ceramic packages. In this case, the main problem is the level of outgassing that can be accepted at the 400–450°C temperature range of lid-sealing. The production of polyimide die attach adhesives reached a maximum of 5 metric tons per year, equally shared between the open and captive markets. It has been at the same level for the last 10 years, despite the fact that the die attach market strongly increased during this period of time. It is most unlikely that heat-resistant polymers will have a prominent part of the die attach market in the near future because of the problems related to their chemical structure.

A second information is provided by the market of precious metals for electronics applications. In the United States, the largest volume in 1992 was for silver powders and flakes with 249 tons, second were silver–palladium alloys with 31 tons, followed by palladium, gold, and platinum with 6.2, 3.1, and 0.8 tons, respectively. The paste and ink formulations loaded with 60–80% of precious metals included a metal value exceeding $180 million, and the added value for the producers of powders and flakes was approximately $30–32 million. The companies making the conductive pastes added approximately $100 million to the final products. This means that in the US alone the total market was over $300 million for conductive adhesives and pastes, inferring a worldwide production worth over 1 billion dollars. The most important manufacturers of multilayer ceramic capacitors and other discrete devices using large quantities of conductive materials are located in Japan and South-East Asia.

As previously stated, conductive adhesives represent only a small part of the electrically conductive polymers when compared to conductive inks, electromagnetic interference shields, and many other applications. However, most of these uses do not refer to very high electrical conductivity and maximum adhesive strength and are thus beyond the scope of this chapter. Considering the overall production in volume, electrically conductive adhesives are mainly used as die attach materials in the semiconductor industry. Two other emerging applications are the bonding of electronic drivers to the LCD devices and the SMT of integrated circuits and discrete devices to printed wiring boards. The new market areas where major growth can be expected comprise two types of adhesives. The first group includes the so called "snap cure" formulations, which are fast curing materials intended to be used directly on rapid moving chip assembly lines. The second category consists of all the new adhesives that are currently evaluated and developed to replace metallic solders for bonding electronic devices to printed wiring boards. With about 60,000 tons per year of solder used worldwide in the assembly of electronic devices, this replacement market potential is huge, but this goal requires the formulation of new adhesives aimed specifically at fulfilling surface-mounting specifications.

References

[1] Lee, H., & Neville, K. (1987). *Handbook of epoxy resins*. McGraw Hill Book, New York.
[2] May, C. A. (Ed.) (1988). *Epoxy resins, chemistry and technology*. Marcel Dekker, New York, 2nd ed.
[3] Ellis, B. (1994). *Chemistry and technology of epoxy resins*. Blackie Academic, Glasgow.
[4] Rabilloud, G. (1997). *High-performance polymers, chemistry and applications, vol. 1, conductive adhesives*. Éditions Technip, Paris.
[5] Asai, S., Saruta, U., & Tobita, M. (1994). *Journal of Applied Polymer Science*, **51**, 1945–1958.
[6] Edwards, M. (1989). Flexible epoxies. In: *Electronic materials handbook, vol. 1, packaging* (pp. 817–821). ASM International, Materials Park, OH.
[7] Hermansen, R. D., & Lau, S. E. (1993). *Adhesives Age*, July, 38–41.
[8] Bauer, R. S. (1993). Application of epoxy resins in electronics. In: C. P. Wong (Ed.), *Polymers for electronic and photonic applications* (pp. 287–331). Academic Press, San Diego.
[9] Sheldal Inc. (1988). *Japan Kokai Tokkyo Koho* JP 63,187,501.
[10] Zahir, S. A. (1982). *Advances in Organic Coatings Science and Technology Series*, **4**, 83–102.
[11] Heise, M. S., & Martin, G. C. (1989). *Macromolecules*, **22**, 99–104.
[12] Sharpe, L. H., Chandross, E. A., & Hartless, R. I. (1983). *Adhesion (London)*, **7**, 1–38.
[13] Lyons, A. M., & Dahringer, D. W. (1974). Electrically conductive adhesives. In: A. Pizzi, & K. L. Mittal (Eds), *Handbook of adhesive technology* (pp. 565–584). Marcel Dekker, New York.
[14] Matejka, L., Lovy, J., Pokorny, S., Bouchal, K., & Dusek, K. (1983). *Journal of Polymer Science, Polymer Chemistry Edition*, **21**, 2873–2885.
[15] Schechter, L., & Wynstra, J. (1956). *Industrial & Engineering Chemistry*, **48**(1), 86–93.

[16] Omoya, K., & Sakurai, W. (1990). *Japan Kokai Tokkyo Koho* JP 02,265,981.
[17] Crivello, J. V., & Bi, D. (1994). *Journal of Polymer Science, Part A: Polymer Chemistry*, **32**, 683–697.
[18] Ho, T. H., & Wang, C. S. (1994). *Journal of Applied Polymer Science*, **54**, 13–23.
[19] Wong, C. P. (1989). *ACS Symposium Series*, **407**, 220–229.
[20] Shimp, D. (1994). *Polymer Material Science and Engineering*, **71**, 561–562.
[21] Shimp, D. (1994). *Polymer Material Science and Engineering*, **71**, 623–624.
[22] Nguyen, M. N., Chien, I. Y., Grosse, M. B., Chau, M. M., & Burkhart, D. A. (1995). *IEEE Proceedings of Electronic Component Technology Conference*, **45**, 682–687.
[23] Rabilloud, G. (1999). *High-performance polymers, chemistry and applications, vol. 2, polyquinoxalines and polyimides*. Éditions Technip, Paris.
[24] Rabilloud, G. (2000). *High-performance polymers, chemistry and applications, vol. 3, polyimides in electronics*. Éditions Technip, Paris.
[25] Goodrich, G. B., & Belani, J. G. (1985). US 4,518,735.
[26] Parker, D., Bussink, J., van de Grampel, H. T., Wheatley, G. W., Dorf, E. U., Ostlining, E., & Reinking, K. (1992). High temperature polymers, Ullman's Encyclopedia of Industrial Chemistry, vol. A21, pp. 449–472.
[27] Mundhenke, R. F., & Schwartz, W. T. (1990). *High Performance Polymers*, **2**, 57–66.
[28] Kusy, R. P. (1986). Applications. In: S. K. Bhattacharya (Ed.), *Metal-filled polymers* (pp. 1–142). Marcel Dekker, New York.
[29] Kusy, R. P. (1977). *Journal of Applied Physics*, **48**, 5301–5305.
[30] Bigg, D. M. (1986). Electrical properties of metal-filled polymer composites. In: S. K. Bhattacharya (Ed.), *Metal-filled polymers* (pp. 165–226). Marcel Dekker, New York.
[31] Cumberland, D. J., & Crawford, R. J. (1987). *The packing of particles*. Elsevier, New York.
[32] Feinstein, L. G. (1989). Die attachment methods. *Electronic materials handbook, vol. 1, packaging* (pp. 214–223). ASM International, Materials Park, OH.
[33] Lewis, T. B., & Nielsen, L. E. (1970). *Journal of Applied Polymer Science*, **14**, 1449–1471.
[34] Nielsen, L. E. (1974). *Industrial & Engineering Chemistry, Fundamentals*, **13**, 17–20.
[35] Bolger, J. C., & Morano, S. L. (1994). *Adhesives Age*, June, 17–20.
[36] Chung, K. K. T., Avery, E., Boyle, A., Dreier, G., Koehn, W., Govaert, G., & Theunissen, D. (1990). *International SAMPE Electronics Conference*, **4**, 241–254.
[37] Tsutsumi, N., Takeuchi, N., & Kiyotsukuri, T. (1991). *Journal of Applied Polymer Science, Part B: Polymer Physics*, **29**, 1085–1093.
[38] Li, L., & Chung, D. D. L. (1990). *International SAMPE Electronics Conference*, **4**, 236–240.
[39] Procter, P., & Solc, J. (1991). *IEEE Transactions on Components, Hybrids, Manufacturing Technology*, **14**(4), 708–713.
[40] Bujard, P. (1988). *InterSociety Conference on Thermal Phenomenon Fabric. Operating Electronics*, 41–49.
[41] Bujard, P., & Ansermet, J. P. (1989). *Fifth IEEE SEMI-Therm Symposium*, 126–130.
[42] Mroz, T. J., & Groat, E. A. (1993). *Ceramic Transactions*, **33**, 333–342.
[43] Bjorneklett, A., Halbo, L., & Kristiansen, H. (1992). *Proceedings of International Electronic Packaging Conference*, 509–522.
[44] Bjorneklett, A. (1993). *Polymers and Polymer Composites*, **1**(4), 275–282.
[45] Bjorneklett, A., & Kristiansen, H. (1994). *Hybrid Circuits*, **33**, 28–31.

[46] Bolger, J. C. (1992). *IEEE Proceedings of Electronic Component Technology Conference*, **42**, 219–224.

[47] Weigand, B. L., & Caruso, S. V. (1983). *International Journal of Hybrid Microelectronics*, **6**(1), 387–392.

[48] Swanson, D. W. (1986). *International SAMPE Electronics Conference*, **3**, 1056–1067.

[49] Military standard: test methods and procedures for microelectronics, MIL-STD-883D. Department of Defence, USA, November 1991.

[50] Pandiri, S. M. (1987). *Adhesives Age*, October, 31–35.

[51] Jost, E. M. (1992). *Precious Metals*, **16**, 265–281.

[52] Jenekhe, S. A. (1983). *Polymer Engineering and Science*, **23**, 830–834.

[53] Shaw, S. J. (1994). Additives and modifiers for epoxy resins. In: B. Ellis (Ed.), *Chemistry and technology of epoxy resins* (pp. 117–143). Blackie Academic, Glasgow.

[54] White, M. L. (1982). *IEEE Proceedings of Electronic Component Conference*, **32**, 262–265.

[55] Ireland, J. E. (1982). *International Journal of Hybrid Microelectronics*, **5**(1), 1–4.

[56] Kropp, P. (1994). *International SAMPE Electronics Conference*, **7**, 487–493.

[57] Miric, A. Z. (1994). Problems in processing SMT adhesives. *Adhesives in electronics 94, International Conference on Adhesives Joining Technology Electronics Manufacture*, VDI/VDE-IT.

[58] Li, T. P. L., & Chadderdon, G. D. (1983). *Proceedings of International Symposium on Microelectronics*, 370–374.

[59] Licari, J. J., & Enlow, R.D. (Ed.) (1988). Assembly processes. *Hybrid microcircuit technology handbook*. Noyes Publications, Park Ridge, NJ, pp. 174–190.

[60] Chaffin, R. J. (1981). *IEEE Transactions on Components, Hybrids, Manufacturing Technology*, **4**(2), 214–216.

[61] Anderson, S. P., & Kraus, H. S. (1981). *International Journal of Hybrid Microelectronics*, **4**(2), 190–193.

[62] Estes, R. H., & Kulesza, F. W. (1982). *International Journal of Hybrid Microelectronics*, **5**(2), 336–348.

[63] Bolger, J. C. (1982). *National SAMPE Technical Conference*, **14**, 257–266.

[64] Bolger, J. C. (1984). Polyimide die attach adhesives for LSI ceramic packages. In: K. L. Mittal (Ed.), *Polyimides: synthesis, characterization, and applications*, vol. 2, (pp. 871–887). Plenum Press, New York.

[65] Lee, S. M. (1988). Electrical and electronic applications. In: C. A. May (Ed.), *Epoxy resins chemistry and technology* (pp. 783–884). Marcel Dekker, New York.

[66] Yang, H., Turlik, I., & Murty, K. L. (1993). *Proceedings of Electronics Materials Processing Conference*, **8**, 55–66.

[67] Bolger, J. C. (1982). *International Journal of Hybrid Microelectronics*, **5**(2), 496–499.

[68] Mahadevan, K., Ramabrahmam, R., Yadagari, G., & Sesharadri, B. V. (1988). *International Journal of Hybrid Microelectronics*, **11**(3), 48–55.

[69] Boyle, O., Whalley, D. C., & Williams, D. J. (1993). *Proceedings of the European Hybrid Microelectronics Conference*, **9**, 75–82.

[70] Hu, D. C., & Chen, H. C. (1992). *Journal of Adhesives Science and Technology*, **6**, 527–536.

[71] Kim, K. S., & Kim, J. (1988). *Journal of Engineering Materials Science and Technology*, **11**(3), 266–273.

[72] Loukis, M. J., & Aravas, N. (1991). *Journal of Adhesives*, **35**, 7–22.
[73] Allen, M. G., Nagarkar, P., & Senturia, S. D. (1989). Aspects of adhesion measurement of thin polyimide films. In: C. Feger, M. M. Khojasteh, & J. E. McGrath (Eds), *Polyimides: materials, chemistry and characterization* (pp. 705–717). Elsevier, Amsterdam.
[74] Pan, J. Y., & Senturia, S. D. (1991). *Annual Technical Conference of the Society of Plastics Engineers*, **49**, 1618–1621.
[75] Chu, Y. Z., Durning, C. J., Jeong, H. S., & White, R. C. (1992). The adhesion strength of metal/polyimide and polyimide/silicon interfaces as determined by the blister test. In: K. L. Mittal (Ed.), *Metallized plastics 3: fundamental and applied aspects* (pp. 347–364). Plenum Press, New York.
[76] Jeong, H. S., Chu, Y. Z., Freiler, M. B., Durning, C. J., & White, R. C. (1992). *Materials Research Society Symposium Proceedings*, **239**, 547–552.
[77] Jeong, H. S., Chu, Y. Z., Durning, C. J., & White, R. C. (1992). *Surface and Interface Analysis*, **18**, 282–292.
[78] Khan, M. M. (1989). Environmental testing for commercial and military applications. *Electronic materials handbook, vol. 1, packaging* (pp. 493–503). ASM International, Materials Park, OH.
[79] Gefken, R. M. (1991). *Proceedings of the Electrochemical Society*, **11**, 667–677.
[80] Totta, P. A., & Sopher, R. P. (1969). *IBM Journal of Research and Development*, **13**, 226 p.
[81] Estes, R. H., & Kulesza, F. W. (1995). Conductive adhesive polymer materials in flip chip applications. In: J. H. Lau (Ed.), *Flip chip technologies* (pp. 223–267). McGraw Hill, New York.
[82] Kulesza, F. W., Estes, R. H., & Spanjer, K. (1988). *Solid State Technology*, **31**(1), 135–139.
[83] Messner, G., Turlik, I., Balde, J. W., & Garrou, P. E. (1992). *Thin film multichip modules*. International Society for Hybrid Microelectronics, Reston.
[84] Massénat, M. (1994). *Multichip Modules d'Hier et de Demain*. Polytechnica, Mentor Science Editions.
[85] Licari, J. (Ed.) (1995). *Multichip module design, fabrication and testing*. McGraw Hill, New York.
[86] Doane, D. A., & Franzon, P. D. (1993). *Multichip module technologies and alternatives*. Van Nostrand Reinhold, New York.
[87] Garrou, P. E., & Turlik, I. (1998). *Multichip module technology handbook*. McGraw Hill, New York.
[88] Goldberg, N. (1981). *Proceedings of the International Society for Hybrid Microelectronics Conference*, **4**, 289–295.
[89] Johnson, R. W., Cornelius, M., Davidson, J. L., & Jaeger, R. C. (1986). *Proceedings of the International Society for Hybrid Microelectronics Conference*, **9**, 758–765.
[90] Fillion, R. A., Wojnarowski, R. J., & Daum, W. (1990). *Proceedings of Electronic Component Technology Conference*, **40**, 554–560.
[91] Cole, H. S., Liu, Y. S., Guida, R., & Rose, J. (1988). *Proceedings of the SPIE — International Society for Optical Engineering*, **877**, 92–96.
[92] Cole, H. S., Gorczyca, T., Wojnarowski, R., Gorowitz, B., & Lupinski, J. (1992). *Proceedings of the International Conference on Multichip Modules*, **1**, 412–422.
[93] Müller, H. G., Paredes, A., Buschick, K., & Reichl, H. (1991). *Materials Research Society Symposium Proceedings*, **203**, 369–374.

[94] Lau, J. H., Erasmus, S. J., & Rice, D. W. (1989). Overview of tape automated bonding technology. *Electronic materials handbook, vol. 1, packaging* (pp. 274–294). ASM International, Materials Park, OH.
[95] Pawlowski, W. P. (1991). *Materials Research Society Symposium Proceedings*, **27**, 247–251.
[96] Ulrich, R. J. (1994). *Semiconductor International*, **17**, 1, 101–105.
[97] Sela, U., & Steinegger, H. (1991). *Microelectronics Manufacturing Techniques*, February, 47–52.
[98] Dudderar, T. D., Degani, Y., Spadafora, J. G., Tai, K. L., & Frye, R. C. (1994). *Proceedings of the International Conference on Multichip Modules*, 306–312.
[99] Aschenbrenner, R., Gwiasda, J., Eldring, J., Zakel, E., & Reichl, H. (1995). *International Journal of Microcircuits and Electronic Packaging*, **18**, 2, 154–161.
[100] Kusagaya, T., Kira, H., & Tsunoi, K. (1993). *Proceedings of the International Conference on Multichip Modules*, 238–246.
[101] Basavanhally, N. R., Chang, D. D., Cranston, B. H., & Seger, S. G. (1992). *IEEE Proceedings of Electronic Component Technology Conference*, **42**, 487–491.
[102] Ogunjimi, A. O., Boyle, O., Whalley, D. C., & Williams, D. J. (1992). *Journal of Electronics Manufacturing*, **2**, 109–118.
[103] Chung, K. K. T., Fleishman, R., Bendorovich, D., Yan, M., & Mescia, N. (1992). *Proceedings of International Electronic Packaging Conference*, 678–689.
[104] Chang, D. D., Crawford, P. A., Fulton, J. A., McBride, R., Schmidt, M. B., Simtski, R. E., & Wong, C. P. (1993). *IEEE Transactions on Components, Hybrids, Manufacturing Technology*, **16**, 828–835.
[105] Chung, K. K. T., Boyle, A., & Sager, J. (1991). *Proceedings of International Electronic Packaging Conference*, 167–175.
[106] Gilleo, K., Cinque, T., Corbett, S., & Lee, C. (1993). *Proceedings of International Electronic Packaging Conference*, 232–242.
[107] Li, L., & Morris, J. E. (1994). Structure and selection models for anisotropic conductive adhesive films. *Adhesives in Electronics 94, International Conference on Adhesives Joining Technology Electronics Manufacture*, VDI/VDE-IT.
[108] Shiozawa, N., Isaka, K., & Ohta, T. (1994). Electric properties of connections by anisotropic conductive films. *Adhesives in Electronics 94, International Conference on Adhesives Joining Technology Electronics Manufacture*, VDI/VDE-IT.
[109] Goward, J. M., Williams, D. J., & Whalley, D. C. (1993). *Journal of Electronics Manufacturing*, **3**(4), 179–190.
[110] Jin, S., Tiefel, T. H., Chen, L. H., & Dahringer, D. W. (1993). *IEEE Transactions on Components, Hybrids, Manufacturing Technology*, **16**, 972–977.
[111] Savolainen, P., & Kivilahti, J. (1995). *Journal of Adhesives*, **49**, 187–196.
[112] Nukii, T., Kakimoto, N., Atarashi, H., Matsubara, H., Yamaura, K., & Matsui, H. (1990). *Proceedings of the International Symposium on Microelectronics*, 257–262.
[113] Adachi, K. (1993). *Solid State Technology* **36**(1), 63–71.
[114] Reinke, R. R. (1991). *IEEE Proceedings of Electronic Component Technology Conference*, **41**, 355–361.
[115] Hogerton, P. B., Carlson, K. E., Hall, J. B., Krause, L. J., & Tingerthal, J. M. (1991). *Proceedings of the International Electronic Packaging Conference*, 1026–1033.
[116] Juskey, F. (1988). *Adhesives Age*, March, 41–44.

[117] Gilleo, K. (1989). *Electronic Components and Technology Conference Proceedings*, **39**, 37–44.

[118] Juskey, F. (1989). *IEEE Transactions on Components, Hybrids, Manufacturing Technology*, **11**, 121–125.

[119] Caers, J. F. J. M., van der Reek, J. N. J., & Kessels, F. J. H. (1994). Criteria to assure the reliability of heat seal interconnections — a critical review. *Adhesives in Electronics 94, International Conference on Adhesives Joining Technology Electronics Manufacture*, VDI/VDE-IT.

[120] Van Noort, H. M, Kloos, M. J. H., & Schäfer, H. E. A. (1994). Anisotropic conductive adhesives for chip on glass and other flip chip applications. *Adhesives in Electronics 94, International Conference on Adhesives Joining Technology Electronics Manufacture*, VDI/VDE-IT.

[121] Asai, S., saruta, U., Tobita, M., Takano, M., & Miyashita, Y. (1995). *Journal of Applied Polymer Science*, **56**, 769–777.

[122] Schäfer, H. E. A., & Van Noort, H. M. (1994). Conductive adhesive processing for chip on glass (COG). *Adhesives in Electronics 94, International Conference on Adhesives Joining Technology Electronics Manufacture*, VDI/VDE-IT.

[123] Mochizuki, A., Maeda, M., Higashi, K., & Sugimoto, M. (1992). *Proceedings of the SPIE — International Society for Optical Engineering*, **1847**, 342–347.

[124] Matenzio, L. J., & Unertl, W. N. (1996). Adhesion of metal films on polyimides. In: M. K. Ghosh, & K. L. Mittal (Eds), *Polyimides fundamentals and applications* (pp. 629–696). Marcel Dekker, New York.

[125] Buchwalter, L. P. (1996). Adhesion of polyimides to various substrates. In: M. K. Ghosh, & K. L. Mittal (Eds), *Polyimides fundamentals and applications* (pp. 587–628). Marcel Dekker, New York.

[126] Jensen, R. J., Cummings, J. P., & Vora, H. (1984). *IEEE Proceedings of Electronic Component Technology Conference*, **34**, 73–81.

[127] Estes, R. H. (1984). *Solid State Technology*, **27**(8), 191–197.

[128] Mahalingam, M., Nagarkar, M., Lofgran, L., Andrews, J., Olsen, D. R., & Berg, H. M. (1984). *IEEE Proceedings of Electronic Component Technology Conference*, **34**, 469–478.

[129] Cognard, J. (1988). *Journal of Adhesives*, **26**, 155–169.

[130] Licari, J. J., Perkins, K. L., & Caruso, S. V. (1975). NASA Contract Report, NASA-CR-150160.

[131] Bolger, J. C., Herberg, M. J., & Mooney, C. T. (1987). *International SAMPE Electronics Conference*, **1**, 616–621.

[132] Estes, R. H., & Pernice, R. F. (1989). *Proceedings of the International Symposium on Microelectronics*, 664–669.

[133] Mitchell, C., & Berg, H. (1976). *Proceedings of the International Microelectronics Symposium*, 52–58.

[134] Svitak, J. J., & Williams, A. F. (1977). *Proceedings of the International Symposium on Microelectronics*, 189–196.

[135] Eisenmann, D. E., & Halyard, S. M. (1976). *Thermochimica Acta*, **14**, 87–97.

[136] Lenkkeri, J., & Rusanen, O. (1993). *Journal of Electronics Manufacturing*, **3**, 199–204.

[137] Sorrells, D. L., Oscilowski, A. A., & Dosher, T. (1984). *Proceedings of the International Microelectronics Symposium*, 181–184.

[138] Rusanen, O., & Lenkkeri, J. (1994). Reliability issues of replacing solder with conductive adhesives. *Adhesives in Electronics 94, International Conference on Adhesives Joining Technology Electronics Manufacture*, VDI/VDE-IT.

[139] Rörgren, R., & Tillströn, A. (1993). *Journal of Electronics Manufacturing*, **3**, 169–173.

[140] Howell, J. R. (1981). *Annual Proceedings of Reliability Physics Symposium*, **19**, 104–110.

[141] Vietl, B., & Rösner, B. (1994). Flip chip with polymer bumps on various substrates. *Adhesives in Electronics 94, International Conference on Adhesives Joining Technology Electronics Manufacture*, VDI/VDE-IT.

[142] Dion, J., Borgesen, P., Yost, B., Lilienfeld, D. A., & Li, C. Y. (1994). *Proceedings of the Materials Research Society Symposium*, **323**, 27–32.

[143] Wakizaka, S., Kobayashi, M., & Mizuno, M. (1993). *IEEE Proceedings of Electronic Component Technology Conference*, **43**, 336–340.

[144] Shi, L. T., Saraf, R., & Huang, W. S. (1992). *Proceedings of the Electronic Materials and Processes Congress*, **7**, 149–153.

[145] Nguyen, G. P., Williams, J. R., Gibson, F. W., & Winster, T. (1993). *Proceedings of the International Society for Hybrid Microelectronics Conference*, 50–55.

[146] Liu, J. (1993). *Circuit World*, **19**, 4–11.

[147] Chang, D. D., Fulton, J. A., Ling, H. C., Schmidt, M. B., Sinitski, R. E., & Wong, C. P. (1993). *IEEE Transactions on Components, Hybrids, Manufacturing Technology*, **16**, 836–842.

[148] Hvims, H. L. (1995). *IEEE Transactions on Components, Hybrids, Manufacturing Technology, Part B*, **18**, 284–291.

[149] Schmidt, C. G., Kristiansen, H., Brox, B., Whalley, D. C., Gileo, K. (Ed.) (1992). *Techniques for improved reliability and high yield of surface mount joints*. Chapman & Hall, London.

[150] Keusseyan, R. L., & Dilday, J. L. (1992). *Proceedings of the SPIE — International Society for Optical Engineering*, **1847**, 510–517.

[151] Van den Bosch, A., & Luyckz, G. (1993). *European Hybrid Microelectronics Conference*, **9**, 68–74.

[152] Jagt, J. C., Berris, P. M. J., & Lijten, G. F. C. (1995). *IEEE Transactions of the Components, Packaging and Manufacturing Technology*, **18**, 292–298.

[153] Nguyen, G. P., Williams, J. R., & Gibson, F. W. (1992). *Proceedings of the SPIE — International Society for Optical Engineering*, **1847**, 510–517.

[154] Klosterman, D., Rak, S., Wille, S., Dubinski, D., & Desai, P. (1994). An investigation of the conductive metal interfaces in Ag-filled adhesives. *Adhesives in Electronics 94, International Conference on Adhesives Joining Technology Electronics Manufacture*, VDI/VDE-IT.

[155] Gaynes, M. A., Russell, H. L., Saraf, R. F., & Roldan, J. M. (1995). *IEEE Transactions of the Components, Packaging and Manufacturing Technology, Part B*, **18**, 299–304.

[156] Estes, R. H., Kulesza, F. W., Buczek, D., & Riley, G. (1993). *International Electronics, Packaging Conference*, **1**, 328–342.

[157] Bolger, J. C., & Gilleo, K. (1994). *Proceedings of the Multichip Module Conference*, 77–82.

[158] Gaynes, M., & Lewis, R. (1994). *International SAMPE Electronics Conference*, **7**, 69–78.

[159] Li, L., Morris, J. E., Liu, J., Lai, Z., Ljungkrona, L., & Li, C. (1995). *IEEE Proceedings of Electronic Component Technology Conference*, **45**, 114–120.

[160] Yoshigara, H., Sagami, Y., Nose, S., & Burkhart, A. (1990). *International SAMPE Electronics Conference*, **4**, 255–266.

[161] O'Grady, P., Stam, F., & Barett, J. (1994). Characterization and reliability study of anisotropic conductive adhesives for fine pitch package assembly. *Adhesives in Electronics 94, International Conference on Adhesives Joining Technology Electronics Manufacture*, VDI/VDE-IT.

[162] Lindner, K., & Reinders, T. G. (1994). Pulse heat bonding method for interconnections with anisotropic conductive adhesive foils and heat seal connectors. *Adhesives in Electronics 94, International Conference on Adhesives Joining Technology Electronics Manufacture*, VDI/VDE-IT.

[163] Hampel, B. (1994). Experiences with heat seal connections and anisotropic conducting materials for the car industry. *Adhesives in Electronics 94, International Conference on Adhesives Joining Technology Electronics Manufacture*, VDI/VDE-IT.

[164] Vanfleteren, J., De Baets, J., Van Calster, A., Dravet, A., Deckelmann, K., Wiese, J., Schmitt, W., Allaert, K., Vetter, P., Schols, G., & Cortès, E. (1994). Anisotropic conductive adhesives (ACAs) for high density interconnection in liquid crystal displays (LCDs). *Adhesives in Electronics 94, International Conference on Adhesives Joining Technology Electronics Manufacture*, VDI/VDE-IT.

[165] Liu, J., & Rörgren, R. (1993). *Journal of Electronics Manufacturing*, **3**, 205–214.

[166] Tong, H. M., & Nguyen, L.T. (Ed.) (1990). *New characterization techniques for thin polymer films*. Wiley, New York.

[167] Chen, W. T., & Nelson, C. W. (1979). *IBM Journal of Research and Development*, **23**, 2, 179–188.

[168] Suhir, E. (1987). *IEEE Proceedings of Electronic Component Technology Conference*, **37**, 508–517.

[169] Riemer, D. E. (1994). *Proceedings of the SPIE — International Society for Optical Engineering*, **2369**, 390–396.

[170] Okabe, T., Kusuhara, A., Mizuno, M., & Horiuchi, K. (1988). *IEEE Proceedings of Electronic Component Technology Conference*, **38**, 468–472.

[171] Lee, C. J. (1993). Polyimidesiloxanes: chemistries and applications. In: C. P. Wong (Ed.), *Polymers for electronic and photonic applications* (pp. 249–286). Academic Press, New York.

[172] Bhattacharya, S. K. (Ed.) (1986). *Metal-filled polymers*. Marcel Dekker, New York.

[173] Bauer, M., & Schneider, J. (1974). Adhesives in the electronic industry. In: A. Pizzi, & K. L. Mittal (Eds), *Handbook of adhesive technology* (pp. 587–598). Marcel Dekker, New York.

[174] *ASM engineered materials handbook series. Vol. 3: adhesives and sealants*. American Technical Publishers, Herts, 1990.

[175] Livesay, B. R., & Nagarkar, N.D. (Ed.) (1990). *New technology in electronic packaging*. American Technical Publishers, Herts.

[176] Mackay, C. (Ed.) (1992). *Electronic packaging: materials and processes to reduce package cycle time and improve reliability*, ASM Conference Book. American Technical Publishers, Herts.

[177] Bolger, J. C. (1989). Conductive adhesives. In: I. Skeist (Ed.), *Handbook of adhesives*. Van Nostrand Reinhold, New York, 3rd edition, 711 p.

Index